Graduate Texts in Contemporary Physics

Series Editors:

R. Stephen Berry
Joseph L. Birman
Jeffrey W. Lynn
Mark P. Silverman
H. Eugene Stanley
Mikhail Voloshin

Springer

New York
Berlin
Heidelberg
Hong Kong
London
Milan
Paris
Tokyo

Graduate Texts in Contemporary Physics

Rabindra N. Mohapatra

Unification and Supersymmetry

The Frontiers of Quark-Lepton Physics

Third Edition

With 60 Illustrations

Springer

Rabindra N. Mohapatra
Department of Physics and Astronomy
University of Maryland
College Park, MD 20742
USA
rmohapatra@umdhep.umd.edu

Series Editors:

R. Stephen Berry
Department of Chemistry
University of Chicago
Chicago, IL 60637
USA

Joseph L. Birman
Department of Physics
City College of CUNY
New York, NY 10031
USA

Jeffrey W. Lynn
Department of Physics
University of Maryland
College Park, MD 20742
USA

Mark P. Silverman
Department of Physics
Trinity College
Hartford, CT 06106
USA

H. Eugene Stanley
Center for Polymer
  Studies
Physics Department
Boston University
Boston, MA 02215
USA

Mikhail Voloshin
Theoretical Physics Institute
Tate Laboratory of Physics
University of Minnesota
Minneapolis, MN 55455
USA

Library of Congress Cataloging in Publication Data
Mohapatra, R. N. (Rabindra Nath)
    Unification and supersymmetry: the frontiers of quark-lepton physics
Rabindra N. Mohapatra.—3rd ed.
        p. cm. (Graduate texts in contemporary physics)
Includes bibliographies and index.
  ISBN 978-1-4419-3042-2   e-ISBN 978-0-387-22736-8
    1. Grand unified theories (Nuclear physics)   2. Supersymmetry.
3. Particles (Nuclear physics)   I. Title.   II. Series.
QC794.6.G7 M64   2002
539.7′25—dc21   2002030271

Printed on acid-free paper.

Printed in the United States of America.

9 8 7 6 5 4 3 2 1

www.springer-ny.com

Springer-Verlag   New York Berlin Heidelberg
A member of BertelsmannSpringer Science+Business Media GmbH

To
Manju
Pramit and Sanjit

"... Where tireless striving stretches it arms towards perfection:

Where the clear stream of reason has not lost its way into the dreary desert sand of dead habit, ... "

GITANJALI, RABINDRA NATH TAGORE,
NOBEL LAUREATE IN LITERATURE

Preface to the Third Edition

The new millennium has brought new hope and vigor to particle physics. The menacing clouds of despair and discontent that enveloped the field following the collapse of SSC have all but vanished. The discovery of neutrino mass has brought the first light of new physics beyond the standard model. The LEP-SLC data has given strong hints of a light Higgs boson, which is widely hoped, will be discovered soon either at the Tevatron of LHC. LEP may quite possibly have missed it by a hair. Many neutrino experiments are either underway or are in the planning stages, and a rough outline of neutrino mixing is appearing on the horizon. There are discussions of pulling resources internationally to build a linear collider after the LHC. Many major breakthroughs in the sister discipline of cosmology have lightened up the sky. Even the job situation in the field is showing signs of improvement after a long plateau.

All this hope and optimism about a bright future for the field seem to be resting on two ideas: unification and supersymmetry. The first is based on the amazing success of the standard model, giving credence to the possibility that the final theory of particle physics could come from gauge theories and string theory, from which the gauge symmetries follow. The belief in supersymmetry arises not only from its beauty and elegance and its ability to truly unify matter and forces but also from the way it embraces gravity into the fold of particle physics. Its hold on the field is almost as pervasive as that of gauge theories. Even though there are many other competing ideas vying for the attention of theorists, the general direction seems to be largely set towards supersymmetry, supergravity, and superstrings. The possibility of extra dimensions playing a role in understanding the new physics is being taken more seriously than ever. Still, there are many unsolved problems remaining: the origin of family replication, the origin of quark and lepton mixings, the origin of supersymmetry breaking, electroweak symmetry breaking, and the possibility of extra dimensions, to name some important ones. There is thus a lot to do, and the new century should be an exciting one.

With this background, I felt it necessary to update the Second Edition of the book by adding new materials that reflect recent developments and hopefully make it more useful to advanced graduate students as well as to researchers in the field. This edition uses the popular $(+, -, -, -)$ metric instead of the Pauli metric used in the first two editions.

I would like to acknowledge the support of the National Science Foundation during the time this book was updated. I would also like to thank Xiang-Dong Ji for carefully reading the beginning chapters on supersymmetry and suggesting many improvements. Last, but not least, I would like to thank those colleagues who have used the book in the their advanced particle physics courses at various universities and to those who have found some use for the book in their own intellectual pursuits.

RABINDRA MOHAPATRA
College Park, Maryland
July 10, 2002

Preface to the Second Edition

Nearly six years have passed since the first edition of *Unification and Supersymmetry* was finished. Many new developments have occurred in both theoretical and experimental areas of particle physics. On the theoretical side, the most notable development was the rise of superstring theories as a candidate theory of everything. A great deal of euphoria swept the theory community that the "end" of theoretical physics was in sight. The years 1987 and 1988 saw the excitement peak. Limitations to the superstring approach have since led to a more sober reassessment of the prospects of string theories. While fewer people now believe the string theories to be the panacea for all the "ailments" of the standard model, it is regarded as a significant theoretical development. I have therefore added an extra chapter dealing with the salient features of string theories and the Calabi–Yau type compactification.

On the experimental side, parameters of the standard model are now much more precisely determined thanks to LEP experiments; the top quark, of course, still remains undiscovered, but it has now a higher, lower limit on its mass, leading to speculation that it may have a key role in understanding the electroweak symmetry breaking.

Finally, many great developments have taken place in neutrino physics—the two most outstanding ones being the apparent confirmation of the existence of a solar neutrino deficit and the possible existence of a 17 keV neutrino. Both of these results will be tested in several planned and ongoing experiments and are the subject of separate books. I have, therefore, only briefly touched on the first topic in Chapter 6. These may yet prove to be the first clue to new physics.

There is no direct evidence yet for supersymmetry, while the recent precise measurement of $\sin^2 \theta_W$ has prompted some, prematurely, to conclude that it is evidence for supersymmetry. These physicists assume that coupling constants must eventually unify. (Although it is an appealing assumption, now there are fewer reasons than ever to require grand unification of gauge couplings, i.e., one can explain both charge quantization and baryon asymmetry without requiring grand unification.) Secondly, even if one assumed unification of couplings, any kind of intermediate scale [such as the ones in SO(10)] would equally well predict $\sin^2 \theta_W$ to match observations precisely.

Apart from several extensive changes in the text, I have also added exercises at the end of each chapter to stimulate further discussion of the issues. I hope that the key theoretical ideas discussed in this book will receive some confirmation in future experiments; but in any case, I will consider my efforts worthwhile if the book inspires even one more graduate student to choose a career as a particle physicist.

Finally, I wish to acknowledge several colleagues, especially C. Kalman and E. Golowich, who have pointed out several typographical errors in the first edition and to Rachel Needle for carefully typing the needed pages for the second edition. The support of the National Science Foundation during the period of the preparation of the second edition is gratefully acknowledged.

RABINDRA MOHAPATRA
College Park, Maryland
September 1, 1991

Preface to the First Edition

Preface to the First Edition

The theoretical understanding of elementary particle interactions has undergone a revolutionary change during the past one and a half decades. The spontaneously broken gauge theories, which in the 1970s emerged as a prime candidate for the description of electroweak (as well as strong) interactions, have been confirmed by the discovery of neutral weak currents as well as the W- and Z-bosons. We now have a field theory of electroweak interactions at energy scales below 100 GeV—the Glashow–Weinberg–Salam theory. It is a renormalizable theory that enables us to do calculations without encountering unnecessary divergences. The burning question now is: What lies ahead at the next level of unification? As we head into the era of supercolliders and ultrahigh-energy machines to answer this question, many appealing possibilities exist: left–right symmetry, technicolor, compositeness, grand unification, supersymmetry, supergravity, Kaluza–Klein models, and, most recently, superstrings that even unify gravity along with other interactions. Experiments will decide if any one or any combination of these is to be relevant in the description of physics at the higher energies. As an outcome of our confidence in the possible scenerios for elementary particle physics, we have seen our understanding of the early universe improve significantly. Such questions as the origin of matter, the creation of galaxies, and the puzzle of the cosmic horizon are beginning to receive plausible answers in terms of new ideas in particle physics. Although a final solution is far from being at hand, reasonable theoretical frameworks for carrying out intelligent discussions have been constructed. Even such difficult questions as the "birth" and "death" of the universe have been discussed.

This book, based on advanced graduate courses offered by me at CCNY (City College of New York) and in Maryland, attempts to capture these exciting developments in a coherent chapter-by-chapter account, in the hope that the frontiers of our understanding (or lack of it) in this exciting field of science can be clearly defined for students as well as beginning researchers. The emphasis has been on physical, rather than technical and calculational, aspects, although some necessary techniques have been included at various points. Extensive references are provided to original works, which the readers are urged to consult in order to become more proficient in the techniques. The prerequisites for this book are an advanced course in quantum field theory (such as a knowledge of Feynman diagrams, the renormalization program, Callan–Symanzik equations, and so forth), group theory (Lie groups at the level of the books by Gilmore or Georgi), basic particle theory such as the quark model, weak interaction, and general symmetry principles (at the level of the books by Marshak, Riazuddin, and Ryan or by Commins), and familiarity with spontaneously broken gauge theories (at the level of the books by J. C. Taylor or Chris Quigg).

The book is divided into two parts: the first eight chapters deal with the introduction to gauge theories, and their applications to standard $SU(2)_L \times U(1) \times SU(3)_C$ models and possible extensions involving left–right symmetry, technicolor, composite models, quarks and leptons, strong and weak CP violation, and grand unification. In the last eight chapters, we discuss global and local supersymmetry, its application to particle interactions, and the possibilities beyond $N = 1$ supergravity. Interesting recent developments in the area of superstrings are only touched on; unfortunately, they could not be discussed as extensively as they ought to be.

I would like to acknowledge many graduate students at CCNY and in Maryland over the past seven years, as well as colleagues at both places who attended the lectures, and helped to sharpen the focus of presentation by their comments. I am grateful to many of my colleagues and collaborators for generously sharing with me their insight into physics. I wish to thank E. F. Redish for suggesting that I compile the lectures into book form, and C. S. Liu and the members of the editorial board of this lecture series for their support at many crucial stages in the production of this book. Finally, I wish to thank Mrs. Rachel Olexa for her careful and prompt typing of the manuscript, and J. Carr for reading several chapters and suggesting language improvements.

I wish to acknowledge the support of the U.S. National Science Foundation during the time this book was written.

RABINDRA MOHAPATRA

College Park, Maryland

March 1986

Contents

Index   **419**

1
Important Basic Concepts in Particle Physics

§1.1   Introduction

Forces observed in nature can be classified into four categories according to their observed strength at low energies: strong, electromagnetic, weak, and gravitational. Their strengths are roughly of the following orders of magnitude:

		Orders of magnitude
Strong	$g_{NN\pi}^2/4\pi \simeq 15$	10^1
Electromagnetic	$e^2/4\pi \simeq 1/137 \cdot 035982$	10^{-2}
Weak	$G_F \simeq 1.16632 \times 10^{-5}\,\mathrm{GeV}^{-2}$	10^{-5}
Gravitational	$G_N \simeq 6 \cdot 70784 \times 10^{-39}\,\mathrm{GeV}^{-2}$	10^{-40}

Furthermore, the ranges of electromagnetic and gravitational interactions are infinite, whereas the weak and strong interactions are known to have very short range. Since, in these lectures, we will study the nature of these forces [1] as they act between known "elementary" (only at our present level of understanding) particles, we will start our discussion by giving a broad classification of the known "elementary" particles (Fig. 1.1) according to first, their spin statistics, and then, whether they participate in strong interactions or not. Leptons and nonhadronic bosons are supposed to participate only in weak and electromagnetic interactions, whereas baryons and mesons have strong and weak (as well as electromagnetic) interactions. Another important point of difference between the nonhadronic (leptons, γ,

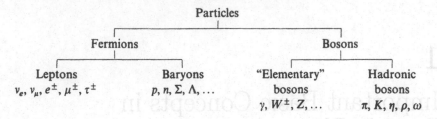

Figure 1.1.

$W^{\pm}, Z \ldots$) and hadronic (baryons and mesons) particles is that all observed nonhadronic particles exhibit pointlike structure to a good approximation (to the extent that experimentalists can tell), whereas the hadronic particles exhibit finite-size effects (or form factors). It is, therefore, convenient to assume that the former are more elementary than the latter, and this has led to the postulate [2] that there exists a more elementary layer of matter (called quarks), of which baryons and mesons are made. Thus, according to present thinking, there are really two sets of basic building blocks of matter: the basic fermions (leptons and quarks, on which the forces like weak, electromagnetic, and strong act) and a basic set of "elementary" bosons such as photon, W, Z, color gluons (V), which are the carriers of the above forces. According to these ideas, the quarks participate in strong, as well as weak, interactions, and it is the binding force of strong interactions mediated by a color octet of gluons that generates the baryons and mesons.

There are, of course, certain other differences between strong interactions and other forms of interactions, the most important being the existence of manifest approximate global symmetries associated with strong interactions like, isospin, SU(3), and so on, which are shared by neither the leptons nor the nonhadronic bosons. Lack of availability of any reliable way to solve the bound-state problems in relativistic physics has given significance to the above symmetries in studying the particle spectra and interactions among the baryons and mesons. The basic strong interactions of the quarks must, of course, respect these symmetries. As far as the nonstrong interactions are concerned, although there does not exist any manifest global symmetry, strong hints of underlying hidden symmetries were detected as early as the late 1950s and early 1960s. Gauge theories [3], in a sense, have opened the door to this secret world of nonmanifest symmetries that have been instrumental in building mathematically satisfactory theories for weak and electromagnetic interactions and strong interactions.

Understanding the symmetries of the quark–lepton world is crucial to our understanding the nature of the various kinds of forces. To understand symmetries, we need to identify the basic degrees of freedom on which the symmetries operate. It is currently believed that the quarks carry two independent degrees of freedom (known as flavor and color), the flavor being a manifest degree of freedom that is responsible for the variety and richness

Table 1.1.

	Color			Lepton
F	u_1	u_2	u_3	ν_e
L	d_1	d_2	d_3	e^-
A	c_1	c_2	c_3	ν_μ
V	s_1	s_2	s_3	u^-
O	t_1	t_2	t_3	ν_τ
R	b_1	b_2	b_3	τ^-

of spectrum in the baryon–meson world, whereas color [4] is a hidden "co-ordinate" that is responsible for the binding of quarks and antiquarks in appropriate combination to baryons and mesons, as well as for the way in which the baryons and mesons interact. Leptons, on the other hand, carry only a flavor degree of freedom. It is, however, a remarkable fact (perhaps indicative of a deeper structure) that the flavor degrees of freedom of quarks and leptons are identical—a fact that is known as quark–lepton symmetry [5]. The existence of a new kind of quark, the charm quark, and a new set of hardrons called charmhadrons was first inferred on the basis of this symmetry. In Table 1.1, we display these quark–lepton degrees of freedom.

The current experience is that the electroweak interactions operate along the flavor degree of freedom, whereas the strong interactions operate along the color degree of freedom. A more ambitious program is, of course, to introduce new kinds of interactions operating between quarks and leptons [6], thus providing a framework for the unified treatment of quarks and leptons. No evidence for such new interactions has yet been uncovered, but when that happens, we would have taken a significant new stride in our understanding of the subatomic world.

To study the structure and forces of the quark–lepton world, we will start by giving a mathematical formulation of symmetries and their implications in the sections that follow. We will work with the assumption that all physical systems are described by a Lagrangian, which is a local function of a set of local fields $\phi_i(x)$ and their first derivative $\partial_\mu \phi_i$, respecting invariance under proper Lorentz transformations. The requirement of renormalizability further restricts the Lagrangian to contain terms of total mass dimension 4 or less. [To count dimensions, we note that bosonic fields have dimension 1 ($d = 1$); fermionic fields, $d = 3/2$; momentum, $d = 1$; and angular momentum, $d = 0$; and so on.]

A few words about the convention, metric, etc. We use the metric $g_{\mu\nu} = diag(+1, -1, -1, -1)$. For a complex scalar field $\phi(x)$, the Lagrangian is

$$\mathcal{L} = \partial^\mu \phi^* \partial_\mu \phi - m^2 \phi^* \phi. \tag{1.1.1}$$

For a fermion field $\psi(x)$, we need to define the γ-matrices:

$$\gamma^i = \begin{pmatrix} 0 & \sigma_i \\ -\sigma_i & 0 \end{pmatrix}, \gamma^0 = \begin{pmatrix} 0 & \mathbf{I} \\ \mathbf{I} & 0 \end{pmatrix}, \tag{1.1.2}$$

where the σ_i are the Pauli matrices.

The Lagrangian for the free Dirac fermion field with mass m is given by

$$\mathcal{L} = \left[i\bar{\psi}\gamma_\mu \partial^\mu \psi - m\bar{\psi}\psi \right]. \tag{1.1.3}$$

From the above equations, the standard canonical quantization method leads to the following propagators:

for bosons:

$$i\Delta_F(k) = \frac{i}{k^2 - m^2} \tag{1.1.4}$$

and for fermions

$$iS_F(p) = \frac{i}{\gamma \cdot p - m}. \tag{1.1.5}$$

§1.2 Symmetries and Currents

Associated with any given symmetry is a continuous or discrete group of transformations whose generators commute the Hamiltonian. Examples of symmetries abound in classical and quantum mechanics, e.g., momentum and angular momentum, parity, symmetry, or antisymmetry of the wave function for many body systems. In the following paragraphs, we will discuss the implications of continuous symmetries within the framework of Lagrangian field theories. To describe the behavior of dynamical systems, we will start by writing down the action S as a functional of the fields ϕ_i, $\partial_\mu \phi_i$

$$S = \int d^4x \, \mathcal{L}(\phi_i, \partial_\mu \phi_i) \tag{1.2.1}$$

($\mathcal{L}$ being the Lagrange density). Under any symmetry transformation

$$\phi \to \phi + \delta\phi, \qquad \partial_\mu \phi \to \partial_\mu \phi + \delta(\partial_\mu \phi)$$

and

$$\mathcal{L} \to \mathcal{L} + \delta\mathcal{L}. \tag{1.2.2}$$

First, we will show that associated with any symmetry transformation there is a current, called the Noether current, J_μ, which is conserved if the symmetry is exact, i.e.,

$$\delta\mathcal{L} = 0. \tag{1.2.3}$$

It is sufficient to have $\delta\mathcal{L} = \partial^\mu 0_\mu$ (i.e., four-divergence of a four-vector) so that the action is invariant under symmetry, i.e., $\delta S = 0$. This is actually the case for supersymmetry transformations (see Chapter 9). To obtain J_μ, we look at the variation of $\mathcal{L}$ under the transformations given in eq. (1.2.2):

$$\delta\mathcal{L} = \sum_i \left[\frac{\delta\mathcal{L}}{\delta\phi_i}\delta\phi_i + \frac{\delta\mathcal{L}}{\delta(\partial_\mu\phi_i)}\delta(\delta_\mu\phi_i) \right]. \tag{1.2.4}$$

Physical fields are those that satisfy the Euler–Lagrange equations resulting from extremization of the action

$$\frac{\delta\mathcal{L}}{\delta\phi_i} = \partial_\mu\frac{\delta\mathcal{L}}{\delta(\partial_\mu\phi_i)}. \tag{1.2.5}$$

Substituting this in to eq. (1.2.4), we get [assuming $\delta(\partial_\mu\phi_i) = \partial_\mu(\delta\phi_i)$]

$$\delta\mathcal{L} = \sum_i \partial^\mu \left[\frac{\delta\mathcal{L}}{\delta(\partial_\mu\phi_i)}\delta\phi_i \right] \tag{1.2.6}$$

$$\equiv \sum_i i\partial^\mu(\varepsilon_a J_\mu^a), \tag{1.2.7}$$

where ε_a is the parameter characterizing the transformation. J_μ^a is the Noether current we were looking for and it satisfies the equation

$$i\varepsilon_a\partial^\mu J_\mu^a = \delta\mathcal{L}. \tag{1.2.8}$$

Thus, if $\delta\mathcal{L} = 0$, since ε is arbitrary, $\partial^\mu J_\mu^a = 0$. It is obvious that the corresponding charge

$$Q^a = + \int d^3x\, J_0^a(x) \tag{1.2.9}$$

satisfies the equation

$$\frac{dQ^a}{dt} = 0 \quad \text{if } \delta\mathcal{L} = 0$$

leading to conservation laws for the corresponding charge.

Let us give some examples of symmetries present in the world of elementary particles. The simplest one is the conservation of electric charge Q, which is a symmetry associated with arbitrary phase transformation of complex field. To give an example, consider the Dirac field ψ transforming as $\psi \to e^{iq_i\theta}\psi$ that keeps the following Lagrangian invariant:

$$\mathcal{L} = [i\bar{\psi}\gamma_\mu\partial^\mu\psi - m\bar{\psi}\psi]. \tag{1.2.10}$$

The corresponding conserved current is $J_\mu = \bar{\psi}\gamma_\mu\psi$. The associated charge operator that generates these symmetries is

$$Q = + \int d^3x\, J_0(x) = \int d^3x\, \psi^\dagger\psi. \tag{1.2.11}$$

Using equal-time commutation relations between ψ and $\psi^\dagger$, we can verify that $e^{iQ\theta}$ indeed generates the phase transformations on the Dirac field ψ.

Isospin

Charge independence of the nuclear forces (pp, np, nn) led Heisenberg to introduce the concept of isospin, according to which proton and neutron are two states (isospin up and down) of the same object, the nucleon. An immediate implication of this hypothesis is that, if we define an internal space, where p and n form the coordinates, any rotation in that space (which, of course, mixes proton and neutron) will leave the nuclear forces invariant. In the language of the Lagrangian, the Lagrangian describing a system of neutrons and protons, interacting only via nuclear forces, will remain invariant under any rotation in that space. So there must be Noether currents, and the charges and conservation laws associated with them. To discuss this question we write a Lagrangian

$$\mathcal{L} = \mathcal{L}_0(\psi, \partial_\mu \psi),$$

where

$$\psi = \begin{pmatrix} p \\ n \end{pmatrix}. \tag{1.2.12}$$

A general rotation in this space is given by

$$\psi \to e^{i\varepsilon \cdot \tau/2} \psi$$

or

$$\delta\psi = \tfrac{i}{2}\varepsilon \cdot \tau \psi + O(\varepsilon^2)$$

for an infinitesimal parameter ε_a, which is a constant independent of space and time. From (1.2.7) we can easily read off the corresponding conserved Noether currents (of which there are now three, corresponding to the three Pauli matrices, τ^i)

$$J_\mu^i = +\tfrac{1}{2}\bar{\psi}\gamma_\mu \tau^i \psi. \tag{1.2.13}$$

Again, using equal-time commutation relations between the fields ψ, we can verify that the corresponding charges Q_i satisfy the algebra

$$[Q_i, Q_j] = i\varepsilon_{ijk}Q_K, \tag{1.2.14}$$

which is the same as the familiar angular momentum algebra of SU(2). The Casimir operator of SU(2), which will be used to designate irreducible representations of SU(2) algebra, has eigenvalues $I(I+1)$, where $t = 0$, $1/2, 1, 3/2, \ldots$, etc.

Isospin is actually an approximate symmetry rather than an exact one, since proton and neutron differ in their electric charge, and electromagnetic interactions do not respect isospin invariance. To see this, let us write the electromagnetic interaction of the proton–neutron system in isospin notation

$$\mathcal{L}_{\mathrm{em}}(p, n, A_\mu) = e\bar{p}\gamma^\mu p A_\mu = \frac{e}{2}\bar{\psi}\gamma^\mu(1+\tau_3)\psi A_\mu. \tag{1.2.15}$$

Note the appearance of the factor τ_3, which makes $\mathcal{L}_{em}$ noninvariant under isospin. A more realistic Lagrangian for the proton–neutron system includes both $\mathcal{L}_0$ and $\mathcal{L}_{em}$.

It is worth pointing out that, even though the isospin symmetry is not respected by $\mathcal{L} \equiv \mathcal{L}_0 + \mathcal{L}_{em}$, the charges Q_i (which are now time-dependent) satisfy the SU(2) algebra

$$[Q_i(t), Q_j(t)] = i\varepsilon_{ijk}Q_K(t). \tag{1.2.16}$$

It was pointed out by Gell-Mann [7] that important physical information can be gained by assuming the algebra of time-dependent charges (the so-called current algebra) even if the symmetry is broken. This idea extended to the chiral symmetry group SU(2) × SU(2) and was exploited by Adler and Weisberger [8] to give a satisfactory explanation of the axial vector coupling constant renormalization (g_A/g_V) in β-decay.

Equation (1.2.15) contains the germ of another important concept in particle physics, i.e., a formula relating electric charge Q to isospin:

$$Q = I_3 + \frac{B}{2}, \tag{1.2.17}$$

where B is the baryon number and I_3 is the third component of isospin. The discovery of V-particles by Rochester and Butler [and its subsequent analysis in terms of a new quantum number, the strangeness (S)] added a new dimension to the discussion of internal symmetries such as electric charge and isospin.

The formula for electric charge now becomes

$$Q = I_3 + \frac{B+S}{2}, \tag{1.2.18}$$

this is the Gell-Mann–Nishijima formula. Gell-Mann and Neeman combined isospin and strangeness (or hypercharge $Y \equiv B+S$) into the larger symmetry SU(3), which now describes both nonstrange and strange baryons and mesons. In the language of quarks SU(3) is the transformation that mixes u, d, and s quarks. With the quark spectrum given in Table 1.1, the new approximate global flavor symmetry is SU(6), and this symmetry is not only broken by electromagnetic interactions but also by different masses of the quarks.

In subsequent sections, we will note that broken symmetries are not the "monopoly" of the hadronic world alone; they exist in the leptonic world too. It is in the way they are hidden that makes them less obviously discernible. The symmetries of the hadronic world can be detected by looking at the level spectrum of particles and approximate mass degeneracies, but such studies do not reveal any symmetry for leptons since ν_e, e^-, μ^-, ..., etc., have so vastly different masses in their scale. Their symmetries, on the other hand, are manifest in their interactions and are consequences of local symmetries introduced in the next section.

§1.3   Local Symmetries and Yang–Mills Fields

In Section 1.2 we discussed global symmetries, where symmetry transformation on the fields is identical at all space–time points. We now introduce the concept of local symmetries, where the symmetry transformations can be arbitrarily chosen at different space–time points. Aesthetically, local symmetries appear much more plausible than global symmetries, where symmetry transformation must be implemented in an identical manner at all points no matter how far their separation is or even whether they are causally connected. The local symmetries are implemented by making the group parameters dependent on space–time. Let G be an arbitrary symmetry group whose elements are expressed in terms of its generators θ_a as follows:

$$G = e^{i\varepsilon^a \theta^a}, \tag{1.3.1}$$

where θ^a satisfy the Lie algebra of the group, i.e.,

$$[\theta^a, \theta^b] = i f^{abc} \theta^c, \tag{1.3.2}$$

where f^{abc} are the structure constants. If $\varepsilon_a(x)$ is a function of space–time x, then we call G a local symmetry. We see below that demanding invariance of a Lagrangian under the local symmetry $G(x)$ requires the introduction of a new set of spin 1 fields, A_μ^a as dynamical variables, and furthermore, its coupling to other fields in the Lagrangian is uniquely fixed by the requirement of gauge invariance. The fields A_μ^a will be called gauge fields. Thus, local symmetries dictate dynamics and therefore provide a more powerful theoretical tool for studies of particle interactions. Moreover, in contrast with global symmetries, the current in the case of local symmetries participates in the interactions and is therefore a physical quantity that, in principle, can be measured (e.g., the electric charge). To state it more pedagogically, a particle carrying a charge of a local symmetry has a field surrounding it (e.g., the electric field), whereas no such thing happens for particles carrying global symmetry charge.

To construct the Lagrangian invariant under an arbitrary local symmetry transformation [9, 10], $G(x)$, we consider a set of spin 1/2 matter fields $\psi(x)$ transforming as an irreducible representation of $G(x)$ as follows:

$$\psi(x) \rightarrow S(x)\psi(x). \tag{1.3.3}$$

Under this transformation, the kinetic energy term in the Lagrangian changes as follows:

$$\bar{\psi}\gamma^\mu \partial_\mu \psi \rightarrow \bar{\psi}\gamma^\mu \partial_\mu \psi + \bar{\psi}\gamma^\mu S^{-1} \partial_\mu S\psi. \tag{1.3.4}$$

To construct an invariant Lagrangian, we must therefore introduce a spin-1 field $B_\mu(x)$ represented as a matrix in the space of the column vector $\psi(x)$ and transforming under $G(x)$ as follows:

$$B_\mu(x) \rightarrow S B_\mu S^{-1} + S \partial_\mu S^{-1}. \tag{1.3.5}$$

If we now write a modified kinetic energy term in the Lagrangian as

$$\mathcal{L} = \bar{\psi}\gamma^\mu(\partial_\mu + B_\mu)\psi, \tag{1.3.6}$$

it is invariant under the transformations of the group $G(x)$.

To make B_μ into a dynamical field that can propagate, we have to write a G-invariant kinetic energy term. To do this we note that the function

$$F_{\mu\nu} = \partial_\mu B_\nu - \partial_\nu B_\mu + [B_\mu, B_\nu] \tag{1.3.7}$$

transforms covariantly under $G(x)$, i.e., $F_{\mu\nu} \overset{G}{\to} S F_{\mu\nu} S^{-1}$. Then the quantity $\operatorname{Tr} F_{\mu\nu}^2$ is G-invariant. We can therefore write full gauge invariant Lagrangians involving gauge and matter fields as

$$\mathcal{L} = +\frac{1}{4g^2} \operatorname{Tr} F^{\mu\nu} F_{\mu\nu} + i\bar{\psi}\gamma^\mu(\partial_\mu + B_\mu)\psi. \tag{1.3.8}$$

To make connection with the gauge field A_μ^a, we have to project out the independent components from the gauge field matrix B_μ. To count the number of independent A_μ^a note that there is one gauge field for each group parameter; therefore, the number of gauge fields is the same as the number of group parameters. We can therefore write the G_μ matrix as follows:

$$B_\mu = -ig\theta^a A_\mu^a. \tag{1.3.9}$$

Using eqs. (1.3.2) and (1.3.9), eq. (1.3.8) can be rewritten as:

$$\mathcal{L} = -\tfrac{1}{4} f^{a,\mu\nu} f_{\mu\nu}^a + i\bar{\psi}\gamma^\mu(\partial_\mu - ig\theta^a A_\mu)\psi, \tag{1.3.10}$$

where

$$f_{\mu\nu}^a = \partial_\mu A_\nu^a - \partial_\nu A_\mu^a + g f^{abc} A_\mu^b A_\nu^c. \tag{1.3.11}$$

Here we have assumed θ^a to be normalized generators satisfying the condition $\operatorname{Tr}\theta^a\theta^b = \delta^{ab}$.

A feature of Yang–Mills theories of great physical significance can be gleaned immediately from eq. (1.3.10), i.e., there is only one coupling constant describing interactions of all matter fields with the gauge field A_μ^a as well as the self-interactions of gauge fields. Thus, Yang–Mills theories are potentially strong candidates for describing interactions such as weak interactions, which, as we will see, have the feature that their interactions respect universality of coupling. There is, however, a potential hurdle in the path of a such applications, since gauge invariance demands that the gauge fields A_μ^a be massless.

While the Lagrangian obtained in the above manner is, of course, correct, we present an alternative derivation for this that emphasizes the role of the Noether current as a dynamical current participating in the gauge interactions. To see this, we write the gauge transformations for the ψ and A_μ^a for infinitesimal values of the group parameters $\varepsilon_a(x)$:

$$\psi \to \psi + i\theta\varepsilon(x)\psi,$$

$$\mathbf{A}_\mu \to \mathbf{A}_\mu + \frac{1}{g}\partial_\mu \boldsymbol{\varepsilon} + \mathbf{A}_\mu \times \boldsymbol{\varepsilon} + O(\varepsilon^2). \tag{1.3.12}$$

We have denoted the adjoint representation of group G by a vector symbol (boldface) for simplicity, and $\mathbf{A} \times \boldsymbol{\varepsilon}$ stands for $f^{abc}A_\mu^b \varepsilon^c$. Let us consider an arbitrary Lagrangian $\mathcal{L}(\psi, \partial_\mu \psi, \mathbf{A}_\mu, \partial_\mu \mathbf{A}_\nu)$ and try to determine its form by requiring invariance under eq. (1.3.12). The variation of the Lagrangian under eq. (1.3.12) is [using eq. (1.3.12)]

$$\delta\mathcal{L} = \frac{\delta\mathcal{L}}{\delta\psi}i\boldsymbol{\theta}\boldsymbol{\varepsilon}\psi + \frac{\delta\mathcal{L}}{\delta(\partial_\mu\psi)}i\boldsymbol{\theta}\partial_\mu(\boldsymbol{\varepsilon} \times \psi) + \frac{\delta\mathcal{L}}{\delta\mathbf{A}_\mu}\left(\frac{1}{g}\partial_\mu\boldsymbol{\varepsilon} + \mathbf{A}_\mu \times \boldsymbol{\varepsilon}\right)$$

$$+ \frac{\delta\mathcal{L}}{\delta(\partial_\mu\mathbf{A}_\nu)}\left(\frac{1}{g}\partial_\mu\partial_\nu\boldsymbol{\varepsilon} + \partial_\nu\mathbf{A}_\mu \times \boldsymbol{\varepsilon} + \mathbf{A}_\mu \times \partial_\nu\boldsymbol{\varepsilon}\right). \tag{1.3.13}$$

To obtain the Noether current generating this symmetry transformation, we can choose a particular form for the gauge parameter functions $\varepsilon(x)$, i.e., $\varepsilon(x) = $ const and get

$$-\mathbf{J}^\mu = i\frac{\delta\mathcal{L}}{\delta(\partial_\mu\psi)}\boldsymbol{\theta}\psi + \frac{\delta\mathcal{L}}{\delta(\partial_\mu\mathbf{A}_\nu)} \times \mathbf{A}_\nu, \tag{1.3.14}$$

$\mathbf{J}_\mu$ is conserved, i.e., $\partial^\mu \mathbf{J}_\mu = 0$. Using this, we can rewrite eq. (1.3.13) as

$$\delta\mathcal{L} = -\partial^\mu\boldsymbol{\varepsilon}\mathbf{J}_\mu + \frac{1}{g}\frac{\delta\mathcal{L}}{\delta\mathbf{A}_\mu}\partial_\mu\boldsymbol{\varepsilon} + \frac{1}{g}\frac{\delta\mathcal{L}}{\delta(\partial_\nu\mathbf{A}_\mu)}\partial_\mu\partial_\nu\boldsymbol{\varepsilon}. \tag{1.3.15}$$

For $\delta\mathcal{L} = 0$, we must have

$$\mathbf{J}_\mu = \frac{1}{g}\frac{\delta\mathcal{L}}{\delta\mathbf{A}_\mu} \tag{1.3.16}$$

and

$$\frac{\delta\mathcal{L}}{\delta(\partial_\nu\mathbf{A}_\mu)} = \frac{\delta\mathcal{L}}{\delta(\partial_\mu\mathbf{A}_\nu)}. \tag{1.3.17}$$

We conclude from eq. (1.3.16) that the Lagrangian must be a function of the symmetry current in the case of local symmetries. Equation (1.3.17) dictates an antisymmetric form for the kinetic energy term in agreement with our previous observations.

To obtain the exact form for $\mathbf{f}_{\mu\nu}$ derived earlier, we can write eq. (1.3.17) to imply the following general form:

$$\frac{\delta\mathcal{L}}{\delta(\partial^\mu\mathbf{A}^\nu)} = \partial_\mu\mathbf{A}_\nu - \partial_\nu\mathbf{A}_\mu + \mathbf{g}_{\mu\nu}(\mathbf{A}_\alpha). \tag{1.3.18}$$

Using eqs. (1.3.16) and (1.3.14), we get

$$\frac{1}{g}\frac{\delta\mathcal{L}}{\delta\mathbf{A}^\mu} = -(\partial_\mu\mathbf{A}_\nu - \partial_\nu\mathbf{A}_\mu + \mathbf{g}_{\mu\nu}) \times \mathbf{A}^\nu. \tag{1.3.19}$$

We can solve for $\mathbf{g}_{\mu\nu}$, by iteration. Let us assume $\mathbf{g}_{\mu\nu} = 0$; then eq. (1.3.19) implies, along with (1.3.18), that

$$\mathcal{L} = -\tfrac{1}{2}g(\partial_\mu \mathbf{A}_\nu - \partial_\nu \mathbf{A}_\mu) \cdot \mathbf{A}^\nu \times \mathbf{A}^\mu - \tfrac{1}{4}(\partial_\mu \mathbf{A}_\nu - \partial_\nu \mathbf{A}_\mu)^2. \qquad (1.3.20)$$

Equations (1.3.18) and (1.3.20) then determine the following unique form for $\mathcal{L}$:

$$\mathcal{L} = -\tfrac{1}{4}(\partial_\mu \mathbf{A}_\nu - \partial_\nu \mathbf{A}_\mu + g\mathbf{A}_\mu \times \mathbf{A}_\nu)^2. \qquad (1.3.21)$$

Thus, for local symmetries, the Noether theorem is even powerful enough to fix the dynamics.

§1.4 Quantum Chromodynamic Theory of Strong Interactions: An Application of Yang–Mills Theories

In the previous section, we saw how powerful the implications of exact local symmetries are. We would therefore like to search for areas of physics where it may be applicable. It is well known that electromagnetic interactions are invariant under a local U(1) symmetry associated with phase transformation of complex fields. The associated massless spin-1 boson is the photon. The only other force in nature that is known to be long range is gravitation, but to consistently describe the properties of gravity we need a massless spin-2 boson, the graviton, which, as we will see in Chapter 14, can be thought of as the gauge field associated with local translation symmetry. So if we want to apply the beautiful idea of non-abelian local symmetries to nature, either (i) we must find a way to give mass to the gauge bosons without spoiling the local symmetry of the Lagrangian; or (ii) we must explain the absence of long-range forces arising from the exchange of massless gauge particles. The first method will be chosen in applying gauge theories to the study of electroweak interactions, whereas the second alternative will be chosen in applications to strong interactions, and this is the focus of this section.

Let us briefly discuss the features of strong interactions that suggest its description in terms of a non-abelian gauge theory. The first important step in the understanding of nuclear forces was the success of SU(3) symmetry for hadrons, which led Gell-Mann and Zweig to introduce a quark picture of hadrons. According to this picture, the quarks are assumed to have spin 1/2 and baryon number 1/3 and transform as SU(3) triplets denoted by (u, d, s); (u, d) carry isospin and s carries strangeness. This picture provides a clear understanding of the spectroscopy of low-lying mesons and baryons in terms simple S- and P-wave bound states of quarks (u, d, s) or quark and antiquarks. The baryon spectroscopy is understood in terms of three-quark bound states, whereas meson spectroscopy arises from nonrelativistic quark–antiquark bound states. Accepting quarks as the constituents of hadrons, we have to search for a field theory that provides the binding force between the quarks.

In trying to understand the Fermi statistics for baryons (such as Ω^-), it became clear that if they are S-wave bound states, then the space part of their wave function is totally symmetric; since a particle such as Ω^- consists of three strange quarks, and has spin 3/2, the spin part of its wave function is symmetric. If there were no other degree of freedom, this would be in disagreement with the required Fermi statistics. A simple way to resolve this problem is to introduce [11] a threefold degree of freedom for quarks, called color (quarks being color triplets) and assume that all known baryons are singlet under this new SU(3). Since an SU(3)$_C$-singlet constructed out of three triplets is antisymmetric in the interchange of indices (quarks), the total baryon wave function is antisymmetric in the interchange of any two constituents as required by Fermi statistics.

It is now tempting to introduce strong forces by making SU(3)$_C$ into a local symmetry. In fact, if this is done, we can show that exchange of the associated gauge bosons provides a force for which the SU(3)$_C$ color singlet is the lowest-lying state; and triplet, sextet, and octet states all have higher mass. By choosing this mass gap large, we can understand why excited states corresponding to the color degree of freedom have not been found.

While this argument in favor of an SU(3)$_C$ gauge theory of strong interaction was attractive, it was not conclusive. The most convincing argument in favor of SU(3)$_C$ gauge theory came from the experimental studies of deep inelastic neutrino and electron scattering off nucleons. These experiments involved the scattering of very-high-energy (E) electrons and neutrinos with the exchange of very high momentum transfers (i.e., q^2 large). It was found that the structure functions, which are analogs of form factors for large q^2 and E, instead of falling with q^2, became scale-invariant functions depending only on the ratio $q^2/2mE$. This was known as the phenomenon of scaling [12]. Two different theoretical approaches were developed to understand this problem. The first was an intuitive picture called the parton model suggested by Feynman [13] and developed by Bjorken and Paschos [14], where it was assumed that, at very high energies, the nucleon can be thought of as consisting of free pointlike constituents. The experimental results also showed that these pointlike constituents were spin-1/2 objects, like quarks, and the scaling function was simply the momentum distribution function for the partons inside the nucleon. These partons could be identified with quarks, thus providing a unified description of the nucleon as consisting of quarks at low, as well as at high, energies. The main distinction between these two energy regimes uncovered by deep inelastic scattering experiments is that at low energies the forces between the quarks are strong, whereas at high energies the forces vanish letting the quarks float freely inside the nucleons. If g described the strong interaction coupling constant for the quarks

$$g(Q^2) \xrightarrow[Q^2 \to \infty]{} 0,$$

where

$$g(Q^2) \xrightarrow[Q^2 \to 0]{} \text{large.} \qquad (1.4.1)$$

The same picture emerged in trying to understand the scaling phenomena from a field-theoretic point of view. It was noted by Wilson and others [15] that the scaling region, i.e., $Q^2 \to \infty$, $E \to \infty$ with $x = Q^2/2mE =$ finite, corresponded in coordinate space to the behavior of current-operator products on the light cone, and revealed that only if current products on the light cone have free field singularities [15], would we get the observed scaling phenomena. Again, this implied that the strong interaction coupling among quarks must vanish for large Q^2 values.

The big challenge to theoretical physics of the early 1970s was to find a field theory that had this property of $g(Q^2) \to 0$ for large Q^2 or "asymptotic freedom." The discovery of the Callan–Symanzik equation [16] in 1970–1971 had given a way to probe the behavior of coupling constants at different momentum transfers. It is given by the following equation:

$$\frac{dg}{dt} = \beta(g), \qquad (1.4.2)$$

where $t = \ln Q^2$ and $\beta(g)$ is a function of the coupling constants, which can be obtained by studying the coupling constant renormalization in a field theory. Asymptotic freedom would result if $\beta(g) \xrightarrow[g \to 0]{} 0$ with $d\beta/dg < 0$.

It was discovered by Gross and Wilczek and by Politzer [17] that only non-abelian gauge theories have this property if the number of fermion flavors is less than 16. More quantitatively, they calculated the function $\beta(g)$ for an SU(N) non-abelian gauge theory with N_f fermion species each transforming as N-dimensional representation under SU(N) and found

$$\beta = -\frac{g^3}{16\pi^2} \left[\tfrac{11}{3}N - \tfrac{2}{3}N_f \right]. \qquad (1.4.3)$$

Thus, in the SU(3)$_C$ theory of quarks, the coupling constant has the desired property of asymptotic freedom and no other theory (scalar, pseudoscalar interactions of quarks, etc.) has this attractive property. Thus, unbroken SU(3)$_C$ gauge theory became the accepted theory of strong interactions and came to be known as quantum chromodynamics (QCD). The associated SU(3)$_C$ gauge bosons are called gluons. In this picture, all observed nuclear forces are residual effects of gluon exchanges between the quarks that are bound inside the nucleon.

An immediate problem was to understand why long-range forces resulting from single gluon exchange have not been seen. Similarly, since gluons are massless, they should be copiously produced in all strong interaction processes. Furthermore, since at high energies quarks become free, they should have been seen because their masses are expected to be less than the nucleon mass. To resolve this conundrum, a new principle, known as "confinement," was introduced [18]. According to this principle, all unbroken non-abelian gauge theories with the property of asymptotic freedom should confine the gauge nonsinglet states, and only observable sectors of the Hilbert space correspond to states that are singlets under this local symmetry. This would then explain why quarks and gluons, and indeed any other nonsinglet bound hadron state, have not been seen. In the last 10 years, many plausible arguments and techniques have been developed to prove confinement and it has, today, become an accepted principle.

We close this subsection by writing down the QCD Lagrangian for different quark flavors:

$$\mathcal{L}_{\text{QCD}} = \sum_{a=flavors} -\bar{q}_a \gamma^\mu (\partial_\mu - \tfrac{i}{2} g_2 \vec{\lambda}_c \mathbf{V}_\mu) q_a - \tfrac{1}{4} f^{i,\mu\nu} f^i_{\mu\nu}, \tag{1.4.4}$$

where $q_a = (q_a^1, q_a^2, q_a^3)$, for three colors and $\vec{\lambda}_C$ are Gell-Mann's matrices for $SU(3)_C$ generators.

§1.5 Hidden Symmetries of Weak Interactions

In the preceding section we presented arguments for a non-abelian gauge theory of strong interactions. In this section we would like to look for clues that will enable us to apply the ideas of gauge theories in the domain of weak interactions. Weak interaction processes known to date are of three kinds: (a) those that involve only leptons such as $\nu_e, e^-, \nu_\mu, \mu^-$ (to be called leptonic); (b) those that involve only hadrons such as $\Sigma, \Lambda, p, \pi^-, K$ (to be called hadronic); and (c) those that involve both hadrons and leptons (to be called semileptonic). (For a detailed study of weak interactions prior to the advent of gauge theories, we refer to several excellent texts [1].) Some typical weak processes are (the lifetimes are given in parentheses)

$$
\begin{aligned}
\mu^- &\rightarrow e^- \bar{\nu}_e \nu_\mu & (2.19 \times 10^{-6}\,\text{s}), \\
n &\rightarrow p e^- \bar{\nu}_e & (898\,\text{s}), \\
K^+ &\rightarrow \pi^0 e^+ \nu_e & (1.2371 \times 10^{-8}\,\text{s}), \\
\Lambda &\rightarrow p \pi^- & (\sim 3.7 \times 10^{-10}\,\text{s}), \\
\Sigma &\rightarrow p \pi^- & (\sim 1.6 \times 10^{-10}\,\text{s}), \\
K^0 &\rightarrow 2\pi & (0.89 \times 10^{-10}\,\text{s}).
\end{aligned}
$$

The first important point to note is that even though the various decay half-lives vary by orders of magnitude from 10^3 to 10^{-10} s, if we try to understand them in terms of a Hamiltonian involving four fermion operators, i.e.,

$$H_{\text{wk}} = \frac{G_F}{\sqrt{2}} \bar{\psi}_1 O \psi_2 \bar{\psi}_3 O \psi_4, \tag{1.5.1}$$

we find that for β-decay as well as μ-decay the associated couplings G_β and G_μ are almost the same ($\sim 10^{-5}/m_p^2$), and for the decays involving strange particles, the coupling strength is $G_\mu \sin\theta_C$ where θ_C, is known as the Cabibbo angle whose value is determined to be [19]

$$\sin\theta_C = 0.231 \pm 0.003. \qquad (1.5.2)$$

So even though the particles taking part in weak interactions have widely varying masses (and would thus not indicate any obvious symmetry), their interaction strength is universal. Because gauge theories lead to universal coupling strengths, as we saw in Section 1.3, this may be the first reason to suspect the relevance of gauge symmetries to weak interactions.

To proceed further, we need to know the form of O in eq. (1.5.1). When Fermi originally wrote down the interaction in eq. (1.5.1), he chose the most general form for O (i.e., 1, γ_5, γ_μ, $\gamma_\mu\gamma_5$, $\sigma_{\mu\nu}$) consistent with Lorentz invariance and parity symmetry. In 1956, in order to understand the $\tau - \theta$ puzzle, Lee and Yang proposed that weak interactions may not conserve parity, an idea that was experimentally confirmed by Wu, Ambler, Hayward, and Hobson in the decay of polarized ^{60}Co. Then in 1957, Marshak, Sudarshan, Feynman, Gell-Mann, and Sakurai proposed the $V - A$ theory of weak interaction that uniquely gave the form for the operator O to be

$$O = \gamma_\mu(1 + \gamma_5). \qquad (1.5.3)$$

The fact that the operator involved in (1.5.1) is a vector–axial-vector type implies that currents are responsible for weak interactions. This is another indication that gauge theories may be relevant to weak interactions, since, as we argued in Section 1.3, gauge symmetries imply that the interaction Lagrangian *must* involve the Noether current that generates the symmetry.

Having said this much, we can now look for the kind of symmetries by simply looking at the current involved in weak processes and its equal-time commutation relations to get the Lie algebra of the weak symmetries.

To make this discussion more transparent we will depart from the historical order of events and proceed to express the weak Hamiltonian and the weak currents in terms of the basic constituents of matter, the quarks (rather than p, n, Λ, Σ, ...) and leptons. The weak Hamiltonian $H_{\text{wk}}^{\text{cc}}$ (cc stands for charged current) can be written in terms of weak charged currents as

$$H_{\text{wk}}^{\text{cc}} = \frac{G_F}{\sqrt{2}} J^\mu J_\mu^+, \qquad (1.5.4)$$

where

$$J_\mu = \bar{u}\gamma_\mu(1 + \gamma_5)(\cos\theta_C d + \sin\theta_C s) + \bar{\nu}_e\gamma_\mu(1 + \gamma_5)e^- + \bar{\nu}_\mu\gamma_\mu(1 + \gamma_5)\mu^-.$$
$$(1.5.5)$$

We will soon generalize this to include all generations. To uncover the hidden symmetry of weak interactions, we go to the limit of $\theta_C = 0$ and ignore the muon and electron pieces and look at the $\bar{u}\gamma_\mu(1 + \gamma_5)d$ piece alone. The charge corresponding to this current is

$$Q_L^- = \int d^3x \, u^+(1+\gamma_5)d, \tag{1.5.6}$$

and using equal-time canonical commutation relations for quark fields we get

$$[Q_L^-, Q_L^+] = 2Q_L^3, \tag{1.5.7}$$

where

$$Q_{3L} = \int d^3x \{ u^+(1+\gamma_5)u - d^+(1+\gamma_5)d \}. \tag{1.5.8}$$

Thus, the weak charges have the potential to generate a weak symmetry; however, if this symmetry is to be local, then the Q_{3L} piece should also be experimentally measured with the same strength as charged currents, i.e., the general form of the full weak Hamiltonian should be

$$H_{wk} = 4\frac{G_F}{\sqrt{2}}[J_L^{+\mu}J_{\mu L}^- + J_L^{3\mu}J_{\mu L}^3], \tag{1.5.9}$$

where

$$J_{\mu L}^3 = \tfrac{1}{4}\big[\bar{u}\gamma_\mu(1+\gamma_5)u - \bar{d}\gamma_\mu(1+\gamma_5)d + \bar{\nu}_e\gamma_\mu(1+\gamma_5)\nu_e - \bar{e}\gamma_\mu(1+\gamma_5)e$$
$$+ \bar{\nu}\gamma_\mu(1+\gamma_5)\nu_\mu - \bar{u}\gamma_\mu(1+\gamma_5)\mu + \cdots \big]. \tag{1.5.10}$$

Again, ignoring the chronological order of events, the processes arising from the neutral-current piece of the Hamiltonian are those where charge does not change at each current vertex, i.e.,

$$\nu + p \to \nu + p,$$
$$\nu_\mu + e \to \nu_\mu + e,$$
$$\nu_e + e \to \nu_e + e. \tag{1.5.11}$$

These processes have been observed [20] and they confirm eq. (1.5.9) as the correct structure for the weak Hamiltonian up to a piece, which we denote by H'_{wk}:

$$H'_{wk} = -\frac{4G_F}{\sqrt{2}}k\bar{\nu}\gamma^\mu(1+\gamma_5)\nu J_\mu^{em}. \tag{1.5.12}$$

If we temporarily ignore eq. (1.5.12), we find that the weak interaction Hamiltonian has all the right properties to incorporate an underlying local weak SU(2) symmetry. Under this weak SU(2) symmetry, the left-handed fermions must transform as doublets as follows:

$$\begin{pmatrix} \nu_{e_L} \\ e_L^- \end{pmatrix} \begin{pmatrix} u_L \\ d_L\cos\theta_C + s_L\sin\theta_C \end{pmatrix}$$
$$\begin{pmatrix} \nu_{\mu_L} \\ \mu_L^- \end{pmatrix}. \tag{1.5.13}$$

If we write down eq. (1.5.9) using the doublets given in eq. (1.5.13), the charged-current piece of the weak Hamiltonian is correctly reproduced. But as far as the neutral-current piece goes, however, it gives rise to interactions that change strangeness, i.e.,

$$H^{\Delta Q=0}_{\Delta S=1} = \frac{G_F}{\sqrt{2}} \cos\theta_C \sin\theta_C \bar{d}\gamma^\alpha (1+\gamma_5)s\bar{u}\gamma_\alpha(1+\gamma_5)\mu + \cdots. \quad (1.5.14)$$

This will lead to the process $K_L^0 \to \mu^+\mu^-$ with a lifetime comparable to $K^+ \to \mu^+\nu$; but, experimentally, it is known that

$$\frac{\Gamma(K_L^0 \to \mu^+\mu^-)}{\Gamma(K^+ \to \text{all})} \simeq (9.1 \pm 1.9) \times 10^{-9}. \quad (1.5.15)$$

This implies that the strangeness-changing neutral-current piece must be absent from the weak Hamiltonian. It was suggested by Glashow, Illiopoulos, and Maiani [5] that a fourth quark (called charm quark, c) with properties similar to the up quark should be introduced and that it must belong to a new weak doublet

$$\begin{pmatrix} c_L \\ -d_L \sin\theta_C + s_L \cos\theta_C \end{pmatrix}.$$

Then the $\Delta S = 1$, $\Delta Q = 0$ piece of the Hamiltonian disappears and the $K_L \to \mu^+\mu^-$ puzzle is solved. In fact, in 1974, the charm quark was discovered [21] and it is now well established that the weak interactions of the charm quark are given by the doublet structure given above.

With the discovery of two new leptons (ν_τ, τ^-) [22] and a new quark b [23], the complete weak $SU(2)_L$ doublet structure, including the effects of all three generations, can be written as follows:

$$\begin{pmatrix} u_L \\ d_L' \end{pmatrix} \begin{pmatrix} c_L \\ s_L' \end{pmatrix} \begin{pmatrix} t_L \\ b_L' \end{pmatrix},$$

$$\begin{pmatrix} \nu_{eL} \\ e_L^- \end{pmatrix} \begin{pmatrix} \nu_{\mu L} \\ u_L^- \end{pmatrix} \begin{pmatrix} \nu_{\tau L} \\ \tau_L^- \end{pmatrix}, \quad (1.5.16)$$

where

$$\begin{pmatrix} d_L' \\ s_L' \\ b_L' \end{pmatrix} = U_{KM}(\theta_1, \theta_2, \theta_3, \delta) \begin{pmatrix} d_L \\ s_L \\ b_L \end{pmatrix}. \quad (1.5.17)$$

U_{KM} is a unitary matrix, known as the Kobayashi–Maskawa [24] matrix, and is parametrized by three real angles and a phase:

$$U_{KM} = \begin{pmatrix} c_1 & -s_1 c_3 & -s_1 s_3 \\ s_1 c_2 & c_1 c_2 c_2 - s_2 s_3 e^{i\delta} & c_1 c_2 s_3 + s_2 c_3 e^{i\delta} \\ s_1 s_2 & c_2 s_2 c_3 + c_2 s_3 e^{i\delta} & c_1 s_2 s_3 - c_2 c_3 e^{i\delta} \end{pmatrix}. \quad (1.5.18)$$

The discovery of the t quark by the Fermilab CDF [25] and DO [26] collaborations completed the third generation of quarks and leptons. The recent DONUT experiment [27] has also given explicit evidence for the τ neutrino. Thus all the matter fermions of the standard model have now been discovered.

The masses of the various quarks and leptons are $M_u = 5.6 \pm 1.1$ MeV; $M_d = 9.9 \pm 1.1$ MeV; $M_s \simeq 199 \pm 33$ MeV; $M_c \simeq 1.35 \pm 0.5$ GeV; $M_b \simeq 5$ GeV; $M_t \simeq 175 \pm 5$ GeV; $M_e = 0.5109990 \pm 0.00000015$ MeV; $M_\mu = 105.65839 \pm 0.00006$ MeV; $M_\tau = 1784.1 \{ ^{+2.7}_{-3.6}$ MeV.

Before closing this section we note that, while having convincingly argued for describing weak interactions by a local $SU(2)_L$ symmetry theory, we realize that these interactions are short range, and therefore, we must give mass to the gauge bosons without destroying the gauge invariance of the Lagrangian. Note that we cannot use confinement arguments, as in the case of QCD, because ν, e^- are observed particles. This will be dealt with in the next chapter, where we seek different realizations of the symmetries present in a Lagrangian.

Exercises

1.1. Write down explicitly the weak hardronic charges including the GIM mechanism, and show that the neutral charge conserves both strangeness and charm.

1.2. Explain, by using dimensional analysis, why the beta decay of the neutron and muon decay have so very different lifetimes, even though the strength of weak interactions is universal (almost) between these two processes.

1.3. Assuming I_3 to be the diagonal generator of weak isospin, write a formula analogous to eq. (1.2.18) that is valid for both hadrons and leptons.

References

[1] For a discussion of weak interactions, see
    R. E. Marshak, Riazuddin, and C. Ryan, *Theory of Weak Interactions in Particle Physics*, Wiley, New York, 1969;
    E. D. Commins, *Weak Interactions*, McGraw-Hill, New York, 1973;
    J. J. Sakurai, *Currents and Mesons*, University of Chicago Press, Chicago, 1969.

[2] M. Gell-Mann, *Phys. Lett.* **8**, 214 (1964);
    G. Zweig, CERN preprint 8182/Th. 401 (1964), reprinted in *Developments in Quark Theory of Hadrons*, Vol. 1, 1964–1978 (edited by S. P. Rosen and D. Lichtenberg), Hadronic Press, MA., 1980.

[3] Some general references on recent developments in gauge theories are:
J. C. Taylor, *Gauge Theories of Weak Interactions*, Cambridge University Press, Cambridge, 1976;
C. Quigg, *Gauge Theories of the Strong, Weak, and Electromagnetic Interactions*, Benjamin-Cummings, New York, 1983;
R. N. Mohapatra and C. Lai, *Gauge Theories of Fundamental Interactions*, World Scientific, Singapore, 1981;
A. Zee, *Unity of Forces in Nature*, World Scientific, Singapore, 1983;
M. A. Beg and A. Sirlin, *Phys. Rep.* **88**, 1(1982); and *Ann. Rev. Nucl. Sci.* **24**, 379 (1974);
E. S. Abers and B. W. Lee, *Phys. Rep.* **9C**, 1 (1973);
T. P. Cheng and L. F. Li, *Gauge Theory of Elementary Particle Physics*, Oxford University Press, New York, 1984;
P. Langacker, *Phys. Rep.* **72C**, 185 (1981);
G. G. Ross, *Grandunified Theories*, Benjamin-Cummings, New York, 1985;
L. B. Okun, *Leptons and Quarks*, North-Holland, Amsterdam, 1981;
R. N. Mohapatra, *Fortsch. Phys.* **31**, 185 (1983);
A. Masiero, D. Nanopoulos, C. Kounas and K. Olive, *Grandunijication and Cosmology*, World Scientific, Singapore, 1985.

[4] O. W. Greenberg, *Phys. Rev. Lett.* **13**, 598 (1964);
M. Y. Han and Y. Nambu, *Phys. Rev.* **139**, B1006 (1965);
H. Fritzsch, M. Gell-Mann, and H. Leutweyler, *Phys. Lett.* **478**, 365 (1973).

[5] A. Gamba, R. E. Marshak, and S. Okubo, *Proc. Nat. Acad. Sci. (USA)* **45**, 881 (1959);
J. D. Bjorken and S. L. Glashow, *Phys. Lett.* **11**, 255 (1965);
S. L. Glashow, J. Illiopoulos, and L. Maiani, *Phys. Rev.* **D2**, 1285 (1970).

[6] J. C. Pati and A. Salam, *Phys. Rev.* **D10**, 275 (1974);
H. Georgi and S. L. Glashow, *Phys. Rev. Lett.* **32**, 438 (1974).

[7] M. Gell-Mann, *Physics* **1**, 63 (1964).

[8] S. L. Adler, *Phys. Rev. Lett.* **14**, 1051 (1965);
W. I. Weisberger, *Phys. Rev. Lett.* **14**, 1047 (1965).

[9] C. N. Yang and R. L. Mills, *Phys. Rev.* **96**, 191 (1954).

[10] R. Shaw, Problem of Particle Types and Other Contributions to the Theory of Elementary Particles Ph.D. Thesis, Cambridge University, 1955.

[11] O. W. Greenberg, *Phys. Rev. Lett.* **13**, 598 (1964);
M. Y. Han and Y. Nambu, *Phys. Revs.* **139**, B1006 (1965).
For a review and references on the subject, see
O. W. Greenberg and C. A. Nelson, *Phys. Rep.* **32**, 69 (1977);
W. Marciano and H. Pagels, *Phys. Rep.* **36C**, 137 (1978).

[12] J. D. Bjorken, *Phys. Rev.* **179**, 1547 (1969).

[13] R. Feynman, *Photon–Hadron Interactions*, Benjamin, Reading, MA., 1972.

[14] J. D. Bjorken and E. A. Paschos, *Phys. Rev.* **185**, 1975 (1969).

[15] K. Wilson, *Phys. Rev.* **179**, 1499 (1969);
R. Brandt and G. Preparata, Nucl. Phys. **27B**, 541 (1971);
H. Fritzsch and M. Gell-Mann, *Proceedings of the Coral Gables Conference*, 1971;
Y. Frishman, *Phys. Rev. Lett.* **25**, 966 (1970).

[16] C. G. Callan, *Phys. Rev.* **D2**, 1541 (1970);
K. Symanzik, *Comm. Math. Phys.* **18**, 227 (1970).

[17] D. Gross and F. Wilczek, *Phys. Rev. Lett.* **30**, 1343 (1973);
H. D. Politzer, *Phys. Rev. Lett.* **30**, 1346 (1973).

[18] K. Wilson, *Phys. Rev.* **D3**, 1818 (1971).

[19] For a review, see
L. L. Chau, *Phys. Rep.* **95C**, 1(1983).

[20] F. Hasert et al., *Phys. Lett.* **46B**, 121 (1973);
B. Aubert et al., *Phys. Rev. Lett.* **32**, 1457 (1974);
A. Benvenuti et al., *Phys. Rev. Lett.* **32**, 800 (1974);
B. Barish et al., *Phys. Rev. Lett.* **32**, 1387 (1974).

[21] J. J. Aubert et al., *Phys. Rev. Lett.* **33**, 1404 (1974);
J.-E. Augustin et al., *Phys. Rev. Lett.* **33**, 1406 (1974);
G. Goldhaber et al., *Phys. Rev. Lett.* **37**, 255 (1976);
I. Peruzzi et al., *Phys. Rev. Lett.* **37**, 569 (1976);
E. G. Cazzoli et al., *Phys. Rev. Lett.* **34**, 1125 (1975).

[22] For a recent survey, see
M. Perl, *Physics in Collision*, Vol. 1 (edited by W. P. Trower and G. Bellini), Plenum, New York, 1982.

[23] S. W. Herb et al., *Phys. Rev. Lett.* **39**, 252 (1977);
W. R. Innes et al., *Phys. Rev. Lett.* **39**, 1240 (1977);
C. W. Darden et al., *Phys. Lett.* **76B**, 246 (1978);
Ch. Berger et al., *Phys. Lett.* **76B**, 243 (1978).

[24] M. Kobayashi and T. Maskawa, *Prog. Theor. Phys.* **49**, 652 (1973).

[25] F. Abe et al., *Phys. Rev. Lett.* **74**, 2626 (1995).

[26] S. Abachi et al., *Phys. Rev. Lett.* **74**, 2632 (1995).

[27] DONUT collaboration, *Nucl. Phys. Proc. Supplement*, **77**, 259 (1999).

2

Spontaneous Symmetry Breaking, Nambu–Goldstone Bosons, and the Higgs Mechanism

§2.1 Symmetries and Their Realizations

A Lagrangian for a physical system may be invariant under a given set of symmetry [1] transformations; but how the symmetry is realized in nature depends on the properties of the ground state. In field theories the ground state is the vacuum state. We will, therefore, have to know how the vacuum state responds to symmetry transformations.

Let $U(\varepsilon)$ denote the unitary representations of the symmetry group in the Fock space of a given field theory. Under the symmetry transformations by parameters ε, a field ϕ transforms as follows:

$$\phi_a \rightarrow \phi'_a = U(\varepsilon)\phi_a\,U^{-1}(\varepsilon) = C_{ab}\phi_b \tag{2.1.1}$$

and

$$\mathcal{L}(\phi', \partial_\mu\phi') = \mathcal{L}(\phi, \partial_\mu\phi). \tag{2.1.2}$$

The vacuum state $|0\rangle$ may or may not be invariant under $U(\varepsilon)$. Let us consider both these cases.

Case (i): $U(\varepsilon)|0\rangle = |0\rangle$. This is known as the Wigner–Weyl mode of symmetry realization. We can show that, in this case, there exists mass degeneracy between particles in a supermultiplet as well as relations between coupling constants and scattering amplitudes. We will illustrate the technique of obtaining mass degeneracy.

The mass of a particle is defined as the position of the pole in the variable p^2 in the propagation function $\Delta_{aa}(p)$ for the corresponding field ϕ_a defined as follows:

$$i\Delta_{aa}^{F}(p) = \int e^{-ip\cdot x} d^4x \langle 0|T(\phi_a(x)\phi_a(0))|0\rangle. \qquad (2.1.3)$$

We can represent the function near its pole as

$$i\Delta_{aa}^{F}(p) = \frac{iZ_a}{p^2 - m_a^2} + \text{terms regular at } p^2 = m_a^2. \qquad (2.1.4)$$

Similarly, for the field ϕ_b we can write

$$i\Delta_{bb}^{F}(p) = \frac{iZ_b}{p^2 - m_b^2} + \text{terms regular at } p^2 = m_b^2, \qquad (2.1.5)$$

where $Z_{a,b}$ denote the wave function renormalization. If ϕ_a and ϕ_b belong to the same supermultiplet, they can be transformed into one another by a symmetry transformation $U(\varepsilon)$, i.e.,

$$U(\varepsilon)\phi_a\,U^{-1}(\varepsilon) = \phi_b. \qquad (2.1.6)$$

Since $U(\varepsilon)|0\rangle = |0\rangle$, this implies $\Delta_{aa}(p) = \Delta_{bb}(p)$, i.e., the two functions are identical. This in turn implies that $m_a = m_b$ and $Z_a = Z_b$.

Similar techniques can be used to relate coupling constants and scattering amplitudes that are derived by looking at the poles of three- and four-point functions of the fields ϕ_a.

Case (ii): $U(\varepsilon)|0\rangle \neq |0\rangle$. This is known as the Nambu–Goldstone realization of the symmetry. We will show that, in this case, the spectrum of particles in the theory must contain a zero-mass particle, known as the Nambu–Goldstone boson.

To prove this we choose an infinitesimal symmetry transformation $U(\varepsilon)$, which can be expanded as follows:

$$U(\varepsilon) = 1 + i\varepsilon_j Q_j, \qquad (2.1.7)$$

Q_j are the generators of symmetry. Noninvariance of vacuum then implies

$$Q_j|0\rangle \neq 0 \qquad \text{(for at least one } j). \qquad (2.1.8)$$

Noether's theorem tells us that

$$\frac{dQ_j}{dt} = 0. \qquad (2.1.9)$$

Equation (2.1.8) implies that there exists at least one state $|m\rangle$ in the Hilbert space for which

$$\langle m|Q_j|0\rangle \neq 0. \qquad (2.1.10)$$

Equation (2.1.9) leads to

$$\langle m|\frac{dQ_j}{dt}|0\rangle = 0. \qquad (2.1.11)$$

Using translation invariance we can rewrite eq. (2.1.11) as

$$E_m \delta^3(\mathbf{p}_m)\langle m|Q_j|0\rangle = 0. \tag{2.1.12}$$

Equation (2.1.12) implies that

$$\lim_{\mathbf{p}_m \to 0} E_m = 0. \tag{2.1.13}$$

This means that there is a massless, spin-0 particle in the theory. Let us now recapitulate the three conditions required to prove this fundamental theorem of particle physics:

(a) Conserved current $\partial_\mu J_\mu = 0$, corresponding to an exact symmetry of the Lagrangian.

(b) $Q_j|0\rangle \neq 0$. This condition can also be written as

$$\langle 0|[Q, \phi]|O\rangle \neq 0, \tag{2.1.14}$$

where ϕ is a local bosonic operator in the theory or as $\langle \phi \rangle \neq 0$.

(c) Lorentz invariance or locality. For instance, without the locality property, we cannot obtain eq. (2.1.12) from eq. (2.1.11).

We will not go into any more detailed discussion of the Goldstone theorem, except to note that the number of Goldstone bosons is precisely the same as the number of charges that do not annihilate the vacuum (or the number of broken-symmetry generators).

Example 2.1 We can give classical, quantum-mechanical, as well as field-theoretical examples of systems for which the Lagrangian respects a symmetry whereas the ground state does not. A classical example is a bead sliding on a frictionless vertical circular wire rotating about its vertical diameter. Above a certain angular velocity the equilibrium (or ground state) corresponds to the bead in a position with azimuthal angle $\theta \neq 0$. This breaks the discrete symmetry corresponding to $\theta \to -\theta$. We leave it as an exercise to the reader to work out the Hamiltonian for this system and prove our assertion.

A well-known quantum-mechanical example is the configuration of the ground state of the ozone (O_3) molecule. Instead of the three oxygen atoms occupying the vertices of an equilateral triangle, they occupy those of an isosceles triangle.

Let us now discuss a field-theoretical example, where a continuous U(1) symmetry is broken. Consider a scalar field ϕ with a nonzero U(1) charge

$$\phi \xrightarrow{\mathrm{U}(1)} e^{i\theta}\phi. \tag{2.1.15}$$

The Lagrangian

$$\mathcal{L} = (\partial^\mu \phi)(\partial_\mu \phi^*) - \mu^2 \phi^* \phi - \lambda(\phi^* \phi)^2 \tag{2.1.16}$$

is invariant under the above U(1) transformations. Let us choose $\mu^2 > 0$, $\lambda > 0$ as required for the Hamiltonian to have a lower bound. The minimum of the Hamiltonian corresponds to

$$\frac{\partial V}{\partial \phi} = 0, \tag{2.1.17}$$

which implies

$$\langle \phi \rangle^2 = \frac{\mu^2}{2\lambda} \neq 0. \tag{2.1.18}$$

This implies that the vacuum state is not invariant under the U(1) symmetry. Therefore, there must be a zero-mass particle in the theory. If we write

$$\phi = \sqrt{\frac{\mu^2}{2\lambda}} + R + iG, \tag{2.1.19}$$

we can easily check that

$$M_G = 0,$$
$$M_R = \sqrt{2}\mu = 2\sqrt{\lambda}\langle \phi \rangle. \tag{2.1.20}$$

Therefore, $G \equiv \text{Im } \phi$ is the Nambu–Goldstone boson corresponding to spontaneous breaking of the U(1) symmetry.

§2.2 Nambu–Goldstone Bosons for an Arbitrary Non-abelian Group

Let G be an n-parameter group with generators θ_A, $A = 1, \ldots, n$. Let ϕ be an irreducible representation of G. Let $V(\phi)$ be the potential for a physical system that is invariant under the group G. To obtain the ground state, we have to look for minima of the potential

$$\frac{\partial V}{\partial \phi_i}\bigg|_{\phi_i = \lambda_i} = 0 \tag{2.2.1}$$

and

$$\frac{\partial^2 V}{\partial \phi_i \partial \phi_j}\bigg|_{\phi = \lambda} \equiv M_{ij}^2, \tag{2.2.2}$$

such that M^2 has positive or zero eigenvalues. The minima generally consist of a manifold of points in the ϕ-space obtained by application of the symmetry transformation G, i.e., $\lambda' = G\lambda$. Invariance of V under the group transformations implies

$$\frac{\partial V}{\partial \phi_i}(\theta_A)_{ij}\phi_j = 0. \tag{2.2.3}$$

Equation (2.2.3) is valid for all values of the field ϕ. Differentiating this equation with respect to k and setting $\phi = \lambda$, and using eqs. (2.2.1) and (2.2.2), we obtain

$$M_{ki}^2 (\theta_A)_{ij} \lambda_j = 0. \qquad (2.2.4)$$

This is the key equation in the study of properties of Nambu–Goldstone bosons. We see that, for those generators for which $(\theta_A)_{ij}\lambda_j \neq 0$, the M^2 matrix has a zero eigenvalue. Thus, the Goldstone boson eigenstates are given by $(\theta_A)_{ij}\lambda_j$. Those generators that satisfy $(\theta_A)_{ij}\lambda_j = 0$ represent the unbroken part of the symmetry group G and generate a symmetry group H which leaves both the vacuum and the Lagrangian invariant. The broken generators belong to the coset space G/H. (Coset space consists of those elements c of group G that satisfy $g = h_i c_j$ for $g \subset G$, $h_i \subset H$. What we have given is the right coset space; we could also define a left coset space in an obvious manner.)

We give a simple O(3) example that illustrates the connection between the zero-mass particles and the broken generators. The generators of the O(3) group can be written as

$$(\theta_A)_{ij} = i\varepsilon_{Aij}. \qquad (2.2.5)$$

Consider a triplet representation of O(3), i.e.,

$$\phi = \begin{pmatrix} \phi_1 \\ \phi_2 \\ \phi_3 \end{pmatrix}. \qquad (2.2.6)$$

Choosing the potential $V(\varepsilon)$ as follows:

$$V(\phi) = -\mu^2 \phi^2 + \lambda(\phi^2)^2; \qquad (2.2.7)$$

we see that for $\mu^2 > 0$ the vacuum state corresponds to $\langle \phi \rangle \neq 0$. In general, we could choose

$$\langle \phi \rangle = \begin{pmatrix} v_1 \\ v_2 \\ v_3 \end{pmatrix}. \qquad (2.2.8)$$

But as we remarked before, all $\langle \phi \rangle$'s obtained from (2.2.8) by applications of O(3) transformations correspond to minima of the potential and are physically equivalent. So we can choose

$$\langle \phi \rangle = \begin{pmatrix} 0 \\ 0 \\ v \end{pmatrix}. \qquad (2.2.9)$$

Using eq. (2.2.5) we can now conclude that

$$\theta_1\langle \phi \rangle \neq 0 \quad \text{and} \quad \theta_2\langle \phi \rangle \neq 0,$$

but

$$\theta_3 \langle \phi \rangle = 0. \tag{2.2.10}$$

Thus, there are two broken generators that reduce the O(3) group to O(2). Working out the mass matrix it is easy to see that $M_{\phi_1} = M_{\phi_2} = 0$; thus, ϕ_2- and ϕ_1-fields are the Nambu–Goldstone boson eigenstates corresponding to the broken generators θ_1 and θ_2.

§2.3   Some Properties of Nambu–Goldstone Bosons

It was generally believed for a long time that there may be no real Nambu–Goldstone bosons in nature because their existence would give rise to long-range forces in classical physics and would lead to new effects in scattering and decay processes. However, starting in 1980 when it was pointed out [2] that any possible nonrelativistic classical long-range forces arising from the existence of massless Goldstone particles are spin dependent and would therefore be extremely difficult to observe [3], the trend of thinking, on the physical reality of Goldstone bosons, has changed and serious studies have been conducted to detect long-range forces in nature [4]. In this section we present some general properties of the coupling of Nambu–Goldstone bosons to matter.

We would like to discuss two specific properties: decoupling of Goldstone bosons from low-energy particles, and spin dependence of long-range forces. We will study the situation when the scale of symmetry breaking is much larger than the highest scales of low-energy physics ($\sim m_W$), i.e., $\Lambda \gg m_W$. (It is also possible [5] to have theories where the symmetry-breaking scale is much lower than m_W, without conflicting with observations.)

To study the physics of decoupling phenomena we consider a simple model with a U(1)′ symmetry under which bosons (ϕ) and fermions (ψ_i) transform nontrivially. The Noether current corresponding to this symmetry is given by

$$J_\mu = i(\phi^* \partial_\mu \phi - \phi \partial_\mu \phi^*) + \bar{\psi}_i \gamma_\mu O_{ij} \psi_j. \tag{2.3.1}$$

We assume that the symmetry is spontaneously broken, i.e., $\langle \phi \rangle = \frac{\Lambda}{\sqrt{2}} \neq 0$. We can then write

$$J_\mu = +\Lambda \partial_\mu G + \bar{\psi}_i \gamma_\mu O_{ij} \psi_j + \cdots, \tag{2.3.2}$$

where $G = -\sqrt{2}\,\mathrm{Im}\,\phi$ is the Nambu–Goldstone boson. Using the fact that $\partial^\mu J_\mu = 0$ and taking the fermion matrix element of (2.3.1), we find that the coupling of fermions to G, i.e., $f_{\psi\psi G}$ is given by

$$f_{\psi\psi G} \simeq \frac{q^\mu}{\Lambda} \bar{u}_i \gamma_\mu O_{ij} u_j. \tag{2.3.3}$$

Since $\bar{u}_i \gamma_\mu \, O_{ij} \, u_j q_\mu$ is a low-energy parameter and is at most of order m_W, we find that $f_{\psi\psi G} \approx m_W/\Lambda \ll 1$. This is the phenomenon of decoupling, i.e., as $\Lambda \to \infty$, $f_{\psi\psi G} \to 0$.

As a specific aspect of the decoupling property we may ask the following question: The potential $V(\phi)$ involves all scalar fields and because the Goldstone boson G is made out of scalar fields, is it not obvious that G will not appear in $V(\phi)$ with an enhanced coupling to other physical scalar bosons of the theory? Below we will show that in the simple case of U(1) symmetries, $V(\phi)$ is indeed independent of G [6].

First, we consider the simple case discussed in eq. (2.1.16). Let us parametrize the complex scalar field ϕ as follows:

$$\phi = \frac{1}{\sqrt{2}}(\Lambda + \rho)e^{iG/\Lambda}, \tag{2.3.4}$$

where $\Lambda = \sqrt{\mu^2/2\lambda}$. We then see that the potential V is independent of G, proving our assertion.

Let us now study a more-complicated situation that involves a U(1) symmetric theory with several scalar fields ϕ_i with U(1) charges Q_i. The Lagrangian in this case can be written as

$$\mathcal{L} = \sum_i +(\partial^\mu \phi_i^*)(\partial_\mu \phi_i) - V(\phi_i), \tag{2.3.5}$$

$V(\phi_i)$ consists of products of monomials of fields ϕ_i. Let us consider a typical term

$$V_x = \prod_i \phi_i^{X_i}, \tag{2.3.6}$$

where X_i are numbers satisfying the relation following from group invariance

$$\sum_i Q_i X_i = 0. \tag{2.3.7}$$

Let the symmetry be broken by various fields ϕ_i acquiring vacuum expectation values (v.e.v.) Λ_i. If we parametrize $\phi_i = (1/\sqrt{2})(\Lambda_i + \rho_i)\exp(i\xi_i/\Lambda_j)$, then eq. (2.2.4) implies that the Nambu–Goldstone boson G is given by

$$G = \sum_i \Lambda_i \xi_i Q_i \Big/ \left(\sum_i \Lambda_i^2 \right)^{1/2}. \tag{2.3.8}$$

Substituting ϕ_i into eq. (2.3.6) we find that

$$V_x = \prod_i \frac{(\rho_i + \Lambda_i)}{\sqrt{2}} \exp\left(i \sum_j X_j \xi_j/\Lambda_j \right). \tag{2.3.9}$$

Using (2.3.7) it is easy to see that the combinations of ξ_i present in V_x are orthogonal to G. This holds for each term in the potential. Thus, V is independent of G.

Let us now discuss the spin dependence of long-range forces resulting from the exchange of real Goldstone bosons. To study this let us look at eqs. (2.3.2) and (2.3.3). We can see using the Dirac equation that

$$f_{\psi_i \psi_j G} = \frac{1}{\Lambda} \bar{\psi}_i (M_i O_{ij} - \varepsilon O_{ij} M_j) \psi_j, \qquad (2.3.10)$$

where $\varepsilon = +1$ if γ_μ commutes with O (i.e., it has no γ_5) and -1 if it anticommutes with O (i.e., it has γ_5 in it). We see that, if $\varepsilon = +1$, only the off-diagonal part of O_{ij} survives, i.e., the Nambu–Goldstone boson G couples two different fermions. If $\varepsilon = -1$, we can have diagonal couplings of G that could lead to long-range forces between matter. The off-diagonality, or the γ_5 nature of the Nambu–Goldenstone boson coupling to fermions, could be inferred from other considerations, too; for instance, note that Nambu–Goldstone bosons satisfy the property that $G \to G + C$ (where C is a constant) which keeps the Lagrangian invariant. Therefore, the only way they can appear is through $\bar{\psi}_i \gamma_5 \psi_j G (i \neq j)$ coupling since, on translation $G \to G + C$, each of these interactions contributes a four-divergence to the Lagrangian and therefore vanishes.

As we noted, the diagonal couplings arise when $\varepsilon = -1$ corresponding to γ_5 couplings and it is well known from standard textbooks that the nonrelativistic limit of γ_5 gives spin-dependent forces as follows:

$$V_G(r) = \frac{f_{\psi\psi G}^2}{m^2 r^3} (\boldsymbol{\sigma}_1 \cdot \boldsymbol{\sigma}_2 - 3(\boldsymbol{\sigma}_1 \cdot \hat{r})(\boldsymbol{\sigma}_2 \cdot \hat{r})). \qquad (2.3.11)$$

The importance of the spin dependence is that for big "chunks" of matter the net spin is infinitesimal compared to the number of nucleons ($\sim$ Avagadro number) leading to tiny unobservable corrections to gravitational effects. The best bound on $f_{\psi\psi G}$ comes from atomic physics where shifts of energy levels can occur due to the presence of these long-range forces with large strength. From an experiment of Ramsey and Code [7] we deduce a bound on $f_{\psi\psi G} \leq 10^{-3}$.

§2.4 Phenomenology of Massless and Near-Massless Spin-0 Bosons

Since 1980 several interesting theoretical models have been proposed that contain massless [2, 4, 8] or near-massless [9] particles. One of the ways they could be detected is through the long-range forces generated by them. If they are Nambu-Goldstone bosons, then their low-energy effect would be to induce spin-dependent forces. On the other hand, combining with CP-violation, the γ_5-couplings could, in principle, change to pure scalar

Table 2.1. Limits on the strengths of new long-range forces from Eotvos, Cavendish, and atomic experiments are summarized in Ref. [3].

N	Limit on λ^{SI}	Limit on Λ^{SD}	Limit on λ^{T}
1	10^{-45} (Ref. 10)		10^{-16} (Ref. 7)
	10^{-47} (Ref. 11)		
2	10^{-23} (Ref. 10)	10^{-8} (Ref. 12)	10^{-11} (Ref. 7)
	10^{-20} (Ref. 11)		
3	10^{-2} (Ref. 10)	10^{6} (Ref. 12)	10^{-6} (Ref. 7)
	10^{7} (Ref. 11)		
	10^{-12} (Ref. 12)		
4			10^{-1} (Ref. 7)
5			10^{4} (Ref. 7)

interactions that would then lead to spin-independent, long-range forces. It may therefore be useful to summarize the present experimental limits on various long-range forces [3].

Spin-Independent Forces

In this case we parametrize the potentials as follows:

$$V^{\mathrm{SI}} = \frac{\lambda_N^{\mathrm{SI}}}{r} \left(\frac{r_0}{r}\right)^{N-1}, \qquad r_0 = 200 \text{ MeV}^{-1}. \qquad (2.4.1)$$

Spin-Dependent Forces

$$V^{\mathrm{SD}} = \frac{\lambda_N^{\mathrm{SD}}}{r} \left(\frac{r_0}{r}\right)^{N-1} \mathbf{S}_1 \cdot \mathbf{S}_2. \qquad (2.4.2)$$

Tensor Forces

$$V^{T} = \frac{\lambda_N^{\mathrm{T}}}{r} \left(\frac{r_0}{r}\right)^{N-1} (\mathbf{S}_1 \cdot \mathbf{S}_2 - 3\mathbf{S}_1 \cdot \hat{r}\mathbf{S}_2 \cdot \hat{r}). \qquad (2.4.3)$$

Finally, we wish to point out that the existence of Nambu–Goldstone bosons or ultralight bosons can affect astrophysical considerations by contributing new mechanisms for energy loss from stars. Consider, for instance, a typical Nambu–Goldstone boson with diagonal γ_5-coupling. Then the Feynman diagram in Fig. 2.1 provides a new mechanism for energy loss. It has been argued [13] that the effective $\psi\psi G$ coupling must be limited by a cosmological known rate of energy loss from red giant stars [14] to be $f_{\psi\psi G} \leq 10^{-12}$, where $\psi = u, d, e$.

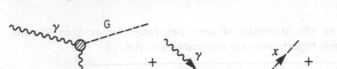

Figure 2.1.

§2.5 The Higgs–Kibble Mechanism in Gauge Theories

In this section, we wish to study field theories where local symmetries are spontaneously broken by vacuum. We will see that, in these cases, the zero-mass particles do not appear in the physical spectrum of states, but rather provide longitudinal modes to the gauge fields, which then become massive. This is, therefore, an extremely welcome situation where two kinds of massless particles, undesirable for physical purposes, combine to give a massive vector-meson state that may mediate short-range forces such as the weak forces. This is known as the Brout–Englert–Higgs–Guralnik–Hagen–Kibble mechanism and it forms the basis for the construction of unified gauge theories.

Before proceeding to study the physics of this mechanism, let us illustrate it with the help of a U(1) model. We consider a Lagrangian for a complex scalar field ϕ with nonzero U(1) charge invariant under local U(1) transformations

$$\phi \to e^{ie\theta(x)}\phi. \qquad (2.5.1)$$

As we discussed in Chapter 1, we need to introduce a real spin-1 field A_μ, which transforms under the gauge transformations as follows:

$$A_\mu \to A_\mu + \partial_\mu\theta. \qquad (2.5.2)$$

The invariance Lagrangian is then given by

$$\mathcal{L} = -\tfrac{1}{4}F^{\mu\nu}F_{\mu\nu} + (D^\mu\phi)^*(D_\mu\phi) + \mu^2\phi^*\phi - \lambda(\phi^*\phi)^2, \qquad (2.5.3)$$

where

$$F_{\mu\nu} = \partial_\mu A_\nu - \partial_\nu A_\mu,$$
$$D_\mu\phi = (\partial_\mu - ieA_\mu)\phi. \qquad (2.5.4)$$

If we look at the potential in eq. (2.5.3) we see that the ground state breaks the U(1) symmetry, i.e.,

$$\langle\phi\rangle = \frac{v}{\sqrt{2}} = \sqrt{\frac{\mu^2}{2\lambda}} \neq 0. \qquad (2.5.5)$$

If we parametrize

$$\phi = \frac{1}{\sqrt{2}}(v + \rho)e^{iG/v}, \qquad (2.5.6)$$

then,

$$D_\mu \phi = \frac{1}{\sqrt{2}}e^{iG/v}(\partial_\mu \rho - ie(\rho + v)B_\mu), \qquad (2.5.7)$$

where

$$B_\mu = A_\mu - \frac{1}{ev}\partial_\mu G. \qquad (2.5.8)$$

We can rewrite $\mathcal{L}$ as

$$\mathcal{L} = -\tfrac{1}{4}B^{\mu\nu}B_{\mu\nu} - \tfrac{1}{2}(\partial_\mu \rho)^2 - \tfrac{1}{2}e^2(\rho^2 + 2\rho v)B_\mu^2 - \tfrac{1}{2}M_B^2 B_\mu^2 - V(\rho), \quad (2.5.9)$$

where $B_{\mu\nu} = \partial_\mu B_\nu - \partial_\nu B_\mu$.

It is clear, from eq. (2.5.9), that the massless fields A_μ and G have disappeared and a new massive vector field B_μ has appeared in their place. This phenomenon occurs in more general cases with non-abelian symmetries.

At this stage we digress to understand the physics behind this mechanism. We saw in Section 2.1 that there are three conditions needed to prove the Goldstone theorem, i.e., current conservation, vacuum noninvariance, and Lorentz invariance. Once we couple the gauge fields, to study the theory, we must fix gauge and, of course, there are various ways to fix gauge: e.g., radiation gauge, i.e., $\nabla \cdot \mathbf{A} = 0$ or Lorentz gauge $\partial_\mu A_\mu = 0$, etc. If we work in radiation gauge, then gauge fixing destroys explicit Lorentz invariance. Therefore, strictly, the third condition needed to prove the Goldstone theorem does not exist and thus no Nambu–Goldstone bosons appear. On the other hand, if we work in Lorentz gauge, all three conditions for the Goldstone theorem are satisfied. Therefore, there must be a massless particle in the theory. However, in this case, the Hilbert space contains more states than are physically acceptable, for instance, states with zero or negative norm. So, the physical subspace of the Hilbert space has to be projected out. It can be shown that the zero-mass particle, as well as the massless gauge fields, remain in the unphysical sector of the Hilbert space; only their linear combination (which gives B_μ) has excitations in the physical subspace.

We illustrate this with the following very simple model involving a real scalar field ϕ and gauge field A_μ:

$$\mathcal{L} = -\tfrac{1}{4}F^{\mu\nu}F_{\mu\nu} + \tfrac{1}{2}(\partial_\mu \phi - gA_\mu)^2. \qquad (2.5.10)$$

This Lagrangian is invariant under the following local symmetry:

$$\begin{aligned}
\phi &\to \phi + g\lambda(x), \\
A_\mu &\to A_\mu + \partial_\mu \lambda(x).
\end{aligned} \qquad (2.5.11)$$

The current that generates this symmetry is

$$J_\mu = \partial_\mu \phi - g A_\mu, \qquad (2.5.12)$$

which is conserved as a result of the field equations. Because the canonical momentum $\pi_\phi = i J_4$, equal-time canonical commutation relations imply that

$$[Q, \phi] = 1, \qquad (2.5.13)$$

i.e., the $\langle 0 | [Q, \phi] | 0 \rangle = 1$ and the symmetry is spontaneously broken. The Higgs phenomena at the Lagrangian level is obtained by introducing the new vector-meson field, $B_\mu = (A_\mu - (1/g)\partial_\mu \phi)$, which now has mass g.

To illustrate the disappearance of the Goldstone boson in Lorentz gauge, we add Lagrangian multiplier terms to $\mathcal{L}$, i.e.,

$$\mathcal{L}' = -A^\mu \partial_\mu B - \tfrac{1}{2} B^2. \qquad (2.5.14)$$

The field equations are

$$A_\mu\text{-variation:} \quad \partial^\mu F_{\mu\nu} = \partial_\nu B - g(\partial_\nu \phi - g A_\nu), \qquad (2.5.15)$$

$$\phi\text{-variation:} \quad \partial^\mu(\partial_\mu \phi - g A_\mu) = 0, \qquad (2.5.16)$$

$$B\text{-variation:} \quad \partial^\mu A_\mu = B. \qquad (2.5.17)$$

From these equations we obtain

$$\Box B = 0 \qquad (2.5.18)$$

and

$$\Box \phi = g B. \qquad (2.5.19)$$

Equation (2.5.18) says that B is a free field and therefore we can make plane-wave decomposition, and the Gupta–Bleuler gauge condition can be imposed as

$$B^{(+)} | \psi_{\text{phys}} \rangle = 0. \qquad (2.5.20)$$

Thus, eq. (2.5.20) defines the physical subspace of the Hilbert space. An operator that does not commute with B will lead to excitations in the unphysical part of the Hilbert space. All we have to show now is that A_μ and the Goldstone boson field ϕ do not commute with B but that $B_\mu = A_\mu - (1/g)\partial_\mu \phi$ does.

To do this, let us write down the canonical commutation relations for various fields

$$\pi_{A_i} = F_{0i}, \qquad \pi_{A_0} = B, \qquad \pi_\phi = J_0. \qquad (2.5.21)$$

From the commutation (equal-time) relations

$$[A_0(x), B(y)]_{x_0 = y_0} = i\delta^3(x - y) \qquad (2.5.22)$$

and

$$[\pi_{A_i}(x), B(y)]_{x_0 = y_0} = 0, \qquad (2.5.23)$$

it follows that

$$[\partial_t A_i(x), B(y)]_{x_0=y_0} = i\delta_i \delta^3(x-y). \tag{2.5.24}$$

This then implies that at arbitrary times

$$[B(x), A_\mu(y)] = -i\partial_\mu D(x-y). \tag{2.5.25}$$

Using eq. (2.5.17) it follows that

$$[B(x), B(y)] = 0. \tag{2.5.26}$$

This condition is important if the physical subspace has to be "stable."
    Furthermore,

$$[\pi_\phi(x), \phi(y)]_{x_0=y_0} = i\delta^3(x-y). \tag{2.5.27}$$

This implies that [using (2.5.15)]

$$\frac{1}{g}[\partial_0 B(x), \phi(y)]_{x_0=y_0} = -i\delta^3(x-y). \tag{2.5.28}$$

For eq. (2.5.19) to be consistent with eq. (2.5.23) we require

$$[B(x), \phi] = -igD(x-y). \tag{2.5.29}$$

We therefore see that neither A_μ nor ϕ are physical operators but that $B_\mu \equiv A_\mu - (1/g)\partial_\mu\phi$ commutes with B and is, therefore, in the physical subspace of states. This proves our assertion.

§2.6   Group Theory of the Higgs Phenomenon

In this section we briefly discuss the Higgs mechanism for the case of general non-abelian local symmetries. Consider the local symmetry group to be G with N generators θ_A. We can choose a basis so that θ_A are imaginary and antisymmetric, i.e.,

$$\theta_A = -\theta_A^{\mathrm{T}} = -\theta_A^* \tag{2.6.1}$$

and satisfy the following commutation relations:

$$[\theta_A, \theta_B] = if_{ABC}\theta_C. \tag{2.6.2}$$

Let V_μ^A be the gauge fields associated with this group and let ϕ be a scalar multiplet used to implement spontaneous symmetry breaking. In the basis we use, ϕ is real. The Lagrangian invariant under gauge transformations can be written as (see Chapter 1)

$$\mathcal{L} = -\tfrac{1}{4}f^{\mu\nu,A}f_{\mu\nu,A} + \tfrac{1}{2}D^\mu\phi^T D_\mu\phi - V(\phi^T\phi), \tag{2.6.3}$$

where

$$D_\mu\phi = (\partial_\mu - i\theta_A V_{\mu,A})\phi.$$

By appropriate choice of $V(\phi)$ we can have a minimum of V that breaks the symmetry G

$$\langle \phi \rangle \neq 0. \tag{2.6.4}$$

Writing

$$\phi = \langle \phi \rangle = +\phi' \tag{2.6.5}$$

and substituting it into eq. (2.6.3) we see that the gauge fields acquire a mass matrix

$$(M^2)_{AB} = +(\langle \phi \rangle^{\mathrm{T}}, \theta_A \theta_B \langle \phi \rangle). \tag{2.6.6}$$

Note that for those generators that annihilate $\langle \phi \rangle$, the mass matrix has zeros. Diagonalizing this matrix we can find the massive and massless gauge bosons. For each symmetry generator that is broken by vacuum, the corresponding gauge field becomes massive, as is clear from (2.6.6). Thus, the Higgs phenomenon generalizes to arbitrary non-abelian groups.

Here we state one group-theoretical result due to Bludman and Klein [15], which throws light on the nature of the symmetry breaking in simple instances (like simple groups). Suppose a Higgs field has n real components and let us suppose v of them acquire vacuum expectation values (v.e.v.). Then the Bludman–Klein theorem states that

$$n - \nu = \text{dimension of the coset space} = N - M \tag{2.6.7}$$

if M is the number of generators of the subgroup that remains unbroken.

§2.7   Renormalizability and Triangle Anomalies

All the techniques for unified gauge theories have been laid out. The important reason for the attractiveness of spontaneously broken gauge theories for model building is the property of renormalizability. We saw in Chapter 1 that gauge invariance provides a natural explanation for universality of coupling strengths, as is observed in the case of weak interactions. The Higgs–Kibble mechanism provides a way to generate masses of the gauge bosons while at the same time maintaining the freedom to gauge transform the fields. This freedom would be lost if the masses for gauge fields were put in from outside. It was shown by 't Hooft [16] that this gauge freedom can be exploited to pass to a gauge where the propagator for massive gauge fields W can be written as

$$i\Delta_{\mu\nu}^W(k,\xi) = \frac{-i}{(2\pi)^4} \left[\frac{g_{\mu\nu} - k_\mu k_\nu/k^2}{k^2 - M^2 - i\varepsilon} + \frac{k_\mu k_\nu}{k^2(k^2/\xi - M^2)} \right], \tag{2.7.1}$$

where ξ is the gauge parameter.

As a result

$$\Delta_{\mu\nu}(k) \xrightarrow[k \to \infty]{} \frac{1}{k^2}, \tag{2.7.2}$$

which leads to a renormalizable theory if we keep in the Lagrangian only terms with mass dimension less than or equal to four. Recall that for a normal massive vector boson

$$i\Delta_{\mu\nu}^{W}(k) = \frac{-i}{(2\pi)^4} \left(\frac{g_{\mu\nu} - k_\mu k_\nu/M^2}{k^2 - M^2 - i\varepsilon} \right). \tag{2.7.3}$$

We see that, $\Delta_{\mu\nu}^{W} \xrightarrow[k\to\infty]{}$ constant. Therefore, the theory is apparently non-renormalizable by power counting arguments. The important point is that for $\xi \to \infty$,

$$\Delta_{\nu\mu}^{W}(k,\xi) \xrightarrow[\xi\to\infty]{} \Delta_{\mu\nu}^{w}(k). \tag{2.7.4}$$

Thus, the gauge freedom enables us to express the massive vector boson propagator in a canonical form (the U-gauge). Because, in the R_ξ-gauge, renormalizability can be explicitly checked, gauge invariance implies that the theory in the U-gauge must also be renormalizable, despite appearance, due to the cancellation of different terms [17].

Another point, important in the construction of renormalizable theories, is the absence of triangle anomalies [18], because if there is an exact local chiral symmetry in a Lagrangian (at the tree level), the current is exactly conserved, i.e., $\partial^\mu J_\mu = 0$. However, at the one-loop level, appearance of triangle graphs lead to breakdown of current conservation, i.e.,

$$\partial_\mu J^\mu = \frac{g^2}{16\pi^2} \varepsilon^{\mu\nu\lambda\sigma} f_{\mu\nu} f_{\lambda\sigma}. \tag{2.7.5}$$

The breakdown of Ward identities [eq. (2.7.5)] is equivalent to the introduction of dimension five terms into the effective Lagrangian, which therefore spoils renormalizability. Since the weak currents are chiral, in building weak interaction models using gauge theories, we must ensure that by appropriate choice of the fermion sector or gauge group G the current conservation is maintained, i.e., anomalies vanish. If θ_A's are the generators of the gauge group (in the space of chiral fermions), the condition for the disappearance of anomalies is that

$$T_r[\theta_A\{\theta_B, \theta_C\}]_L - T_r[\theta_A\{\theta_B, \theta_C\}]_R = 0. \tag{2.7.6}$$

This is an additional constraint on unified gauge models [19].

Exercises

2.1. Prove that the Higgs potential is independent of the Goldstone boson field for a simple non-abelian group group such as SU(2).

2.2. In the abelian example of Section 2.3, decompose the scalar field $\phi = (1/\sqrt{2}) \times (\Lambda + R + iG)$ as in eq. (2.1.19). When ϕ, in this form, is substituted in $V(\phi)$, it depends explicitly on the Goldstone boson field

G. Show that the scattering amplitudes involving only R fields and at least one G field vanish in the limit of zero external momenta for particles. This proves that Goldstone boson effects decouple in the limit of $\Lambda \to \infty$.

2.3. Show that if any representation of a group is pseudo-real, that representation is automatically anomaly free.

2.4. In four dimensions, only interactions of dimension 4 or less are renormalizable. What is the maximum dimension of renormalizable interactions in $D = 2$, 3, and 5? Prove this by analyzing the degree of divergence of an arbitrary loop diagram with an arbitrary number of external fermions and boson legs in the above dimensions.

2.5. What is the analog of the axial anomaly equation in dimension 6 and 2? What kind of one loop diagram will lead to anomalies in the case of $d = 6$ and $d = 2$?

2.6. In four dimensions, there can also be mixed anomalies involving gauge and gravitational vertices. Write the general form of this anomaly and derive the constraints it imposes on the representations of the gauge.

References

[1] For an excellent discussion of and references on symmetries and spontaneously broken symmetries, see
S. Weinberg, Brandeis Lectures, 1970;
M. A. B. Beg, Lectures Notes in Mexico, 1971;
G. Guralnik, C. R. Hagen, and T. W. B. Kibble, *Advances in High-Energy Physics* (edited by R. Cool and R. E. Marshak), Wiley, New York, 1969.
R. Gatto, *A Basic Course in Modern Weak Interaction Theory*, Bologna preprint (1979) (unpublished).

[2] Y. Chikashige, R. N. Mohapatra, and R. Peccei, *Phys. Lett.* **98B**, 265 (1981).

[3] For a survey of known limits on long-range forces, see
G. Feinberg and J. Sucher, *Phys. Rev.* **D20**, 1717 (1979).

[4] G. Gelmini, S. Nussinov, and T. Vanagida, *Nucl. Phys.* **B219**, 31 (1983);
H. Georgi, S. L. Glashow, and S. Nussinov, *Nucl. Phys.* **B193**, 297 (1981);
J. Moody and F. Wilczek, *Phys. Rev.* **D30**, 130 (1984).

[5] G. Gelmini and M. Roncadelli, *Phys. Lett.* **99B**, 411 (1981).

[6] R. Barbieri, R. N. Mohapatra, D. V. Nanopoulos, and D. Wyler, *Phys. Lett.* **107B**, 80 (1981).

[7] N. Ramsey and R. F. Code, *Phys. Rev.* **A4**, 1945 (1971).

[8] R. Barbieri and R. N. Mohapatra, *Z. Phys.* **C.11**, 175 (1981);
F. Wilczek, *Phys. Rev, Lett.* **49**, 1549 (1982);
D. Reiss, *Phys. Lett.* **115B**, 217 (1982).

[9] J. E. Kim, *Phys. Rev. Lett.* **43**, 103 (1979);
     M. Dine, W. Fischler, and M. Srednicki, *Phys. Lett.* **101B**, 199 (1981);
     D. Chang, R. N. Mohapatra, S. Nussinov, *Phys. Rev. Lett.* **55**, 2835 (1985).

[10] R. V. Eotvos, D. Pekar, and E. Fekele, *Ann. Phys.* **68**, 11 (1922).

[11] V. B. Braginsky and V. I. Panov, *Sov. Phys. JETP* **34**, 464 (1972);
     For other related experiments, see
     H. J. Paik, *Phys. Rev.* **D19**, 2320 (1979);
     H. J. Paik, H. A. Chan, and M. Moody, *Proceedings of the Third Marcel Grossmann Meeting on General Relativity*, 1983, p. 839;
     R. Spero, J. K. Hoskins, R. Newman, J. Pellam, and J. Schultz, *Phys. Rev. Lett.* **44**, 1645 (1980).

[12] D. R. Long, *Phys. Rev.* **D9**, 850 (1974);
     Y. Fujii and K. Mima, *Phys Lett.* **79B**, 138 (1978);
     Nature **260**, 417 (1976).

[13] D. Dicus, E. Kolb, V. Teplitz, and R. Wagoner, *Phys. Rev.* **D18**, 1829 (1978).

[14] M. Fukugita, S. Watamura, and M. Yoshimura, *Phys. Rev. Lett.* **18**, 1522 (1982).

[15] S. Bludman and A. Klein, *Phys. Rev.* **131**, 2363 (1962).

[16] G. 't Hooft, *Nucl. Phys.* **33B**, 173 (1971).

[17] For a detailed discussion of renormalizability of Yang–Mills theories, see
     E. S. Abers and B. W. Lee, *Phys. Rep.* **9C**, 1 (1973);
     G.'t Hooft and M. Veltman, *Nucl. Phys.* **B44**, 189 (1972);
     H. Kluberg-Stein and J. B. Zuber, *Phys. Rev.* **D12**, 467, 482, 3159 (1975);
     C. Becchi, A. Rouet, and R. Stora, *Commun. Math. Phys.* **42**, 127 (1975);
     J. C. Taylor, *Nuel. Phys.* **B33**, 436 (1971);
     J. Zino-Justin, Lecture Notes, Bonn, 1974.

[18] S. Adler, *Phys. Rev.* **177**, 2426 (1969);
     J. Bell and R. Jackiw, *Nuovo Cimento.* **51A**, 47 (1969);
     W. Bardeen, *Phys. Rev.* **184**, 1848 (1969).

[19] D. Gross and R. Jackiw, *Phys. Rev.* **D6**, 477 (1972);
     C. Bouchiat, J. Illiopoulos, and Ph. Meyer, *Phys Lett.* **38B**, 519 (1972).

3
The $SU(2)_L \times U(1)$ Model

§3.1 The $SU(2)_L \times U(1)$ Model of Glashow, Weinberg, and Salam

In this section we will apply the ideas of spontaneously broken gauge theories to construct the first successful model of the electroweak interaction of quarks and leptons. As we discussed in the Chapter 1 the observed universality of the four-Fermi coupling of weak-decay processes suggests the existence of a hidden symmetry of weak interactions, and the symmetry manifests itself not through the existence of degenerate multiplets but through broken local symmetries. The $SU(2)_L \times U(1)$ model of Glashow, Weinberg, and Salam [1] provides a realization of this idea in the framework of a renormalizable field theory, and the discovery of $W^{\pm}$- and Z-bosons in the proton–antiproton collider experiments [2] has proved the correctness of these ideas and given a boost to the study of spontaneously broken non-abelian gauge theories as the way to probe further into the structure of quark–lepton interactions. This will be explored in the subsequent sections.

To discuss the $SU(2)_L \times U(1)$ model, we will start by giving the assignment of quarks and leptons to representations of the gauge group. We will denote, by subscript a, the various generations; and by i, the $SU(3)_c$ index on quarks. (We will consider three generations of fermions.)

To implement the spontaneous breaking of the $SU(2)_L \times U(1)_Y$ symmetry, we have to include scalar bosons (the Higgs bosons) into the model. We choose the Higgs bosons ϕ to transform as the simplest nontrivial representation, i.e., doublets under the gauge group SU(2)$_L$ × U(1)$_Y$ × SU(3)$_C$

$$\phi \equiv \begin{pmatrix} \phi^+ \\ \phi^0 \end{pmatrix}, \qquad (2,+1,1). \tag{3.1.1}$$

	SU(2)$_L$ × U(1)$_Y$ × SU(3)$_c$ representation
$\psi_{a,L}^{(0)} \equiv \begin{pmatrix} \nu_a \\ e_a^- \end{pmatrix}_L$,	$(2,-1,1)$,
$e_{a,R}^{-(0)}$,	$(1,-2,1)$,
$Q_{a,iL} \equiv \begin{pmatrix} u_a^{(0)} \\ d_a^{(0)} \end{pmatrix}_{iL}$,	$(2,\frac{1}{3},3)$,
$u_{a,iR}^{(0)}$,	$(1,\frac{4}{3},3)$,
$d_{a,iR}^{(0)}$,	$(1,-\frac{2}{3},3)$.

$$\tag{3.1.2}$$

(The superscript zero denotes the fact that the quarks are eigenstates of weak interactions but not of the mass matrices to be generated subsequent to spontaneous breaking.)

Except for the Higgs potential part and the Higgs–Yukawa couplings the rest of the Lagrangian is completely dictated by the requirements of gauge invariance and renormalizability, i.e., it is simply the gauge invariant kinetic term for the above fields $\psi_L^{(0)}$, $E_R^{(0)}$, $Q_L^{(0)}$, $U_R^{(0)}$, $d_R^{(0)}$, ϕ:

$$\mathcal{L} = \mathcal{L}_{\text{kin}}^{\text{matt}} + \mathcal{L}_{\text{kin}}^{\text{gauge}} - V(\phi) + \mathcal{L}_Y, \tag{3.1.3}$$

where

$$\mathcal{L}_{\text{kin}}^{\text{matt}} = -\bar{Q}_L^{(0)} \gamma^\mu \left(\partial_\mu - \frac{ig}{2}\boldsymbol{\tau} \cdot \mathbf{W}_\mu - \frac{ig'}{6} B_\mu \right) Q_L^{(0)}$$
$$- \bar{\psi}_L^{(0)} \gamma^\mu \left(\partial_\mu - \frac{ig}{2}\boldsymbol{\tau} \cdot \mathbf{W}_\mu + \frac{ig'}{2} B_\mu \right) \psi_L^{(0)}$$
$$- \bar{E}_R^{(0)} \gamma^\mu (\partial_\mu + ig' B_\mu) E_R^{(0)} - \bar{U}_R^{(0)} \gamma_\mu \left(\partial_\mu - \frac{2i}{3} g' B_\mu \right) U_R^{(0)}$$
$$- \bar{d}_R^{(0)} \gamma^\mu \left(\partial_\mu + \frac{i}{3} g' B_\mu \right) d_R^{(0)} - \left| \left(\partial_\mu \phi - \frac{ig}{2}\boldsymbol{\tau} \cdot \mathbf{W}_\mu \phi - \frac{ig'}{2} B_\mu \phi \right) \right|^2,$$

$$\tag{3.1.4}$$

where the color and generation indices are suppressed (and are summed over)

$$\mathcal{L}_{\text{kin}}^{\text{gauge}} = -\frac{1}{4} \mathbf{W}^{\mu\nu} \cdot \mathbf{W}_{\mu\nu} - \frac{1}{4} B^{\mu\nu} B_{\mu\nu}, \tag{3.1.5}$$

where

$$\mathbf{W}_{\mu\nu} = \partial_\mu \mathbf{W}_\nu - \partial_\nu \mathbf{W}_\mu + g\mathbf{W}_\mu \times \mathbf{W}_\nu.$$

Before discussing weak interactions, let us study the breakdown of SU(2)$_L$ × U(1)$_Y$ symmetry to U(1)$_{\text{em}}$, which is the only observed exact local symmetry in nature. For this purpose we choose the Higgs potential $V(\phi)$ as follows:

$$V(\phi) = -\mu^2 \phi^+ \phi + \lambda(\phi^+ \phi)^2, \qquad \mu^2 > 0. \tag{3.1.6}$$

As noted earlier, the minimum of V corresponds to

$$\langle \phi \rangle = \frac{1}{\sqrt{2}} \binom{0}{v}, \tag{3.1.7}$$

where

$$v = \frac{\mu}{\sqrt{\lambda}}. \tag{3.1.8}$$

It is easily checked that

$$\tfrac{1}{2}(\tau_3 + Y)\langle \phi \rangle = 0, \tag{3.1.9}$$

which, therefore, is the unbroken generator. We will call this "electric charge" and we get the formula

$$Q = I_{3W} + \frac{Y}{2}. \tag{3.1.10}$$

Y is a free parameter of the theory and we have adjusted it appropriately [anticipating eq. (3.1.10)] so that the electric charges of the quarks and leptons come out correctly.

All other electroweak generators are broken by vacuum. It therefore follows from our previous discussion (Chapter 2) that the gauge bosons corresponding to those generators, to be called $W^\pm$ and Z, pick up mass. We now calculate their masses.

Substituting eq. (3.1.7) in the last term of eq. (3.1.4) we isolate the following mass terms involving the gauge bosons:

$$\mathcal{L}_{\text{mass}} = -\tfrac{1}{4}g^2 v^2 W^{+\mu} W^-_\mu - \tfrac{1}{8}v^2(gW_{3\mu} - g'B_\mu)^2. \tag{3.1.11}$$

This implies that

$$m_W = \frac{gv}{2}, \tag{3.1.12}$$

and one linear combination of the two neutral gauge bosons, i.e.,

$$\frac{1}{\sqrt{g^2 + g'^2}}(gW_{3\mu} - g'B_\mu) \equiv Z_\mu \tag{3.1.13}$$

picks up the mass

$$m_Z = \tfrac{1}{2}(g^2 + g'^2)^{1/2}v. \tag{3.1.14}$$

The massless orthogonal combination

$$A_\mu = \frac{g'W_{3\mu} + gB_\mu}{\sqrt{g^2 + g'^2}} \tag{3.1.15}$$

associated with the unbroken generator Q is the electromagnetic potential (the photon field). The gauge coupling associated with A_μ is, of course, the electric charge, which can now be expressed in terms of g and g' as follows.

Look at the gauge interactions involving W_3 and B, i.e., keeping the SU(2)$_L$×U(1)$_Y$ generators in place of currents and dropping Lorentz indices

$$\mathcal{L}_{\text{int}}(W_3, B) \equiv -\frac{i}{2}(g\tau_3 W_3 + g'YB). \tag{3.1.16}$$

Using eqs. (3.1.13) and (3.1.15) we can rewrite $\mathcal{L}_{\text{int}}$ as follows (using $I_{3W} = \frac{1}{2}\tau_3$):

$$\mathcal{L}_{\text{int}} = -i\left[\left(\frac{g^2 I_{3W} - (g'^2/2)Y}{\sqrt{g^2 + g'^2}}\right)Z + \frac{gg'}{\sqrt{g^2 + g'^2}}QA\right]. \tag{3.1.17}$$

The second term implies that the magnitude of the electric charge of the positron is

$$e = \frac{gg'}{\sqrt{g^2 + g'^2}}. \tag{3.1.18}$$

The first term of eq. (3.1.17) predicts the structure of the neutral current interaction.

At this point it is convenient to introduce a reparametrization of g and g' in terms of an angle θ_W (known as the Weinberg angle) and the electric charge of the positron, e:

$$\tan\theta_W \equiv \frac{g'}{g}. \tag{3.1.19}$$

Equation (3.1.18) then implies that

$$g = e\,\text{cosec}\,\theta_W, \tag{3.1.20a}$$
$$g' = e\sec\theta_W. \tag{3.1.20b}$$

The W- and Z-masses can be rewritten as

$$m_W = \frac{ev}{2\sin\theta_W}, \tag{3.1.21a}$$
$$m_Z = \frac{ev}{2\sin\theta_W\cos\theta_W}. \tag{3.1.21b}$$

This leads to a relation between the W- and Z-boson masses

$$m_W = m_Z\cos\theta_W \tag{3.1.21c}$$

or if we define a parameter $\rho_W = m_W^2/(m_Z\cos\theta_W)^2$, the Glashow–Weinberg–Salam model predicts $\rho_W = 1$ at the tree level. This feature depends not just on the gauge structure of the model but also on the fact that symmetry breaking is implemented by the doublet Higgs bosons. Using eqs. (3.1.20a,b) in the first term of eq. (3.1.17), i.e., the neutral-current interaction involving the Z-boson can be written as

$$\mathcal{L}_{\text{N.C.}} = -i\frac{e}{\sin\theta_W \cos\theta_W}Z \cdot (I_{3W} - \sin^2\theta_W Q) \tag{3.1.22}$$

[where I_{3W} and Q symbolize the fermionic currents with respective SU(2)$_L$ × U(1) transformation properties]. Equation (3.1.22) predicts the form of the interaction of Z for any kind of matter multiplet, once their representation content is known. For arbitrary fermions we can write the $\mathcal{L}_{\text{N.C.}}$ as (restoring the Lorentz indices)

$$\mathcal{L}_{\text{N.C.}} = -i\frac{e}{\sin\theta_W \cos\theta_W}Z^\mu \sum_i \bar{\psi}_i\gamma_\mu(I_{3W} - Q\sin^2\theta_W)\psi_i. \tag{3.1.23}$$

From this we can read off the neutral-current interaction of any particle. This is an important prediction of the SU(2)$_L$ × U(1) model and has been confirmed to a great degree of accuracy by neutral-current experiments and provides the first curcial test of the ideas of the gauge theories. We will discuss this in somewhat greater detail in the next section. Here we consider the charged-current aspects of the model.

We have seen that because the massive gauge bosons and their antiparticles are charged, this will generate the conventional charged-current weak interaction, which has the following form:

$$\mathcal{L}_{\text{wk}} = -\frac{ig}{2\sqrt{2}}W^{+\mu}[\bar{u}_a^{(0)}\gamma_\mu(1+\gamma_5)d_a^{(0)} + \bar{\nu}_a^{(0)}\gamma_\mu(1+\gamma_5)e_a^{-(0)}] + \text{h.c.} \tag{3.1.24}$$

From this we can conclude that the Fermi coupling can be expressed as

$$\frac{G_F}{\sqrt{2}} = \frac{g^2}{8m_W^2}. \tag{3.1.25}$$

Using eq. (3.1.20a) we can write

$$m_W^2 = \frac{\pi\alpha}{G_F \sin^2\theta_W} \cdot \frac{1}{\sqrt{2}}. \tag{3.1.26}$$

Once the Weinberg angle is determined independently from neutral-current interactions, the mass of the charged W boson can be predicted and vice versa. The m_W and $\sin^2\theta_W$, determined from independent experiments, do not agree with eq. (3.1.26) until radiative corrections are included. It is important to note that this feature depends only on the gauge structure of the model and not on the detailed nature of symmetry breaking or on the fermionic structure of the model.

In eq. (3.1.24) we have not displayed the mixing between the various generations. We will address this equation in a subsequent section. However, the universality of weak four-Fermi coupling (such as that between muon decay and β decay) already follows from eq. (3.1.24).

§3.2   Neutral-Current Interactions

In this section we will study the properties of the new kind of interactions induced by electroweak gauge unification that do not involve the exchange of charged gauge bosons, the neutral-current interactions. Its properties are dictated by the interaction Lagrangian in eq. (3.1.23). The low-energy four-Fermi interaction induced by eq. (3.1.23) is given by

$$\mathcal{L}_{N.C.} = -\frac{1}{2}\frac{e^2}{\sin^2\theta_W \cos^2\theta_W} \cdot \frac{1}{m_Z^2} J^{N.C.\mu} J_\mu^{N.C.}, \qquad (3.2.1)$$

where

$$J_\mu^{N.C.} = \sum_{i=\nu,e,u,d} \bar{\psi}_i \gamma_\mu (I_{3W} - \sin^2\theta_W \cdot Q)\psi_i. \qquad (3.2.2)$$

If we define a parameter $\rho_W = (m_W/m_Z \cos\theta_W)^2$, then we can write

$$H_{N.C.} = -\frac{4}{\sqrt{2}} G_F \rho_W J_\mu^{N.C.} J_\mu^{N.C.}. \qquad (3.2.3)$$

From eq. (3.2.3) we can write the various neutral-current interactions between various particles in order to compare with experiments

$$H_{N.C.} = II_{N.C.}^\nu + H_{N.C.}^{eH} + H_{N.C.}^{c\mu}, \qquad (3.2.4)$$

$$H_{N.C.}^\nu = -\frac{G_F}{\sqrt{2}} \bar{\nu}\gamma^\mu(1+\gamma_5)\nu \sum_f [\varepsilon_L(f)\bar{f}\gamma_\mu(1+\gamma_5)f$$

$$+ \varepsilon_R(f)\bar{f}\gamma_\mu(1-\gamma_5)f], \qquad (3.2.5)$$

where

$$\varepsilon_L(u) = \rho_W\left(\tfrac{1}{2} - \tfrac{2}{3}\sin^2\theta_W\right),$$
$$\varepsilon_R(u) = \rho_W\left(-\tfrac{2}{3}\sin^2\theta_W\right),$$
$$\varepsilon_L(d) = \rho_W\left(-\tfrac{1}{2} + \tfrac{1}{3}\sin^2\theta_W\right),$$
$$\varepsilon_R(d) = \rho_W\left(+\tfrac{1}{3}\sin^2\theta_W\right),$$
$$\varepsilon_L(e) = \rho_W\left(-\tfrac{1}{2} + \sin^2\theta_W\right),$$
$$\varepsilon_R(e) = \rho_W\left(\sin^2\theta_W\right). \qquad (3.2.6)$$

The parameters $\varepsilon_{L,R}$ can be extracted from the various neutrino neutral-current experiments conducted at Fermilab and CERN and applying the parton model for high-energy scattering of leptons off nucleons. The theoretical prediction of them using $\rho_W = 1$ and $\sin^2\theta_W = 0.233$ are noted on the right of Table 3.1. Here we have written $g_{V,A}^f \equiv \varepsilon_L(f) \pm \varepsilon_R(f)$. We have not included further radiative corrections, which multiply the $\sin^2\theta_W$ terms in each expression by a factor κ that is very close to one and adds to the whole expression in each line a term λ_f, which is close to zero.

Table 3.1.

Neutral-current parameters	Values extracted from experiments (Ref. 4)	Predictions of the Glashow, Weinberg, and Salam model $\sin^2 \theta_W = 0.233$
$\varepsilon_L(u)$	0.339 ± 0.033	0.345
$\varepsilon_L(d)$	-0.424 ± 0.026	-0.423
$\varepsilon_R(u)$	-0.179 ± 0.019	-0.155
$\varepsilon_R(d)$	-0.016 ± 0.058	$+0.077$
$g_V(e)$	-0.043 ± 0.063	-0.034
$g_A(e)$	-0.545 ± 0.056	-0.5
$\rho_W = 1.001 \pm 0.21$		

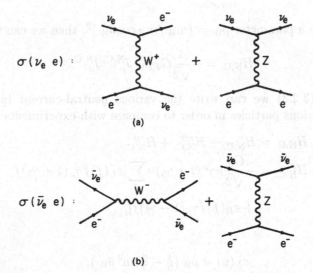

Figure 3.1. Tree-level Feynman diagrams contributing to $\overset{(-)}{\nu_e} e$ scattering in the standard model.

The agreement between theory and experiment is very good. The value of the ρ_W parameter is actually better known than indicated by the neutrino data alone.

An important test of the standard model can be performed by improved $\nu_e e \rightarrow \nu_e e$, $\bar{\nu}_e e \rightarrow \bar{\nu}_e e$ scattering experiments. The reason is that, unlike the neutrino–hadron neutral-current experiments, the $\overset{(-)}{\nu_e} e$ scattering experiments involve interference between charged and neutral boson exchange diagrams in the standard model (see Fig. 3.1). The interference term in the standard model is proportional to $I_3(e_L) + \sin^2 \theta_W$, which is equal to $\approx -\frac{1}{2} + 0.233 \approx -0.267$ leading to destructive interference. In variants of standard model [5], the situation will, in general, be different; for instance, if there exist flavor-changing neutral current involving ν_e, then additional pieces that do not interfere with the charged-current piece will appear. This experiment could therefore provide a sensitive test of (or reveal new physics beyond) the standard model. A main difficulty with this experiment is its low cross section compared to neutrino–hadron scattering.

Let us now look at the other pieces of the neutral-current Hamiltonian. We parametrize $H_{N.C.}^{eH}$ as follows:

$$\mathcal{L}_{N.C.}^{eH} = -\frac{G_F}{\sqrt{2}} \sum_i (c_{1i}\bar{e}\gamma^\mu\gamma_5 e\bar{q}_i\gamma_\mu q_i + c_{2i}\bar{e}\gamma^\mu e\bar{q}_i\gamma_\mu\gamma_5 q_i), \quad (3.2.7)$$

$$C_{1u} = \frac{2}{\rho_W}g_A^e g_V^u, \qquad C_{1d} = \frac{2}{\rho_W}g_A^e g_V^d,$$

$$C_{2u} = \frac{2}{\rho_W}g_V^e g_A^u, \qquad C_{2d} = \frac{2}{\rho_W}g_V^e g_A^d. \qquad (3.2.8)$$

Determination of g_V^f and g_A^f from neutrino neutral-current data can be used to predict C_{1i} and C_{2i}, which can then be used to predict parity-violating effects in electronucleon interactions. Such effects can be measured by looking for parity-violating effects in atomic physics, as well as looking for differences in left- and right-handed electron scattering off nucleons. By now, both of these kinds of experiments have been carried out and the results agree with the $SU(2)_L \times U(1)$ model to within the experimental accuracy.

A. Deep Inelastic Electron–Hadron Scattering

The SLAC experiment [6] on polarized electron–deep inelastic scattering $eD \rightarrow e + X$ measures the parity-violating asymmetry

$$A = \frac{d\sigma(+) - d\sigma(-)}{d\sigma(+) + d\sigma(-)}. \qquad (3.2.9)$$

Using the parton model approximations, and neglecting the effect of antiquarks in nucleons for an isosinglet target such as deuteron, we can predict

$$\frac{A_D}{Q^2} = \frac{3G_F}{5\sqrt{2}\pi\alpha}\left[(C_{1u} - \tfrac{1}{2}C_{1d}) + F(y)(C_{2u} - \tfrac{1}{2}C_{2d})\right], \qquad (3.2.10)$$

where

$$F(y) = \frac{1 - (1-y)^2}{1 + (1-y)^2},$$

where y is the dimensionless variable in parton models given by

$$y = \frac{E_i - E_f}{E_i}.$$

In the Glashow–Weinberg–Salam model

$$C_{1u} - \tfrac{1}{2}C_{1d} = -\tfrac{3}{4} + \tfrac{5}{3}\sin^2\theta_W,$$

$$C_{2u} - \tfrac{1}{2}C_{2d} = 3(\sin^2\theta_W - \tfrac{1}{4}), \qquad (3.2.11)$$

and

$$\frac{3G_F}{5\sqrt{2}\pi\alpha} = 2.16 \times 10^{-4}\,\mathrm{GeV}^{-2}. \qquad (3.2.12)$$

SLAC data [6] yield

$$\frac{A_D}{Q^2} = [(-9.7 \pm 2.6) + (4.9 \pm 8.1)F(y)] \times 10^{-5},$$

which implies

$$C_{1u} - \tfrac{1}{2}C_{1d} = -0.45 \pm 0.12$$

and

$$C_{2u} - \tfrac{1}{2}C_{2d} = 0.23 \pm 0.38. \qquad (3.2.13)$$

Putting $\sin^2\theta_W = 0.233$ yields $C_{1u} - \tfrac{1}{2}C_{1d} = -0.365$ and $C_{2u} - \tfrac{1}{2}C_{2d} = -0.051$, both values are in agreement with data.

B. Atomic Parity Violation

The dominant contribution to atomic parity violation in heavy nuclei comes from the C_{1i} term of eq. (3.2.7). The reason for this is that in the nonrelativistic limit $\bar{q}\gamma_\mu q \to q^+ q$ (which counts the number of quarks in a nucleus) whereas $\bar{q}\gamma_5\gamma_\mu q \xrightarrow[\mathrm{N.R.}]{} q^+\boldsymbol{\sigma} q$ (which measures the spin of the nucleus), which is a number of order 1 or 2 regardless of how heavy a nucleus is. For a typical nucleus, the effective weak parity-violating electron nucleus interaction can be written as

$$H_{\mathrm{wk}} = \frac{G_F}{\sqrt{2}}Q_w\boldsymbol{\sigma}_e \cdot \nabla\delta^3(\mathbf{r}), \qquad (3.2.14)$$

where

$$Q_W(N,Z) = -2[C_{1u}(2Z+N) + C_{1d}(Z+2N)], \qquad (3.2.15)$$

where N and Z denote the number of protons and neutrons in a nucleus (note that the proton has two up-quarks and one down-quark, and vice versa for the neutron). For bismuth and thallium, on which experiments [7] have been carried out,

$$Q_w^{\mathrm{Bi}}(126,83) = -584C_{1u} - 670C_{1d} \approx -120.8,$$
$$Q_w^{\mathrm{Tl}}(123,81) = -570C_{1u} - 654C_{1d} \approx -118.$$

The results of these experiments are

$$Q_w^{\mathrm{Bi}} = -126 \pm 45$$

and

$$Q_w^{\mathrm{Tl}} = -155 \pm 63,$$

which are in agreement with the standard model. More recently, very precise measurements have been made of Q_w for cesium[1]

$$Q_w^{Cs} = -71.04 + 1.58 \pm [0.88]$$

to be compared with $Q_w^{Cs}(\text{THEORY}) = -73.1$.

C. Front–Back Asymmetry in $e^+e^- \to \mu^+\mu^-$

We now focus on the neutral weak interaction between electrons and muons. This interaction can be parametrized as follows:

$$\begin{aligned} H_{N.C.}^{eh} = -\frac{G_F}{\sqrt{2}} [&h_{VV}(\bar{e}\gamma_\lambda e + \bar{\mu}\gamma_\lambda\mu)(\bar{e}\gamma_\lambda e + \bar{\mu}\gamma_\lambda\mu) \\ &- 2h_{VA}(\bar{e}\gamma_\lambda e + \bar{\mu}\gamma_\lambda\mu)(\bar{e}\gamma_\lambda\gamma_5 e + \bar{\mu}\gamma_\lambda\gamma_5\mu) \\ &+ h_{AA}(\bar{e}\gamma_\lambda\gamma_5 e + \bar{\mu}\gamma_\lambda\gamma_5\mu) \times (\bar{e}\gamma_\lambda\gamma_5 e + \bar{\mu}\gamma_\lambda\gamma_5\mu)]. \end{aligned} \quad (3.2.16)$$

The forward–backward asymmetry measures the difference between the number of μ^-'s in the forward and backward hemisphere, with e^+e^- colliding direction as the polar axis, and is defined as follows:

$$A_{\mu\mu} = \frac{\int d\Omega[(d\sigma/d\Omega)(\theta) - (d\sigma/d\Omega)(\pi - \theta)]}{\int d\Omega[(d\sigma/d\Omega)(\theta) + (d\sigma/d\Omega)(\pi - \theta)]}. \quad (3.2.17)$$

The dominant contribution to the denominator is from electromagnetic interactions, which cancel in the numerator, leading to the result

$$A_{\mu\mu} = \frac{3}{16} \frac{1}{\sqrt{2}\sin^2\theta_W} \cdot \frac{S}{S - m_Z^2}. \quad (3.2.18)$$

It predicts $A_{\mu\mu} \simeq -9\%$ at the highest available e^+e^- energies right now, i.e., at the e^+e^- machine at DESY and this parameter has been measured by the various groups with the combined result [7a]:

$$A_{\mu\mu}^{\text{expt}} = -10.4 \pm 1.4.$$

Again, there is good agreement between theory and experiment.

§3.3 Masses and Decay Properties of W and Z Bosons

In this section we will consider the predictions for W- and Z-boson masses in the standard model. This model has three free parameters (two gauge coupling constants, g and g', and the spontaneous symmetry breaking scale, v), which can be reexpressed in terms of: e, the electric charge of positron;

[1]M. C. Noecker et al., *Phys. Rev. Lett.* **61**, 310 (1988).

$\sin \theta_W$, the weak mixing angle; and G_F ($\equiv G_\mu$), the Fermi coupling constant. We can therefore predict the W- and Z-boson masses. Let us discuss this step-by-step [8]. At the tree level, we have the following relation [eq. (3.1.26)]:

$$m_W^0 = \left[\frac{\pi \alpha^0}{\sqrt{2} G_\mu^0 \sin^2 \theta_W} \right]. \qquad (3.3.1)$$

We have denoted by a superscript zero the fact that no radiative corrections are applied. Let us first give physical argument that the radiative corrections are indeed important. The point is that radiative corrections will be defined after subtracting out the infinite contributions that introduce a subtraction point, μ, and will depend on it. In eq. (3.3.1), the left-hand side must be defined at $\mu = m_W$; but conventional values of α^0 are at $\mu = 0$ because it is obtained from Compton scattering of real photons. Also, G_μ is obtained from μ-decay, where $\mu \simeq 0.1\,\text{GeV}$. To get the tree level prediction for m_W we use:

$$(\alpha^0)^{-1} = 137 \cdot 035963(15), \qquad (3.3.2)$$
$$G_\mu = (1.16634 \pm 0.00002) \times 10^{-5}\,\text{GeV}^{-2}. \qquad (3.3.3)$$

Using this we get

$$m_W^0 = \frac{37.2810 \pm 0.0003}{\sin \theta_W^0}\,\text{GeV}. \qquad (3.3.4)$$

The uncorrected tree level value of $\sin \theta_W^0$ can be obtained[3] from ν_μ-hadron scattering data:

$$\sin \theta_W^0 = 0.227. \qquad (3.3.5)$$

This leads to

$$m_W^0 = 78.3\,\text{GeV} \qquad \text{(lowest-order)},$$
$$m_Z^0 = 2 \left(\frac{\pi \alpha^0}{\sqrt{2} G_\mu^0} \right) \frac{1}{\sin 2\theta_W^0} = 89.0\,\text{GeV} \qquad \text{(lowest-order)}. \qquad (3.3.6)$$

These tree level predictions are in clear disagreement with experiments and one has to take into account radiative corrections [8–10] and renormalization effects to obtain the physical values of m_W and m_Z. These will involve the physical value of $\sin^2 \theta_W$. The procedure generally adpoted is to fix the renormalization scheme and using either m_W or m_Z as input determine $\sin^2 \theta_W$ from various processes and check for its consistency. The various renormalization schemes have been summarized along with their results in [11].

For instance, in a scheme known as the on-shell scheme, the tree level formula $\sin^2 \theta_W = 1 - \frac{m_W^2}{m_Z^2}$ is used to define the $\sin^2 \theta_W$ to all orders in perturbation theory. In this scheme, we have:

$$m_W = \frac{37.2810 \pm 0.0003 \,\text{GeV}}{\sin\theta_W(1-\Delta r)^{1/2}} \tag{3.3.7a}$$

and

$$M_Z = \frac{74.562 \pm 0.0006 \text{GeV}}{\sin 2\theta_W(1-\Delta r)^{1/2}}. \tag{3.3.7b}$$

where, Δr denotes the radiative corrections including the top quark effects. For $m_t = 175 \pm 5$ GeV and $M_H - m_Z$, one finds $\Delta r = 0.0349 \pm 0.0019 \pm \pm 0.0007$. Using the current values of $m_W = 80.41 \pm 0.10$ GeV and $m_Z = 98.186 \pm 0.0020$ GeV, one gets $\sin^2\theta_W = 0.2231 \pm 0.0005$ and 0.2228 ± 0.0006 respectively. These values agree very well with the value of $\sin^2\theta_W$ derived from various neutral current data compared with experiments in the same renormalization scheme.

Another useful renormalization scheme is called modified minimal substraction scheme $\bar{M}S$-scheme, which turns out to be more suitable for comparing with the predictions of grand unification. In this scheme, the expressions for $m_{W,Z}$ are

$$m_W = \frac{37.28}{s_Z(1-\Delta\tilde{r})^{1/2}},$$
$$m_Z = \frac{m_W}{\rho^{1/2}c_Z}, \tag{3.3.8}$$

where $c_Z = \cos\theta_W$ in this scheme. The experimental values of $m_{W,Z}$ lead to $\sin^2\theta_W = 0.231$ that is to be compared with $\sin^2\theta_W = 0.23124 \pm 0.00017$, which is the average value when all neutral-current data are taken into account.

The W and Z bosons were discovered a year ago at the CERN $p\bar{p}$ colliding facility by the UA1 and UA2 experiments [12] and the values for their masses were taken from Ref. 11.

Let us now look at the decay properties of W and Z. The decay modes of W and Z are of two types: hadronic and leptonic:

$$W^+ \to u\bar{d}, \bar{e}\nu_e, c\bar{s}, \bar{\mu}\nu_\mu, \tau\nu_\tau, \tag{3.3.9}$$
$$Z \to u\bar{u}, d\bar{d}, c\bar{c}, s\bar{s}, b\bar{b}, \nu_e\bar{\nu}_e, \tag{3.3.10}$$

and

$$\nu_\mu\bar{\nu}_\mu, \nu_\tau\bar{\nu}_\tau, e\bar{e}, \mu\bar{\mu}, \tau\bar{\tau}.$$

From the weak couplings given in previous sections we predict

$$\Gamma(W \to \text{hadrons}) = \frac{G_\mu m_W^3}{\sqrt{2}\pi} \times \left(1 + \frac{\alpha_s(m_W)}{\pi}\right)$$
$$\simeq 1.37\,\text{GeV}, \tag{3.3.11}$$

where three colors have been taken into account, QCD corrections have been included and quark masses other than m_t have been ignored.

$$\Gamma(W \to l\bar\nu_l) = \frac{G_\mu m_W^3}{6\sqrt{2}} \simeq 0.23 \, \text{GeV} . \tag{3.3.12}$$

The total width of W is therefore about 2.8 GeV. Electroweak correction to these results are small.

Coming to the Z decay, its coupling can be inferred from eq. (3.1.23) to be

$$\mathcal{L}_{Z \to f\bar f} = -i \frac{M_Z}{\sqrt{2}} \left(\frac{G_F}{\sqrt{2}} \right)^{1/2} Z^\mu \bar f \gamma_\mu (v_f - a_f \gamma_5) f, \tag{3.3.13}$$

where

$$|a_f| = 1,$$
$$|v_f| = (1 - 4Q \sin^2 \theta_W). \tag{3.3.14}$$

Using this we can calculate

$$\Gamma(Z_0 \to l\bar l) = \frac{G_\mu M_Z^3}{12\sqrt{2}\pi}(1 - 4\sin^2 \theta_W + 8\sin^2 \theta_W) \simeq 0.092 \, \text{GeV},$$
$$\tag{3.3.15}$$

$$\Gamma(Z_0 \to \nu_l \bar\nu_l) = \frac{G_\mu M_Z^3}{12\sqrt{2}\pi} \simeq 0.166 \, \text{GeV}, \tag{3.3.16}$$

$$\Gamma(Z_0 \to \text{hadrons}) = \frac{G_\mu M_Z^3}{4\sqrt{2}\pi} f\left(\sin^2 \theta_W, \frac{m_t^2}{m_t^2}\right)\left(1 + \frac{\alpha_s}{\pi}\right), \tag{3.3.17}$$

where

$$f\left(\sin^2 \theta_W, \frac{m_t^2}{M_Z^2}\right) = 5 - \frac{28}{3}\sin^2 \theta_W + \frac{80}{9}\sin^4 \theta_W; \tag{3.3.18}$$

$$\Gamma(Z_0 \to \text{hadrons}) \simeq 1.74 \, \text{GeV} . \tag{3.3.19}$$

This gives a total width for Z_0 of about 2.85 GeV. We also see from here that each additional massless neutrino, with properties of similar neutrinos, will add 0.18 GeV to the width. Thus, measurement of Z-width will limit the number of neutrinos of type $m_\nu \ll M_Z$. Especially relevant is the invisible width that gets a contribution only from neutrinos. Present results for the invisible width of Z is $\Gamma_Z(\text{inv.}) = 498.3 \pm 4.2$ MeV [8]. This corresponds precisely to three light neutrinos, as have been confirmed by other observations and is in remarkable agreement with the presence of only three generations of fermions.

There is a very clear signature for pure $V - A$ (or $V + A$) structure of W coupling to fermions in high-energy $p\bar p$ collision. There is an asymmetry in the $e^-(e^+)$ production with respect to the direction of collision that results from pure $V - A$ (or $V + A$) coupling. This can be seen as follows. We

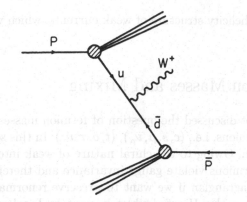

Figure 3.2. The Drell–Yan diagram for the production of W-bosons in $p\bar{p}$ collision.

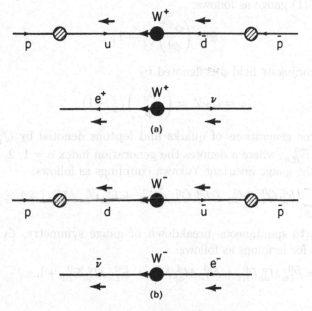

Figure 3.3. Asymmetry in $e^{\pm}$ distribution resulting from pure $V - A$ structure of weak currents.

assume that W production results from the collision o f a quark (q), from proton and antiquark ($\bar{q}$), from antiproton, via the diagram in Fig. 3.2.

If W coupling to quarks is left handed, then to antiquark it is right handed. Consider Fig. 3.3(a). The u quark must be left handed and the $\bar{d}$ must be right-handed giving rise to W^+ polarized toward the left. Because, by electric charge conservation $W^+ \to e^+\nu$, helicity conservation implies that for $V - A$ coupling e^+ must be right-handedly polarized and therefore emerge in the beam direction of the antiproton. By similarly analyzing Fig. 3.3(b) we conclude that e^- must emerge in the beam direction of the incident proton. This feature has been confirmed experimentally, thus

proving the pure helicity structure of weak currents, which we can take as
$V - A$ type.

§3.4   Fermion Masses and Mixing

So far we have not discussed the question of fermion masses or the higher
generations of fermions, i.e., (c, s, μ, ν_μ), (t, b, τ, ν_τ). In this section we consider this question. Owing to the chiral nature of weak interactions, bare
mass terms for fermions violate gauge invariance and therefore cannot be
included in the Lagrangian if we want to preserve renormalizability. We,
therefore, use the scalar Higgs doublet ϕ used to break gauge symmetries in Section 3.2 for this purpose. Recall that ϕ transforms under the
SU(2)$_L$ × U(1) gauge as follows:

$$\phi = \begin{pmatrix} \phi^+ \\ \phi^0 \end{pmatrix} (\tfrac{1}{2}, +1). \tag{3.4.1}$$

Its charge conjugate field $\tilde{\phi}$ is denoted by

$$\tilde{\phi} = i\tau_2\phi^* = \begin{pmatrix} \phi_0^* \\ -\phi^- \end{pmatrix} (\tfrac{1}{2}, -1). \tag{3.4.2}$$

For the three generations of quarks and leptons denoted by $Q^0_{L,a}$, $U^0_{R,a}$,
$d^0_{R,a}$, $\psi^0_{L,a}$, $E^0_{R,a}$, where a denotes the generation index $a = 1, 2, 3$, we can
now write the gauge invariant Yukawa couplings as follows:

$$\mathcal{L}_Y = \sum_{a,b} (h^d_{ab}\bar{Q}^0_{La}\phi d^0_{Rb} + h^u_{ab}\bar{Q}^0_{La}\tilde{\phi}U^0_{Rb} + h^e_{ab}\bar{\psi}^0_{La}\phi E^0_{Rb}) + \text{h.c.} \tag{3.4.3}$$

Subsequent to spontaneous breakdown of gauge symmetry, $\mathcal{L}_Y$ leads to
mass terms for fermions as follows:

$$\mathcal{L}_m = \bar{P}^0_{La}M^u_{ab}P^0_{Rb} + \bar{N}^0_{La}M^d_{ab}N^0_{Rb} + \bar{E}^0_{La}M^e_{ab}E^0_{Rb} + \text{h.c.}, \tag{3.4.4}$$

where

$$M^P = \frac{1}{\sqrt{2}}h^P v, \qquad p = u, d, e \tag{3.4.5}$$

denotes the mass matrices for up- and down-quarks and negatively charged
leptons. Note that neutrinos are massless and will never acquire mass in
higher orders because its chiral partner ν_R does not exist in the theory.

These mass matrices mix the weak eigenstates (denoted by superscript
zero) of different generations and give rise to mixing angles such as the
Cabibbo angle. Owing to off-diagonal mixing terms in the mass matrix,
the quanta of the weak eigenstates are not eigenstates of the Hamiltonian
(or mass). To get mass eigenstates we must diagonalize the mass matrices
by means of biunitary transformations as given below

$$U^{(p)}_L M^{(p)} U^{(p)\dagger}_R = M^{(p)}_{\text{diag}}, \tag{3.4.6}$$

where

$$M_{\text{diag.}}^{(u)} = (m_u, m_c, m_t, \ldots), \tag{3.4.7}$$

$$M_{\text{diag.}}^{(d)} = (m_d, m_s, m_b, \ldots), \tag{3.4.8}$$

$$M_{\text{diag.}}^{(e)} = (m_e, m_\mu, m_\tau, \ldots). \tag{3.4.9}$$

The mass eigenstates can be written as

$$P_{L,R} = U_{L,R}^{(u)} P_{L,R}^0,$$

$$N_{L,R} = U_{L,R}^{(d)} N_{L,R}^0,$$

$$E_{L,R} = U_{L,R}^{(e)} E_{L,R}^0. \tag{3.4.10}$$

where

$$P = (u, c, t, \ldots),$$

$$N = (d, s, b, \ldots),$$

$$E = (e^-, \mu^-, \tau^-, \ldots). \tag{3.4.11}$$

We will also call the mass eigenstate basis the flavor basis. If we now rewrite the gauge boson interactions given in (3.1.22) and (3.1.24) in the flavor basis, we find that the neutral current interaction of the Z boson remains diagonal, i.e., different flavors do not mix. This is an extremely desirable feature of the standard model because it implies that any neutral-current process that changes flavor, like strangeness, can arise only at the one or higher loop level and must, therefore, be suppressed. This agrees with observations such as the suppression of $K_L^0 \to \mu\bar{\mu}$ decay (compared to $K^+ \to \mu^+\nu_\mu$ decay), $K_L - K_S$ mass difference, etc. We return to these questions in a subsequent section.

Turning to the interaction Lagrangian of the $W^\dagger$ with fermions, eq. (3.1.24) can be written as

$$\mathcal{L}_{\text{gauge}} = -\frac{ig}{2\sqrt{2}} W^{\dagger\mu} \bar{P} \gamma_\mu U_L^{(u)\dagger} U_L^{(d)} (1 + \gamma_5) N + \text{h.c.} \tag{3.4.12}$$

We can write $U_{CKM} = U_L^{(u)\dagger} U_L^{(d)}$ as the matrix that mixes different generations and is responsible for such phenomena as strangeness-changing weak processes (e.g., $\Sigma \to p\pi^-$, $K \to \pi e\nu$, $K \to e\nu$, etc.). Since $U^{(p)}$'s are unitary matrices, U_{CKM}, is a unitary matrix. To give an example for two generations

$$U_{CKM} = \begin{pmatrix} \cos\theta_C & \sin\theta_C \\ -\sin\theta_C & \cos\theta_C \end{pmatrix}, \tag{3.4.13}$$

where θ_C is the Cabibbo angle. This matrix for two generations was first introduced by Glashow, Illiopoulos, and Maiani [13], and the absence of flavor-changing neutral current in lowest-order is known as the Glashow–Illiopoulos–Maiani mechanism. Two other points about mixings are worth pointing out. The mixing matrices corresponding to U_R never appear in the final theory involving flavor eigenstates. Similarly, the left-handed mixing matrix for the charged leptons is also not observable because, in the charged current, we can redefine the neutrino states and absorb $U_L^{(l)}$; since neutrinos are massless, all rotated bases are equivalent.

§3.5   Higher-Order-Induced Flavor-Changing Neutral-Current Effects

As discussed in the previous section the lowest-order neutral-current couplings of Z^0 conserve all flavors such as strangeness, charm, etc. However, in higher orders these effects are induced by the charged currents. Since these effects are not present at the tree level and the theory is renormalizable, the magnitude of induced flavor-changing neutral-current effects can be computed and used to test the standard model. These calculations have been carried out by Gaillard and Lee [14]. We briefly note the results for the two processes $K_L^0 \to \mu\bar{\mu}$ and $K_1 - K_2$ mass difference.

$K_L^0 \to \mu\bar{\mu}$ Decay

Experimentally the branching ratio for this decay is around 9×10^{-9} leading to an amplitude whose strength is $\sim G_F \times 10^{-4}$. There are two different sets of contributions to this process: (a) The 2γ intermediate state involving $K_L^0 \to 2\gamma \to \mu\bar{\mu}$. The absorptive part of this process involves the two photons or mass shell and can be predicted using known branching ratio for $K_L^0 \to 2\gamma$ decay [15] to be of the order $G_F\alpha^2$, which is of the same order as the experimental observation (Fig. 3.4). (b) The pure weak-interaction contribution involving the exchange of virtual $W^\pm$ bosons [Fig. 3.4(a)]. The divergent parts of the diagram cancel between the up- and charm-quark intermediate states (the Glashow–Illiopoulos–Maiani cancellation) due to the unitary (or orthogonal) characters of the mixing matrices leading to a finite answer given by

$$A(K_L^0 \to \mu\bar{\mu}) \simeq \frac{G_F\alpha}{\pi} \frac{(m_c^2 - m_u^2)}{m_W^2 \sin^2\theta_W} \ln\left(\frac{m_W^2}{m_c^2}\right) \bar{d}\gamma_\mu\gamma_5 s\bar{\mu}\gamma_\mu\gamma_5\mu. \quad (3.5.1)$$

This explains the suppression of $K_L^0 \to \mu\bar{\mu}$ decay compared to $K \to \mu\nu$ decay and it also indicates that $m_c \ll m_W$.

$K_L - K_S$ Mass Difference

This arises from the higher-order-induced $\Delta S = 2$ matrix element in the standard model and is given by the contributions of the graphs shown in Fig. 3.5. The expression for the $\Delta S = 2$ matrix element in the two-

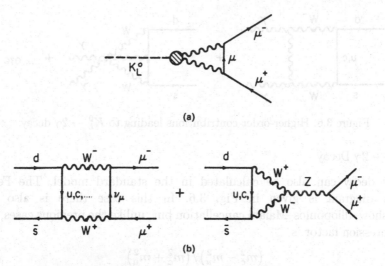

Figure 3.4. (a) Contribution of the two-photon intermediate state to $K_L^0 \to \mu\bar\mu$ decay. (b) Higher-order weak corrections to $K_L^0 \to \mu\bar\mu$ decay.

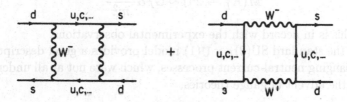

Figure 3.5. Higher order corrections leading to $\Delta S = 2$, $K^0 - \bar{K}^0$ mixing.

generation model is given by

$$H_{\Delta S=2} = \frac{G_F^2}{4\pi^2} M_W^2 (\sin\theta_C \cos\theta_C)^2 \left(\frac{m_c^2}{M_W^2}\right) \bar{d}_L \gamma^\mu s_L \bar{d}_L \gamma_\mu s_L. \qquad (3.5.2)$$

To obtain the value of the $K_1 - K_2$ mass difference, we have to evaluate the value of the $\langle K^0 | \bar{d}_L \gamma_\mu s_L \bar{d}_L \gamma_\mu s | \bar{K}^0 \rangle$ matrix element. It can be done using the vacuum saturation approximation [16] to obtain

$$\Delta m_K \simeq \frac{G_F^2}{6\pi^2} M_W^2 f_K^2 M_K \left(\frac{m_c^2}{M_W^2}\right) \cos^2\theta_C \sin^2\theta_C \simeq 3.1 \times 10^{-15}\,\text{GeV}.$$

$$(3.5.3)$$

This can be compared with the experimental value of 3.5×10^{-15} GeV. This can be regarded as a successful prediction of the standard model, although many simplistic assumptions have been made such as keeping only the box graph that reflects only the short-distance contribution and vacuum dominance of the matrix element. In any case, suppression of the $K_L - K_S$ mass difference is well understood.

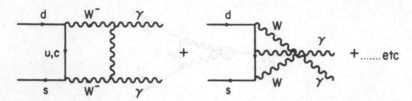

Figure 3.6. Higher-order contributions leading to $K_L^0 \to 2\gamma$ decay.

$K_L^0 \to 2\gamma$ Decay

This decay can also be calculated in the standard model. The Feynman diagram is given in Fig. 3.6. In this case there is also the Glashow–Illiopoulos–Maiani cancellation but, unlike the previous cases, the suppression factor is

$$\left(m_c^2 - m_u^2\right) / \left(m_c^2 + m_u^2\right)$$

and we get

$$M(K_L^0 \to 2\gamma) \sim G_F \alpha \frac{\Delta m^2}{m_c^2}. \tag{3.5.4}$$

Again, this is in accord with the experimental observations.

Thus, the standard SU(2)$_L$ × U(1) model provides a good description of flavor-changing neutral-current processes, which were not at all understood prior to the advent of gauge theories.

§3.6   The Higgs Bosons

As we saw in Section 3.2, the gauge symmetry was spontaneously broken by means of an explicit scalar multiplet belonging to a doublet representation of the gauge group

$$\phi = \begin{pmatrix} \phi^+ \\ \phi^0 \end{pmatrix}. \tag{3.6.1}$$

We can write the complex field ϕ^0 in terms of real fields, i.e., $\phi^0 = (1/\sqrt{2})(\sigma + i\chi)$. Subsequent to spontaneous breakdown of the gauge symmetry, $\langle\sigma\rangle = v$, we can shift the σ field. The $\phi^\pm$ and χ fields, then get absorbed as the longitudinal components of the $W^\pm$ and Z fields, respectively, leaving the shifted σ-field as a physical, real, scalar boson. We should therefore study the implications of the existence of this particle for physical processes and look for ways to detect it. Important for this purpose are the mass and couplings of the physical Higgs bosons, σ. To study its mass, we have to look at the Higgs potential

$$V = -\mu^2 \phi^+ \phi + \frac{\lambda}{2} \left(\phi^+ \phi\right)^2. \tag{3.6.2}$$

On substituting $\phi^0 = (1/\sqrt{2})(v + \sigma + i\chi)$, and setting $\phi^\pm$ and χ to zero in the unitary gauge, we get the mass of the Higgs boson as

$$m_\sigma^2 = 2\lambda v^2. \tag{3.6.3}$$

This formula for the mass of the physical Higgs bosons is not valid for all λ; for $\lambda \simeq 10^{-4}$, the radiative contributions to the Higgs boson mass become important. In fact, it has been pointed out by Linde [17] and by Weinberg [17] that radiative corrections provide a lower bound on m_σ. To illustrate the way to obtain the lower bound, we take the effective Higgs potential including the radiative corrections à la Coleman and E. Weinberg [18]

$$V(\sigma) = -\frac{\mu^2}{2}\sigma^2 + \frac{\lambda}{4}\sigma^4 + \frac{B}{4}\sigma^4 \ln\frac{\sigma^2}{V^2}, \tag{3.6.4}$$

where

$$B = \frac{3e^4(1 + \frac{1}{2}\sec^4\theta_W)}{64\pi^2 \sin^4\theta_W} \simeq 6.4 \times 10^{-4}. \tag{3.6.5}$$

For $\mu^2 < 0$ we get

$$\frac{\partial V}{\partial\sigma} = 0 \tag{3.6.6}$$

leading to symmetry breaking, i.e., $\langle\sigma\rangle \neq 0$ only if $\lambda < -B/2$ in which case we get $\mu^2 = (\lambda + B/2)v^2$. There are then two minima of the potential: one at $\langle\sigma\rangle = 0$ and another at $\langle\sigma\rangle = v$. Because

$$V|_{\langle\sigma\rangle=v} < V|_{\langle\sigma\rangle=0} \tag{3.6.7}$$

implies that

$$|\lambda| < B.$$

Thus

$$\frac{B}{2} < |\lambda| < B. \tag{3.6.8}$$

Taking the second derivative of V with respect to σ, we find

$$m_\sigma^2 = (3B - 2|\lambda|)v^2 \geq Bv^2. \tag{3.6.9}$$

Using the value of B given above we find $m_\sigma \geq 6.6\,\text{GeV}$.

The above discussion omits the effects of the fermion as well as the Higgs boson loops. Including these effects, B modifies to the form

$$B = \frac{3e^4(1 + \frac{1}{2}\sec^4\theta_W)}{64\pi^2 \sin^4\theta_W} + \frac{9\lambda^2 - 3h_f^4}{128\pi^2}. \tag{3.6.5a}$$

Similarly, there is also an upper bound on the mass of the Higgs boson [19] which can be derived from many considerations. A naive way to see this is to look at the formula in eq. (3.6.3) and observe that for the theory to remain perturbative we require $\lambda^2/4\pi \leq 1$. This would imply that

$$m_\sigma^2 \leq 8\pi v^2. \qquad (3.6.10)$$

This implies $m_\sigma \geq 1$ TeV. Another way to obtain a bound is to look at the radiative corrections to the ρ_w parameter. We find

$$\rho_w = 1 + \frac{\alpha}{4\pi} G \left(\frac{m_\sigma^2}{M_Z^2}, \cos^2 \theta_W = c^2, \sin^2 \theta_W = s^2 + \cdots \right), \qquad (3.6.11)$$

where

$$G(\xi, c^2) = \frac{3\xi}{4s^2} \left[\frac{\ln(c^2/\xi)}{c^2 - \xi} + \frac{\ln \xi}{c^2(1 - \xi)} \right]$$

$$\xrightarrow[\xi \to \text{large}]{} -\frac{3}{4} \left(\frac{\ln \xi}{c^2} + \frac{\ln c^2}{s^2} \right). \qquad (3.6.12)$$

The experimental fact that $\Delta\rho_w \leq 0.01$ implies that $m_\sigma \leq 10^2 M_Z \simeq 9$ TeV. Thus we expect the physical Higgs boson mass to remain within 7 GeV to 1 TeV.

Let us now turn to its coupling to fermions. It is easy to see from eq. (3.4.6) that

$$\mathcal{L}\sigma = \frac{\sigma}{v} \left(\bar{P} M^u P + \bar{N} M^d N + \bar{E} M^c E \right). \qquad (3.6.13)$$

The heavier a fermion is, the stronger is its coupling to the Higgs boson σ. Therefore, in its decay, heavy fermions play an important role. Its decay modes to various final states can be calculated [20]

$$\Gamma(\sigma \to f\bar{f}) = \frac{(3)G_F}{4\pi\sqrt{2}} m_f^2 M_\sigma \left(1 - \frac{4m_f^2}{M_\sigma^2} \right)^{1/2}, \qquad (3.6.14)$$

where factor 3 will appear only for quarks. For $M_\sigma > 2M_Z$

$$\Gamma(\sigma \to W^+ W^-) = \frac{G_F M_\sigma^3}{8\pi\sqrt{2}} \left(1 - \frac{4M_W^2}{M_\sigma^2} \right)^{1/2} f \left(\frac{M_W}{M_\sigma} \right) \qquad (3.6.15)$$

and

$$\Gamma(\sigma \to Z^0 Z^0) = \frac{G_F M_\sigma^3}{16\sqrt{2}} \left(1 - \frac{4M_Z^2}{M_\sigma^2} \right)^{1/2} f \left(\frac{M_Z}{M_\sigma} \right), \qquad (3.6.16)$$

where

$$f(x) = (1 - 4x^2 + 12x^4). \qquad (3.6.17)$$

Clearly the coupling to the $W^+ W^-$ is the strongest. This would provide a clean signature of the Higgs boson by looking at two high transverse momentum leptons.

The production for heavy Higgs bosons in proton–proton and proton–antiproton collisions proceeds through the gluon fusion mechanism given in

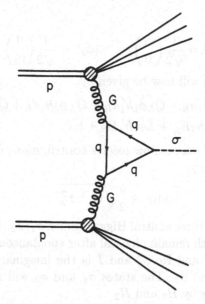

Figure 3.7. Gluon fusion diagram for the production of the Higgs boson in the standard model.

Fig. 3.7. This process has been calculated [21]. This cross section is about 10^{-38} cm^2.

In e^+e^- collision, the dominant process for Higgs search is $e^+e^- \to Z+\sigma$ with $b\bar{b}$ decay mode being the Higgs signature and leptonic decay modes of Z being its signature. At present, the LEP experiments using this search mode have provided a lower limit on the Higgs boson mass of 115 GeV.

The Higgs boson is expected to be the biggest prize for the next generation of machines such as the LHC at CERN.

§3.7   SU(2)$_L$ × U(1) Model with Two Higgs Doublets

Until the previous section we have discussed the minimal Higgs model. Often, from a theoretical point of view (such as the strong CP problem to be discussed in Chapter 4) it is necessary to consider models with two Higgs doublets. In this section we present this model with a brief discussion of it. We keep all the features of the SU(2)$_L$ × U(1) model except that we include two Higgs doublets ϕ_1 and ϕ_2 and let both of them participate in the spontaneous breakdown of the gauge symmetry

$$\phi_1 = \begin{pmatrix} \phi_1^{0*} \\ \phi_1^- \end{pmatrix} \quad \text{and} \quad \phi_2 = \begin{pmatrix} \phi_2^+ \\ \phi_2^0 \end{pmatrix}, \tag{3.7.1}$$

with

$$\langle\phi_1\rangle = \frac{1}{\sqrt{2}}\begin{pmatrix}v_1\\0\end{pmatrix}, \qquad \langle\phi_2\rangle = \frac{1}{\sqrt{2}}\begin{pmatrix}0\\v_2\end{pmatrix}. \tag{3.7.2}$$

The Yukawa coupling will now be given by

$$\mathcal{L}_Y = \bar{Q}_L\phi_1 h_1 u_R + \bar{Q}_L\tilde{\phi}_1 h_1' d_R + \bar{Q}_L\phi_2 h_2 d_R + \bar{Q}_L\tilde{\phi}_2 h_2' d_R$$
$$+ \bar{L}\phi_2 h_3 E_R^- + \bar{L}\tilde{\phi}_1 h_3' E_R + \text{h.c.} \tag{3.7.3}$$

The W- and Z-boson masses now receive contributions from both vacuum expectation values, i.e.,

$$m_W = \frac{g}{2}\sqrt{v_1^2 + v_2^2}. \tag{3.7.4}$$

In this case, there are three neutral Higgs bosons, σ_1, σ_2, I, and one charged Higgs boson, $\phi^\pm$, which remain physical after spontaneous breakdown. The σ_1 and σ_2 are Re ϕ_1^0 and Re ϕ_2^0 and I is the imaginary part of a linear combination of ϕ_1 and ϕ_2. The states σ_1 and σ_2 will mix. If we denote their mass eigenstates by H_1 and H_2

$$H_1 = \sigma_1 \cos\phi + \sigma_2 \sin\phi,$$
$$H_2 = -\sigma_1 \sin\phi + \sigma_1 \cos\phi, \tag{3.7.5}$$

their couplings to fermions are given by (denoting $v_1 = v \cos\alpha$ and $v_2 = v \sin\alpha$)

$$\mathcal{L}_Y = \frac{1}{\sqrt{2}}\sigma_1[\bar{P}_L f_1^{(u)} P_R + \bar{N}_L f_1^{(d)} N_R + \bar{E}_L f_1^{(e)} E_R]$$
$$+ \frac{1}{\sqrt{2}}\sigma_2[\bar{P}_L f_2^{(u)} P_R + \bar{N}_L f_2^{(d)} N_R + \bar{E}_L f_2^{(e)} E_R]$$
$$+ \frac{i}{\sqrt{2}}I[\bar{P}_L f_3^{(u)} P_R + \bar{N}_L f_3^{(d)} N_R + \bar{E}_L f_3^{(e)} E_R]$$
$$+ H^-[\bar{d}_L f_4 u_R + \bar{d}_R f_5 u_L] + \text{h.c.}, \tag{3.7.6}$$

where

$$\tilde{f}_1^{(u)} = (h_1 \cos\phi + h_2' \sin\phi),$$
$$\tilde{f}_1^{(d)} = (h_2 \sin\phi - h_1' \cos\phi),$$
$$\tilde{f}_1^{(e)} = (h_3 \sin\phi - h_3' \cos\phi),$$
$$\tilde{f}_2^{(u)} = (-h_1 \sin\phi - h_2' \cos\phi),$$
$$\tilde{f}_2^{(d)} = (h_2 \cos\phi + h_1' \sin\phi),$$
$$\tilde{f}_2^{(e)} = (h_3 \cos\phi + h_3' \sin\phi),$$
$$\tilde{f}_3^{(u)} = (h_1 \sin\alpha + h_2' \cos\alpha),$$
$$\tilde{f}_3^{(d)} = (h_2 \cos\alpha - h_1' \sin\alpha),$$
$$\tilde{f}_3^{(e)} = (h_3 \cos\alpha - h_1' \sin\alpha),$$

$$f_4 = U_L^{(d)} (h_1 \sin \alpha - h_2 \cos \alpha) U_R^{(u)\dagger},$$

$$f_5 = U_L^{(u)} (h_2 \cos \alpha - h_1 \sin \alpha) U_R^{(d)\dagger}, \qquad (3.7.7)$$

where terms with and without a tilde are related as follows:

$$U_L^{(p)} \tilde{f}_i^{(p)} U_R^{(p)\dagger} = f_i^{(p)}, \qquad p = u, d, e; \; i = 1, 2, 3. \qquad (3.7.8)$$

The fermion mass matrices in this case are given by the formula

$$M_{u-} = \frac{v}{\sqrt{2}} (h_1 \cos \alpha + h_2' \sin \alpha),$$

$$M_{d-} = \frac{v}{\sqrt{2}} (h_2 \sin \alpha - h_1' \cos \alpha),$$

$$M_{e-} = \frac{v}{\sqrt{2}} (h_3 \sin \alpha - h_3' \cos \alpha).$$

Looking back at the Higgs couplings of σ_1 and σ_2 we find that, unlike the single Higgs doublet case, the Yukawa couplings of the physical quarks (mass eigenstates) are not flavor diagonal. This implies that at the tree level this model will lead to flavor-changing neutral currents. This difficulty can be avoided if we set the primed Yukawa couplings h_1', h_2', and h_3' equal to zero. It is then easy to convince oneself that all σ_1-couplings are flavor diagonal and parity conserving, whereas those of I are parity violating and flavor conserving. For the couplings in this case we get

$$\begin{aligned} \mathcal{L}_Y = 2^{1/4} G_F^{1/2} [&\sec \alpha \bar{P} M_u P (\cos \phi \sigma_1 - \sin \phi \sigma_2) \\ &+ \operatorname{cosec} \alpha \bar{N} M_d N (\sin \phi \sigma_1 + \cos \phi \sigma_2) \\ &+ i I (\tan \alpha \bar{P} \gamma_5 M_u P + \cot \alpha \bar{N} \gamma_5 M_d N) \\ &+ H^+ (\tan \alpha \bar{P}_R M_u U_{KM} N_L + \cot \alpha \bar{P}_L U_{KM} M_d N_R)] + \text{h.c.} \qquad (3.7.9) \end{aligned}$$

We do not discuss this model any further here. We only note that this model has an extra chiral U(1) symmetry in the gauge as well as the Yukawa coupling Lagrangian, which is the U(1)$_{PQ}$ symmetry used as one possible solution to the strong CP problem in Chapter 4. The advantage of this model over the single Higgs doublet minimal model is that here couplings to quarks need not be suppressed due to their mass factor. The model also has a richer phenomenology [22].

§3.8 Puzzles of the Standard Model

In spite of the impressive success of the standard model in correlating all observed low-energy data in terms of a very few parameters, it is still very unsatisfactory since it builds on many assumptions and leaves many fundamental questions unanswered. Most of the successes of the standard model pertain only to the gauge sector of the theory where, with the help of only one free parameter, $\sin^2 \theta_W$, numerous neutral-current data are successfully understood. But in the fermionic sector, all masses and mixings are unexplained. Below we give a list of the major puzzles of the standard model.

(a) Gauge Problem

There are many aspects to the gauge problem and we enumerate them one by one:

(i) The standard model has three different gauge groups postulated from phenomenological considerations. They are associated with three different gauge coupling constants, which are arbitrary. In a more satisfactory theory, one should have a way of not only understanding the three gauge groups but also the origin of the three gauge couplings. In such a theory, one would be able to predict the value of $\sin^2 \theta_W$ as well as the color gauge coupling, α_s. The most popular approach in addressing these problems has been via the program of grand unification (Chapters 5 and 7), where simple non-abelian groups such as SU(5) and SO(10) are chosen as the grand unification groups.

(ii) Another aspect of the gauge problem is the choice of the fermion representation under the gauge group, which is done mainly to fit observations such as the nature of beta decay, values of electric charges, etc. In the context of the standard model, the question would be: Why assign left-handed fermions to doublets and how does one fix the values of hypercharge quantum numbers? In fact, the choice of hypercharge is related to the question of the quantization of electric charges of fermions. We will argue later that constraints of gauge anomaly combined with other reasonable requirements may help to explain the values of hypercharges of various fermion multiplets and hence their electric charges.

(b) Fermion Problem

This is the most unsatisfactory aspect of the standard model. First of all, what dictates that the fermions must be assigned to doublets, except for the a posteriori justification that it fits data? Second, why are there three generations, or are there more? Furthermore, the fermion masses and mixings seem to exhibit a hierarchical pattern, i.e., $m_t, m_b \gg m_c, m_s \gg m_u, m_d$, and $m_\tau \gg m_\mu \gg m_e$, and $V_{ud} \approx V_{cs} \gg V_{us} \approx V_{cd} \approx V_{bc} \gg V_{ub}, V_{td}$. Finally, the chirality of fermions is also put in by hand although this problem is somewhat less awkward in left–right symmetry models where weak interactions involve both left- and right-handed helicities. Then there are questions of CP violation and the QCD θ-problem all related to the fermionic sector. Finally, in the standard model, the neutrino has zero mass. The convincing evidence for neutrino mass from recent neutrino oscillation experiments imply that one will most certainly need an extension of the standard model, the most likely scenario being the existence of a left–right symmetric model (Chapter 6) at some scale, since this is the only class of model where small neutrino masses receive a plausible explanation.

(c) Higgs Problem

The Higgs boson is the most obscure aspect of the standard model. Not only that, its choice and its vacuum expectation value is decided ad hoc, but more importantly, its mass receives quadratically divergent quantum loop corrections, making its natural value of the order of Planck mass, unless there is some new physics at the TeV scale where the Higgs mass should be. The standard model certainly provides no clue to this new physics. Two promising scenarios are supersymmetry (Chapters 9–16) and technicolor (Chapter 8). We will consider the supersymmetric approach in a subsequent section.

(d) Gravity Problem

Again there are several aspects to this problem. First, and the most obvious one, is that Einstein's gravity is totally outside the framework of gauge theories. In fact, adding gravity to gauge theories destroys the attractive property of renormalizability. Second, all of physics, except gravity, seems to follow from the principle of local gauge invariance, whereas, to understand gravity, one must use an additional principle—the principle of equivalence. These two problems are addressed within the framework of supergravity and Kaluza–Klein (Chapters 14–17) theories, and perhaps "most satisfactorily" in the superstring theories, once these theories are understood.

(e) Quantization of Electric Charge

One of the fundamental mysteries of nature is the fact that all observed electric charges appear in quantized units, which are multiples of the electric charge of the positron. In the hadronic sector, the quark model implies that the smallest unit of charge may be $-\frac{1}{3}$ for the down-quark. One may hope that a fundamental theory of the forces of nature will explain the quantization of electric charges. For a long time, it was believed that the existence of the abelian U(1)-group as part of the standard model would leave the electric charges of quarks and leptons arbitrary. It has recently been realized [24,25,25a] that constraints of anomaly freedom (required for renormalizability) and nonvanishing fermion masses lead to change quantization in the standard model. However, for various reasons outlined above, it is obvious that one must go beyond the standard model, and the interesting question is to see if the above result holds, once one goes beyond the standard model. For instance, if a right-handed neutrino is included in the standard electric charge, becomes a free parameter. One possibility is to consider grand unified theories based on simple groups, in which case charge quantization follows automatically. Recently, an alternative possibility has been noted [25] according to which in gauge models, the charge quantization follows from the requirements of anomaly freedom and the Majorana neutrino. The deeper reason for this is that inclusion of the right-handed neutrino elevates $B - L$ to a hidden local symmetry, which then causes dequantization of electric charge. But once the neutrino is a Majorana particle, $B - L$ is no more exact and charge quantization is restored. An important implication of this result is that, to resolve the puzzles of the standard models, one need not invoke physics near the Planck scale. The physics in the TeV scale (which should be testable in the near future) may indeed be sufficient.

§3.9   Outline of the Various Scenarios

Over the past fifteen years, various scenarios have been advocated which address the various puzzles of the standard model. All of them involve new physics beyond the standard model and lead to testable predictions, whose pursuit is the major concern of many collider and nonaccelerator experiments. In this section, we wish to present a brief overview of the various scenarios, the various puzzles of the standard models that they address, and their possible tests in a qualitative language. We discuss these in more detail in the subsequent chapters. We present these ideas in the form of a table (Table I). In this table we have presented interesting new ideas rather than extensions of the standard model that increase the number of Higgs doublets or the inclusion of a fourth generation. Except for a restricted class of multi-Higgs doublet models that are well motivated by supersymmetric models, the other models (multi-Higgs or the fourth generation of fermions) are at this moment not so well motivated. I will therefore not discuss them.

§3.10   Beyond the Standard Model

The discovery of the $W^{\pm}$ and Z bosons in the CERN collider experiments (with masses M_W and M_Z in striking agreement with the predictions of the standard model) has provided confirmation not only of the theoretical approach based on spontaneously broken gauge theories but also of the SU(2)$_L$ × U(1) as the low-energy electroweak (gauge?) symmetry. This model, however, introduces a number of free parameters into the theory to explain everything from the masses of the gauge bosons to the masses of leptons and quarks, as well their mixings. This is a highly unsatisfactory state of affairs and it is hoped that, while the standard model provides a good starting point, a lot of physics remains to be uncovered as we move into regimes beyond the energy scales explored in the standard model, i.e., $E > 100\,\text{GeV}$ or so. In the remaining chapters we present various theoretical schemes that extend our knowledge into this unexplored regime, each one attempting to cure, in one way or another, the shortcomings of the $SU(2)_L \times U(1)$ theory. It is, however, quite possible that the true reality may be a combination of some of these ideas or, even worse, outside their scope completely. It may, therefore, be useful to analyze ways to discover this new physics in a way that does not rely on any specific theoretical framework.

To this end, we must focus on aspects of physics that are unambiguously predicted in the standard model. The strategy would then be to look for deviations from these predictions. A number of papers [26] have focused on this question and below we provide a short summary of these ideas. Before proceeding to this discussion we give a summary (in Table 3.2) of the experimental situation [27] for rare processes and the theoretical expectations for them based on the standard model.

Clearly, observation of any of the above rare processes would indicate the existence of new physics. The burden on the theorist, however, is the following: Once any indication of new physics appears, what can we say about the nature of new physics? To answer this question we note that from the observed magnitude of new physics a general idea can be had about the mass scale of the new physics (or new interaction). To say anything beyond that we must resort to a particular model. In this section we will give a rough description of how the first part of the answer can be given, focusing on a few specific examples.

(a) Neutrino Mass

The simplest (yet, extremely profound) example of new physics has been the discovery of nonzero neutrino mass. To see how and what we can learn about possible new physics from here, we remind ourselves that in the standard model $m_\nu = 0$. However, using the fields present in the standard model, i.e., the lepton doublet $\psi_L \equiv \binom{\nu_e}{e^-}_L$ and the Higgs doublet $\phi \equiv \binom{\phi^+}{\phi^-}$, we can write the following $SU(2)_L \times U(1)$ invariant operator:

$$L^{(1)} = \frac{\lambda}{M}\psi_L^T C^{-1}\tau_2\tau\psi_L \cdot \phi^T\tau_2\tau\phi. \tag{3.10.1}$$

The appearance $1/M$ is to make the operator have mass dimension 4 so it can be a legitimate Lagrangian. The $L^{(1)}$ must be $SU(2)_L \times U(1)_Y$ invariant because we are exploring physics at a higher mass scale $M \gg M_W$ where electroweak symmetry is exact. After $SU(2)_L \times U(1)_Y$ symmetry is broken,

TABLE I.

Scenario and symmetry (global if within square brackets)	The puzzles of the standard model addressed	Testable predictions	Present status
(i) Axions SU(3)$_C$ × SU(2)$_L$ × U(1)$_Y$ [U(1)$_{PQ}$]	Strong CP-problem	Axion ≡ light pseudoscalar particle coupling to all fermions.	Visible axion ruled out
(ii) Majorons, familons SU(3)$_c$ × SU(2)$_L$ × U(1)$_Y$ × [G] $G = $ U(1)$_{B-L}$ or family symmetry	Small neutrino Majorana mass or possible quark mixings	Majoron (χ) emissible in $(\beta\beta)_{0\nu}$ decay, measurement of Z width Familon: (f) $K \rightarrow \pi + f$ $\mu \rightarrow e + f$	
(iii) Left-right symmetric models	Origin of parity and CP violation, quark mixings, strong CP problem, possible small neutrino mass	New gauge bosons: $W_R^{\pm} Z$ in the few hundred to TeV range; $(\beta\beta)_{0\nu}$ decay, muonium–antimuonium oscillation, large neutron electric dipole moment ($\approx 10^{-26}$ ecm level), heavy Majorana right-handed neutrino, etc.	
(iv) Supersymmetric model SU(3)$_c$ × SU(2)$_L$ × U(1)$_Y$ × SUSY	(a) Higgs mass problem; (b) Automatic inclusion of gravity	New set of fermions of bosonic partners with identical electroweak properties to bosonic and fermionics of the standard model: gravitino	

(v) Technicolor $SU(3)_c \times SU(2)_L \times U(1)_Y \times G_{\text{Hypercolor}}$ and compositeness	Higgs mass problem	Low-mass ($m \leq M_W$) neutral and charged Higgs bosons	
(vi) Grand unification $SU(5)$, $SC(10)$, E_6, etc.	Unification of gauge couplings	Proton decay, neutron-antineutron oscillation measurement	Minimal $SU(5)$ ruled out by proton decay and $\sin^2 \theta_W$
(vii) Higher-dimensional theories (Kaluza–Klein type)	Unifies gauge and equivalence principle into a larger equivalence principle	Without monopolelike configurations, these theories predict vectorlike weak interactions and are therefore ruled out	
(viii) Superstrings	Unification of gravity with all other fundamental interactions in a locally supersymmetric higher-dimensional theory; explains fermion generations	Can be ruled out by discovery of doubly charged Higgs bosons or processes such as muonium–antimuonium oscillation at a significant level ($\gtrsim G_F \times 10^{-3}$)	Plagued by the theoretical problem of nonunique vacuum, at the moment

$L^{(1)}$ can lead to an effective mass for neutrino as

$$m_{\nu_e} \simeq \frac{4m_W^2}{g^2 M} \lambda. \tag{3.10.2}$$

If m_{ν_e} is in the electron volt range and if λ is 10^{-6}–10^{-1}, the value of the new scale M is 10^6 GeV to 10^{11} GeV. Since λ is an unknown parameter, in the absence of a detailed theory, the precise value of M cannot be determined; however, the existence of a new scale is clearly established.

The same kind of analysis can be carried out for most of the processes listed in Table 3.2. We give two more examples.

(b) Proton Decay

The strategy here is to consider the lowest-dimensional operator invariant under the SU(2)$_L$ × U(1) × SU(3)$_c$ group that changes the baryon number. This question has been studied in detail in Ref. [23]. Here we simply present an illustrative example

$$\mathcal{L}_1^{\Delta B \neq 0} = \frac{\varepsilon_{ijk}}{M^2} u_{R_i}^T C^{-1} d_{R_j} Q_{Lk}^T C^{-1} \tau_2 \psi_L. \tag{3.10.3}$$

Another operator of this type is

$$\mathcal{L}_2^{\Delta B \neq 0} = \frac{\varepsilon_{ijk}}{M^2} Q_{L_i}^T \tau_2 C^{-1} Q_{L_j} U_{R_k}^T C^{-1} e_R. \tag{3.10.4}$$

Other operators of this type can be constructed, and it turns out that all these operators satisfy the selection rule $\Delta B = \Delta L$ and lead to decays of type

$$p \rightarrow e^+ \pi^0$$
$$\rightarrow \bar{\nu} \pi^+. \tag{3.10.5}$$

Observation of any of these processes would again indicate new physics with mass scale M. The present lower limits on $\tau_{p \rightarrow e^+ \pi^0}$ imply $M \geq 10^{15}$ GeV or so.

(c) Neutron–Antineutron Oscillation

The $\Delta B \neq 0$ operator that causes $N - \bar{N}$ transition must necessarily involve six quark fields and will therefore have dimension nine. A typical SU(2)$_L$ × U(1) × SU(3)$_c$ invariant operator of this type is

$$\mathcal{L}_{(3)}^{\Delta B = 2} = \frac{\varepsilon_{ijk}}{M^5} \varepsilon_{i'j'k'} u_{R_i}^T C^{-1} d_{R_{i'}} u_{R_j}^T C^{-1} d_{R_{j'}} d_{R_k}^T C^{-1} d_{R_{k'}}^R. \tag{3.10.6}$$

This involves a high power of the mass scale M and, therefore, is particularly interesting. The reason is that observation of $N - \bar{N}$ oscillation with

Table 3.2.

Physical process	Expectation based on the standard $SU(2)_L \times U(1)$ model	Experimental situation
Higgs boson	One neutral with parity conserving fermion couplings	None observed
$\frac{B(Z \to e^+e^-\gamma)}{B(Z \to e^+e^-)}$	2–3%	
$\frac{B(W \to e\nu\gamma)}{B(W \to e\nu)}$	2–3%	None observed
Baryon nonconserving processes such as $p \to e^+\pi^0, \ldots, N \leftrightarrow \bar{N}$, etc.	0	$\tau_{p \to e^+\pi^0} > 7 \times 10^{33}$ yr $\tau_{N-\bar{N}} > 8.6 \times 10^7$ s
Lepton number non-conserving processes		
(a) Neutrinoless double β-decay		$\tau_{\beta\beta0\nu} \geq 1.25 \times 10^{25}$ yr from enriched ^{76}Ge
(b) $B(K^+ \to \pi^-e^+e^+)$	0	$< 10^{-8}$
(c) $B(\mu^- + N \to e^+ + N')$	0	$< 1.7 \times 10^{-12}$
Lepton flavor violation		
(a) $B(\mu \to e\gamma)$	0	$< 1.2 \times 10^{-11}$
$B(\mu \to e\gamma\gamma)$	0	$< 7 \times 10^{-11}$
Strength of $\mu^+e^- \to \mu^-e^-$ transition	0	$\leq G_F \times 3 \times 10^{-3}$
(b) $B(\mu \to e + f)$ f = massless particle	0	$< 6 \times 10^{-6}$
(c) $B(\mu \to 3e)$	0	$\leq 10^{-12}$
Other rare processes		
(a) $B(K^0_L \to \mu\bar{e})$	0	$< 6 \times 10^{-6}$
(b) $B(K^0_L \to \bar{e})$	10^{-13}	$< 2 \times 10^{-7}$
(c) $B(K^0_L \to \pi^+\mu^e)$	0	$< 7 \times 10^{-9}$
(d) $B(K^0_L \to \pi^+\mu^{\bar{\nu}})$	$\geq 4 \times 10^{-10}$	$< 1.4 \times 10^{-7}$
Electric charge violation		
(a) Photon mass, m_γ	0	$< 3 \times 10^{-27}$ eV
(b) $\tau(e^+ \to \nu + \gamma)$	0	$> 2 \times 10^{22}$ yr
Neutrino masses		
m_{ν_e}	0	< 2.4 eV
m_{ν_μ}	0	< 0.16 MeV
m_{ν_τ}	0	< 17.2 MeV
Limits on parameters in μ-decay*		
ρ	3/4	0.7518 ± 0.0026
η	0	-0.007 ± 0.013
ξ	1	$1.0027 \pm 0.0079 \pm 0.0030$
δ	3/4	$0.7486 \pm 0.0026 \pm 0.0028$

* The parameters are defined in terms of the electron spectrum in μ-decay as follows:

$$dNe/x^2 \, dx \, d(\cos\theta) = \text{const.}[(3 - 2x) + (4/3\rho - 1)(4x - 3) + 12(m_e/m_\mu)(1 - x)/x\eta$$
$$+ \{(2x - 1) + (4/3\delta - 1)(4x - 3)\}\xi P_\mu \cos\theta],$$

where $x = 2E_e/m_\mu$, θ is the angle between the electron momentum and polarization vector of the muon, and P_μ is the muon-polarization vector.

mixing time of order ~ 1 yr would imply (see Chapter 6) $M \simeq 10^6$ GeV, indicating new physics in a nearby regime.

Exercises

3.1. Assigning arbitrary values of weak hypercharge to the quarks and leptons:

    (a) obtain relations between them resulting from anomaly cancellation;

    (b) show that, if Dirac mass constraints on weak hypercharges are included, electric charges of all fermions are quantized;

    (c) show that the charge quantization is lost once the right-handed neutrino is included in the standard model;

    (d) charge quantization is again restored if the neutrino is assumed to be a Majorana particle.
    [See Ref. 25.]

3.2. Show that the general formula for the ρ_w parameter [defined after eq. (3.1.21c)], if the gauge symmetry breaking is implemented by a Higgs multiplet with weak isospin I and hypercharge Y, is

$$\rho_W^{(0)} = \frac{\sum_a \left[I^a(I^a + 1) - I_3^{a^2} \right] (V^a)^2}{2 \sum_a \left(I_3^a \right)^2 V_a^2}.$$

Show that if all I_3^a components of an arbitrary irreducible Higgs representation contribute equal v.e.v., then $\rho_W^0 = 1$.

3.3. Include the effect of fermion doublets in the effective potential expression eq. (3.6.4), and discuss its impact on the lower bound on the Higgs boson mass.

3.4. (a) Consider the two Higgs model with the requirement of natural flavor conservation in neutral weak interactions,

    (b) Write down the one-loop induced effective Higgs potential in this model and discuss the spectrum and interactions of neutral Higgs bosons.

3.5. (a) What are the symmetries of the standard model in the absence of all Yakawa couplings?

    (b) Are baryon and lepton numbers separate symmetries of the standard model? If the answer is yes, explain; if the answer is no, is any combination of B and L a symmetry of the model?

    (c) Could that combination have been gauged?

References

[1] S. L. Glashow, *Nucl. Phys.* **22**, 579 (1961);
A. Salam and J. C. Ward, *Phys. Lett.* **13**, 168 (1964);
S. Weinberg, *Phys. Rev. Lett.* **19**, 1264 (1967);
A. Salam, in *Elementary Particle Theory* (edited by N. Svartholm),
Almquist and Forlag, Stockholm, 1968;
For an excellent review, see
E. S. Abers and B. W. Lee, *Phys. Rep.* **9C**, 1 (1973).

[2] UA1 Collaboration, G. Arnison et al., *Phys. Lett.* **122B**, 103 (1983);
UA2 Collaboration, M. Banner et al., *Phys. Lett.* **122B**, 476 (1983).

[3] Particle Data Tables, 2000.

[4] See, for instance,
M. Jonker et al. *Phys. Lett.* **99B**, 265 (1981).

[5] B. Kayser, E. Fishbach, S. P. Rosen, and H. Spivack, *Phys. Rec.* **D20**, 87 (1979).

[6] C. Y. Prescott et al., *Phys. Lett.* **77B**, 347 (1978).

[7] P. Bucksbaum, E. Commins, and L. Hunter, *Phys. Rev. Lett.* **46**, 640 (1981);
L. M. Barkov, M. Zolotorev, and I. Khriplovich, *Sov. Phys. Usp.* **23**, 713 (1980);
M. A. Bouchiat et al., *Phys. Lett.* **117B**, 358 (1982);
J. Hollister et al., *Phys. Lett.* **46**, 643 (1981).

[7a] For a review, see
Albrecht Bohm, *Proceedings of the SLAC Summer Institute* (edited by M. Zipf et al.), Stanford, 1983.

[8] W. J. Marciano and A. Sirlin, *Phys. Rev.* **D29**, 945 (1984);
W. J. Marciano and A. Sirlin, *Phys. Rev.* **D22**, 2695 (1980);
A. Sirlin and W. J. Marciano, *Nucl. Phys.* **B189**, 442 (1981);

[9] M. Böhm, W. Hollik, and H. Spiesberger, *Fortsch. Phys.* **34**, 687 (1986);
D. Kennedy and B. W. Lynn, *Nucl. Phys.* **B322**, 1 (1989).

[10] F. Antonelli, M. Consoli, and G. Corbo, *Phys. Lett.* **91B**, 90 (1980);
C. Llewellynsmith and J. Wheater, *Phys. Lett.* **105B**, 486 (1981);
J. Wheater and C. Liewellynsmith, *Nucl. Phys.* **B208**, 27 (1982);

[11] For a summary and references, see J. Erler and P. Langacker in *Particle Data Group, Europhys. J.* **C 3**, 1 (1998).

[12] G. Arnison et al., *Phys. Lett.* **126B**, 398 (1983);
P. Bagnaia et al., *Phys. Lett.* **129B**, 130 (1983);
G. Arnison et al., *Phys. Lett.* **129B**, 273 (1983).

[13] S. L. Glashow, J. Lliopoulos, and L. Maiani, *Phys. Rev.* **D2**, 1285 (1970).

[14] M. K. Gaillard and B. W. Lee, *Phys. Rev.* **D10**, 897 (1974).

[15] B. R. Martin, E. de Rafael, and J. Smith, *Phys. Rev.* **D1**, (1970).

[16] R. N. Mohapatra, J. Subbarao, and R. E. Marshak, *Phys. Rev.* **171**, 1502 (1968).

[17] S. Weinberg, *Phys. Rev. Lett.* **36**, 294 (1976);
      A. Linde, *JETP Lett.* **23**, 73 (1976).

[18] S. Coleman and E. Weinberg, *Phys. Rev.* **D7**, 1888 (1973).

[19] D. A. Dicus and V. S. Mathur, *Phys. Rev.* **D7**, 3111 (1973);
      B. W. Lee, C. Quigg, and H. Thacker, *Phys. Rev.* **D16**, 1519 (1977);
      M. Veltman, *Acta Phys. Polon.* **B8**, 475 (1977).

[20] T. Rizzo, *Phys. Rev.* **D22**, 722 (1980).

[21] H. Georgi, S. L. Glashow, A Machachek, and D. Nanopoulos, *Phys. Rev. Lett.* **40**, 692 (1978);
      See also H. Gordon, W. Marciano, F. E. Paige, P. Grannis, S. Naculich, and H. H. Williams, *Proceedings of the 1982 DPF Summer Study on Elementary Particle Physics*, Snowmass, 1982, p. 161.

[21a] See review talk by
      A. Jawahery, *Int. Conference on High Energy Physics*, 1990, Singapore.

[22] L. Hall and A Wise, *Nucl. Phys.* **B187**, 397 (1981);
      M. Barnett, G. Senjanovic, and D. Wyler, ITP Santa Barbara preprint, (1984);
      J. A Frere, A Gavela, and J. Varmaseren, *Phys. Lett.* **125**, 275 (1983).

[23] For recent reviews and references for Higgs boson effects, see
      Higgs Hunters' Guide by H. Haber, J. Gunion, G. Kane, and S. Dawson (to be published).
      M. Sher, *Phys. Rep.* **179**, 274 (t989).

[24] N. G. Deshpande, Oregon Preprint (1981);
      R. Foot, G. C. Joshi, H. Lew, and R. Volkas, Mod. *Phys. Lett.* **A5**, 95 (1990).

[25] K. S. Babu and R. N. Mohapatra, *Phys. Rev. Lett*, **63**, 938 (1989).

[25a] For further discussion of change quantization from anomaly constraints, see
      C. Geng and R. Marshak, *Phys. Rev.* **D13**, 693 (089);
      P. Ramond, J. Minahan, and R. Warner, *Phys. Rev.* **D41**, 715 (1990);
      S. Rudaz, *Phys. Rev.* **D41**, 2619 (1990);
      E. Golowich and P. Pal. *Phys. Rev.* **D41**, 3537 (1990).

[26] S. Weinberg, *Phys. Rev. Lett.* **43**, 1566 (1979);
      F. Wilczek and A. Zee, *Phys. Rev. Lett.* **43**, 1571 (1979);
      A. H. Weldon and A. Zee, *Nucl. Phys.* **B173**, 269 (1980);
      R. N. Mohapatra, *Proceedings of the First Workshop on Grand Unification* (edited by P. Frampton, H. Georgi, and S. L. Glashow), Math Sci. Press, Brookline, MA, 1980.
      For excellent recent reviews, see
      J. D. Vergados, *Phys. Rep.* (1986) (to appear).
      G. Costa and M. Zwirner, *Rev. Nuovo. Cim.* (1886) (to appear).
      C. Burges and H. Schnitzer, *Nucl. Phys.* **B228**, 464 (1983).
      M. K. Gaillard, *Proceedings of the Workshop on Intense Medium Energy Sources of Strangeness* (edited by T. Goldman, H. Haber, and H. F. Sadrozinski), (AIP), New York 1983, p. 54.

[27] The latest experimental situation in weak interaction has recently been
summarized in
G. Barbiellini and C. Santoni, *Rev. Nuovo Cim.* **9N2**, 1 (1986).
Particle Data Group, *Rev. Mod. Phys.* **56**, S1 (1984).

References 73

4

CP Violation: Weak and Strong

§4.1 CP Violation in Weak Interactions

The phenomenon of CP-violation was discovered in 1964 by Christenson,
Cronin, Fitch, and Turlay in K^0 decays. For a long time kaon systems re-
mained the only place where breakdown of CP invariance was observed.
Recently, CP violation has been confirmed in the B-system in the CDF[1],
BABAR[1], and BELLE [1] experiments. Since Lorentz invariant local field
theories are CPT invariant, breakdown of CP symmetry implies breakdown
of time-reversal invariance. This fact is used in experimental search for CP
violation by looking for kinematic effects odd under time-reversal symmetry. In the $K^0 - \bar{K}^0$ system, however, CP-violating interference effects
appear that are experimentally measurable. The CP-violating phenomenology in K^0 decays has been extensively described in many places [1a] and
we do not repeat it here, except to note some salient points.

In the absence of weak interactions, $K^0 - \bar{K}^0$ are separate eigenstates
of the Hamiltonian and they do not mix due to strangeness conservation.
Once CP-conserving weak interactions are turned on, strangeness is no
longer a good symmetry and this implies that the $K^0 - \bar{K}^0$ mix, and in the
rest frame of the kaons the mass matrix looks like the following:

$$M^{(+)}_{K^0-\bar{K}^0} = \begin{pmatrix} m_{K^0} & m_{K^0\bar{K}^0} \\ m_{K^0\bar{K}^0} & m_{\bar{K}^0} \end{pmatrix}, \qquad (4.1.1)$$

with $m_{K^0} = m_{\bar{K}^0}$. The equality of off-diagonal elements follows from CP
invariance. The eigenstates of this matrix define the physical K mesons K_1

and K_2

$$K_1^0 = \frac{K^0 - \bar{K}^0}{\sqrt{2}},$$

$$K_2^0 = \frac{K^0 + \bar{K}^0}{\sqrt{2}}. \tag{4.1.2}$$

If we assume $C|K^0\rangle = |\bar{K}^0\rangle$, $K_1^0(K_2^0)$ is *CP*-even (odd) and $m_{K_1^0} - m_{K_2^0} = 2m_{K^0\bar{K}^0}$. Also note that since K_1 and K_2 are not stable particles, $m_{K_{1,2}}$ are not real. If we now turn to the *CP*-violating interactions, then, in the $K_1 - K_2$ basis, the mass matrix is no longer diagonal and it looks as follows:

$$M_{K_1-K_2} = \begin{pmatrix} m_{K_1} & \delta m_{K_1^0 - K_2^0} \\ -\delta m_{K_1^0 - K_2^0} & m_{K_2} \end{pmatrix}. \tag{4.1.3}$$

CPT invariance has been assumed in writing eq. (4.1.3). Let us write

$$m_{K_1^0} = m_1 - \frac{i}{2}\gamma_1,$$

$$m_{K_2^0} = m_2 - \frac{i}{2}\gamma_2,$$

$$\delta m_{K_1^0 - K_2^0} = i\left(m' - \frac{i}{2}\gamma'\right). \tag{4.1.4}$$

Then $M_{K_1^0 - K_2^0}$ can be diagonalized, leading to the short- and long-lived eigenstates K_S and K_L, i.e.,

$$|K_S\rangle = \frac{|K_1^0\rangle + \rho|K_2^0\rangle}{1 + |\rho|^2},$$

$$|K_L\rangle = \frac{|K_2^0\rangle + \rho|K_1^0\rangle}{1 + |\rho|^2}, \tag{4.1.5}$$

where

$$\rho \simeq \frac{-i(m' - i\gamma'/2)}{[(m_1 - m_2) - (i/2)(\gamma_1 - \gamma_2)]}. \tag{4.1.6}$$

In terms of hadronic matrix elements

$$m' = \text{Im}\, M_{K_1^0 K_2^0}$$

and

$$\gamma' = \text{Im}\, \Gamma_{K_1^0 K_2^0}. \tag{4.1.7}$$

So far, we have calculated the *CP* impurity in K_L and K_S systems. In order to apply this discussion to $K \to 2\pi$ decays, we have to realize that there are phases arising from the two-pion final states as follows:

$$\langle(2\pi); I|H_{\text{wk}}|K^0\rangle = A_I e^{i\delta_I}$$

and

$$\langle(2\pi); I|H_{\text{wk}}|\bar{K}^0\rangle = \bar{A}_I e^{i\delta_I} \tag{4.1.8}$$

where $\langle(2\pi); I|$ denotes the definite isospin combination of the $\pi^+\pi^-$ and $\pi^0\pi^0$ states. The CPT theorem implies that $\bar{A}_I = -A_I^*$, which can be used to calculate that

$$2^{-1/2}\langle(2\pi); I|H_{\text{wk}}|K_2^0\rangle = ie^{i\delta_I}\operatorname{Im} A_I$$

and

$$2^{-1/2}\langle(2\pi); I|H_{\text{wk}}|K_1^0\rangle = e^{i\delta_I}\operatorname{Re} A_I. \tag{4.1.9}$$

Let us now define

$$\eta_{ij} = \langle\pi_i\pi_j|H_{\text{wk}}|K_L^0\rangle/\langle\pi_i\pi_j|H|K_S^0\rangle, \tag{4.1.10}$$

where $i, j = (+, -)$ or $(0, 0)$.

We can then write

$$\eta_{+-} = \varepsilon + \frac{\hat{\varepsilon}'}{1 + \hat{\omega}/\sqrt{2}},$$

$$\eta_{00} = \varepsilon - \frac{\hat{\varepsilon}'}{1 - \sqrt{2}\hat{\omega}}, \tag{4.1.11}$$

where

$$\varepsilon = \rho + i\frac{\operatorname{Im} A_0}{\operatorname{Re} A_0},$$

$$\varepsilon' = -\frac{i}{\sqrt{2}}\left\{\left(\frac{\operatorname{Im} A_0}{\operatorname{Re} A_0}\right) - \left(\frac{\operatorname{Im} A_2}{\operatorname{Re} A_2}\right)\right\}\hat{\omega},$$

$$\hat{\omega} = \left(\frac{\operatorname{Im} A_2}{\operatorname{Re} A_0}\right) \cdot e^{i(\delta_2 - \delta_0)}. \tag{4.1.12}$$

From eqs. (4.1.6) and (4.1.12) we can calculate the phase of η_{+-} and η_{00} without reference to any models. Using $\Delta m = m_2 - m_1$ and $\gamma_1 \gg \gamma_2$ and $\gamma' \ll m'$, we find (using the fact that $\gamma'/\gamma_1 \approx \operatorname{Im} A_0/\operatorname{Re} A_0$, and $\gamma_1/2 \simeq \Delta m$)

$$\varepsilon \simeq \frac{i(m'/\Delta m) + i(\operatorname{Im} A_0/\operatorname{Re} A_0)}{1 + (i/2)\gamma_1/\Delta m}$$

$$\approx \frac{(m'/\Delta m) + \operatorname{Im} A_0/\operatorname{Re} A_0}{\gamma_1/2\Delta m - i}. \tag{4.1.13}$$

Experimentally, $\gamma_1 \simeq 2\Delta m \simeq 0.70 \times 10^{-14}$ GeV implies that

$$\varepsilon \simeq \frac{e^{i\pi/4}}{\sqrt{2}}\left(\frac{m'}{\Delta m} + \frac{\operatorname{Im} A_0}{\operatorname{Re} A_0}\right). \tag{4.1.14}$$

Furthermore, $\delta_2 - \delta_0 + \pi/2 \approx \pi/4$, which can be used to show that ε'/ε is real. Furthermore, since $\Delta I = 3/2K \to 2\pi$, the amplitudes are known to be about twenty times smaller than the $\Delta I = 1/2$ amplitudes, and as we would expect, $\varepsilon' \ll \varepsilon$. This would imply that $|\eta_{+-}| \approx |\eta_{00}|$. Present experimental results [2] bear this out as we see below

$$|\eta_{+-}| = (2.27 \pm 0.3) \times 10^{-3},$$
$$\text{Arg}\,\eta_{+-} = 44.6 \pm 1.2°,$$
$$(\eta_{00}/\eta_{+-})^2 = 1.028 \pm 0.032 \pm 0.014 \quad \text{(Ref. 3)},$$
$$\varepsilon = (2.280 \pm 0.013) \cdot 10^{-3}. \tag{4.1.15}$$

The parameter ε'/ε was measured in 1999 in the KTeV experiment at Fermilab and NA48 experiment at CERN [2a] giving the present world average value:

$$\frac{\varepsilon'}{\varepsilon} = (21.4 \pm 4.6) \times 10^{-4}. \tag{4.1.16}$$

There are other evidences for CP violation in the K-system if one assumes the validity of CPT theorem. For example, there is now a measurement at CPLEAR of the time asymmetric observable

$$A = \frac{P(K^0 \rightarrow \bar{K}^0, t) - P(\bar{K}^0 \rightarrow K^0, t)}{P(K^0 \rightarrow \bar{K}^0, t) + P(\bar{K}^0 \rightarrow K^0, t)}. \tag{4.1.17}$$

The value of A is measured to be $(6.6 \pm 1.3 \pm 1.0) \times 10^{-3}$. As was noted long ago by Kabir [2b], this is a test of T violation. However, if one assumes CPT conservation, this also is a signal of CP violation and can be predicted from the observations of the ε parameter. One then gets $A = (6.63 \pm 0.06) \times 10^{-3}$, in good agreement with observations.

The KTeV collaboration has also observed a large T-odd effect in the angular distribution between the $e^+ e^-$ and $\pi^+ \pi^-$ decay planes in the K_L^0 decay mode $K_L^0 \rightarrow \pi^+ \pi^- e^+ e^-$. The magnitude of the observed asymmetry is $(13.6 \pm 2.5 \pm 1.2)\%$, in agreement with theoretical predictions. A similar number has also been found by the NA48 collaboration.

Another important experiment relevant for the understanding of CP violation is the search for an electric dipole moment of the electron and the neutron. The latest results are:

$$d_n = (-1.4 \pm 0.6) \times 10^{-25} \text{ ecm [3a]}$$
$$(-0.3 \pm 0.5) \times 10^{-25} \text{ ecm [3b]}$$
$$d_e = (-1.5 \pm 5.5 \pm 1.5) \times 10^{-26} \text{ ecm [3c]}$$

Any model of CP-violation must explain the approximate equality of η_{+-} and η_{00} or the extreme smallness of ε'/ε as well as the small values of dipole moments. A phenomenological model, proposed by Wolfenstein [4], known as the superweak model, predicts $\varepsilon' = 0$ and negligible dipole moments. The basic feature of this model is that all CP-violating interactions are $\Delta S = 2$ type and can, therefore, lead to $m' \neq 0$, but since Im A_0 and Im A_2 arise from $\Delta S = 1$ transitions, they must be zero. This phenomenological model can be embedded into gauge theories in many ways [5]. This model is now ruled out by the measurement of ε'/ε.

§4.2 CP Violation in Gauge Models: Generalities

In order to introduce CP violation into gauge models, we must make one or more parameters of the theory complex. Furthermore, this "complexity" must be complex enough so that, by all possible phase redefinitions of the fields in the theory, we should not be able to remove it. This phase can reside in two sectors of the theory: (i) gauge interactions; and/or (ii) Higgs boson interactions of the fermions. Moreover, the CP violation can be intrinsic to the parameters of the original Lagrangian prior to spontaneous symmetry breaking, or it could be of spontaneous origin in the sense that all parameters of the theory are real, but the vacuum expectation values of the Higgs fields are complex. Thus, it appears that, in general, there could be lots of arbitrariness in the discussion of CP violation in gauge theories. However, a great deal of simplification can occur under certain circumstances. For instance, imagine a theory in which all parameters are real prior to spontaneous symmetry breaking and subsequent to spontaneous breaking only one Higgs field acquires a complex vacuum expectation value. Then there will only be one phase parametrizing all CP-violating phenomena. Another circumstance was discovered by Kobayashi and Maskawa [6] in 1973. It was known earlier [7] that if we consider only left-handed currents participating in weak interactions, then for two generations, we cannot have a nontrivial CP phase. In fact, it was suggested [7] that we may invoke right-handed currents to generate nontrivial CP-violating effects. Kobayashi and Maskawa pointed out that if we have only left-handed currents going to three generations of quarks (i.e., six flavors), only *one* nontrivial CP phase appears in the theory for arbitrary values of the parameters of the Lagrangian.

To prove this statement, it is important to discuss the origin of the CP-violating phase in gauge models. First, we realize that, since the adjoint representation of the gauge group to which the gauge fields belong is a real representation, the gauge coupling must be real. We can therefore make the Yukawa couplings complex. Subsequent to spontaneous symmetry breaking, this gives rise to a complex mass matrix M for quarks. To identify the physical quark states which are mass eigenstates, the mass matrix must be diagonalized. In general, it requires a *bi*-unitary transformation to do this

$$U_L M U_R^\dagger = M_{\text{diag}}. \tag{4.2.1}$$

If we denote the original quark fields, which are the weak eigenstates by $Q^0 \equiv (q_1^0, q_2^0, \ldots)$, then the mass eigenstates are denoted by

$$Q_{L,R} = U_{L,R} Q_R^0. \tag{4.2.2}$$

The $U_{L,R}$ matrices, in general, involve complex phases which are potential sources of CP violation.

If we consider the standard model, then the gauge group is $SU(2)_L \times U(1)$ and the fermions transform as follows (see Chapter 3):

$$\text{doublet:} \quad \begin{pmatrix} u_a^0 \\ d_a^0 \end{pmatrix}_L ;$$

$$\text{singlets:} \quad u_{aR}^0, d_{aR}^0, \qquad a = 1, \ldots, Ng \text{ for generations.}$$

Let us denote by $P^0(u_1^0, u_2^0, \ldots)$ the up-quark vector, and by $N^0 \equiv (d_1^0, d_2^0, \ldots)$ the down-quark vector $(1, \ldots, Ng$ denote generations).

As discussed in Chapter 3, the gauge interactions for quarks can be written as follows:

$$\mathcal{L}_{\text{wk}} = \frac{-ig}{\sqrt{2}} W^{+\mu} \bar{P}_L^0 \gamma_\mu N_L^0$$

$$+ \frac{-ig}{\cos\theta_W} Z^\mu \left[\tfrac{1}{2} (\bar{P}_L^0 \gamma_\mu P_L^0 - \bar{N}_L^0 \gamma_\mu N_L^0) \right.$$

$$\left. - \sin^2\theta_W \left(\tfrac{2}{3} \bar{P}^0 \gamma_\mu P^0 - \tfrac{1}{3} \bar{N}^0 \gamma_\mu N^0 \right) \right]. \quad (4.2.3)$$

Using eq. (4.2.2) to define physical quarks that are mass eigenstates, i.e., $P_{L,R} \equiv U_{L,R} P_{L,R}^0$ and $N_{L,R} \equiv V_{L,R} N_{L,R}^0$, we find the following expression for $\mathcal{L}_{\text{wk}}$:

$$\mathcal{L} = \frac{-ig}{\sqrt{2}} W^{\dagger\mu} \bar{P}_L \gamma_\mu U_L V_L^\dagger N_L$$

$$+ \frac{-ig}{2\cos\theta_W} Z^\mu \left[\bar{P}_L \gamma_\mu P_L - \bar{N}_L \gamma_\mu N_L \right.$$

$$\left. - \tfrac{4}{3} \sin^2\theta_W (\bar{P}\gamma_\mu P - \tfrac{1}{2}\bar{N}\gamma_\mu N) \right]. \quad (4.2.4)$$

$U_L V_L^\dagger$ is the charged current mixing matrix and will, in general, contain some complex phases. For N_g generations, $U_L V_L^\dagger \equiv U_{\text{KM}}$ is an $N_g \times N_g$ unitary matrix and contains N_g^2 parameters, out of which $N_g(N_g - 1)/2$ are real angles that parametrize the $N_g \times N_g$ orthogonal matrix. So, the starting number of complex phases is

$$\eta_p^0 = (N_g^2) - \frac{N_g(N_g - 1)}{2}$$

$$= \frac{N_g^2}{2} + \frac{N_g}{2} = \tfrac{1}{2}(N_g)(N_g + 1). \quad (4.2.5)$$

Now, the phases of both the N- and P-fields could be separately redefined and thereby $2N_g - 1$ phases could be removed. This, then, leaves us with n_p genuine *CP* phases

$$n_p = \frac{N_g^2}{2} + \frac{N_g}{2} - 2N_g + 1$$

$$= \tfrac{1}{2}(N_g - 1)(N_g - 2). \quad (4.2.6)$$

We see that for $N_g = 1$ or 2, $n_p = 0$ and for three generations

$$n_p = 1. \tag{4.2.7}$$

Thus, in pure left-handed models, we need three generations to get *CP* violation. This is the Kobayashi–Maskawa model, which we discuss in the next section.

We will show in a subsequent section that if we include right-handed charged currents the phase counting argument is different [7a], and we can obtain nontrivial phases even for two generations. We will show in Section 4.4 that this leads to a very interesting alternative model of *CP* violation.

§4.3   The Kobayashi–Maskawa Model

It is the simplest extension of the standard one-generation model described in the previous chapter that can accommodate *CP* violation. The Lagrangian describing weak interactions in this model is given in eq. (4.2.4) where $U_{\rm KM} = U_L V_L^\dagger$ contains both generation mixing and *CP*-violating effects.

To study the implications of this model, we write down $U_{\rm KM}$, which is parametrized in terms of three real angles and a phase

$$U_{\rm KM} = \begin{pmatrix} c_1 & -s_1 c_3 & -s_1 s_3 \\ +s_1 c_2 & c_1 c_2 c_3 - s_2 s_3 e^{i\delta} & c_1 c_2 s_3 + s_2 c_3 e^{i\delta} \\ +s_1 s_2 & c_1 s_2 c_3 - c_2 s_3 e^{i\delta} & c_1 s_2 s_3 - c_2 c_3 e^{i\delta} \end{pmatrix}, \tag{4.3.1}$$

where $c_i = \cos\theta_i$, $s_i = \sin\theta_i$, $i = 1, 2, 3$. We will restrict all the angles θ_i to the first quadrant, i.e., $0 \le \theta_i \le \pi/2$ and allow δ to vary between 0 and 2π. The detailed implications of this model have been studied in a number of recent papers and have been extensively reviewed in several references [8].

To study the implications of this model for weak processes, we must know the values of θ_i and δ. Before doing this we can give an alternative representation for $U_{\rm KM}$ as follows:

$$U_{\rm KM} = \begin{pmatrix} V_{ud} & V_{us} & V_{ub} \\ V_{cd} & V_{cs} & V_{cb} \\ V_{td} & V_{ts} & V_{tb} \end{pmatrix}. \tag{4.3.2}$$

The various elements of $U_{\rm KM}$ have been determined from low-energy weak-interaction data: the muon-decay rate defines the Fermi coupling constant G_F [9]:

$$G_F = (1.16638 \pm 0.00002) \times 10^{-5}\,{\rm GeV}^{-2}. \tag{4.3.3}$$

From $0^+ \to 0^+$ allowed Fermi transition and using eq. (4.3.3) we can determine [10]

$$|V_{ud}| = 0.9738 \pm 0.025. \tag{4.3.4}$$

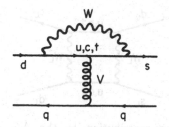

Figure 4.1. Penguin diagram that gives dominant contribution to the $\Delta I = \frac{1}{2} K \to 2\pi$ amplitudes.

V_{us} can be determined [10] using semileptonic decays of hyperon and Ke_3 decays to be

$$|V_{us}| = 0.2205 \pm 0.0018. \tag{4.3.5}$$

Recent information [11] on the lifetime and leptonic branching ratio can be used to determine V_{cb} and put an upper limit on V_{ub}:

$$|V_{cb}| = 0.040 \pm 0.002$$

and

$$|V_{ub}| = (3.56 \pm 0.56) \times 10^{-3}. \tag{4.3.6}$$

A more convenient parametrization of the CKM matrix has been suggested by Wolfenstein [12] in terms of a small parameter $\lambda \equiv \sin\theta_c \equiv V_{us}$:

$$U_{\mathrm{KM}} = \begin{pmatrix} 1 - \frac{\lambda^2}{2} & \lambda & A\lambda^3(\rho - i\eta) \\ -\lambda & 1 - \frac{\lambda^2}{2} & A\lambda^2 \\ A\lambda^3(1 - \rho - i\eta) & -A\lambda^2 & 1 \end{pmatrix}, \tag{4.3.7}$$

where $\lambda \simeq 0.22$, $A = 0.826 \pm 0.041$; $-0.15 \leq \rho \leq 0.35$ and $+0.20 \leq \eta \leq +0.45$.

To calculate ε we use eq. (4.1.12) and calculate the imaginary part of the $\Delta S = 2K^0 - \bar{K}^0$ matrix element to find ρ, and the imaginary part of the $\Delta I = \frac{1}{2} K \to 2\pi$ decay amplitude to find Im A_0/Re A_0. The latter receives dominant contribution from the gluon-mediated penguin diagrams shown in Fig. 4.1. As far as the evaluation of ρ is concerned, reliable estimates can only be made of the short-distance contribution coming from Fig. 4.2. To obtain an expression for ρ and the short-distance contribution to $K_L - K_S$ mass difference, let us define a parameter λ_i, $i = u, c, t$

$$\lambda_i = V_{id}^* V_{is} \tag{4.3.8}$$

and note that

$$m_{K_L} - m_{K_S} = 2\mathrm{Re}\, M_{K^0 - \bar{K}^0}. \tag{4.3.9}$$

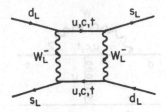

Figure 4.2. Short-distance contribution to the $\Delta S = 2$ processes in the $SU(2)_L \times U(1)$ model.

The box graph of Fig. 4.2 makes the following contribution to $m_{K_L} - m_{K_S}$ [13]:

$$(m_{K_L} - m_{K_S})_{\text{box}} = \frac{G_F^2}{6\pi^2} M_W^2 f_K^2 m_K B F(X_i, \theta_j), \qquad (4.3.10)$$

where

$$\begin{aligned}
F(x_i, \theta_j) = &[(\text{Re}\,\lambda_c)^2 (\text{Im}\,\lambda_c)^2] \eta_1 f(x_c) \\
&+ [(\text{Re}\lambda_t)^2 - (\text{Im}\,\lambda_t)^2] \eta_2 f(x_t) \\
&+ 2(\text{Re}\lambda_c)\text{Re}\lambda_t - \text{Im}\,\lambda_c \text{Im}\lambda_t) \eta_3 f(x_t, x_c), \qquad (4.3.11)
\end{aligned}$$

where

$$\begin{aligned}
f(x_i) &= x_i[\tfrac{1}{4} + \tfrac{9}{4}(1-x_i)^{-1} - \tfrac{3}{2}(1-x_i)^{-2}] + \frac{3}{2}\left[\frac{x_i}{x_i-1}\right]^3 \ln x_i, \\
f(x_i, x_j) &= x_i x_j \left\{ [\tfrac{1}{4} + \tfrac{3}{2}(1-x_j)^{-1} - \tfrac{3}{4}(1-x_j)^{-2}] \frac{\ln x_j}{x_j - x_i} \right. \\
&\qquad\qquad \left. + (x_j \leftrightarrow x_i) - \tfrac{3}{4}\big[(1-x_j)(1-x_i)\big]^{-1} \right\}. \qquad (4.3.12)
\end{aligned}$$

$x_i = m_i^2/M_W^2$ and η_i are the various QCD corrections given [14] by $\eta_i = 0.85$, $\eta_2 \simeq 0.6$, and $\eta_3 \simeq 0.39$. The parameter B denotes the matrix element $\langle K^0 | d\gamma^\mu(1+\gamma_5)s\bar{d}\gamma_\mu(1+\gamma_5)s | \bar{K}^0 \rangle$. This is an unknown parameter whose value is expected [15] to be between 0.33 and 1; putting in the values of G_F, $f_K = 160$ MeV, $M_W = 82$ GeV, we find (using $m_c \simeq 1.4$ GeV)

$$\begin{aligned}
(\Delta M)_{\text{box}} &\simeq 1.92 \times 10^{-10} B F(x_i, \theta_j) \text{ GeV} \\
&\simeq 0.22 \times 10^{-14} B \text{ GeV}. \qquad (4.3.13)
\end{aligned}$$

Thus, we need $B \simeq 1.5$ to understand the $K_L - K_S$ mass difference as coming entirely from short-distance effects. It has, however, been argued that there may be significant long-distance effects in $K_L - K_S$ mass difference, which are hard to estimate.

Let us now turn to the prediction of the Kobayashi–Maskawa model for ε and ε'. Again, we will ignore any possible long-distance effects. From eq. (4.1.14) we see that we have to find m', which is the imaginary part of $M_{K^0 \bar{K}^0}$, and $\xi \equiv \text{Im}\,A_0/\text{Re}\,A_0$: using the box graph in Fig. 4.2 we can write

$$|\varepsilon| = \frac{G_F^2}{12\sqrt{2}\pi^2}\frac{BM_W^2 f_K^2}{\Delta M}m_K D(x_i, \theta_j) + \xi, \qquad (4.3.14)$$

where

$$D(x_i, \theta_j) = 2\,\mathrm{Im}\,\lambda_t\{-\mathrm{Re}\,\lambda_c f(x_c)\eta_1 + \mathrm{Re}\,\lambda_t f(x_t)\eta_2$$
$$+ (\mathrm{Re}\,\lambda_c - \mathrm{Re}\,\lambda_t)f(x_t, x_c)\eta_3\}. \qquad (4.3.15)$$

All the symbols in eq. (4.3.15) have already been explained.

$$|\varepsilon| \simeq 0.4s_2 s_3 s_\delta B\left\{-\eta + \frac{\mathrm{Re}\,\lambda_t}{\mathrm{Re}\,\lambda_c}\eta_2\frac{f(x_t)}{f(x_c)}\right.$$
$$\left. + \eta_3\left(1 - \frac{\mathrm{Re}\,\lambda_t}{\mathrm{Re}\,\lambda_c}\right)f(x_t, x_c)/f(x_c)\right\} + \xi. \; (4.3.16)$$

In terms of Wolfenstein's parametrization, one simply has to replace $s_2 s_3 s_\delta$ by $A^2\lambda^4\eta$. For the measured value of m_t and using values of Λ and λ and value of $B \simeq 0.8$, one gets the correct value for ε. We will see in the following discussion that $\xi \sim A^2\lambda^4\eta \ll 2 \times 10^{-3}$ so that second term in 4.3.14 makes a negligible contribution to ε.

Let us now study the predictions for ε' in this model starting from (4.1.12). As already mentioned, the value of ε'/ε has recently been measured. Therefore modulo the hadronic matrix element uncertainties, this should provide a crucial test of the KM model because there are no more free parameters in the theory to adjust.

The contribution to the ε' comes from penguin diagrams of the type in Fig. 4.1. There are both gluon and Z penguins. The effective $\Delta S = 1$ Hamiltonian is generally written in terms of four-Fermi operators denoted by Q_i with $i = 1, 2, \ldots, 11$.

$$H^{\Delta S=1} = \frac{G_F}{\sqrt{2}}V_{ud}V_{us}^* \sum_i C_i(\mu)Q_i(\mu) \qquad (4.3.17)$$

where $C_i(\mu) = z_i(\mu) + \tau y_i(\mu)$ with $z_i(\mu), y_i(\mu)$ are the Wilson coefficients, which depend on how extrapolation between the scale m_W where the operator was first evaluated and μ where its value is measured in ε'. In eq. (4.3.17), $\tau = -V_{td}V_{ts}^*/V_{ud}V_{us}^*$. The operators $Q_i(\mu)$ are the following:

$$Q_1 = (\bar{s}_\alpha u_\beta)_{V-A}(\bar{u}_\beta d_\alpha)_{V-A}, \qquad (4.3.18)$$

$$Q_2 = (\bar{s}u)_{V-A}(\bar{u}d)_{V-A}, \qquad (4.3.19)$$

$$Q_{3,5} = (\bar{s}d)_{V-A}\sum_q(\bar{q}q)_{V\pm A}, \qquad (4.3.20)$$

$$Q_{4,6} = (\bar{s}_\alpha d_\beta)_{V-A}\sum_q(\bar{q}_\beta q_\alpha)_{V\pm A}, \qquad (4.3.21)$$

$$Q_{7,9} = \frac{3}{2}(\bar{s}d)_{V-A}\sum_q e_q(\bar{q}q)_{V\pm A}, \qquad (4.3.22)$$

$$Q_{8,10} = \frac{3}{2}(\bar{s}_\alpha d_\beta)_{V-A} \sum_q e_q (\bar{q}_\beta q_\alpha)_{V\pm A}, \qquad (4.3.23)$$

where α, β are color indices; $e_u = 2/3; e_d = e_s = -1/3$. The color indices for color singlet operators are omitted. The $V \pm A$ refer to the Dirac matrices $\gamma_\mu(1 \pm \gamma_5)$. The operator Q_1 originates from the lowest-order W-exchange graph whereas Q_2 arises from the gluon correction to the W-exchange graph. The rest are from penguin-type graphs in Fig. 4.1.

We have omitted in our discussion the so-called chromomagnetic operators of the form $\bar{q}\sigma_{\mu\nu} \cdot G^{\mu\nu}q$, because they emerge with suppressed coefficients in the standard model. These operators are, however, very important in supersymmetric models.

To get a reasonably believable estimate of ε'/ε, one needs an estimate of the hadronic matrix elements of the form $\langle 2\pi|Q_i|K\rangle$ and a great variety of methods have been tried leading to values from 10^{-4} to 2.6×10^{-3} [16]. To get a feeling for how these numbers emerge, let us assume that the relevant four-Fermi operators have same value. To apply Eq. (4.1.12), let us estimate Im A_0:

$$\text{Im } A_0 \simeq \frac{G_F}{\sqrt{2}} \frac{\alpha_s}{4\pi} \text{Im } \lambda_t \langle Q_i \rangle$$

$$\sim \frac{G_F}{\sqrt{2}} A\lambda^5 \eta \langle Q_i \rangle. \qquad (4.3.24)$$

Since $|A_0| \sim \frac{G_F}{\sqrt{2}}\lambda$, the ratio, $\frac{\text{Im } A_0}{\text{Re } A_0} \sim \frac{\alpha_s}{4\pi} A\lambda^4 \eta \sim 5 \times 10^{-5}$. Using the fact that $\omega \simeq 1/22$ and $\varepsilon \sim 2.3 \times 10^{-3}$, one can get a rough estimate of $\frac{\varepsilon'}{\varepsilon} \approx 2 \times 10^{-3}$, which is of the right order of magnitude. Thus whether the standard model contribution can completely explain the observed value of $\frac{\varepsilon'}{\varepsilon}$ remains an open question.

An important test of CKM model in the B system is to use the unitarity of the CKM matrix especially the bd element:

$$U_{ub}^* U_{ud} + U_{cb}^* U_{cd} + U_{tb}^* U_{td} = 0. \qquad (4.3.25)$$

Dividing the equation by the real term in the sum, i.e., $V_{bc}^* V_{cd}$, we get an equation for a triangle one of whose sides is along the x-axis and has length one. Denoting the base angles by β and γ, the triangle relation gives the vertex α to be $\alpha = 2\pi - \beta - \gamma$. The power of this triangle is that all the angles under certain assumptions can be related to observable physical processes. For instance, the process $B_d \to J/\psi K_S^0$ measures $\sin 2\beta$. This process has been searched for in the CDF experiment at Fermilab and also at ALEPH and OPAL collaborations. The CDF collaboration has published a number $\sin 2\beta = 0.79^{+0.41}_{-0.44}$. There are two dedicated experiments, the BABAR collaboration at SLAC and the BELLE collaboration at KEK, which have also measured the same parameter and their numbers are:[1]

$$\sin 2\beta = 0.75 \pm 0.09 \pm 0.04 \text{BABAR}$$
$$\sin 2\beta = 0.82 \pm 0.12 \pm 0.05 \text{BELLE} \qquad (4.3.26)$$

§4.4 Left–Right Symmetric Models of CP Violation

Soon after the CP violation models [5, 6] based on $SU(2)_L \times U(1)$ models were constructed, it was pointed out by Mohapatra and Pati [17] that extending the gauge group to $SU(2)_L \times SU(2)_R \times U(1)_{B-L}$ helps the introduction of CP-violation into gauge models even with two generations of quarks. The fact that with right-handed currents two generations are enough for CP violation was actually noted in Ref. 7. The $SU(2)_L \times SU(2)_R \times U(1)_{B-L}$ extends this idea and makes it much more appealing and phenomenologically acceptable. By that time (1974) even charm quarks were not discovered, let alone the third generation of fermions. As we will see below, an important aesthetic feature of the left–right symmetric model of CP violation is that CP and P violation are linked to each other. In this section, we would like to discuss this model and its implications in brief.

The left–right symmetric electroweak models will be discussed in detail in Chapter 6. Here we briefly introduce those aspects of the model needed for the discussion of CP violation. The quarks and leptons will be assigned in a left–right symmetric manner under the gauge group. Denoting $Q = (u, d)$ and $\psi = (\nu, e^-)$ we assign [the numbers within parentheses stand for $SU(2)_L$, $SU(2)_R$, and $U(1)_{B-L}$ quantum numbers]:

$$Q_L \equiv \left(\tfrac{1}{2}, 0, \tfrac{1}{3}\right), \qquad Q_R \equiv \left(0, \tfrac{1}{2}, \tfrac{1}{3}\right),$$
$$\psi_L \equiv \left(\tfrac{1}{2}, 0, -1\right), \qquad \psi_R \equiv \left(0, \tfrac{1}{2}, -1\right). \qquad (4.4.1)$$

There are two sets of gauge bosons $\mathbf{W}_L$, $\mathbf{W}_R$, and B whose coupling to fermions is uniquely given by the requirement of gauge invariance. The details of gauge symmetry breaking will be discussed in Chapter 6. It is sufficient to note that we will choose the following pattern:

$$SU(2)_L \times SU(2)_R \times U(1)_{B-L}$$
$$\downarrow M_{W_R}, M_{Z_R}$$
$$SU(2)_L \times U(1)_Y$$
$$\downarrow M_{W_L}$$
$$U(1)_{\text{em}}.$$

[1]CDF: T. Affolder et al. *Phys. Rev.* **D 61**, 072005 (2000); BaBar collaboration: B. Aubert et al. *Phys. Rev. Lett.*, **86**, 2515 (2001); BELLE collaboration, K. Abe et al. *Phys. Rev. Lett.* **87**, 161601 (2001).

The first stage can be implemented by the choice of either left–right symmetric Higgs doublets (χ_L, χ_R) or triplets (Δ_L, Δ_R). The second stage is implemented by the mixed doublet $\phi(2, 2, 0)$, which acquires, in general, the following vacuum expectation value:

$$\langle \phi \rangle = \begin{pmatrix} \kappa & 0 \\ 0 & \kappa' e^{i\alpha} \end{pmatrix}. \tag{4.4.2}$$

This has the property of maintaining $\rho_W = 1$, a fact that is experimentally well established. An important point is that for both κ, $\kappa' \neq 0$ the W_L and W_R mix with a CP-violating phase $e^{i\alpha}$.

To discuss CP violation we have to study the fermion masses. They arise from Yukawa coupling between Q and ϕ. We will consider the minimal set of Higgs bosons, i.e., one ϕ and its charge conjugate field $\tilde{\phi} \equiv \tau_2 \phi^* \tau_2$. Under $SU(2)_L \times SU(2)_R$ transformations we have

$$\phi \to U_L \phi U_R^+$$

and

$$\tilde{\phi} \to U_L \tilde{\phi} U_R^+. \tag{4.4.3}$$

The most general Yukawa coupling involving these fields, the quark fields Q, and leptonic fields ψ is

$$\mathcal{L}_Y = h_{ij} \bar{Q}_{Li} \phi Q_{Rj} + \tilde{h}_{ij} \bar{Q}_{Li} \tilde{\phi} Q_{Rj} + \text{h.c.}$$
$$+ h'_{ij} \bar{\psi}_{Li} \phi \psi_{Rj} + h''_{ij} \bar{\psi}_{Li} \tilde{\phi} \psi_{Ri} + \text{h.c.} \tag{4.4.4}$$

We will consider the general case of complex Yukawa couplings. Under left–right symmetry transformation we have

$$\phi \leftrightarrow \phi^+,$$
$$\tilde{\phi} \leftrightarrow \tilde{\phi}^+. \tag{4.4.5}$$

Thus, $\mathcal{L}_Y$ will be left–right symmetric provided [18]

$$h_{ij} = h^*_{ji}, \tag{4.4.6}$$
$$\tilde{h}_{ij} = \tilde{h}^*_{ji}, \tag{4.4.7}$$

and similarly for h' and h''.

The quark mass matrices are obtained from eq. (4.4.4) on substituting vacuum expectation values for ϕ and $\tilde{\phi}$ in eq. (4.4.2).

The first entry in eq. (4.4.2) can be made real by a gauge transformation. It is worth emphasizing that for the most general Higgs potential and Yukawa couplings, eq. (4.4.2) is the only allowed minimum. On the other hand, if in addition to parity invariance, Lagrangian is chosen to respect CP invariance, there exists a range of parameters for which $\alpha = 0$ naturally. The interesting point in this case is that the mass matrices are real and symmetric with equal left- and right-handed quark mixing angles.

If we require the weak Lagrangian to conserve both P and CP prior to spontaneous breakdown, two cases arise:

(i) $\alpha = 0$. In this case, the mass matrices are real and symmetric leading to equal left- and right-fermion mixing matrices, i.e.,

$$U_L = U_R. \tag{4.4.8}$$

This has been called (in literature) manifest left–right symmetry [19].

(ii) $\alpha \neq 0$. This case turns out to be the most interesting because here the Yukawa couplings are real but CP symmetry is spontaneously broken by vacuum leading to

$$M_u = M_u^T, \qquad M_d = M_d^T. \tag{4.4.9}$$

This implies [19]

$$U_R = U_L^* K, \tag{4.4.10}$$

where K is a diagonal unitary matrix. This case has an important bearing on CP violation. It is important to emphasize here that with the straightforward definition of left–right transformation [eq. (4.4.5)] of the ϕ fields, we always have the real mixing angles (for both left- and right-hand sectors) equal. Thus, even though a priori the presence of right-handed charged currents could have brought in a whole new set of mixing angles, the symmetries of the theory in the simplest model do not permit this. The only new parameters in weak interactions therefore remain m_{W_R}, $W_L - W_R$ mixing angle ζ, and new complex phases in right-handed currents.

In the case of $\alpha \neq 0$ with arbitrary Yukawa couplings the mass matrices are no more Hermitian and, therefore, left- and right-mixing angles will be different, their difference being proportional to α.

The phase counting in left–right symmetric models is different due to the presence of right-handed charged currents. The situation is, of course, dependent on phase convention in one of which the number of significant phases (after arbitrary rotation of quark fields as in Section 4.3 in the left- and right-handed charged currents) is given by

$$N_L = \frac{(N-1)(N-2)}{2} \tag{4.4.11}$$

and

$$N_R = \tfrac{1}{2} N(N+1). \tag{4.4.12}$$

Note that this reduces to eq. (4.2.6) in the absence of right-handed currents.

We find that for two generations, $N_L = 0$, $N_R = 3$; for three generations, $N_L = 1$, $N_R = 6$. Thus, for two generations it is possible to introduce CP violation for two generations in the left–right symmetric models as claimed earlier.

It is now clear that all CP violation resides in the W_R sector, in the limit of $m_{W_R} \to \infty$, $\eta_{+-} \to 0$, which implies that

$$\eta_{+-} = \left(\frac{m_{W_L}}{m_{W_R}}\right)^2 \sin\delta. \qquad (4.4.13)$$

To present the details we denote $P \equiv (u, c)$ and $N \equiv (d, s)$, and write the weak Lagrangian as

$$\mathcal{L}_{\text{wk}} = \frac{g}{2\sqrt{2}} [\bar{P}_L \gamma^\mu U_L N_L W_{L\mu}^+ + \bar{P}_R \gamma^\mu U_R N_R W_{R\mu}^+] + \text{h.c.}, \qquad (4.4.14)$$

where

$$U_L = \begin{pmatrix} \cos\theta & \sin\theta \\ -\sin\theta & \cos\theta \end{pmatrix},$$

$$U_R = e^{i\tau_3\alpha_1} U_L e^{-\tau_3\alpha_2} e^{i\beta} \quad [\tau_3 \text{ is the diagonal Pauli matrix}].(4.4.15)$$

Ignoring Higgs and $W_L - W_R$-mixing effects, we can write the $\Delta S = 1$ nonleptonic weak Hamiltonian in this model as $[\eta \equiv (M_{W_L}/m_{W_R})^2]$

$$H_{\text{wk}}^{\text{P.V.}} = \frac{G_F}{2\sqrt{2}} \sin 2\theta O_1^{(+)} (1 - +\eta e^{i\delta}) + \text{h.c.},$$

$$H_{\text{wk}}^{\text{P.C.}} = \frac{G_F}{2\sqrt{2}} \sin 2\theta O_2^{(+)} (1 + \eta e^{i\delta}) + \text{h.c.}, \qquad (4.4.16)$$

where $\delta = -2\alpha_2$ and $O_i^{(+)}$ include both the long-distance and the penguin contributions to the effective nonleptonic weak Hamiltonian. This form of the weak Hamiltonian satisfies the relations

$$[I_3, H_{\text{wk}}^{(+)}] \simeq i(m_{W_L}/m_{W_R})^2 \sin\delta H_{\text{wk}}^{(-)} \qquad (4.4.17)$$

for both parity-conserving and parity-violating parts. This was called the isoconjugate relation. It leads to

$$\frac{M(K_2^0 \to \pi_i \pi_j)}{M(K_1^0 \to \pi_i \pi_j)} = i \left(\frac{m_{W_L}}{m_{W_R}}\right)^2 \sin\delta \qquad (i, j = +, -, \text{ or } 00) \qquad (4.4.18a)$$

and

$$\frac{M(K_1^0 \to \pi_i \pi_j \pi_k)}{M(K_2^0 \to \pi_i \pi_j \pi_k)} = -i \left(\frac{m_{W_L}}{m_{W_R}}\right)^2 \sin\delta \qquad (i, j, k = +, -, 0 \text{ or } 0, 0, 0).$$

$$\qquad (4.4.18b)$$

From these we obtain

$$\eta_{+-} = e^{i\pi/4} \left[\frac{\text{Im} M_{K^0\bar{K}^0}}{\Re M_{K^0\bar{K}^0}} + \left(\frac{m_{W_L}}{m_{W_R}}\right)^2 \sin\delta\right] \qquad (4.4.19)$$

and

$$\eta_{+-} - \eta_{+-0} = 2 \left(\frac{n_{W_L}}{m_{W_R}} \right)^2 \sin \delta. \qquad (4.4.20)$$

Thus, in the approximation that we ignore Higgs and $W_L - W_R$-mixing effects, we obtain $\varepsilon' = 0$ and $\eta_{+-} - \eta_{+-0} = 2\eta \sin \delta$, the former coinciding with the superweak result and the latter potentially different from superweak prediction. To discuss the second prediction, which could be tested in experiments involving intense kaon beams, we have to study $M_{K^0 \bar{K}^0}$. In addition to the $W_L - W_L$ contribution discussed earlier, there are $W_L - W_R$ and Higgs exchange graphs all of which are potentially large and can be written symbolically as

$$M_{K^0 \bar{K}^0} = f_{LL} + [1 + \eta e^{i\delta} A_{LR} + e^{i\delta} 2 A_H] \dots . \qquad (4.4.21)$$

The absolute values of A_{LR} and A_H are much larger than 1, e.g., in the vacuum-saturation approximation [20], $A_{LR} \simeq -430$ as we will discuss in Chapter 6. The complexion of the CP-violating effect will depend on whether A_{LR} and A_H cancel each other in $M_{K\bar{K}}$ [Case (a) or not Case (b)].

Case (a). In this case, $\varepsilon \simeq \eta \sin \delta$ and $|\eta_{+-} - \eta_{+-0}| \simeq \eta_{+-}$, a prediction that can be tested by the next generation of kaon beam experiments.

Case (b). In this case, A_{LR} dominates η_{+-} and $A_{LR} \gg 1$, and we obtain $\eta \simeq \eta A_{LR} \sin \delta \approx \sin \delta$ and $|\eta_{+-} - \eta_{+-0}| \leq |\eta_{+-}|$ [10^{-3} to 10^{-2}] which is like superweak prediction.

Note that, in either case, CP and P violations are related to each other, which is a unique feature of this model. In fact, if we are to take the model seriously as the only source of CP violation in K-meson decays, W_R cannot be too heavy. This gives an upper bound on m_{W_R}:

$$m_{W_R} \leq 35 \text{ TeV}. \qquad (4.4.22)$$

Including QCD corrections [21], the upper limit goes up from 60 TeV to 80 TeV. Thus, the right-handed W-boson could be observable in machines with energies in the multi-TeV range.

Another noteworthy feature of this kind of model is the possible large CP violation for semileptonic D-meson decays (for light W_R), a distinguishing feature from the Kobayashi–Maskawa and Higgs models.

In the discussion of CP violation so far, we have ignored the $W_L - W_R$-mixing parameter. If we also ignore the Higgs-induced CP-violating effects by assuming heavy physical Higgs boson masses, we obtain

$$\varepsilon' = 0 \quad \text{and} \quad d_n^e = 0, \qquad (4.4.23)$$

where d_n^e is the electric dipole moment of the neutron. Once the $W_L - W_R$ mixing effect is taken into account this leads to both ε' and d_n^e, by the graphs in Figs. 4.3 and 4.4, respectively [22].

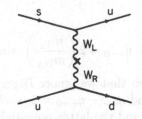

Figure 4.3. $W_L - W_R$-mixing contribution to ε'.

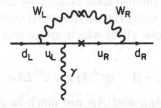

Figure 4.4. $W_L - W_R$-mixing contribution to d_n^e.

$$\varepsilon' \approx \frac{1}{20} \left(\frac{\kappa'}{\kappa} \right) \left(\frac{m_{W_L}}{m_{W_R}} \right)^2 \sin \delta \qquad (4.4.24)$$

and

$$d_n^e \approx 10^{-21} \sin \delta \left(\frac{\kappa'}{\kappa} \right) \left(\frac{m_{W_L}}{m_{W_R}} \right)^2 A, \qquad (4.4.25)$$

where A is a model-dependent parameter that can be of order 1 to 1/10. Then we obtain the following relation between ε' and d_n^e:

$$d_n^e \approx A \quad 10^{-20} \varepsilon' \text{ e.c.m.} \qquad (4.4.26)$$

Using the experimental bound that $\varepsilon'/\varepsilon \leq 2 \times 10^{-3}$, we get $d_n^e \leq 10^{-27} - 10^{-26}$ e.c.m. Thus, if ε' is measured at the level of 1%, then, for this model to be valid, the electric dipole moment of the neutron should be $\approx 10^{-27} - 10^{-26}$ e.c.m.

Looking back for a moment at the prediction of $\varepsilon = (m_{W_L}/m_{W_R})^2$ $430 \sin \delta$, we realize that, owing to the box graph enhancement mentioned, which requires m_{W_R} in the TeV region, the smallness of CP-violation requires the phase δ to be extremely small, i.e., $\delta \leq 10^{-3}$. In fact, an amusing choice is $\eta = \delta = 1/430$ which reproduces the observed value of ε. It is then legitimate to question:

How natural is it to choose δ so small? Actually, it turns out that with models with soft CP violation and gauge hierarchies we can show [23] that

$$\delta \approx \left(\frac{m_{W_L}}{m_{W_R}} \right)^2. \qquad (4.4.27)$$

The basic argument behind eq. (4.4.27) is the following. In the limit of $m_{W_R} \to \infty$, the effective low-energy theory is the standard model with only one Higgs doublet. In this model the minimum of the Higgs potential is CP-conserving. Thus, the soft CP phase $\delta \to 0$ as $m_{W_R} \to \infty$. But, since it is nonvanishing when m_{W_R} is finite, we must have eq. (4.4.27).

This model has recently been extended [24] to the six-quark model by the following observation: isoconjugate relations hold if the mixing matrices obey the form

$$U_L = O$$

and

$$U_R = K_1 O K_2, \tag{4.4.28}$$

where K_1 and K_2 are diagonal unitary matrices and O is an orthogonal matrix. This kind of mixing matrix results if the matrices are such that $MM^+ =$ real but M^+M not. The relation between CP and P remains valid in this model and so does the rest of the phenomenology. A natural implementation of this model, however, requires some ad hoc discrete symmetries in the starting Hamiltonian.

Another Way to Relate CP and P Violation

It has been noted by Chang [25] that left–right symmetric models provide another way to realize the connection between P and CP violation, if the theory is CP conserving prior to spontaneous breakdown. The key to this observation is in eq. (4.4.7), where we noted that if $\kappa' = 0$ (i.e., $W_L - W_R$ mixing vanishes), all CP violation disappears. Since $\alpha \neq 0$ is the only source of CP violation, all phases in the weak currents must be expressible in terms of α. The left-right model in this case becomes a single–phase theory of CP violation.

It has been shown that for the case of two generations the phases in eq. (4.4.15) are

$$-4\beta = 2\alpha_2 = \alpha_1 = \frac{m_c}{m_s}\left(\frac{\kappa'}{\kappa}\right)\sin\alpha. \tag{4.4.29}$$

For the parameters ε and ε' we obtain [25]

$$\varepsilon \simeq \frac{430}{2}\eta\left(\frac{\kappa'}{\kappa}\right)\frac{m_c}{m_s}\sin\alpha, \tag{4.4.30}$$

$$\varepsilon' \simeq \frac{\chi}{20\sqrt{2}}4\eta\left(\frac{\kappa'}{\kappa}\right)\sin\alpha, \tag{4.4.31}$$

where χ is a strong interaction parameter of order 10.

First point to note here is the connection between CP and P violation as in the isoconjugate model. Second, the phase α must also be very small to explain the observed magnitude. Finally, we obtain, $\varepsilon'/\varepsilon \approx \frac{1}{430} \approx 2\times 10^{-3}$, which could be tested in the next round of experiments searching for ε'.

The model can be extended to three generations. Now, of course, there are seven phases all of which, in principle, are expressible in terms of α- and CP-violating phenomena that receive not only contributions from right-handed currents but also from the usual Kobayashi–Maskawa phase in the left-handed currents. The interesting point is that the box graph enhancement factor of 430 makes the right-handed current contributions dominate over the Kobayashi–Maskawa effect as the source of CP violation, thus preserving the relation between CP and P violation.

These models have been extensively studied in recent papers [26], where it has been emphasized that in the low energy limit, i.e., $\mu \ll M_{W_R}$, this class of models reduces to standard model with a KM phase, which is roughly of order ~ 0.1; this will therefore lead to small CP violating effects in the B system and could therefore be used as a test.

Before closing, we point out that the difference between the isoconjugate model and the model of Ref. 26 is that ε'/ε can be arbitrarily small in isoconjugate models but not so in the other one. Thus, if the upper limit on ε'/ε is stretched to its maximum possible value of 10^{-4}, the model of Ref. 26 will be ruled out. However, in the framework of the isoconjugate model it will simply mean that the $W_L - W_R$ mixing of κ'/κ is very small.

Thus, right-handed currents provide a very appealing framework for describing CP-violating interactions, and can be soon tested by improved experimental search for ε and neutron electric dipole moment as well as CP violation in the B system.

§4.5   The Higgs Exchange Models

The idea, in these models, is to have CP violation reside only in the Higgs sector while keeping the gauge sector CP conserving as in the two-generation $\mathrm{SU}(2)_L \times \mathrm{U}(1)$ model. These models are of two types: (i) where the neutral Higgs couplings to quarks is flavor violating; and (ii) where it is flavor conserving. In the first case, $K_L - K_S$ mass difference implies that the masses of the Higgs bosons must be very heavy (typically $\alpha m_f^2/m_H^2 \leq 10^{-12}$ GeV, which implies $m_H \geq 10^2$ TeV or so). Such heavy masses are barely compatible with unitarity. Therefore, they can lead to superweak-type models for CP violation as discussed in Ref. 27. In case (ii) CP violation can arise through the exchange of charged Higgs bosons. Therefore, it can lead to milliweak ($\sim G_F \times 10^{-3}$ as the strength of CP violating interaction) type models of CP-violation. These have been discussed by Weinberg [28]. An important point has been made by Branco [29] that, for an arbitrary number of generations, the requirement of natural flavor conservation leads to CP-conserving gauge interactions in the $\mathrm{SU}(2)_L \times \mathrm{U}(1)$ model. In this model the only source of CP violation will be from the Higgs exchanges. This gives a sort of uniqueness to the Higgs

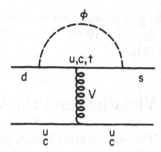

Figure 4.5. The Higgs–penguin graph that contributes to ξ.

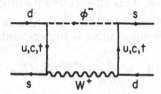

Figure 4.6. New contribution to $\Delta S = S$ processes in the Higgs exchange models of CP violation.

exchange models of CP violation which otherwise would be complicated by the presence of phases in weak currents.

Higgs interactions with quarks in this class of models can be written as

$$\mathcal{L}_\phi = \frac{1}{\lambda_1^*} \bar{N}_R M^d U_{\text{KM}}^+ P_L \phi_1^- - \frac{1}{\lambda_2} \bar{P}_R M^u U_{\text{KM}} N_L \phi_2^+ + \text{h.c.}, \quad (4.5.1)$$

where $P = (u, c, t, \ldots)$; $N = (d, s, b, \ldots)$; U_{KM} is the Cabibbo–Kobayashi–Maskawa mixing matrix; $\phi_{1,2}$ are the physical charged Higgs bosons; and $M^{u,d}$ are the diagonal mass matrices for the up- and down-quarks. CP violation arises due to the complex propagator

$$\Delta_\phi = \frac{\langle 0|T(\phi_1^- \phi_2^+)|0\rangle_{\text{F.T.}}}{\lambda_1^* \lambda_2}, \quad (4.5.2)$$

where F.T. stands for Fourier transform. It has been shown in Refs. 30 and 31 that these models predict too large a value for ε'/ε and are in conflict with observations. To see this we note from eq. (4.1.12) that

$$\frac{\varepsilon'}{\varepsilon} = -\frac{1}{20\sqrt{2}} \left(\frac{\xi}{\varepsilon + \xi} \right). \quad (4.5.3)$$

The dominant graph contributing to ξ and ε now involves Higgs bosons as shown in Figs. 4.5 and 4.6 and it has been found that $\varepsilon \ll \xi$. Therefore, $\varepsilon'/\varepsilon \simeq -\frac{1}{20}$ which is in conflict with experiments. Again, if long-distance effects [32], which strictly are not calculable, are assumed to be big the result could be weakened.

This model also tends to give a large contribution to the neutron electric dipole moment [33], i.e., $d_n > 3 \times 10^{-26}$ e.c.m. and therefore could be in conflict with experiments [34].

§4.6   Strong CP Violation and the θ-Problem

So far we have discussed CP-violation arising from weak interactions. It was observed in 1976 by Callan, Dashen, and Gross [35], and Jackiw and Rebbi [36] that quantum chromodynamics (QCD) leads to new CP-violating effects in strong interactions. This effect owes its origin to the fact that, in QCD, the axial baryon number current $A_\mu \equiv \sum_a \bar{q}_a \gamma_\mu \gamma_5 q_a$, where a is the flavor index and sum over color index is understood, has an anomaly, i.e.,

$$\partial^\mu A_\mu = \frac{g^2}{32\pi^2} \varepsilon^{\mu\nu\alpha\beta} G_{\mu\nu} G_{\alpha\beta}; \qquad (4.6.1)$$

this leads to a degenerate multiple vacuum [37] structure (labeled by $|\nu\rangle$, $\nu = 0, 1, \ldots$) for QCD. This implies that the true vacuum must be a superposition of all $|\nu\rangle$ vacua with an arbitrary phase $e^{i\nu\theta}$; ν is the topological winding number characterizing the different vacuum solutions and is given by

$$\Delta Q_A = \frac{g^2}{32\pi^2} \int \varepsilon^{\mu\nu\alpha\beta} G_{\mu\nu} G_{\alpha\beta} = \nu, \qquad (4.6.2)$$

Q_A is the axial charge.

This is equivalent to adding to the QCD Lagrangian a term

$$\mathcal{L}_\theta = \theta \frac{g^2}{32\pi^2} \varepsilon^{\mu\nu\alpha\beta} G_{\mu\nu} G_{\alpha\beta}, \qquad (4.6.3)$$

where $G_{\mu\nu}$ is the gauge covariant antisymmetric gluon field. Note that $\mathcal{L}_\theta$ is P and CP odd. The new CP-violating Lagrangian will lead to observable consequences such as electric dipole moment of the neutron. To calculate this, we note the observation that θ can be expressed in terms of quark fields and for one generation, we can write [38]

$$\mathcal{L}_{CP} = i\theta \frac{m_u m_d}{m_u + m_d} (\bar{u}\gamma_5 u + \bar{d}\gamma_5 d). \qquad (4.6.4)$$

From this, we can estimate that

$$d_n^e = (2.7 - 5.2) \times 10^{-16} \theta \text{ e.c.m.}$$

Using present limits on d_n^e we conclude that $\theta \leq 10^{-9}$. Thus, θ must either be zero or an extremely small number. The question now arises as to how to understand this small parameter in a natural manner. This is known as the strong CP problem and, below, we discuss some of the proposed mechanisms to solve this problem. Before we discuss them it is important to point out that the additional term $\mathcal{L}_\theta$ in the Lagrangian is generated if we make a color and flavor singlet axial (or axial baryon number) transformation in the Lagrangian by an amount θ. This is because of Noether's theorem, which says that if

$$q_j \to e^{i\theta\gamma_5} q_j, \tag{4.6.5}$$

then

$$\mathcal{L} \to \mathcal{L} + \delta\mathcal{L}, \tag{4.6.6}$$

where

$$\delta\mathcal{L} = i\theta\partial^\mu A_\mu. \tag{4.6.7}$$

Using eq. (4.6.1) we see that strong CP-parameter θ is the parameter of chiral transformations. Therefore, if no other interactions are present and any of the quarks are massless, then, by a chiral transformation, θ can be removed. This fact is reflected in eq. (4.6.4) in that the physical effect of θ vanishes if either m_u or $m_d \to 0$.

Let us now briefly discuss whether the models of weak CP-violation described in this chapter lead to an acceptable θ. As a prototype example let us consider the Kobayashi–Maskawa model. In this model CP violation is introduced through complex dimension-four Yukawa couplings. As a result, we would expect the quark mass matrices to have phases at the tree level that will receive infinite corrections in higher orders. Of course, not all the phases present in the mass matrix are related to the strong CP problem. It is only the phase of the Det $M_u \cdot M_d$ that changes under axial baryon number transformation (4.6.5) and is therefore connected with θ. So, if in a theory Det$(M_u M_d)$ is finite to all orders, the θ will be finite. Unfortunately, in the Kobayashi–Maskawa model, Det$(M_u M_d)$ receives infinite contributions at the fourteenth order [38a]. Therefore, the Kobayashi–Maskawa model does not provide a natural framework for understanding the smallness of θ and extensions of this model are needed for this purpose.

Peccei–Quinn Mechanism

Perhaps the most popular solution to this problem is the one proposed by Peccei and Quinn [39], who proposed that the full Lagrangian of the world, i.e., $\mathcal{L} = \mathcal{L}_{\text{QCD}} + \mathcal{L}_{\text{wk}}$, be invariant under a chiral $U(1)_A$ symmetry [to be called $U(1)_{\text{PQ}}$ from now on] operating on the quark flavors and Higgs bosons. This symmetry can then be used to "rotate" θ away. A very nontrivial aspect of this interesting proposal is to show that quark masses, which arise subsequent to spontaneous breakdown, remain real as the $U(1)_{\text{PQ}}$ transformation is made to remove the θ. It was shown [39] that for the simple model considered there this indeed happens. Thus, after the strong CP violation is removed by a chiral transformation, there remains no trace of CP violation in the theory if there was no phase in weak interactions.

We now proceed to show how these ideas can be incorporated into the standard electroweak model. We would like the weak interaction model to have realistic weak *CP* violation. Let Q_{La} $(a = 1, 2, 3$ for generations) denote the left-handed quark doublets and u_{Ra} and d_{Ra} be right-handed singlet fields. We define the action of U(1)$_{PQ}$ on them as follows:

$$Q_{La} \to e^{i\alpha} Q_{La},$$
$$(u)_{Ra}, (d)_{Ra} \to e^{-i\alpha}(u_{Ra}, d_{Ra}). \tag{4.6.8}$$

To generate fermion masses, Yukawa couplings of quarks must also be U(1)$_{PQ}$ invariant. If we have one doublet, called ϕ_d, such that it couples to Q_L and d_R, the coupling is $\bar{Q}_L \phi_d d_R$ where $\phi_d \equiv \begin{pmatrix} \phi_d^+ \\ \phi_d^0 \end{pmatrix}$; the local U(1)$_Y$ invariance demands that $Y(\phi_d) = +1$ because U$(Q_L) = \frac{1}{3}$ and $Y(d_R = -\frac{2}{3})$. Equation (4.6.8) then requires that under U(1)$_{PQ}$

$$\phi_d \xrightarrow{\text{U}(1)_{PQ}} e^{2i\alpha} \phi_d. \tag{4.6.9}$$

Now U(1)$_{PQ}$ forbids θ_d couplings to $\bar{Q}_L$ and u_R. Thus we must have another Higgs doublet ϕ_u with $Y(\phi_u) = -1$ or $\phi_u = \begin{pmatrix} \phi_u^0 \\ \phi_u^- \end{pmatrix}$ and

$$\phi_u \xrightarrow{\text{U}(1)_{PQ}} e^{2i\alpha} \phi_u. \tag{4.6.10}$$

The full gauge invariant $U(1)_{PQ}$ symmetric Yukawa coupling can now be written as

$$\mathcal{L}_Y = \sum_{a,b}(h_{ab}^u \bar{Q}_{L,a} \phi_u u_{R,b} + h_{ab}^d \bar{Q}_{La} \phi_d d_{R,b}) + \text{h.c.} \tag{4.6.11}$$

$h_{ab}^{u,d}$ are complex couplings that enable us to introduce *CP* violation into the weak currents as in the Kobayashi–Maskawa model. The Higgs potential must be chosen in an U(1)$_{PQ}$ invariant manner, yet such that the minimum of the potential will correspond to $\langle\phi_u^0\rangle = v_u \neq 0$ and $\langle\phi_d^0\rangle = v_d \neq 0$ leading to W, Z, and quark masses. In this theory the strong *CP*-violating parameter θ can be rotated to give zero.

The Axion

Subsequent to the work of Peccei and Quinn (who, in fact, proposed a model different from the one just described), it was shown by Weinberg [40] and Wilczek [40] that since U(1)$_{PQ}$ symmetry is broken by the minimum of the potential, there must be a zero-mass particle in the theory—called the axion. They then pointed out that the axion actually picks up a very tiny mass from the instanton effects that break U(1)$_{PQ}$ symmetry. To see how this happens we recall a result first obtained by 't Hooft which says that the instanton effects can be written in terms of an effective Lagrangian involving the quark fields as follows:

$$\mathcal{L}_{\text{eff}} \approx c e^{i\theta} \text{Det} \|\bar{Q}_L Q_R\| + \text{h.c.} \tag{4.6.12}$$

$\bar{Q}_L Q_R$ in eq. (4.6.12) symbolically denotes a matrix whose elements are quark bilinears and whose rows and columns denote the various flavors: for instance, for one flavor u, Det $\|\bar{Q}_L Q_R\| = \bar{u}_L u_R$; for two flavors u and d,

$$\text{Det} \|\bar{Q}_L Q_R\| = \text{Det} \begin{vmatrix} \bar{u}_L u_R & \bar{u}_L d_R \\ \bar{u}_R u_L & \bar{u}_L d_R \end{vmatrix}; \tag{4.6.13}$$

etc. Thus for N flavors, Det is a monomial in quark fields of degree $2N$. This, being a determinant, is not invariant under the axial baryon number transformation, and will therefore give mass to the axion that has been estimated by various authors [40,41] to be the following:

$$m_a \approx \frac{N m_\pi f_\pi}{2(m_u + m_d)^{1/2}} \left[\frac{m_u m_d m_s}{m_u m_d + m_d m_s + m_s m_u} \right]^{1/2} \frac{2^{1/4} G_F^{1/2}}{\sin 2\alpha}$$

$$\approx 23 \text{Kev} \times \frac{N}{\sin 2\alpha}, \tag{4.6.14}$$

where $\tan \alpha = (v_d/v_u)$ and N is the number of quark flavors. For six flavors we expect $m_a \approx 138/\sin 2\alpha$ keV, i.e., less than 1 MeV. We now wish to study its coupling to matter, to look for processes, where it can be looked for. The axion is a mixture of the Higgs particles Im ϕ_u^0, Im ϕ_d^0, and the hadrons π^0 and η and can be written as

$$a = N \{ \xi_{\pi^0} \pi^0 + \xi_\eta \eta^0 + \chi(\sin \alpha \, \text{Im} \, \phi_u + \cos \alpha \, \text{Im} \, \phi_d) \}. \tag{4.6.15}$$

The combination of the last two terms is obtained from orthogonality with the Higgs–Kibble particle which is fixed by spontaneous breaking of gauge symmetry. The admixture of π^0 comes entirely from the fact the m_u and m_d are nonzero and has been calculated [40] to be

$$\xi_\pi = \xi \left[\frac{3m_d - m_u}{m_d + m_u} \tan \alpha - \frac{3m_u - m_d}{m_u + m_d} \cot \alpha \right],$$

$$\xi_\eta = \xi \left[\sqrt{3} \tan \alpha + \frac{1}{\sqrt{3}} \cot \alpha \right],$$

$$\xi = \frac{2^{1/4}}{4} G_F^{1/2} f_\pi, \tag{4.6.16}$$

x is fixed by normalization. The axion couplings to heavy quarks and leptons are easily noted to be of γ_5-type since it is a pseudo-Goldstone boson (see Chapter 2) and is given by

$$\mathcal{L} = i 2^{1/4} G_F^{1/2} A [m_c \tan \alpha \, \bar{c} \gamma_5 c + m_b \cot \alpha \, \bar{b} \gamma_5 b$$

$$+ (m_e \bar{e} \gamma_5 e + m_\mu \bar{\mu} \gamma_5 \mu)(\tan \alpha \text{ or } \cot \alpha)]. \tag{4.6.17}$$

As far as the light quarks are concerned, we can find their coupling to nucleons from the π^0 and η admixture.

It is now possible to discuss the possible decay and production characteristics of the axion. If it has mass below $2m_e$, it will decay chiefly by processes $\alpha^0 \rightarrow 2\gamma$. If $m_a > 2m_e$, we have a decay mode $\alpha^0 \rightarrow e^+e^-$. The axions could be produced in decays of excited nuclei [42] and has indeed been looked for in such experiments [43] without success, although, another experiment by the Aachen group [44] has seen effects that can be interpreted as an axionlike object. The present consensus appears to be that an axion with mass less than 1 MeV is inconsistent with experiments.

The apparent failure of the experimental searches for the axion has prompted recent theoretical speculation that the $U(1)_{PQ}$ symmetry may be a symmetry at ultrahigh ($\sim 10^9$ GeV) energies [45] with breaking scale $V_{PQ} \gg n_W$. This hypothesis has the consequence that both the axion mass and its matter couplings become highly suppressed, i.e.,

$$m_a \approx \frac{f_\pi^2}{V_{PQ}} \qquad (4.6.18)$$

and

$$g_{ffa} \approx \frac{f_\pi}{V_{PQ}}. \qquad (4.6.19)$$

This particle is similar to the Majoron [46] discussed in Chapter 2 and therefore the astrophysical constraints from red giants imply [47] that $V_{PQ} \geq 10^9$ GeV. This implies that $m_a \approx 10^{-2}$ eV and $g_{ffa} \simeq 10^{-10}$. Because of this weak strength, this particle, like the Majoron, would escape experimental detection. Further experimental implications of this idea have been studied in a number of papers [48]. This idea is, however, criticized on the ground that to explain small θ we introduce a very high scale $V_{PQ} \leq 10^9$ GeV that requires additional fine tuning of parameters. However, if this concept is successfully accommodated into a grand unified theory, where the fine tuning is needed to study gauge hierarchies, then it appears less unnatural. Nevertheless, all these additional theoretical inputs, coupled with lack of the conventional light axion to appear, has made it more attractive to consider alternative solutions to the strong *CP* problem, which we briefly discuss in the next section.

§4.7   Solutions to the Strong *CP* Problem without the Axion

The basic idea of the previous section was to use a continuous symmetry that removes the θ terms. Several alternative solutions use the idea of replacing the continuous symmetry by appropriate discrete symmetries so that it leads to $\theta = 0$ in the Lagrangian. Once these discrete symmetries are spontaneously broken, we will not have any light axionlike particle. However, the number of possibilities for such theories is limited by the fact that because the discrete symmetries are broken, a finite 0 will arise as a result of quantum effects and only those theories will be acceptable that have $\theta_{\text{loop}} \leq 10^{-9}$. We discuss below four different possibilities that solve the strong CP-problem without a $U(1)_{\text{PQ}}$ symmetry.

The general strategy in all these models consists of the following:

(a) Choose $\theta = 0$ using some symmetry (or by appropriate choice of the gauge symmetry) and its particle content.

(b) After spontaneous symmetry breaking we obtain quark mass matrices $M_{u,d}$. On diagonalization these can also lead to nonvanishing θ_{QFD}. Thus, the theory must be such that subsequent to spontaneous breakdown

$$\theta_{\text{QFD}} \equiv \operatorname{Arg} \operatorname{Det}(M_u M_d) = 0. \tag{4.7.1}$$

Combination of conditions (a) and (b) guarantee that $\theta_{\text{tree}} = 0$ naturally. This implies that any contribution to 0 arising in higher orders must be finite. To estimate these effects we calculate the radiative correction $\delta M_{u,d}$ to the mass matrices. The finite contribution to θ is then

$$\delta\theta_{\text{QFD}} = \operatorname{Arg} \operatorname{Det}(M_u + \delta M_u)(M_d + \delta M_d). \tag{4.7.2}$$

(i) The first series of such models [49] was constructed using the left–right symmetric models where parity is a good symmetry of the Lagrangian prior to spontaneous breakdown. Under parity $\theta \to -\theta$; thus, in left–right symmetric models, $\theta = 0$ naturally. Furthermore, if the model has manifest left–right symmetry subsequent to spontaneous breakdown (see Chapter 6), i.e.,

$$M_{u,d} = M_{u,d}^+, \tag{4.7.3}$$

this satisfies eq. (4.7.1) automatically. In Ref. 49 it has been shown that nonvanishing, finite contributions to θ arise at the two-loop level keeping $\theta \leq 10^{-9}$ without any unnatural adjustment of parameters.

Similar strategies have been employed outside the left–right symmetric models, by requiring additional discrete symmetries such as CP, which are then softly broken [50] to make the model realistic.

(ii) A second class of models, that do not use $U(1)_{\text{PQ}}$ symmetry, employs the observation that, in theories where CP violation arises entirely out of vacuum expectation values, the value of the phases in the observable sector of quarks and leptons depends on whether there are high mass scales in the theory [51]. For instance, consider a theory with two mass scales μ and M with $\mu \ll M$. If, in the limit of $M \to \infty$, the effective low-energy theory is such that it is CP conserving, then any spontaneous CP-violating phase δ associated with the heavy scale M will manifest in the low-energy sector in such a way that

$$\delta \approx \left(\frac{\mu}{M}\right)^2. \tag{4.7.4}$$

Because θ_{tree} is zero in this theory because of CP invariance any θ arising at the tree level will be $\approx (\mu/M)^2$. Choosing M appropriately we could make θ naturally small. How this can work in a realistic model has been demonstrated in Ref. 51 where $\mu = m_{W_L}$ and $M = M_{W_R}$. This would require $M_{W_R} \geq 10^6$ GeV. It would be interesting to implement this idea in other theories to test the generality of such models.

(iii) A third class of models recently proposed [52] considers models with softly broken Peccei–Quinn symmetry. If the soft-breaking terms are properly chosen, they will lead to a theory where, if $\theta_{\text{tree}} = 0$, θ_{loop} will be computable and finite.

A simple realization of this idea is to consider the $\text{U}(1)_{\text{PQ}}$ symmetric Lagrangian described in Section 4.6 and add the following soft-breaking gauge invariant term to the Lagrangian such as

$$\mathcal{L}_B = -\mu^2 \phi_u \phi_d + \text{h.c.} \tag{4.7.5}$$

This class of models is devoid of a light axion and leads to nonvanishing 0 at the three-loop level. It has also been shown [52] that these theories arise naturally in $N = 1$ broken supergravity models.

(iv) Finally, a new class of models, that lead to finite and small θ without Peccei–Quinn symmetry, has recently been proposed [53]. Here we consider the CP-invariant Lagrangian (so that $\theta_{\text{QCD}} = 0$) with heavy fermions that belongs to real representations under the gauge group (i.e., $C + \bar{C}$ where C is a complex representation). If we denote the light fermions of the model by F, then to solve the strong CP problem the following constraints must be imposed on the theory:

(a) no $\bar{C}F$- or CC-type mass or Yukawa interaction terms at the tree level;

(b) CP-violating phases only in coupling of F to $C + \bar{C}$ together.

The most general fermion mass matrix in these models is such that its determinant is real leading to vanishing θ_{QFD}. A very simple realization of this class of models is provided by [54] an extension of the left–right symmetric models that includes three pairs of vectorlike quarks.

§4.8 Summary

In summary we wish to note that CP violation in gauge theories has two aspects to it: strong and observed weak CP violation. The first is a theoretical problem that must be solved in any realistic model of weak CP violation. It is worth pointing out that the Kobayashi–Maskawa model does not provide a solution to the strong CP problem. If we calculate a higher loop, at the fourteenth order, an infinite contribution [38a] to θ appears. So θ must be renormalized and is therefore an arbitrary parameter. In fact, the third proposal [52], in Section 4.7, provides the simplest modification of the Kobayashi–Maskawa model which solves the strong CP-problem.

Exercises

4.1. All CP-violating effects in the Kobayashi–Maskawa model are proportional to the quantity J [defined by C. Jarlskog, *Phys. Rev. Lett.* **55**, 1039 (1985)].

$$J \equiv s_1^2 s_2 s_3 c_1 c_2 c_3 s_\delta.$$

Show that J (the Jarlskog invariant) is related to the determinant c

$$c = [M_u M_u^+, M_d M_d^+]$$

as follows:

$$J = -\frac{1}{2} \cdot \frac{\det c}{(m_t^2 - m_c^2)(m_c^2 - m_u^2)(m_u^2 - m_t^2)(m_b^2 - m_s^2)(m_s^2 - m_d^2)(m_d^2 - m_b^2)}.$$

4.2. Show that in the left–right symmetric model, the CP-violating vacuum could either be parametrized by eq. (4.4.2) or, equivalently, by

$$\langle \phi \rangle = e^{i\delta} \begin{pmatrix} \kappa & 0 \\ 0 & \kappa' \end{pmatrix}.$$

Show that in the limit that $\kappa' \to 0$, $\alpha \to 0$.

4.3. In the two-generation left–right symmetric model with spontaneous CP violation, the CP-violating phases α_1, α_2, and β of eq. (4.4.15) can be related to κ'/κ and the quark masses. Derive those relations in the approximation $\kappa' \ll \kappa$.

4.4. Give a parametrization of a quark mixing matrix with CP violation in the case of four generations of quarks.

4.5. In the gauge models of CP violation, effective dimension-6 CP-violating gluon operators of type $f_{abc}\varepsilon^{\mu\nu\alpha\beta} G_{\mu\nu}^a G_{\alpha\gamma}^b G_\beta^{c\gamma}$ are induced at the two-loop level. Show that these operators lead to a large electric dipole moment of the neutron in models with Higgs exchanges [S. Weinberg, Austin Preprint 1989], whereas their effects in the case of left–right models lead to $d_n^e \simeq 10^{-26}$–10^{-27} e.c.m.

4.6. Show that in theory with a light- ($\sim M_W$) and heavy- ($M_H \gg M_W$) mass scale, spontaneous breaking CP invariance is impossible without additional fine tuning of parameters in the Higgs potential (i.e., in addition to that required to obtain gauge hierarchy). (See Ref. 51.)

References

[1] CDF collaboration, F. Abe et al. *Phys. Rev. Lett.* **81**, 5513 (1998); BABAR collaboration, hep-ex/0011024; BELLE collaboration, hep-ex/0010018.

[1a] R. E. Marshak, Riazuddin, and C. P. Ryan, *Theory of Weak Interactions in Particle Physics*, Wiley, New York, 1969;
L. Wolfenstein, *Theory and Phenomenology in Particle Physics* (edited by A. Zichichi), Academic Press, New York, 1969;
E. Paul, in *Elementary Particle Physics*, Springer Tracts in Modern Physics, vol. 79.
P. K. Kabir, *CP Puzzle*, Academic Press, New York, 1968;
CP-Violation (edited by C. Jarlskog), World Scientific, Singapore, 1988;
G. Branco, L. Lavoura and J. de Silva, *CP Violation*, (Oxford University Press), 1999;
I. I. Bigi and A. Sanda, *CP Violation*, Cambridge University Press (2000).

[2] Particle Data Group, *Rev. Mod. Phys.* **56**, S1 (1984).

[2a] KTeV collaboration: A. Alavi-Harati et al. *Phys. Rev. Lett.*, **83**, 22 (1999); V. Fanti et al. NA48 collaboration, *Phys. Lett.* **B 465**, 335 (1999).

[2b] P. K. Kabir, *Phys. Rev.* **D2**, 510 (1970).

[3] B. Weinstein, Talk Given at XIth International Neutrino Conference, Dortmund, West Germany, June, 1984.

[3a] 1. Altarev et al., *JETP Lett.* **44**, 460 (1986).

[3b] K. F. Smith et al., *Phys. Lett.* **234B**, 191 (1990).

[3c] S. Murthy et al., *Phys. Rev. Lett.* **63**, 965 (1989).

[4] L. Wolfenstein, *Phys. Rev. Lett.* **13**, 562 (1964).

[5] R. N. Mohapatra, J. C. Pati, and L. Wolfenstein, *Phys. Rev.* **D11**, 3319 (1975);
P. Sikivie, *Phys. Lett.* **65B**, 141 (1976);
A. Joshipura and L. Montvay, *Nucl. Phys.* **B196**, 147 (1982);
S. M. Barr and P. Langacker, *Phys. Rev. Lett.* **42**, 1654 (1979).

[6] M. Kobayashi and T. Maskawa, *Prog. Theor. Phys.* **49**, 652 (1973).

[7] R. N. Mohapatra, *Phys. Rev.* **D6**, 2023 (1972).

[7a] R. N. Mohapatra and D. P. Sidhu, *Phys. Rev.* **D17**, 1876 (1978);
D. Chang, *Nucl. Phys.* **B214**, 435 (1983);
P. Herczeg, *Phys. Rev.* **D28**, 200 (1983).

[8] L. L. Chau, *Phys. Rep.* **95C**, 1 (1983); A. Buras, *Lectures at the LAKE LOUISE Winter School*, hep-ph/9905473; I. I. Bigi, *Fundamental Particles and Interactions*, ed. R. S. Panvini and T. Weiler, AIP (1997), p. 3; Y. Nir, *Particle Physics, 1999*, ed. G. Senjanović and A. Smirnov (World Scientific (2000); p. 165.

[9] K. L. Giovanetti et al., *Phys. Rev.* **D29**, 343 (1984).

[10] A. Ali and D. London, hep-ph/9903535; F. Parodi, P. Roudeau and A. Strocci, hep-ex/9903063.

[11] For a review, see D. Cassel, *Fundamental Particles and Interactions*, ed. R. Panvini and T. Weiler, AIP (1997); p. 84

[12] L. Wolfenstein, *Phys. Rev. Lett.* **51**, 1945 (1983).

[13] M. K. Gaillard and B. W. Lee, *Phys. Rev.* **D10**, 897 (1974).

[14] F. J. Gilman and A Wise, *Phys. Rev.* **D27**, 1128 (1983).

[15] J. F. Donoghue, E. Golowich, and B. R. Holstein, *Phys. Lett.* **119B**, 412 (1982);
J. Trampetic, *Phys. Rev.* **D27**, 1565 (1983);
B. Guberina, B. Machet, and E. deRafael, *Phys. Lett.* **128B**, 269 (1983).

[16] For reviews, see A. Buras, hep-ph/9905437; S. Bertolini, M. Fabbrichesi and J. Eeg, hep-ph/9802405.

[17] R. N. Mohapatra and J. C. Pati, *Phys. Rev.* **D11**, 566 (1975).

[18] R. N. Mohapatra, F. E. Paige, and D. P. Sidhu, *Phys. Rev.* **D17**, 2642 (1978).

[19] M. A. B. Beg, R. V. Budny, R. N. Mohapatra, and A. Sirlin, *Phys. Rev. Lett.* **38**, 1252 (1977).

[20] G. Beall, M. Bender, and A. Soni, *Phys. Rev. Lett.* **48**, 848 (1982).

[21] I. I. Bigi and J. M. Frere, University of Michigan preprint (1983).

[22] G. Beall and A. Soni, *Phys. Rev. Lett.* **47**, 552 (1981);
G. Ecker, W. Grimus, and H. Neufeld, *Nucl. Phys.* **B229**, 421 (1983).

[23] A. Masiero, R. N. Mohapatra, and R. D. Peccei, *Nucl. Phys.* **B192**, 66 (1981).

[24] G. Branco, J. Frere, and J. Gerard, *Nucl. Phys.* **B221**, 317 (1983).

[25] D. Chang, *Nucl. Phys.* **B214**, 435 (1983).
R. N. Mohapatra, *Phys. Lett.* **159B**, 374 (1985).

[26] For recent discussions of this model, see G. Barenboim, J. Bernabeu and M. Raidal, *Phys. Rev. Lett.* **80**, 4625 (1998);
G. Barenboim, *Phys. Lett.* **B443**, 317 (1998);
G. Barenboim et al., *Phys. Rev.* **D60**, 016003 (1999); P. Ball, J. M. Frere and P. Matias, *Nucl. Phys.* **B562**, 3 (2000);
P. Ball and R. Fleisher, *Phys. Lett.* **B475**, 111 (2000).

[27] P. Sikivie, *Phys. Lett.* **65B**, 141 (1976).

[28] S. Weinberg, *Phys. Rev. Lett.* **37**, 657 (1976).

[29] G. Branco, *Phys. Rev. Lett.* **44**, 504 (1980).

[30] N. G. Deshpande, *Phys. Rev.* **D23**, 2654 (1981).

[31] A. I. Sanda, *Phys. Rev.* **D23**, 2647 (1981).

[32] D. Chang, *Phys. Rev.* **D25**, 1381 (1982).

[33] G. Beall and N. G. Deshpande, Oregon preprint (1983).

[34] W. Dress et al., *Phys. Rev.* **D15**, 9 (1977);
I. Altarev et al., *Phys. Lett.* **102B**. 13 (1981).

[35] C. Callan, R. Dashen, and D. Gross, *Phys Lett.* **63B**, 334 (1976).

[36] R. Jackiw and C. Rebbi, *Phys. Rev. Lett.* **37**, 172 (1976).

[37] G. 't Hooft, *Phys. Rev.* **D14**, 3432 (1976);
A. Belavin, A. M. Polyakov, A. Schwartz, and Yu S. Tyupkin, *Phys. Lett.* **59B**, 85 (1975).

[38] V. Baluni, *Phys. Rev.* **D19**, 2227 (1979);
R. Crewther, P. di Vecchia, G. Veneziano, and E. Witten, *Phys. Lett.* **88B**, 123 (1979).

[38a] J. Ellis and M. K. Gaillard, *Nucl. Phys.* **B150**, 141 (1979).

[39] R. D. Peccei and H. Quinn, *Phys. Rev. Lett.* **38**, 1440 (1977) *Phys. Rev.* **D16**, 1791 (1977).

[40] S. Weinberg, *Phys. Rev. Lett.* **40**, 223 (1978);
F. Wilczek, *Phys. Rev. Lett.* **40**, 279 (1978).

[41] W. Bardeen and S. H. H. Tye. *Phys. Lett.* **74B**, 229 (1978);
J. Kandaswamy, J. Schecter, and P. Salomonson, *Phys. Lett.* **74B**, 377 (1978).

[42] J. Barosso and N. Mukhopadhyaya, SIN preprint (1980).

[43] M. Zender, SIN preprint (1981).

[44] H. Faissner et al. Aachen preprint (1983).

[45] J. E. Kim, *Phys. Rev. Lett.* **43**, 103 (1979);
M. Shifman, A. Vainstein, and V. Zakharov, *Nucl. Phys.* **B166**, 493 (1980);
M. Dine, W. Fischler, and M. Srednicki, *Phys. Lett.* **104B**, 199 (1981).

[46] Y. Chikashige, R. N. Mohapatra, and R. D. Peccei, *Phys. Lett.* **98B**, 265 (1981).

[47] D. A. Dicus, E. Kolb, V. Teplitz, and R. V. Wagoner, *Phys. Rev.* **D18**, 1829 (1978);
M. Fukugita, S. Watamura, and M. Yoshimura, *Phys. Rev. Lett.* **48**, 1522 (1978).

[48] M. Wise, H. Georgi, and S. L. Glashow, *Phys. Rev. Lett.* **47**, 402 (1981);
R. Barbieri, R. N. Mohapatra, D. Nanopoulos, and D. Wyler, *Phys. Lett.* **107B**, 80 (1981);
J. Ellis, M. K. Gaillard, D. Nanopoulos, and S. Rudaz, *Phys. Lett.* **107B**, (1981);

[49] M. A. B. Bég and H. S. Tsao, *Phys. Rev. Lett.* **41**, 278 (1978);
R. N. Mohapatra and G. Senjanovic, *Phys. Lett.* **79B**, 283 (1978).

[50] H. Georgi, *Hadron J.* **1**, 155 (1978);
G. Segre and H. A. Weldon, *Phys. Rev. Lett.* **42**, 1191 (1979);
S. Barr and P. Langacker, *Phys. Rev. Lett.* **42**, 1654 (1979).

[51] A. Masiero, R. N. Mohapatra, and R. D. Peccei, *Nucl. Phys.* **B192**, 66 (1981).

[52] R. N. Mohapatra and G. Senjanovic, *Z. Phys.* **20**, 365 (1983);
R. N. Mohapatra, S. Ouvry, and G. Senjanovic, *Phys. Lett.* **126B**, 329 (1983);

D. Chang and R. N. Mohapatra, *Phys. Rev.* **D32**, 293 (1985);
K. S. Babu, B. Dutta and R. N. Mohapatra, *Phys. Rev.* **D65**, 016005 (2002).

[53] A. Nelson, *Phys. Lett.* **136B**, 387 (1984);
S. Barr, *Phys. Rev. Lett.* **53**, 329 (1984).

[54] K. S. Babu and R. N. Mohapatra, *Phys. Rev.* **D41**, 1286 (1990).

5

Grand Unification and the SU(5) Model

§5.1   The Hypothesis of Grand Unification

We emphasized in the first chapter that the Yang–Mills theories provide a unique framework for describing interactions with universal couplings for different, apparently unrelated, processes such as μ decay and β decay, etc. This led to the successful electroweak unification theories based on $\mathrm{SU}(3)_c \times \mathrm{SU}(2)_L \times \mathrm{U}(1)_Y$ group (G_{321}). The 321-theory has three gauge couplings of different magnitudes and fermions assigned according to convenience, rather than any principle. It also has many parameters that are adjusted arbitrarily to fit different observations such as masses and mixings of quarks, etc. A logical next step to consider is a higher symmetry that unifies all three couplings and also simultaneously offers the possibility of relating the different parameters so that one has a more satisfactory theory than the standard model. The first attempt in this direction was made by Pati and Salam [1], who unified the quarks and leptons within the group $\mathrm{SU}(2)_L \times \mathrm{SU}(2)_R \times \mathrm{SU}(4)_C$ by extending the color gauge group to include the leptons. This model unified the quarks and leptons into a single representation and explained the quantization of electric charge, although it had three coupling constants, g_{2L}, g_{2R}, and g_c. This shortcoming was partially removed in [2] by making the theory left–right symmetric so that the two $SU(2)$ couplings became equal making this theory a two-coupling constant partial unification theory. This kind of theory will be discussed in Chapter 6.

A more ambitious approach was taken independently in the same year by Georgi and Glashow [3] who proposed the rank-4 simple group SU(5) as the grand unification group that not only explains the quantization of electric charge but also leads to the unification of all coupling constants. In this chapter we will discuss the SU(5) model, its predictions, and note that even though this model has now been ruled out by experiments on proton decay, it provides a prototype example with many features that can guide the construction of general grand unification models. We will also present some general results pertaining to more general grand unified models than SU(5).

§5.2   SU(N) Grand Unification

Because $SU(3)_c \times SU(2)_L \times U(1)$ is a group of rank 4, the minimal unifying group is SU(5). One may, however, consider more general SU(N) groups fpr the purpose. We therefore start by giving some requisites for choosing the representations of a general SU(N) group. First, in order to ensure the vector nature of color gauge interactions [i.e., to have only 3 and 3* of color SU(3)], the simplest choice is to assign fermions only to antisymmetric representations. In other words, the irreducible representations to be chosen must be of type $\psi_{ijk...}$ [$i, j, k, ... = 1, ..., N$ for SU(N)], antisymmetric in the interchange of any two indices. We remind the reader that the dimensionality of an mth rank antisymmetric tensor under SU(N) is

$$d(\psi_{i,...,i_m}) = \frac{N!}{m!(N-m)!}. \tag{5.2.1}$$

Furthermore, because we would like to have anomaly-free combinations, let us also give a formula for the anomalies $A_{m,N}$ associated with an antisymmetric representation of SU(N) with m boxes [3a]:

$$A_{m,N} = \frac{(N-3)!(N-2m)}{(N-m-1)!(m-1)!}. \tag{5.2.2}$$

With these two as the minimal criteria, we give some examples of sets of antisymmetric representations that are anomaly free and are therefore suitable candidates for grand unification ([N, m] means SU(N) with m antisymmetric indices)

$$
\begin{aligned}
\text{SU(5)}: &\quad [5, 1] + [5, 3], \\
\text{SU(6)}: &\quad 2[6, 1] + [6, 4], \\
\text{SU(7)}: &\quad [7, 2] + [7, 4] + [7, 6], \\
\text{SU(8)}: &\quad [8, 1] + [8, 2] + [8, 5], \\
\text{SU(9)}: &\quad [9, 2] + [9, 5], \\
\text{or} &\quad [9, 1] + [9, 3] + [9, 5] + [9, 7], \\
\text{SU(10)}: &\quad [10, 3] + [10, 6],
\end{aligned}
$$

etc. We could of course keep going; but it turns out that the smallest anomaly-free combination beyond SU(1O) makes the theory ultraviolet unstable [3b]. The formula for the β function for SU(N) with fermions in $[N, m]$ can be written as

$$\beta(g) = - \left[\tfrac{11}{3} N - \tfrac{1}{3} \sum_m \binom{N-2}{m-1} c_m \right] \frac{g^3}{16\pi^2} + O(g^5). \qquad (5.2.3)$$

It is clear from the above that, if we want to unify only one generation of fermions, SU(5) is satisfactory, because the total number of components is 15 for the representation listed.

§5.3   Sin$^2 \theta_W$ in Grand Unified Theories (GUT)

Before proceeding to a detailed discussion of GUT models, we give a formula for the calculation of $\sin^2 \theta_W$ in the exact symmetry limit in grand unified theories. In this subsection this formula will be derived in terms of the representation content of a particular kind of grand unification [4]. To do this, we note that the grand unification group must contain diagonal generators (or normalized linear combinations of diagonal generators) that can be identified with T_{3_L} and Y, with the corresponding gauge bosons being called W_{3_L} and W_0. The corresponding couplings are not necessarily identical to the grand unifying coupling constant g_U. So let us call them g and g'. Next, if we calculate the fermion contributions to the leading divergences in the self-energies $\pi_{33}(q)$ and $\pi_{00}(q)$, then invariance under the grand unifying group implies that

$$\pi_{33}^{\text{div}}(q) = \pi_{00}^{\text{div}}(q). \qquad (5.3.1)$$

But it is easy to see that the leading divergences of π_{33} and π_{00} are proportional, respectively, to $g^2 \sum_i T_{3i}^2$ and $g'^2 \sum_i Y_i^2$, where i goes over all the fermions in the theory.

Thus we obtain

$$g^2 \sum_i T_{3i}^2 = g'^2 \sum_i Y_i^2. \qquad (5.3.2)$$

But $g'^2/g^2 = \tan^2 \theta_W$, leading to

$$\cot^2 \theta_W = \frac{\sum_i Y_i^2}{\sum_i T_{3i}^2}. \qquad (5.3.3)$$

Using the fact that $\sum_i Q_i^2 = \sum_i \left(T_{3i}^2 + Y_i^2 \right)$, the formula for $\sin^2 \theta_W$ follows:

$$\sin^2 \theta_W = \frac{\sum_i T_{3i}^2}{\sum_i Q_i^2}. \qquad (5.3.4)$$

§5.4   SU(5)

In this section we will discuss the minimal grand unification based on SU(5) and its implications. As discussed in the previous chapter, the anomaly-free combination of SU(5) representations is $\{\bar{5}\} + \{10\}$ if both representations are chosen left-handed. Below we denote the $\{5\}$- and $\{10\}$-dimensional representations of fermions, where the $\{5\}$ is right-handed and the $\{10\}$ is chosen left-handed:

$$\psi = \begin{pmatrix} d_1 \\ d_2 \\ d_3 \\ e^+ \\ \nu^c \end{pmatrix}_R \quad ; \chi = \begin{pmatrix} 0 & u_3^c & -u_2^c & u_1 & d_1 \\ & 0 & u_1^c & u_2 & d_2 \\ & & 0 & u_3 & d_3 \\ & & & 0 & e^+ \\ & & & & 0 \end{pmatrix}_L . \qquad (5.4.1)$$

There are $\{24\}$ gauge bosons associated with SU(5). Under $SU(3)_c \times SU(2)_L \times U(1)$ they decompose as follows:

$$\{24\} = \{8,1,0\} + \{1,3,0\} + \{3,2,\tfrac{5}{3}\} + \{3^*,2,-\tfrac{5}{3}\} + \{1,1,0\}. \qquad (5.4.2)$$

To express this in matrix form we choose the following form for the SU(5) generators:

$$\lambda_i = \begin{pmatrix} & & & 0 & 0 \\ & \lambda_i & & 0 & 0 \\ & & & 0 & 0 \\ 0 & 0 & 0 & 0 & 0 \\ 0 & 0 & 0 & 0 & 0 \end{pmatrix}, \quad i = 1,\ldots,8,$$

$$\lambda_9 = \begin{pmatrix} & & & 1 & 0 \\ & 0 & & 0 & 0 \\ & & & 0 & 0 \\ 1 & 0 & 0 & & \\ 0 & 0 & 0 & & 0 \end{pmatrix}, \quad \lambda_{10} = \begin{pmatrix} & & & -i & 0 \\ & 0 & & 0 & 0 \\ & & & 0 & 0 \\ i & 0 & 0 & & \\ 0 & 0 & 0 & & 0 \end{pmatrix},$$

$$\lambda_{11} = \begin{pmatrix} & & & 0 & 1 \\ & 0 & & 0 & 0 \\ & & & 0 & 0 \\ 0 & 0 & 0 & & \\ 1 & 0 & 0 & & 0 \end{pmatrix}, \quad \lambda_{12} = \begin{pmatrix} & & & 0 & -i \\ & 0 & & 0 & 0 \\ & & & 0 & 0 \\ 0 & 0 & 0 & & \\ i & 0 & 0 & & 0 \end{pmatrix},$$

$$\lambda_{13} = \begin{pmatrix} & & & 0 & 0 \\ & 0 & & 1 & 0 \\ & & & 0 & 0 \\ 0 & 1 & 0 & & \\ 0 & 0 & 0 & & 0 \end{pmatrix}, \quad \lambda_{14} = \begin{pmatrix} & & & 0 & 0 \\ & 0 & & -i & 0 \\ & & & 0 & 0 \\ 0 & i & 0 & & \\ 0 & 0 & 0 & & 0 \end{pmatrix},$$

$$\lambda_{15} = \begin{pmatrix} & & 0 & 0 \\ 0 & & 0 & 1 \\ & & 0 & 0 \\ 0 & 0 & 0 & \\ 0 & 1 & 0 & 0 \end{pmatrix}, \qquad \lambda_{16} = \begin{pmatrix} & & 0 & 0 \\ 0 & & 0 & -i \\ & & 0 & 0 \\ 0 & 0 & 0 & \\ 0 & i & 0 & 0 \end{pmatrix},$$

$$\lambda_{17} = \begin{pmatrix} & & 0 & 0 \\ 0 & & 0 & 0 \\ & & 1 & 0 \\ 0 & 0 & 1 & \\ 0 & 0 & 0 & 0 \end{pmatrix}, \qquad \lambda_{18} = \begin{pmatrix} & & 0 & 0 \\ 0 & & 0 & 0 \\ & & -i & 0 \\ 0 & 0 & i & \\ 0 & 0 & 0 & 0 \end{pmatrix},$$

$$\lambda_{19} = \begin{pmatrix} & & 0 & 0 \\ 0 & & 0 & 0 \\ & & 0 & 1 \\ 0 & 0 & 0 & \\ 0 & 0 & 1 & 0 \end{pmatrix}, \qquad \lambda_{20} = \begin{pmatrix} & & 0 & 0 \\ 0 & & 0 & 0 \\ & & 0 & -i \\ 0 & 0 & 0 & \\ 0 & 0 & i & 0 \end{pmatrix},$$

$$\lambda_{20+j} = \begin{pmatrix} & & 0 & 0 \\ 0 & & 0 & 0 \\ & & 0 & 0 \\ 0 & 0 & 0 & \\ 0 & 0 & 0 & \tau_j \end{pmatrix}, \qquad j = 1,2,3,$$

$$\lambda_{24} = \frac{2}{\sqrt{15}} \begin{pmatrix} 1 & & & & \\ & 1 & & & \\ & & 1 & & \\ & & & -3/2 & \\ & & & & -3/2 \end{pmatrix}. \tag{5.4.3}$$

This gives the following form for the gauge boson matrix:

$$\begin{pmatrix} \frac{1}{\sqrt{2}}\boldsymbol{\lambda}\cdot\mathbf{V}_{\{8\}} + \sqrt{\frac{2}{15}}V_{24} & \begin{matrix} X_1^{-4/3} & Y_1^{-1/3} \\ X_2^{-4/3} & Y_2^{-1/3} \\ X_1^{-4/3} & Y_3^{-1/3} \end{matrix} \\ \begin{matrix} X_1^{4/3} & X_2^{4/3} & X_3^{4/3} \\ Y_1^{1/3} & Y_2^{1/3} & Y_3^{1/3} \end{matrix} & \frac{1}{\sqrt{2}}\boldsymbol{\tau}\cdot\mathbf{W} - \sqrt{\frac{3}{10}}V_{24} \end{pmatrix}. \tag{5.4.4}$$

One of the first things we can do now is to predict the value of the weak-electromagnetic mixing angle $\sin^2\theta_W$. For this purpose recall that θ_W is defined for the $SU(2) \times U(1)$ Glashow–Weinberg–Salam group as follows:

$$\tan\theta_W = \frac{g_{U(1)}}{g_{SU(2)}} \equiv \frac{g'}{g}. \tag{5.4.5}$$

Owing to unification we can predict g'/g at the masses above which unification has occurred. To see this note that λ_{24} corresponds to Y. Let us take its eigenvalue for the e^+ and compare with the $SU(2) \times U(1)$ case:

we have

$$\sqrt{\tfrac{3}{5}} g_{SU(5)} = g'. \tag{5.4.6}$$

But

$$g_{SU(5)} = g. \tag{5.4.7}$$

It therefore follows that $\tan\theta_W = \sqrt{\tfrac{3}{5}}$, leading to the famous prediction $\sin^2\theta_W = \tfrac{3}{8}$. Incidentally, the same result is also obtained if we use eq. (5.3.4).

Next we would like to present the breakdown of the SU(5) group to $SU(3)_c \times U(1)_{cm}$. Within the conventional Higgs picture this is achieved by introducing two Higgs multiplets: one belonging to the {24}-dimensional irreducible representation (denoted by Φ), and the other to the {5}-dimensional (denoted by H) representations. The stages of symmetry breakdown are given by

$$SU(5) \to SU(3)_c \times SU(2)_L \times U(1) \to SU(3)_c \times U(1)_{em}.$$

The first stage is achieved by $\langle\Phi\rangle \neq 0$ and the second stage by $\langle H\rangle \neq 0$ as follows: from the general group theory of spontaneous breakdown we can show that [5]

$$\langle\Phi\rangle = V \begin{pmatrix} 1 & & & & \\ & 1 & & & \\ & & 1 & & \\ & & & -3/2 & \\ & & & & -3/2 \end{pmatrix}, \tag{5.4.8}$$

i.e., the breaking is along the λ_{24} direction. This is responsible for the first stage of the symmetry breakdown. The second stage is caused by

$$\langle H\rangle = \begin{pmatrix} 0 \\ 0 \\ 0 \\ 0 \\ \rho/\sqrt{2} \end{pmatrix}. \tag{5.4.9}$$

In the presence of both H and Φ, the v.e.v. of Φ changes somewhat and looks like the following:

$$\langle\Phi\rangle = V \begin{pmatrix} 1 & & & & \\ & 1 & & & \\ & & 1 & & \\ & & & -3/2-\varepsilon/2 & \\ & & & & -3/2+\varepsilon/2 \end{pmatrix}. \tag{5.4.10}$$

It is then a straightforward matter to compute all the gauge boson masses

Superheavy bosons X, Y: $m_X^2 \approx m_Y^2 = \tfrac{25}{8} g^2 V^2,$ \hfill (5.4.11)

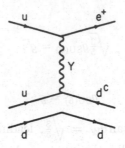

Figure 5.1.

$$W\text{-, } Z\text{-bosons:}\quad m_W^2 \simeq \frac{g^2\rho^2}{4}(1+\varepsilon), \qquad m_Z^2 \simeq \frac{g^2\rho^2}{4\cos^2\theta_W}.\quad(5.4.12)$$

The reason why the X and Y bosons must be heavy can be seen by looking at the part of the gauge Lagrangian that involves X and Y. For one
generation it is given by

$$\mathcal{L}_{X,Y} = \frac{ig_x}{\sqrt{2}} X_{\mu,i}\left(\varepsilon_{ijk}\bar{u}_{kL}^c\gamma^\mu u_{jL} + \bar{d}_i\gamma^\mu e^+\right)$$

$$+ \frac{ig_Y}{\sqrt{2}} Y_{\mu,i}\left(\varepsilon_{ijk}\bar{u}_{kL}^c\gamma^\mu d_{jL} - \bar{u}_{i_L}\gamma^\mu e_L^+ + \bar{d}_{i_R}\gamma^\mu \nu_R^c\right) + \text{h.c.} \quad (5.4.13)$$

We thus see that

$$\bar{X} \to uu \qquad \left(B = \tfrac{2}{3}\right)$$
$$\to \bar{d}e^+ \qquad \left(B = -\tfrac{1}{3}\right),$$

and

$$\bar{Y} \to ud \qquad \left(B = \tfrac{2}{3}\right)$$
$$\to \bar{u}e^+ \qquad \left(B = -\tfrac{1}{3}\right).$$

Thus they can lead to proton decay of type $p \to e^+\pi^0$ via the diagrams
shown in Fig. 5.1 with strength

$$M(p \to e^+\pi^0) \approx \frac{g_{X,Y}^2}{m_{X,Y}^2}$$

leading to a crude estimate for the lifetime of the proton to be about

$$\tau_p \approx \frac{m_X^4}{g^4 m_p^5}. \qquad (5.4.14)$$

The lower limits [6] on the proton lifetime, existing up to 1979, then implied
$m_{X,Y} \geq 10^{15}$ GeV.

Several dedicated searches for proton decay have been going on at various
places around the world since that time [7–12] and we will discuss these
results at the end of this chapter. On the basis of these experiments we
can safely say that $\tau_p \geq 10^{31}$ yr with more stringent limits on particular

decay modes such as $p \rightarrow e^+\pi^0$, etc. We now focus on some theoretical aspects of the SU(5) model, especially the problem of understanding such large masses previously nonexistent in elementary particle physics.

One aspect of the SU(5) model is that there exist two totally different mass scales. The theoretical question then arises as to how, in a theory with divergences (albeit of renormalizable type), such large mass ratios can be maintained. We will argue that, in order to maintain this disparity between mass scales, certain parameters of the theory must be adjusted to one part in 10^{13}. This is the so-called "gauge hierarchy" problem. There is, of course, a related problem that can be called "supposed gauge hierarchy impasse," which we will also mention below.

Gauge Hierarchy Problems

To study this problem we write the potential involving the $\Phi\{24\}$ and $H\{5\}$ Higgs multiplets

$$V = -\tfrac{1}{2}\mu^2 \operatorname{Tr}\phi^2 + \tfrac{1}{4}a(\operatorname{Tr}\phi^2)^2 + \tfrac{1}{2}b\operatorname{Tr}\phi^4 - \tfrac{1}{2}\nu^2 H^+ H + \frac{\lambda}{4}(H^+ H)^2$$
$$+ \alpha H^+ H \operatorname{Tr}\phi^2 + \beta H^+ \phi^2 H. \tag{5.4.15}$$

Substituting the expected minima given in eqs. (5.4.9) and (5.4.10) we obtain

$$V_{\min} - -\tfrac{1}{2}\left(\tfrac{15}{2}V^2 + \tfrac{1}{2}\varepsilon^2 V^2\right) + \frac{a}{4}\left(\tfrac{15}{2}V^2 + \tfrac{1}{2}\varepsilon^2 V^2\right)^2$$
$$+ \frac{b}{2}\left(\tfrac{105}{8}V^4 + \tfrac{27}{4}V^4\varepsilon^2 + \cdots\right) - \tfrac{1}{4}\nu^2\rho^2$$
$$+ \frac{\lambda}{16}\rho^4 + \frac{\alpha}{2}\rho^2\left(\tfrac{15}{2}V^2 + \tfrac{1}{2}\varepsilon^2 V^2\right) + \beta\frac{\rho^2 V^2}{8}(3-\varepsilon)^2. \tag{5.4.16}$$

The conditions for minimum give the following equations:

$$\varepsilon \simeq \frac{3\beta}{20b}(\rho/V)^2, \tag{5.4.17a}$$

$$\mu^2 = \tfrac{15}{2}av^2 + \tfrac{7}{2}bv^2 + \alpha\rho^2 + \tfrac{3}{10}\beta\rho^2 + O(\varepsilon\rho^2), \tag{5.4.17b}$$

$$\nu^2 = \tfrac{1}{2}\lambda\rho^2 + 15\alpha v^2 + \tfrac{9}{2}\beta v^2 - 3\varepsilon\beta v^2 + O(\varepsilon\rho^2). \tag{5.4.17c}$$

From (5.4.17c) we see that if $\nu^2 \approx \rho^2$, then both α, β must be of order ρ^2/V^2 10^{-26} so that ρ can be much smaller than V; the other possibility is that α, β are of order 1 and ν^2 is of order V^2; however, there must then be a cancellation between ν^2 and $15\alpha V^2 + \tfrac{9}{2}\beta V^2$ (to an incredible degree of accuracy) to have a quantity of order $10^{-26}V^2$. Both of these are technically allowed (though extremely unnatural) and may indeed be a symptom of something wrong with such simple-minded grand unification.

Let us now discuss the second point, i.e., the "gauge hierarchy impasse [13]." If we chose α and β at the tree level to an accuracy of 10^{-26}, once we include radiative corrections, α and β "may acquire" a lower bound of $\geq g^4/32\pi^2$ and therefore may make such a choice technically unrealizable.

This, if true, would indeed have been a fatal blow to SU(5) and other such simple grand unification schemes. It has, however, been noted (by explicit computation of the one-loop potential) that no such impasse occurs [14], and indeed (at least up to the one-loop level) it is possible to adjust parameters arbitrarily to obtain any desired gauge hierarchy. From now on we will assume the existence of a gauge hierarchy, which is still an open problem and which has been one of the strongest motivations for considering supersymmetric models.

§5.5   Grand Unification Mass Scale and $\text{Sin}^2 \theta_W$ at Low Energies

The philosophy of grand unification is that all interactions are described by a single coupling constant. However, we know quite well that at low energies (i.e., $E \approx 1$ GeV), strong-interaction couplings are much larger than the weak and electromagnetic couplings. The next problem in the discussion of grand unification is how to reconcile these two ideas. The solution was first suggested by Georgi, Quinn, and Weinberg [15]. Their observation was that coupling constants are scale-dependent quantities whose rate of variation with scale is governed by Callan–Symanzik equations. Therefore, these coupling constants must change with energy and, if the hypothesis of grand unification is to hold, they must be equal to the SU(5) coupling at some mass M_U. We will call M_U the grand unification scale. For $\mu < M_U$, SU(5) breaks down to $SU(3)_c \times SU(2) \times U(1)$ and the separate couplings behave differently. To study this behavior we need to know how the gauge coupling constants change with energy. This is given by a powerful theorem first proved by Appelquist and Carazone [16].

Decoupling Theorem

If a gauge invariant (under group G) Lagrangian field theory contains particles with two very different masses m and M ($m \ll M$) and is described by a renormalizable Lagrangian, the behavior of the light particles in the theory for $E \ll M$ can be described completely by a renormalizable Lagrangian involving only the light particles. The effect of the heavy particles [17] is simply to rescale the coupling constants and renormalization parameters of the theory.

In the context of the SU(5) model this means that if, at mass scale $M \approx V$, the SU(5) symmetry breaks down to $SU(3)_c \times SU(2)_L \times U(1)$, then the behavior of the three gauge couplings g_3, g_2, and g_1 corresponding to these symmetries must evolve according to the β functions corresponding to SU(3), SU(2), and U(1), respectively, with no memory whatsoever of

SU(5). We can therefore write

$$\frac{dg_i}{dt} = \beta_i(g_i) \equiv +b_{0i} \frac{g_i^3}{16\pi^2}, \qquad i = 3, 2, 1, \qquad (5.5.1)$$

where

$$t = \ln \mu.$$

We know that

$$\beta_3(g_3) = -\frac{g_3^3}{16\pi^2} \left(11 - \tfrac{2}{3} N_f - \frac{N_H}{6} \right), \qquad (5.5.2)$$

where N_f is the number of four-component color triplet fermions and N_H is the number of the color triplet Higgs bosons. Similarly,

$$\beta_2(g_2) = -\frac{g_2^3}{16\pi^2} \left(\tfrac{22}{3} - \tfrac{2}{3} N_f - \frac{N_H}{6} \right), \qquad (5.5.3)$$

$$\beta_1(g_1) = +\tfrac{2}{3} N_f \cdot \frac{g_1^3}{16\pi^2} + \text{Higgs boson contributions.} \qquad (5.5.4)$$

Let us first omit the contribution of the Higgs bosons and define $\alpha_i = g_i^2/4\pi$. Then, $d\alpha_i/dt = (1/2\pi) \cdot g_i(dg_i/dt)$. Equation (5.5.1) can then be written as

$$\frac{d\alpha_3}{dt} = -\frac{\alpha_3^2}{2\pi} (11 - \tfrac{2}{3} N_f),$$

$$\frac{d\alpha_2}{dt} = -\frac{\alpha_2^2}{2\pi} (\tfrac{22}{3} - \tfrac{2}{3} N_f),$$

$$\frac{d\alpha_1}{dt} = -\frac{\alpha_1^2}{2\pi} \cdot \tfrac{2}{3} N_f. \qquad (5.5.5)$$

The boundary condition on these differential equations is

$$\alpha_1(M_X) = \alpha_2(M_X) = \alpha_3(M_X) = \alpha_U. \qquad (5.5.6)$$

Using (5.5.5) and (5.5.6) we will find $\alpha_i(\mu)$ for $\mu \approx m_W$ To identify low-energy parameters we note that $\alpha_3(m_W) = \alpha_s(m_W)$, $\alpha_2(m_W) = \alpha_w(m_W)$, and $\tfrac{3}{5}\alpha_1(m_W) = g'^2(m_W)/4\pi$, with $\sin^2 \theta_W = g'^2(m_W)/\{g'^2(m_W) + g^2(m_W)\}$. The solutions for these equations can be written as

$$\frac{1}{\alpha_i(\mu)} = \frac{1}{\alpha_i(M_X)} + \frac{b_{0i}}{2\pi} \ln \frac{M_X}{\mu}. \qquad (5.5.7)$$

One equation that we obtain from this is

$$\frac{1}{\alpha_2(\mu)} - \frac{1}{\alpha_1(M_X)} = -\frac{1}{2\pi} \cdot \frac{22}{3} \ln \frac{M_X}{\mu}, \qquad (5.5.8a)$$

which leads to a connection between $\sin^2 \theta_W$ and the unification scale M_X, i.e.,

$$\sin^2 \theta_W(m_W) = \frac{3}{8} - \frac{55\alpha(m_W)}{24\pi} \ln \frac{M_X}{m_W}. \qquad (5.5.8b)$$

If we include one Higgs doublet, this changes the above expression as follows:

$$\sin^2 \theta_W(m_W) = \tfrac{3}{8} - \alpha(m_W) \cdot \frac{109}{48\pi} \ln \frac{M_X}{m_W}. \qquad (5.5.9a)$$

Another equation that can be obtained from eq. (5.5.7) is [call $\alpha_3(m_W) \equiv \alpha_s(m_W)$]

$$\frac{\alpha(m_W)}{\alpha_s(m_W)} = \frac{3}{8} - \frac{67}{16\pi} \alpha(m_W) \ln \frac{M_U}{m_W}. \qquad (5.5.9b)$$

The experimentally measured value of α is at $\mu = 0$ since it involves physical photons. Therefore we must evaluate $\alpha(m_W)$. To do this we note that α satisfies a renormalization group equation as follows:

$$\frac{\partial e^2}{\partial t} = \frac{1}{6\pi^2} \sum_f Q_f^2 e^4. \qquad (5.5.10)$$

This implies that

$$\frac{1}{\alpha(m_W)} \simeq \frac{1}{\alpha(0)} - \frac{2}{3\pi} \sum_f Q_f^2 \ln \frac{m_W}{m_f} + \frac{1}{6\pi} + \dots, \qquad (5.5.11)$$

where m_f is the fermion mass that includes all quarks and leptons. Putting in $m_t \simeq 40$ GeV we find [18] $\alpha^{-1}(m_W) \simeq 128$. From eq. (5.5.9) we can evaluate M_X by using a value of $\alpha_s(m_W)$.

The fine structure constant for QCD (α_s depends on the value of the QCD scale parameter $\Lambda_{\overline{MS}}$, which roughly denotes the scale at which the chromodynamic interactions become strong) is given by the following formula:

$$\alpha_s(Q^2) = \frac{12\pi}{25 \ln(Q^2/\Lambda_{\overline{MS}}^2)} + \text{higher-order terms}. \qquad (5.5.12)$$

A plausible value [19] for $\Lambda_{\overline{MS}}$ is

$$\Lambda_{\overline{MS}} = 160^{+100}_{-80} \text{ MeV}. \qquad (5.5.13)$$

This value is derived from decays of the $B(\gamma \to \text{hadrons} + \text{photon})/ B(\gamma \to \text{hadrons})$ for which the next order in α_s corrections is small. For a general $\Lambda_{\overline{MS}}$ the predicted value [20] of M_X is

$$M_X = 2.1 \times 10^{14} \times (1.5)^{\pm 1} \left[\frac{\Lambda_{\overline{MS}}}{0.16 \text{ GeV}} \right]. \qquad (5.5.14)$$

This leads to a prediction for $\sin^2 \theta_W(m_W)$ as follows:

$$\sin^2 \theta_W(M_W) = 0.214 \pm 0.003 \pm 0.006 \ln \left[\frac{0.16 \text{GeV}}{\Lambda_{\overline{MS}}} \right]. \qquad (5.5.15)$$

This prediction for $\sin^2 \theta_W$ is in disagreement with the values obtained from the world average value from the study of neutral currents [21] and

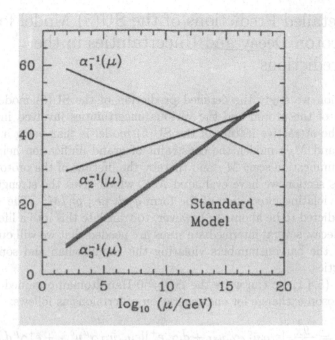

Figure 5.2. This figure shows the lack of unification of gauge couplings with standard model spectrum. α_i^{-1} is plotted against the mass scale and the values at the weak scale are the measured values from LEP and SLC as well as other experiments.

the recent study of W-and Z-masses from SPS [22a], CDF [22b], LEP [22c], and SLC [22d]

$$\sin^2 \theta_W = 0.23161 \pm 0.00018. \tag{5.5.16}$$

This disagreement is clearly described in Fig. 5.2, which gives the evolution the gauge couplings in the SU(5) model. In this figure, we have used the values of the three gauge coupling parameters measured at the M_Z scale in the LEP and SLC experiments, i.e.,

$$\alpha_1^{-1}(M_Z) = 58.97 \pm 0.05, \tag{5.5.17}$$
$$\alpha_2^{-1}(M_Z) = 29.61 \pm 0.05,$$
$$\alpha_3^{-1}(M_Z) = 8.47 \pm 0.22.$$

§5.6   Detailed Predictions of the SU(5) Model for Proton Decay and Uncertainties in the Predictions

In this section we study the detailed prediction of the SU(5) model for the lifetime of the proton and the various uncertainties involved in this estimate. The attractive feature of the SU(5) model is that being a two-scale (m_W and M_X) model, the constraint of grand unification helps to predict the unification scale M_X and thereby the lifetime of the proton. In the previous section we have evaluated M_X, which gives the strength of the baryon-violating interaction of the form $qqql$, i.e., g_U^2/M_X^2. The value of α_U is predicted to be about $\frac{1}{50}$; however, to translate this into a lifetime for proton decay, several intermediate steps are needed that we will outline after giving the baryon numbers violating the Hamiltonian and some of their properties.

Using eq. (5.4.1) we can write the $\Delta B \neq 0$ Hamiltonian obtained from the gauge boson exchange for one generation of fermions as follows:

$$H_{\Delta B \neq 0} = \frac{g_X^2}{2M_X^2}[\varepsilon_{ijk}\bar{u}_{kL}^c\gamma_\mu u_{jL} + \bar{d}_i\gamma_\mu e^+][\varepsilon_{ilm}\bar{u}_{iL}\gamma^\mu u_{mL}^c + \bar{e}^+\gamma^\mu d_i]$$

$$+ \frac{g_Y^2}{2M_Y^2}[\varepsilon_{ijk}\bar{u}_{kL}^c\gamma_\mu d_{jL} - \bar{u}_{iL}\gamma_\mu e_L^+ + \bar{d}_{iR}\gamma_\mu\nu_R^c]$$

$$\times [\varepsilon_{ilm}\bar{d}_{lL}\gamma^\mu u_{mL}^c - \bar{e}_L^+\gamma^\mu u_{iL} + \bar{\nu}_R^c\gamma^\mu d_{iR}]$$

$$+ \frac{1}{M_{H_i}^2}\mathcal{O}_i^\dagger\mathcal{O}_i, \tag{5.6.1}$$

where H_i are the colored Higgs bosons in the five-dimensional Higgs field and the operators $\mathcal{O}_i$ are given by

$$\mathcal{O}_i = f_{11}\{\varepsilon_{ijk}\bar{u}_{kL}^c d_{jR} + \bar{u}_{iL}e_R^+\} \tag{5.6.2}$$

$$+ h_{11}\{\varepsilon_{ijk}u_{jL}^T C^{-1}d_{kL} + \varepsilon_{ijk}d_{jL}^T C^{-1}u_{kL} \tag{5.6.3}$$

$$+ u_{iL}^{cT}C^{-1}e_L^+)\}, \tag{5.6.4}$$

If the Higgs boson masses are of order 10^{13} GeV or so, the Higgs contribution to proton decay can be neglected because the associated Yukawa couplings are $\approx G_F^{+1/2}m_f$.

Looking at this Hamiltonian it is clear that it conserves the quantum number $B - L$, where L is the lepton number. This is also true for the Higgs bosons contributions, as will be discussed in the next section. This can be understood by realizing that even though the gauge bosons X, Y have no definite baryon number (thus leading to proton decay), they have a definite $B - L$ quantum number $2/3$. The dominant decay mode that respects this selection rule is $p \rightarrow e^+\pi^0$. Let us discuss how one can estimate its partial lifetime. There are several steps involved in this estimate:

(a) The effective $\Delta B \neq 0$ Hamiltonian is defined at the grand unification scale M_X, and therefore has to be extrapolated down to the $\mu \simeq 1$ GeV scale by calculating weak as well as strong QCD corrections to the vertex [23]. The QCD correction is given by

$$A_3 \approx \left[\frac{\alpha_3(1\,\text{GeV})}{\alpha_U}\right]^{2/11-2/3f} \tag{5.6.5}$$

The electroweak corrections are

$$A_{21} \approx \left[\frac{\alpha_2(100\,\text{GeV})}{\alpha_U}\right]^{27/(86-8f)} \times \begin{cases} \left[\dfrac{\alpha_1(100\,\text{GeV})}{\alpha_U}\right]^{-69/(6+40f)}, \\[2ex] \left[\dfrac{\alpha_1(100\,\text{GeV})}{\alpha_U}\right]^{-33/(6+40f)}. \end{cases} \tag{5.6.6}$$

The top factor in eq. (5.6.6) applies to the operators involving e_L^- and the bottom factor to that involving ν_L and e_R^-; f is the number of flavors. These factors have the effect of enhancing the matrix element by a factor of 3.5–4, decreasing the proton lifetime by roughly a factor 15.

(b) The next problem is to go from the quark–lepton form of the Hamiltonian to calculate the proton decay matrix elements. We may use [23], [24] the naive nonrelativistic SU(6) wave functions (for the proton and the various mesons) to evaluate the proton decay matrix element or ultrarelativistic bag model [25]. In terms of quark diagrams we can draw the three types shown in Fig. 5.3. Figure 5.3(a) is not expected to be significant due to phase-space suppression. Recent estimates [26] of proton decay have therefore focused on the diagrams in Figs. 5.3(b) (c). All these calculations lead to the value of τ_p as follows:

$$\tau_p = 2.5 \times 10^{28}\,\text{yr to } 1.6 \times 10^{30}\,\text{yr.} \tag{5.6.7}$$

In most models $p \to e^+\pi^0$ is predicted to occur most often with a branching ratio of 40–60%, and the next most frequent being $p \to e^+\omega, e^+\rho^0$, and $\bar{\nu}_e\pi^+$ with branching ratios of 5–20%, 1–10%, and 16–24%, respectively. This gives a maximum value for the partial lifetime of $p \to e^+\pi^0$ decay of about 3×10^{30} yr. The old experimental value for this number comes from the IMB,[1] Kamiokande,[2] and Frejus[3] experiments:

$$\tau_{p \to e^+\pi^0} \geq 6.1 \times 10^{32}\,\text{yr.} \tag{5.6.8}$$

and a recent much higher limit from the Sups-Kamiskande experiment of $\tau_{p \to e^+\pi^0} \geq 6.1 \times 10^{33}$ yr. This result rules out the minimum SU(5) model as the grand unification symmetry.

(c) *Effect of Flavor Mixing on Proton Decay.* This was first studied in Ref. 27 and subsequently in Ref. 28. The basic question, first raised by

---

[1] S. Seidel et al., *Phys. Rev. Lett.* **61**, 2522 (1988).

[2] K. S. Hirata et al., *Phys. Lett.* **B220**, 308 (1989).

[3] Ch. Berger et al., *Nucl. Phys.* **B313**, 509 (1989).

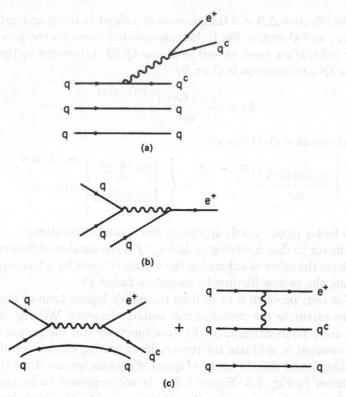

Figure 5.3. Different classes of contributions to $p \to e^+\pi^0$ decay amplitude.

Jarlskog [29], is that in the presence of flavor mixing effects in SU(5) models with several generations, it is *a priori* possible that the mixing matrix that appears in the baryon-number-conserving weak charged currents, the Cabibbo–Kobayashi–Maskawa matrix need not be the same as the one in the baryon-number-violating interactions. This would have profound implications on the predictions for proton decay in the SU(5) model. For instance, if the two mixing matrices are unrelated, we could "rotate" away proton decay by adjusting the mixing angles in the $\Delta B \neq 0$ sector. It was, however, shown in Ref. 27 that in the minimal SU(5) model this is not possible, i.e., both mixing matrices in the gauge interactions of $\Delta B = 0$ and $\Delta B \neq 0$ types are the same. We do not provide a detailed proof of this here and it can be found in Ref. 27. We only note that the key observation that leads to this result is the symmetry of the up quark mass matrix in the minimal model. This result makes it clear that in estimating the prediction of the minimal SU(5) model for proton decay, it is safe to ignore the flavor mixing.

If one includes additional Higgs bosons to the model [say, for instance, those belonging to the $\{45\}$ representation H_{gr}^p, of SU(5)], these results will change. They will generate new kinds of Yukawa couplings involving $\{5^*\}$

and $\{\underline{10}\}$ as well as those involving only the $\underline{10}$ representations of fermions. These interactions will be given by

$$\mathcal{L}'_Y = \sum_{a,b} f'_{ab} \bar{\chi}^a_{pq,L} \psi_{r,R} H^r_{pq} + \sum_{a,b} h'_{ab} \varepsilon^{pqrst} \chi^{Ta}_{pq,L} C^{-1} \chi^b_{lr,L} H^l_{st}. \qquad (5.6.9)$$

The H fields acquire vacuum expectation values as follows: $\langle H^4_{45} \rangle = -3\langle H^i_{i5} \rangle$ (no sum over i). By substituting $\langle H \rangle$ into eq. (5.6.9) it becomes clear that the contribution of $\mathcal{L}_{Y'}$ to the up-quark mass matrix makes it asymmetric and, in principle, proton lifetime can be arbitrarily adjusted.

A source of uncertainty in the proton lifetime, that we have not mentioned has to do with superheavy Higgs boson masses [29a]. The problem is that, owing to uncertainties in $\lambda\phi^4$-type couplings, the masses of the Higgs bosons could easily be uncertain by a factor of 100. This would imply that the evolution of the coupling constants will change leading to changes in M_X. These uncertainties have been estimated in Ref. 29a and could change M_X by a factor of 2–3.

§5.7 Some Other Aspects of the SU(5) Model

In this section we focus our attention on some other aspects of the SU(5) model: this will include predictions for fermion masses, selection rules for baryon nonconservations, and possible extensions of the SU(5) model.

(i) Fermion Masses in the SU(5) Model

In the minimal SU(5) model with only the Higgs multiplets in the {24}- and {5}-dimensional (denoted by H) representations, the down-quark and lepton masses will arise from the following Yukawa couplings [the up-quark mass has been discussed in eq. (5.6.9)]

$$\mathcal{L}_Y = h\bar{\psi}_{pR} H^+_q \chi_{pq} + \text{h.c.} \qquad (5.7.1)$$

Substituting $\langle H_5 \rangle = \rho/\sqrt{2}$, we find equal mass for electron and down-quarks

$$m_e = m_d = h\rho/\sqrt{2}. \qquad (5.7.2)$$

In the presence of all generations of quarks and leptons the generalized mass matrix looks like

$$M^{(l^-)}_{ab} = M^{(d)}_{ab} = h_{ab}\rho/\sqrt{2}. \qquad (5.7.3)$$

Diagonalization of this mass matrix therefore says that all its eigenvalues are equal, i.e.,

$$m_e = m_d, \qquad m_\mu = m_s, \qquad m_\tau = m_b. \qquad (5.7.4)$$

Note, however, that these equations are valid only at mass scales where SU(5) is a good symmetry. But since observed masses correspond to energies of order 1 GeV, these relations must be extrapolated. The extrapolation formula can be written as follows:

$$\ln \frac{m_{f_1}(\mu)}{m_{f_2}(\mu)} = \ln \frac{m_{f_1}(M)}{m_{f_2}(M)} + \int \sum_i [\gamma^i_{f_1} - \gamma^i_{f_2}] \frac{d\mu'}{\mu'}, \qquad (5.7.5)$$

where the notation is self-explanatory except for i, which stands for subgroup G_i. However, if we look at certain mass ratios such as

$$\frac{m_e}{m_\mu} = \frac{m_d}{m_s}, \qquad (5.7.6)$$

without doing any calculation it is easy to argue that they will remain the same. The reason is that, since the SU(2) and U(1) quantum numbers of particles in each ratio are equal, their contributions will cancel. In fact, before calculating any fermion masses, it is quite clear that the above relation is grossly violated by experiments and therefore requires drastic changes in the simple structure of the model.

One way to correct [30], [31] for this is to include a {45}-dimensional H^P_{qs} Higgs boson in the theory. In the presence of this we will have additional Yukawa couplings of the form

$$\mathcal{L}_Y = h' \bar{\psi}_{p,R} H^{p+}_{sq} \chi_{sq,L} + \text{h.c.} \qquad (5.7.7)$$

The property of the {45}-dimensional representation of the Higgs meson is that

$$H^p_{sq} = -H^p_{qs}, \qquad \sum_{p=1}^{5} H^p_{ps} = 0. \qquad (5.7.8)$$

From this it follows that the only allowed nonzero vacuum expectation values of H are

$$\langle H^i_{i5} \rangle = -\tfrac{1}{3} \langle H^4_{45} \rangle = \Lambda. \qquad (5.7.9)$$

Thus, using (5.7.9) will yield lepton–quark mass relations of type $m_q = 3m_L$. Now, using this, let us see how we can fix the lepton–quark mass relations: consider only the $e - \mu$ sector of the mass matrix. Then, by means of suitable discrete symmetries, we can have the following type of Yukawa coupling:

$$\mathcal{L}_Y = h_1(\bar{\psi}^{(1)}_{p,R} H^+_q \chi^{(2)}_{pq,L} + \bar{\psi}^{(2)}_{p,R} H^+_q \chi^{(1)}_{pq,L}) + h_2 \bar{\psi}^{(2)}_{p,R} H^{p+}_{gr} \chi^{(2)}_{gr,L} + \text{h.c.} \qquad (5.7.10)$$

This leads to mass matrices of the form

$$\begin{array}{cc} & \begin{array}{cc} e & \mu \end{array} \\ \begin{array}{c} e \\ \mu \end{array} & \begin{pmatrix} 0 & a \\ a & 3b \end{pmatrix}, \end{array} \qquad \begin{array}{cc} & \begin{array}{cc} d & s \end{array} \\ \begin{array}{c} d \\ s \end{array} & \begin{pmatrix} 0 & a \\ a & b \end{pmatrix}. \end{array} \qquad (5.7.11)$$

This implies

$$\frac{m_e}{m_\mu} = \frac{1}{9}\frac{m_d}{m_s}. \tag{5.7.12}$$

This relation is, of course, quite well satisfied by experiment. But we pay the price of including several Higgs multiplets in the theory.

(ii) $B - L$ Conservation in Proton Decay

Next we want to display a very important property of the minimal SU(5) model that contains only {24}- and {5}-dimensional Higgs multiplets. We will show that even after symmetry breaking there exists an exact global symmetry in the theory which can be identified with the $B - L$ quantum number [32].

To see this, note that only {5}-dimensional Higgs {H} couples to fermions with couplings of the form $\bar{\psi}H^\dagger\chi$, $\chi\chi H$ [ψ and χ defined in eq. (5.4.1)]. These couplings are therefore invariant under a global $\tilde{U}(1)$ transformation with "charges" $\tilde{Q}$

$$\tilde{Q}(\psi) = +1,$$
$$\tilde{Q}(H) = -\tfrac{2}{3},$$
$$\tilde{Q}(\chi) = +\tfrac{1}{3}, \qquad \tilde{Q}(\phi) = 0. \tag{5.7.13}$$

Of course, the gauge interactions and the Higgs potential clearly respect this symmetry. The second stage of SU(5) symmetry breaking is implemented by $\langle H_5 \rangle \neq 0$ (the first stage being implemented by $\langle \phi \rangle \neq 0$ clearly leaves the symmetry intact). Therefore, it breaks $\tilde{U}(l)$; however, it keeps a linear combination with ω still invariant:

$$\omega = -\left(\frac{2}{\sqrt{15}}\lambda_{24} - \frac{3}{5}\tilde{Q}\right) \tag{5.7.14}$$

since $\omega(H_5) = 0$. Looking at ω it is apparent that

$$\omega = B - L. \tag{5.7.15}$$

Thus all baryon and lepton nonconserving processes in SU(5) must respect the $B - L$ quantum number. This is, of course, very important and can be used to distinguish the minimal SU(5) model from other models such as the minimal $SU(2)_L \times SU(2)_R \times U(1)_{B-L}$ model, where the baryon non-conserving processes respect the $\Delta B = 2$ selection rule. A typical allowed $\Delta B \neq 0$ process in SU(5) is $p \to e^+\pi^0$, $n \to e^+\pi^-$, etc., whereas in the latter case it is the $N \leftrightarrow \bar{N}$ oscillation and $N + P \to \pi$'s, etc. Note that in this connection the process $n \to e^-\pi^+$, which obeys the $\Delta(B + L) = 0$ selection rule, is indeed a rare process not allowed in any simple unified model. Another consequence of $B - L$ conservation is that the neutrino cannot acquire Majorana mass due to higher orders and must therefore remain massless to all orders. We wish to note two further points at this stage.

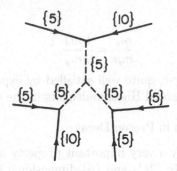

Figure 5.4.

(i) It has been pointed out by Weinberg [33] and by Wilczek and Zee [33] that exact $B - L$ conservation follows for all four-fermion operators (operators of dimension 6) that break the baryon number but which respect $SU(3)_c \times SU(2)_L \times U(1)$ symmetry and is therefore independent of any particular grand unification scheme. Thus, violation of $B - L$ conservation at the level of $\tau_{\Delta b \neq 0} \simeq 10^{30}$ yr must therefore mean that operators of dimension higher than six are involved, and must require the existence of mass scale below 10^{15} GeV to be observable in low-energy processes.

(ii) Another important point worth noting is that if we extend the SU(5) model by including an extra {15}-dimensional Higgs representation, then that leads to $B - L$ violation. To see this, denote the {15}-dimensional Higgs meson by S_{pq}: this allows additional couplings into the theory

$$\mathcal{L}_Y = h_{15}\psi_{pR}^T C^{-1}\psi_{qR}S_{pq}^+ + \lambda_{15}M_{15}H_p^+ H_q^+ S_{pq} + \text{h.c.} \qquad (5.7.16)$$

It is clear that this interaction violates the new $\tilde{U}(1)$ symmetry present in the minimal SU(5) model. To study its effects observe that (due to λ_{15} term, we obtain, regardless of the sign of its mass term)

$$\langle S_{55} \rangle = \kappa \neq 0. \qquad (5.7.17)$$

This leads to the Majorana mass term for the neutrino: $h_{15}\kappa$. A priori, there exists a rather weak restriction of the magnitude of κ arising from the fact that it contributes to the violation of the weak $\Delta I_W = 1/2$ rule (or $\rho_W = 1$) and must therefore satisfy the constraint

$$(g\kappa/m_{W_L})^2 \ll 1. \qquad (5.7.18)$$

This still, however, leaves κ big enough so that to understand the smallness of neutrino mass h_{15} may have to be tuned to a small value.

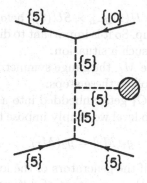

Figure 5.5.

(iii) $N - \bar{N}$ Oscillation

In this model $\Delta B = 2$ transitions, such as $N - \bar{N}$ oscillation, proceed via the Feynman graph shown in Fig. 5.4 and have strength $(h_{15}h_5^2/m_{15}^2 m_5^4)\lambda M$. But {15} also contributes a piece to the proton decay in addition to the usual {5} and X, Y-mediated graphs (Fig. 5.5). The sum total of this contribution is (symbolically) (Fig. 5.5).

$$\frac{h_5^2}{m_5^2} + \frac{h_5 h_{15} \lambda_{15} M_{15} \langle H_5 \rangle}{m_5^2 m_{15}^2}. \tag{5.7.19}$$

Observed stability of proton with $\tau_p > 10^{30}$ yr implies

$$h_5^2/m_5^2 < 10^{-30}\,\text{GeV}^{-2} \qquad \text{and} \qquad \frac{h^{15}}{h_5} \frac{\lambda_{15} M_{15} \langle H_5 \rangle}{m_{15}^2} < 1. \tag{5.7.20}$$

Equation (5.7.19) implies that

$$A_{N \leftrightarrow \bar{N}} \approx \left(\frac{h_{15} \lambda_{15} M_{15}}{m_{15}^2} \right) \left(\frac{h_{15}^2}{m_5^4} \right) < \left(\frac{h_5^4}{m_5^4} \right) \frac{1}{h_5 \langle H_5 \rangle}$$

$$\leq 10^{-60}\,\text{GeV}^{-4} \frac{1}{h_5 \langle H_5 \rangle} \leq 10^{-58}\,\text{GeV}^{-5} \tag{5.7.21}$$

(assuming $h_5 \langle H_5 \rangle$ which is a typical fermion mass ≤ 10 MeV). Such a small $N - \bar{N}$ transition amplitude would lead to $\tau_{N-\bar{N}} \geq 10^{28}$ yr which is insignificant. Thus, unless the model is made considerably complicated, SU(5) will not allow a significant $N - \bar{N}$ oscillation.

§5.8 Gauge Coupling Unification with Intermediate Scales before Grand Unification

An important aspect of grand unification is the possibility that there are intermediate symmetries before the grand unification symmetry is realized. For instance a very well motivated example is the presence of the gauge group $SU(2)_L \times SU(2)_R \times U(1)_{B-L} \times SU(3)_c$ before the gauge symmetry enlarges to the SO(10) group. So it is important to discuss how the evolution equations are modified in such a situation.

Suppose that at the scale M_I, the gauge symmetry enlarges. To take this into account, we need to follow these steps:

(i) If the smaller group G_1 gets embedded into a single bigger group G_2 at M_I, then at the one loop level we simply impose the matching condition:

$$g_1(M_I) = g_2(M_I) \tag{5.8.1}$$

(ii) On the other hand, if the generators of the low-scale symmetry arise as linear combinations of the generators of different high-scale groups as follows:

$$\lambda_1 = \Sigma_b p_b \theta_b, \tag{5.8.2}$$

then the coupling matching condition is

$$\frac{1}{g_1^2(M_I)} = \Sigma_b \frac{p_b^2}{g_b^2(M_I)}. \tag{5.8.3}$$

One can prove this as follows: for simplicity let us consider only the case where $G_2 = \Pi_b U(1)_b$, which at M_I breaks down to a single $U(1)$. Let this breaking occur via the v.e.v. of a single Higgs field ϕ with charges $(q_1, q_2, \ldots)$ under $U(1)$. The unbroken generator is given by:

$$Q = \Sigma p_a Q_a \tag{5.8.4}$$

with $\Sigma_c p_c q_c = 0$. The gauge field mass matrix after Higgs mechanism can be written as

$$M_{ab}^2 = g_a g_b q_a q_b \langle \phi \rangle^2 \tag{5.8.5}$$

This mass matrix has the following massless eigenstate, which can be identified with the unbroken U(1) gauge field:

$$A_\mu = \frac{1}{(\Sigma_b \frac{p_b^2}{g_b^2})^{1/2}} \Sigma \frac{p_b}{g_b} A_{\mu,b} \equiv N \Sigma_b \frac{p_b^2}{g_b^2} A_{\mu,b}. \tag{5.8.6}$$

To find the effective gauge coupling, we write

$$L \sim \Sigma g_b Q_b A_{\mu,b} \tag{5.8.7}$$
$$= \Sigma_b g_b (p_b Q + \ldots) N(\frac{p_b}{g_b} A_\mu + \ldots).$$

Collecting the coefficient of A_μ and using the normalization condition $\Sigma p_b^2 = 1$, we get the result we wanted to prove in eq. (5.8.3).

We will see the usefulness of this equation when we discuss the groups such as $SU(2)_L \times SU(2)_R \times SU(4)_c$ or SO(10) in the subsequent chapters.

To summarize this chapter, a beautiful illustration of the hypothesis of grand unification is the SU(5) model discussed in this chapter. Even though experimental results on proton decay have ruled out this model in its simplest form, the basic concept of grand unification may be right, and indeed, has generated a great deal of activity and excitement. Part of this excitement will be recorded in some of the subsequent chapters.

Exercises

5.1. Discuss the dependence of the grand unification scale, and hence proton lifetime, on the masses of Higgs bosons. Assuming these masses are arbitrary, give a few sample choices of Higgs multiplets that will bring the SU(5) model back into agreement with the proton lifetime as well as with $\sin^2 \theta - W$.

5.2. Taking into account the explicit form of the anomalous dimensions of the quark and lepton mass operators, show that, at the weak scale,

$$\frac{M_b}{M_\tau} \simeq 3.$$

5.3. Discuss CP-violating effects in baryon nonconserving processes. Do you need three generations to have nontrivial CP-violating phases in $\Delta B \neq 0$ processes?

5.4. Consider the SU(8) and SU(9) models as possible models for three and four generations of fermions. Discuss the choice of Higgs bosons that will lead to the correct symmetry-breaking patterns, as well as to fermion masses in these models.

5.5. Consider an alternative assignment of fermions to the five-dimensional representation of SU(5) such that

$$\{\bar{5}\} = \begin{pmatrix} u_1^c \\ u_2^c \\ u_3^c \\ \nu \\ e^- \end{pmatrix} \qquad \text{and} \qquad \{1\} = e_L^+.$$

(a) Show that the gauge group for this model is at least $SU(5) \times U(1)_{\tilde{Y}}$.
(b) Give the assignment of fermions to the $\{10\}$-dimensional representation of SU(5).
(c) Obtain the $\tilde{Y}$ quantum numbers of the different multiplets.
(d) Show that the model is free of anomalies and discuss symmetry breaking in this model.
(e) Discuss fermion masses arising from spontaneous symmetry breaking in this model. Show that one way to solve the problem of neutrion masses in this model is to have a $\{50\}$-dimensional representation participate in symmetry breaking.

References

[1] J. C. Pati and A. Salam, *Phys. Rev.* **D10**, 275 (1974).

[2] R. N. Mohapatra and J. C. Pati, *Phys. Rev.* **D11**, 2558 (1975);
G. Senjanovic and R. N. Mohapatra, *Phys. Rev.* **D12**, 1502 (1975).

[3] H. Georgi and S. L. Glashow, *Phys. Rev. Lett.* **32**, 438 (1974).

[3a] S. Okubo, *Phys. Rev.* **D16**, 3528 (1977);
J. Banks and H. Georgi, *Phys. Rev.* **D14**, 1158 (1976).

[3b] J. Chakrabarti, M. Popovic, and R. N. Mohapatra, *Phys. Rev.* **D21**, 3212 (1980).

[4] J. D. Bjorken, SLAC Summer Institute Lectures, 1976.

[5] L. F. Li, *Phys. Rev.* **D9**, 1723 (1974).

[6] J. Learned, F. Reines, and A. Soni, *Phys. Rev. Lett.* **43**, 907 (1979).

[7] M. Krishnaswamy, M. G. K. Menon, N. Mandal, V. Narasimhan, B. V. Sreekantan, S. Ito, and S. Miyake, *Phys. Lett.* **106B**, 339 (1981); **115B**, 349 (1892).

[8] R. Bionta et al., *Phys. Rev. Lett.* **51**, 27 (1983);
B. Cortez et al., *Phys. Rev. Lett.* **52**, 1092 (1984).

[9] M. Koshiba, *Proceedings of the Leipzig Conference*, 1894.

[10] E. Fiorini, *Proceedings of the Leipzig Conference*, 1984.

[11] D. Cline, Washington APS Meeting, 1984.

[12] M. L. Cherry et al., *Phys. Rev. Lett.* **47**, 1507 (1981).

[13] E. Gildener, *Phys. Rev.* **D14**, 1667 (1976).

[14] R. N. Mohapatra and G. Senjanovic, *Hadron. J.* **1**, 903 (1978);
I. Bars, *Orbis Scientia*, 1979;
K. T. Mahanthappa and D. Unger, *Phys. Lett.* **78B**, 604 (1978);
T. N. Sherry, *Phys. Lett.* **88B**, 76 (1979);
S. Weinberg, *Phys. Lett.* **82B**, 387 (1979).

[15] H. Georgi, H. Quinn, and S. Weinberg, *Phys. Rev. Lett.* **33**, 451 (1974).

[16] T. Appelquist and J. Carazone, *Phys. Rev.* **D11**, 2856 (1975).

[17] For further discussion of heavy particle effects at low energies, see D. Toussaint, *Phys. Rev.* **D18**, 1626 (1978);
G. Senjanovic and A. Sokorac, *Phys. Rev.* **D18**, 2708 (1978);
J. Kubo and S. Sakakibara, Dortmund preprint, 1980;
Y. Kazama and Y. P. Yao, Michigan preprint, 1980;
S. Weinberg, *Phys. Lett.* **91B**, 51 (1980);
L. Hall, *Nucl. Phys.* **B178**, 75 (1981);
N. P. Chang, A. Das, and J. P. Mercader, *Phys. Lett.* **39B**, 137 (1980).

[18] W. Marciano, Brookhaven preprint 34728, 1984;
Phys. Rev. **D20**, 274 (1979).

[19] P. B. MacKenzie and G. P. Lapage, *Phys. Rev. Lett.* **47**, 1244 (1981);
A. Buras, 1981 *International Symposium on Lepton and Photon Interactions*, 1981, p. 636.

[20] For excellent reviews see:
P. Langacker, *Phys. Rep.* **72**, 185 (1981);
J. Ellis, in *21st Scottish University Summer School* (edited by K. C. Bowler and
D. G. Sutherland), Edinburgh, 1981, p. 201;
W. Marciano, in *Fourth Workshop on Grand Unification* (edited by H. A. Weldon et al.), Birkhauser, Boston, 1983, p. 13.

[21] U. Amaldi et al., *Phys. Rev.* **D36**, 1385 (1987);
G. Costa et al., *Nucl. Phys.* **B297**, 244 (t988).

[22a] G. Arnison et al., *Phys. Lett.* **126B**, 398 (1983);
P. Bagnaia et al., *Phys. Lett.* **129B**, 130 (1983).

[22b] P. Sinervo, *Proceedings of The 14th International Symposium on Lepton and Photon Interactions*, Stanford, 1989, World Scientific, Singapore.

[22c] B. Adeva et al., *Phys. Lett.* **231B**, 509 (1989);
D. Decamp et al., *Phys. Lett.* **231B**, 519 (1989);
M. Z. Akrawy et al., *Phys. Lett.* **231B**, 530 (1989);
P. Aarnio et al., *Phys. Lett.* **231B**, 539 (1989).

[22d] G. Feldman et al., *Phys. Rev. Lett.* **63**, 724 (1989).

[23] A. Buras, J. Ellis, M. K. Gaillard, and D. V. Nanopoulos, *Nuel. Phys.* **B135**, 66 (1978).

[24] T. Goldman and D. A. Ross, *Nucl. Phys.* **B171**, 273 (1980);
C. Jarlskog and F. J. Yndurain, *Nuel. Phys.* **B149**, 29 (1979);
M. Machachek, *Nucl. Phys.* **B159**, 37 (1979);
M. B. Gavela, A. LeYouanc, L. Oliver, O. Pene, and J. C. Raynal, Orsay preprint, PTHE 80/6, 1980.

[25] A. Din, G. Girardi, and P. Sorba, *Phys. Lett.* **91B**, 77 (1980);
J. Donoghue, *Phys. Lett.* **92B**, 99 (1980);
E. Golowich, University of Massachusetts preprint, 1980.

[26] For a recent review and references see
P. Langacker, *Proceedings of the Fermilab Conference on Inner Space/Outer Space Connection*, 1984.

[27] R. N. Mohapatra, *Phys. Rev. Lett.* **43**, 893 (1979).

[28] J. Ellis, M. K. Gaillard, and D. V. Nanopoulos, *Phys. Lett.* **80B**, 360 (1979).

[29] C. Jarlskog, *Phys. Lett.* **82B**, 401 (1979).

[29a] G. Cook, K. T. Mahanthapa, and M. Sher, *Phys. Lett.* 90B, 298 (1980).

[30] P. Frampton, S. Nandi, and J. Scanio, *Phys. Lett.* **85B**, 255 (1979).

[31] H. Georgi and C. Jarlskog, *Phys. Lett.* **86B**, 297 (1979).

[32] M. T. Vaughn, Northeastern University preprint, 1979.

[33] S. Weinberg, *Phys. Rev. Lett.* **43**, 1566 (1979);
F. Wilczek and A. Zee, **43**, 1571 (1979).

6

Left–Right Symmetric Models of Weak Interactions and Massive Neutrinos

§6.1   Why Left–Right Symmetry?

While the standard electroweak model, based on the spontaneously broken local symmetry $SU(3)_c \times SU(2)_L \times U(1)_Y$, has been extremely successful in the description of low-energy weak phenomena, it leaves many questions unanswered as discussed in Chapter 3. One of them has to do with understanding of the origin of parity violation in low-energy weak-interaction processes, i.e., while all other forces in nature are parity conserving, why are the weak forces apparently not or are they really parity conserving at a fundamental level and we just do not see it. The second one, of a more phenomenological nature but an urgent one, has to do with the origin of neutrino masses, for which now there is convincing evidence from neutrino oscillation searches. In view of the historical background, where the $V - A$ structure of weak interactions were first motivated from the masslessness of the neutrino, one would be led to suspect that both the above questions are really connected. As we will see in this chapter, in the framework of the left–right symmetric models, both these questions receive a unified answer.

The basic premise of the left–right symmetric models is that the fundamental weak-interaction Lagrangian is invariant under parity symmetry at energy scales much above the standard model scale of ~ 100 GeV and the parity asymmetry observed in nature in processes such as β decay and μ decay arise from vacuum being noninvariant under parity symmetry. This is a fundamentally new approach to weak interactions, which puts the origin of parity violation on a very different footing than the standard model. An immediate consequence of this hypothesis is that there must be right-handed neutrinos in nature and, as a consequence, neutrinos must be massive. Thus neutrino mass and left–right symmetry of weak interactions would seem to go hand in hand.

Within the framework of gauge theories, this idea has found its realization in the $SU(2)_L \times SU(2)_R \times U(1)_{B-L}$ models [1] constructed in 1973–1974. An important feature of these models is that, at low energies, they reproduce all the features of the successful standard model based on the $SU(2)_L \times U(1)$ gauge symmetry; but as we move up in energies new effects associated with parity invariance of the Lagrangian (such as a second neutral Z boson, right-handed charged currents, right-handed neutrino) are expected to appear. The mass scale at which these new effects appear (the mass of W_R and Z') can be related to the neutrino mass, which makes the connection between left–right models and the standard model via the small neutrino mass an elegant one.

There exist several other considerations having to do with weak interactions that make the left–right symmetric models more attractive than the standard model. They are (i) quark–lepton symmetry of the fundamental weak interactions: (ii) a simple electric charge formula in terms of physical quantum numbers such as weak isospin and $B - L$ quantum numbers.

We will return to neutrino masses [2] later in this chapter but here we wish to emphasize as a motivation for left–right symmetric theories that there is now convincing evidence for neutrino masses from observation of the phenomenon of neutrino oscillation from studies of atmospheric neutrinos as well as the neutrinos from the core of the Sun [3]. The basic observation is that there is a deficit in the number of electron neutrinos produced in the core of the Sun as well as a deficit in the upward-going muon neutrinos produced in the upper atmosphere in the collision of the cosmic rays with the atmospheric protons. The simplest interpretation of these deficits is that the neutrinos (ν_e's in the solar case and ν_μ's in the atmospheric case) are oscillating into other species of neutrinos that the present detectors cannot detect for various reasons. It is well known from quantum mechanics that for neutrinos to be able to oscillate, they must have nonzero mass. Evidence for neutrino oscillation is therefore evidence for neutrino mass. This is the first clear evidence for physics beyond the standard model and as we will argue in this chapter, it provides a strong argument for the existence of the left–right symmetric gauge structure of nature at a higher scale. Detailed discussion of neutrinos and their properties is the subject of many books [3] and we do not enter into a detailed discussion of this here. Instead we want to discuss the left–right symmetric models and their implications and constraints for the parameters of the standard model and toward the end of this chapter briefly return to some salient features of neutrinos.

There are several other reasons to suspect that the standard model will eventually become part of a left–right symmetric gauge structure. First, in models where weak-interaction symmetries arise out of a more fundamental substructure of quarks and leptons, and if the forces at the substructure level are assumed to be similar to those operating in nuclear physics, i.e., QCD, then $SU(2)_L \times SU(2)_R \times U(1)_{B-L}$ arises as a more natural weak-interaction symmetry than $SU(2)_L \times U(1)$. Second, fundamental Planck scale theories such as string theories more easily lead to a left–right symmetric gauge structure than the standard model gauge structure.

To return to an earlier point, we note that a deficiency of the standard model is the lack of any physical meaning of the hypercharge $U(1)$ generator, which is arbitrarily adjusted to give whatever charge is desired for a particle. On the other hand, in the left–right symmetric models $U(1)$ generator can be identified as the $B - L$ quantum number [4] so that all of the weak-interaction symmetry generators have a physical meaning. As if suggesting a deeper symmetry structure in the $SU(2)_L \times U(1)_Y$ model once the right-handed neutrinos are included, the only anomaly-free quantum number left ungauged is $B - L$, and once $B - L$ is included as a gauge generator, the weak gauge group becomes $SU(2)_L \times SU(2)_R \times U(1)_{B-L}$, and electric charge is given by [4]

$$Q = I_{3L} + I_{3R} + \frac{B - L}{2}. \qquad (6.1.1)$$

This relation is so powerful that several phenomenological consequences of the left–right model can be deduced simply by analyzing the group theory of this formula, as we will see subsequently.

Furthermore, as we saw in Chapter 4, the parity invariance of these theories helps to provide solutions to the strong CP problem without invoking an axion, a possibility that may become much more essential as searches for the axion continue giving negative results,

In this chapter we will study the left–right symmetric models of weak interactions based on the gauge group $SU(2)_L \times SU(2)_R \times U(1)_{B-L}$ and their various implications for particle physics.

In Section 6.2 we introduce the model, discuss the bounds on the right-handed gauge boson masses, and possible ways of detecting the $W_R^\pm$ and Z_R in high-energy experiments. In Section 6.3 we discuss neutrino masses, lepton numbers violation, and other implications for low-energy experiments. In Section 6.4 we discuss neutrino mass and neutrinoless double β decay. In Section 6.5 we discuss the selection rules for baryon-number-violating processes and higher unification constraints on the scale of right-handed currents. In Section 6.6 we discuss some recent work where parity and $SU(2)_R$ scales are decoupled, and their impact on physics. In Section 6.7 we discuss an alternative realization of the idea of left–right symmetry, where neutrino is a Dirac rather than a Majorana mass.

§6.2 The Model, Symmetry Breaking, and Gauge Boson Masses

The left–right symmetric models of weak interactions are based on the gauge group $SU(2)_L \times SU(2)_R \times U(1)_{B-L}$ with the following assignment of fermions to the gauge group: denoting $Q \equiv \binom{u}{d}$ and $\psi \equiv \binom{\nu_e}{e}$ we have

$$Q_L: \quad (\tfrac{1}{2}, 0, \tfrac{1}{3}), \qquad Q_R: \quad (0, \tfrac{1}{2}, \tfrac{1}{3}),$$
$$\psi_L: \quad (\tfrac{1}{2}, 0, -1), \qquad \psi_R: \quad (0, \tfrac{1}{2}, -1). \tag{6.2.1}$$

As mentioned in the introduction, the $U(1)$ generator corresponds to the $B - L$ quantum numbers of the multiplet. The electric charge formula is given by [4]

$$Q = I_{3L} + I_{3R} + \frac{B - L}{2}. \tag{6.2.2}$$

Owing to the existence of the discrete parity symmetry, the model has only two gauge coupling constants prior to the symmetry breaking, i.e., $g_2 \equiv g_{2L} = g_{2R}$ and g' and, as before, we can define

$$\sin \theta_W = e/g_{2L} \tag{6.2.3}$$

and we will parametrize the neutral-current Hamiltonian in terms of θ_W.

Because the weak-interaction symmetry is $SU(2)_L \times SU(2)_R \times U(1)_{B-L} \times P$, its breaking to the standard model gauge group $SU(2)_L \times U(1)$ model can proceed in one of the following two ways: one can break both parity and $SU(2)_R \times U(1)_{B-L}$ at one stage or one can break parity before one breaks the gauge symmetry [5], i.e., $SU(2)_L \times SU(2)_R \times U(1)_{B-L} \times P \xrightarrow[M_P]{} SU(2)_L \times SU(2)_R \times U(1)_{B-L} \xrightarrow[M_{W_R}]{} SU(2)_L \times U(1)_Y \xrightarrow[M_{W_L}]{} U(1)_{em}$. In the second case, because at the first stage only the parity symmetry is broken and weak gauge symmetry is unbroken the W_L and W_R remain massless; their masses arise at the subsequent stages of symmetry breaking. The parity breaking at the first stage would manifest as different gauge couplings $g_{2L} \neq g_{2R}$ at $\mu \geq M_{W_R}$. For simple choice of Higgs multiplets, both parity and $SU(2)_R$ break at the same scale. If, however, one wanted to break parity separately from $SU(2)_R \times U(1)_{B-L}$ symmetry, it could be achieved by including in the theory a parity odd, neutral scalar field σ, which has a nonzero vacuum expectation value. We discuss this in Chapter 7.

In the bulk of this chapter we will assume that $M_P = M_{W_R}$, which happens with the minimal choice of the Higgs multiplets discussed below.

In order to study the detailed implications of the model, we start by writing down the Lagrangian, which consists of three parts:

$$\mathcal{L} = \mathcal{L}_{kin} + \mathcal{L}_Y - \mathcal{V}(\phi). \tag{6.2.4}$$

$\mathcal{L}_{kin}$ is dictated by the gauge group and the assignment of fermions and Higgs boson to the gauge group. Since we have not yet specified the Higgs boson sector of the theory, we present first only the fermion kinetic energy terms, using eq. (6.2.1),

$$\mathcal{L}_{\text{kin}}^f = -\bar{Q}_L \gamma^\mu \left(\partial_\mu - \frac{ig}{2} \boldsymbol{\tau} \cdot \mathbf{W}_{L\mu} - \frac{ig'}{6} B_\mu \right) Q_L$$

$$- \bar{\psi}_L \gamma^\mu \left(\partial_\mu - \frac{ig}{2} \boldsymbol{\tau} \cdot \mathbf{W}_{L\mu} + \frac{ig'}{2} B_\mu \right) \psi_L$$

$$- \bar{Q}_R \gamma^\mu \left(\partial_\mu - \frac{ig}{2} \boldsymbol{\tau} \cdot \mathbf{W}_{R\mu} - \frac{ig'}{6} B_\mu \right) Q_R$$

$$- \bar{\psi}_R \gamma^\mu \left(\partial_\mu - \frac{ig}{2} \boldsymbol{\tau} \cdot \mathbf{W}_{R\mu} + \frac{ig'}{2} B_\mu \right) \psi_R. \tag{6.2.5}$$

Note that under the interchange $L \leftrightarrow R$, the above Lagrangian is completely invariant. Before discussing the weak interactions in this model, we first want to show how the Higgs mechanism reduces the $SU(2)_L \times SU(2)_R \times U(1)_{B-L}$ gauge group to the standard model. This leads to the following fermion gauge boson interaction:

$$\mathcal{L}_{\text{gauge}} = \frac{ig}{2} [\bar{Q}_L \gamma_\mu \boldsymbol{\tau} Q_L + \bar{\psi}_L \gamma_\mu \boldsymbol{\tau} \psi_L] \cdot \mathbf{W}_{\mu L}$$

$$+ \frac{ig}{2} [\bar{Q}_R \gamma_\mu \boldsymbol{\tau} Q_R + \bar{\psi}_R \gamma_\mu \boldsymbol{\tau} \psi_R] \cdot \mathbf{W}_{\mu R}$$

$$+ \frac{ig}{2} [\tfrac{1}{3} \bar{Q} \gamma_\mu Q - \bar{\psi} \gamma_\mu \psi] B_\mu. \tag{6.2.6}$$

We use the Higgs multiplets [6, 7] dictated by the intuitive dynamical requirement that they be bilinear in the basic fermionic multiplets. Then the unique minimal set required to break the symmetry down to the $U(1)_{\text{em}}$ is

$$\Delta_L(1, 0, +2) + \Delta_R(0, 1, +2)$$

and

$$\psi(\tfrac{1}{2}, \tfrac{1}{2}, 0) \tag{6.2.7}$$

under left–right symmetry $\Delta_L \leftrightarrow \Delta_R$ and $\phi \leftrightarrow \phi^+$.

We now show [6] that, for a range of parameters, an exactly left–right symmetric potential would lead to parity-violating minima that will break the gauge symmetry. Let us ignore the ϕ multiplet, which will be used to break $SU(2)_L \times U(1)_Y$, in the first stage. The most general parity and gauge-invariant potential involving $\Delta_{L,R}$ is

$$V(\Delta_L, \Delta_R) = -\mu^2 Tr(\Delta_L^\dagger \Delta_L + \Delta_R^\dagger \Delta_R)$$

$$+ \rho_1 [(Tr(\Delta_L^\dagger \Delta_L))^2 + (Tr(\Delta_R^\dagger \Delta_R))^2]$$

$$+ \rho_2 [Tr(\Delta_L^\dagger \Delta_L \Delta_L^\dagger \Delta_L) + Tr(\Delta_R^\dagger \Delta_R \Delta_R^\dagger \Delta_R)]$$

$$+ \rho_3[\text{Tr}(\Delta_L^\dagger \Delta_L)\,\text{Tr}(\Delta_R^\dagger \Delta_R)]$$
$$+ \rho_4[\text{Tr}(\Delta_L^\dagger \Delta_L^\dagger)\,\text{Tr}(\Delta_L \Delta_L)$$
$$+ \text{Tr}(\Delta_R^\dagger \Delta_R^\dagger)\,\text{Tr}(\Delta_R \Delta_R)], \qquad (6.2.8)$$

where we write

$$\Delta_{L,R} = \frac{1}{\sqrt{2}}\tau \cdot \delta_{L,R} \equiv \begin{pmatrix} \delta^+/\sqrt{2} & \delta^{++} \\ \delta^0 & -\delta^+/\sqrt{2} \end{pmatrix}_{L,R}. \qquad (6.2.9)$$

By suitable symmetry transformation, we can choose the v.e.v. of $\Delta_{L,R}$ as follows:

$$\langle \Delta_{L,R} \rangle = \begin{pmatrix} 0 & 0 \\ v_{L,R} & 0 \end{pmatrix}. \qquad (6.2.10)$$

The potential V then reduces to

$$V(v_L, v_R) = -\mu^2(v_L^2 + v_R^2) + (\rho_l + \rho_2) + (v_L^4 + v_R^4) + \rho_3 v_L^2 v_R^2. \qquad (6.2.11)$$

Parametrizing $v_L = v\sin\alpha$ and $v_R = v\cos\alpha$ and differentiating V with respect to α, we get the extremum condition

$$[\rho_3 - 2(\rho_1 + \rho_2)]v\sin 2\alpha \cos 2\alpha = 0. \qquad (6.2.12)$$

For $\rho_3 \neq 2(\rho_1 + \rho_2)$, the solutions of eq. (6.2.11) are $\alpha = 0, \pi/4, \pi/2 \ldots$. $\alpha = 0$ corresponds to $v_L = 0$ and $v_R \neq 0$, whereas $\alpha = \pi/4$ corresponds to $v_R = 0$ and $v_L = 0$. We thus see that $\alpha = 0$ corresponds to the desired parity-violating solution. Taking the second derivative of V with respect to α and demanding positivity, we see that, for $\rho_3 > 2(\rho_1 + \rho_2)$, $\alpha = 0$ and $\pi/2$ corresponds to the minimum whereas $\alpha = \pi/4$ is a maximum. Thus, for a range of parameters, we get the correct unbroken symmetry group of the standard model. Once we include the ϕ multiplet, the potential becomes

$$V(\Delta_L, \Delta_R, \Phi) = V(\Delta_L, \Delta_R) - \sum_{i,j} \mu_{ij}^2\,\text{Tr}(\phi_i^\dagger \phi_j)$$

$$+ \sum_{i,j,k,l} \lambda_{ijkl}\,\text{Tr}(\phi_i^\dagger \phi_j)\,\text{Tr}(\phi_k^\dagger \phi_l)$$

$$+ \sum_{i,j,k,l} \lambda'_{ijkl}\,\text{Tr}(\phi_i^\dagger \phi_j \phi_k^\dagger \phi_l)$$

$$+ \sum \alpha_{ij}\,\text{Tr}[\phi_i^\dagger \phi_j]\,\text{Tr}(\Delta_L^\dagger \Delta_L + \Delta_R^\dagger \Delta_R)$$

$$+ \sum \beta_{ij}\,\text{Tr}[\phi_i \phi_j^\dagger \Delta_L^\dagger \Delta_L + \phi_i^\dagger \phi_j \Delta_R^\dagger \Delta_R]$$

$$+ \sum [\gamma_{ij}\,\text{Tr}\,\Delta_L^\dagger \phi_i \Delta_R \phi_j^\dagger + \text{h.c.}], \qquad (6.2.13)$$

where

$$\phi_1 = \begin{pmatrix} \phi_1^0 & \phi_1^\dagger \\ \phi_2^- & \phi_2^0 \end{pmatrix}, \qquad (6.2.14)$$

$$\phi_2 = \tau_2 \phi_1^* \tau_2 = \begin{pmatrix} \phi_2^{0*} & -\phi_2^\dagger \\ -\phi_1^- & \phi_1^{0*} \end{pmatrix}. \tag{6.2.15}$$

Note that under $SU(2)_L \times SU(2)_R$ transformation

$$\Delta_L \to U_L \Delta_L U_L^\dagger,$$
$$\Delta_R \to U_R \Delta_R U_R^\dagger,$$
$$\phi \to U_L \phi W_R^\dagger. \tag{6.2.16}$$

Equation (6.2.13) is an extremely complicated potential, but the minimum of this potential can be studied assuming, for $\langle \Delta_R \rangle$, $\langle \Delta_L \rangle$, $\langle \phi \rangle$, the following general form:

$$\langle \Delta_{R,L} \rangle = \begin{pmatrix} 0 & 0 \\ v_{R,L} & 0 \end{pmatrix},$$

$$\langle \phi \rangle = \begin{pmatrix} \kappa & 0 \\ 0 & \kappa' \end{pmatrix} e^{i\alpha}. \tag{6.2.17}$$

First note that this pattern is true for a large range of values of the parameters in the theory. Further note that both κ and κ' are nonzero and

$$v_L = \left(\frac{\gamma_{12}}{2(\rho_1 + \rho_2) - \rho_3} \right) \frac{\kappa^2}{v_R} \tag{6.2.18}$$

if $\kappa' \ll \kappa$, which we will assume for simplicity from here on. It follows from eq. (6.2.16) that, since $\kappa \ll v_R$, we get $v_L \ll \kappa$. This is a seesaw-like formula for vacuum expectation values and can be useful in understanding, in the gauge theory context, why certain physical parameters are small compared to others.

A further consequence of the general potential (6.2.13) is that, in the limit of κ/v_R small, the spontaneous CP phase α vanishes unless we allow arbitrary fine tuning of parameters in the potential.

Charged Gauge Boson Masses

Let us now study the effect of symmetry breaking on the gauge boson masses of the theory: In the charged gauge boson sector we obtain, for the $W_L^\dagger - W_R^\dagger$ mass matrix,

$$\begin{pmatrix} \frac{1}{2}g^2(\kappa^2 + \kappa'^2 + 2v_L^2) & g^2\kappa\kappa' \\ g^2\kappa\kappa' & \frac{1}{2}g^2(\kappa^2 + \kappa'^2 + 2v_R^2) \end{pmatrix}. \tag{6.2.19}$$

The eigenstates of this matrix denote the physical $W_{1,2}$ bosons

$$W_1 = W_L \cos\zeta + W_R \sin\zeta, \quad M_{W_1}^2 \simeq \tfrac{1}{2}g^2(\kappa^2 + \kappa'^2),$$
$$W_2 = -W_L \sin\zeta + W_R \cos\zeta, \quad M_{W_2}^2 \simeq \tfrac{1}{2}g^2(\kappa^2 + \kappa'^2 + 2v_R^2),$$

with

$$\tan 2\zeta = \frac{2\kappa\kappa'}{v_R^2 - v_L^2}. \tag{6.2.20}$$

The charged-current weak-interactions can now be written as

$$\mathcal{L} = \frac{g}{\sqrt{2}}(\bar{u}_L\gamma_\mu d_l + \bar{\nu}_L\gamma_\mu e_L)(W_1^{\dagger\mu}\cos\zeta + W_2^{\dagger\mu}\sin\zeta)$$

$$+ \frac{g}{\sqrt{2}}(\bar{u}_R\gamma_\mu d_R + \bar{\nu}_R\gamma_\mu e_R)(W_2^{\dagger\mu}\cos\zeta - W_1^{\dagger\mu}\sin\zeta)$$

$$+ \text{h.c.} \tag{6.2.21}$$

We have suppressed the generation index in eq. (6.2.21). Furthermore, in the presence of quark mixings, the weak quark and lepton currents will include mixing angles as well as CP-violating phases.

The effective four-Fermi interaction in the low-energy limit ($q^2 \ll M_{W_{1,2}}^2$) can be written as

$$H_{wk}^{cc} = \frac{4G_F}{\sqrt{2}}\left\{[\cos^2\zeta + \eta\sin^2\zeta]J_L^{\dagger\mu}J_{\mu L}^- + [\eta\cos^2\zeta + \sin^2\zeta]J_R^{\dagger\mu}J_{\mu R}^-\right.$$

$$\left. + \cos\zeta\sin\zeta(1-\eta)(J_L^{\dagger\mu}J_{\mu R}^- + J_R^{\dagger\mu}J_{\mu L}^-)\right\}, \tag{6.2.22}$$

where $J_{\mu L} = \bar{u}_L\gamma_\mu d_L + \bar{\nu}\gamma_\mu e_L$ and similarly for $J_{\mu R}$; $\eta = (M_{W_1}^2/M_{W_2})^2$. Now the observed predominance of $V - A$ current at low energies can be understood if $v_R \gg \kappa, \kappa'$, which implies that both η and ζ are much less than one and eq. (6.2.22) reduces to almost pure $V - A$ theory. Turning this around, we can use the accuracy of low-energy weak-interaction data to obtain a lower bound on the mass of the W_R boson. Also note that for $\zeta \ll 1$, $W_1 \simeq W_L$ and $W_2 \simeq W_R$.

Neutral Gauge Boson Masses

Coming to the neutral weak boson W_{3L}, W_{3R}, and B, their mass matrix is given by

$$\begin{array}{ccc} W_{3L} & W_{3R} & B \end{array}$$
$$\begin{pmatrix} g^2/2(\kappa^2 + \kappa'^2 + 4v_L^2) & -g^2/2(\kappa^2 + \kappa'^2) & -2gg'v_L^2 \\ -g^2/2(\kappa^2 + \kappa'^2) & g^2/2(\kappa^2 + \kappa'^2 + 4v_R^2) & -2gg'v_R^2 \\ -2gg'v_L^2 & -2gg'^2v_R^2 & 2g'^2(v_L^2 + v_R^2) \end{pmatrix}.$$

To obtain the exact eigenstates of this matrix, we work in the following basis (define $\sin\theta_W = e/g$):

$$A = \sin\theta_W(W_{3L} + W_{3R}) + \sqrt{\cos 2\theta_W}B,$$

$$Z_L = \cos\theta_W W_{3L} - \sin\theta_W \tan\theta_W W_{3R} - \tan\theta_W\sqrt{\cos 2\theta_W}B, \tag{6.2.23}$$

$$Z_R = \frac{\sqrt{\cos 2\theta_W}}{\cos\theta_W}W_{3R} - \tan\theta_W B.$$

Here A is the photon that remains massless after symmetry breaking and is an exact eigenstate. The remaining exact eigenstates are

$$Z_1 = Z_L \cos\xi + Z_R \sin\xi,$$
$$Z_2 = -Z_L \sin\xi + Z_R \cos\xi,$$
$$(6.2.24)$$

where

$$\tan 2\xi \simeq 2\sqrt{\cos 2\theta_W}\,(M_{Z_L}^2/M_{Z_R}^2), \tag{6.2.25}$$

$$M_{Z_L}^2 \simeq \frac{g^2}{2\cos^2\theta_W}(\kappa^2 + \kappa'^2 + 4v_L^2),$$

$$M_{Z_R}^2 \simeq \frac{g^2}{2\cos^2\theta_W \cos 2\theta_W}[4v_R^2 \cos^4\theta_W + (\kappa^2 + \kappa'^2)\cos^2 2\theta_W + 4v_L^2 \sin^4\theta_W].$$
$$(6.2.26)$$

As $v_R \gg \kappa, \kappa'$, ξ approaches zero and we have $Z_1 \simeq Z_L$ and $Z_2 \simeq Z_R$. Then using eqs. (6.2.6) and (6.2.23), the neutral-current Lagrangian in the leading order can be written down as

$$\mathcal{L}_{\text{N.C}} = \frac{g}{\cos\theta_W}\left[\left\{J_L^\mu - \frac{\xi}{\sqrt{\cos 2\theta_W}}(\sin^2\theta_W J_L^\mu + \cos^2\theta_W J_R^\mu)\right\}Z_{1\mu}\right.$$

$$\left. + \frac{1}{\sqrt{\cos 2\theta_W}}\{\sin^2\theta_W J_L^\mu + \cos^2\theta_W J_R^\mu\}Z_{2\mu}\right], \tag{6.2.27}$$

where

$$J_{L,R}^\mu = \sum_f \bar{f}\gamma^\mu[I_{3L,R} - Q\sin^2\theta_W]f. \tag{6.2.28}$$

Fermion Masses and Mixings

As in the standard model, fermion masses in the left–right model arise also from the Yukawa coupling between quarks, leptons, and the Higgs bosons responsible for gauge symmetry breaking. The most general gauge-invariant Yukawa coupling is given by

$$\mathcal{L}_Y = \sum_{i,j}(h_{ij}^q \bar{Q}_{Li}\phi Q_{Rj} + \tilde{h}_{ij}^q \bar{Q}_{Li}\tilde{\phi}Q_{Rj})$$

$$+ (h_{ij}^l \bar{\psi}_{Li}\phi\psi_{Rj} + \tilde{h}_{ij}^{(l)} \bar{\psi}_{Li}\tilde{\phi}\psi_{Rj})$$

$$+ f_{ij}(\psi_{Li}^T C^{-1}\tau_2\Delta_L\psi_{ij} + \psi_{Ri}^T C^{-1}\tau_2\Delta_R\psi_{Rj})$$

$$+ \text{h.c.} \tag{6.2.29}$$

On substituting the vacuum expectation values from eq. (6.2.17), we get, for charged fermions, the following mass matrices:

$$M_{ij}^u = h_{ij}^{(q)}\kappa e^{i\alpha} + \tilde{h}_{ij}^{(q)}\kappa' e^{-i\alpha},$$

$$M_{ij}^d = h_{ij}^{(q)}\kappa e^{-i\alpha} + \tilde{h}_{ij}^{(q)}\kappa' e^{i\alpha}, \tag{6.2.30}$$

$$M_{ij}^e = h_{ij}^{(l)}\kappa e^{-i\alpha} + \tilde{h}_{ij}^{(l)}\kappa' e^{i\alpha}.$$

We defer the discussion of neutrino masses to a later section.

The invariance of the Yukawa coupling under parity symmetry (i.e., $Q_L \leftrightarrow Q_R$, $\psi_L \leftrightarrow \psi_R$, $\phi \leftrightarrow \phi^\dagger$) requires that

$$h_{ij}^{(a)} = h_{ji}^{*(a)}, \qquad (6.2.31)$$

where a goes over all the Yukawa couplings involving ϕ and $\tilde{\phi}$. This property leads to interesting constraints on the mass matrices, which were touched upon in Chapter 4 and we repeat them here for completeness.

(i) h complex, $\alpha = 0$

In this case, we have

$$M^{(a)} = M^{(a)\dagger}, \qquad a = u, d, e. \qquad (6.2.32)$$

Since Hermitian matrices are diagonalized by unitary transformations, i.e., $UM^{(a)}U^\dagger = D^{(a)}$, we see that the mixing angles in the left- and right-handed quark sectors are identical. This case is referred to in the literature as manifest left–right symmetry.

(ii) h real, $\alpha \neq 0$

The reality of h is guaranteed by demanding CP invariance of the Lagrangian prior to symmetry breaking. In this case, eq. (6.2.30) implies that the mass matrices are complex and symmetric and are diagonalized by [8]

$$UM^{(a)}U^T K = D^{(a)}, \qquad (6.2.33)$$

where K is a diagonal unitary matrix. As a result, the left- and right-handed mixing angles are equal but the corresponding phases are different phases. This case is called pseudo-manifest.

In these two cases, analyzing weak-interaction processes such as muon and beta decay to derive lower bounds on M_{W_R} becomes easier, since only the CKM mixing angles are present in both the left- and right-handed sectors and existing experiments have given a great deal of information on them.

Another case, which does not emerge in the simplest left–right symmetric models but cannot in general be ruled out, is called nonmanifest left–right symmetry. In this case, the right-handed charged-current mixing angles are completely arbitrary. For this to happen, we must have mass matrix, which is neither symmetric nor Hermitian. This can be obtained if there are two ϕ's (denoted by ϕ_1 and ϕ_3) and under parity $\phi_1 \leftrightarrow \tilde{\phi}_3^\dagger$. It is left as an exercise to the reader to check that, in this case, left- and right-handed mixing angles are indeed different.

§6.3 Limits on M_{Z_R} and m_{W_R} from Charged-Current Weak Interactions

We are now ready to discuss the bounds on the masses of the neutral Z_R and charged W_R gauge bosons, as well as the left–right mixing parameters from weak-interaction data. The most model-independent limits are obtained on M_{Z_R} from analysis of existing neutral-current data, as well as $p\bar{p}$ collider data [9]. The basic strategy is to look at the effect of $Z_L - Z_R$ mixing on the shift of the Z_L mass and therefore on the ρ parameter. Another way is to use the knowledge of the leptonic decay of Z_R and its hardronic coupling to deduce the production probability of

$$p\bar{p} - Z_R \quad + X,$$
$$\hookrightarrow \mu\bar{\mu}$$

and from this derive limits on M_{Z_R}. The best lower bound on M_{Z_R} appears to be $M_{Z_R} \geq 370$ GeV. Stronger limits can be derived for the case of Dirac neutrinos from emission of ν_R's from supernova core. These bounds are of order 2–3 TeV.

A model-independent limit on the mass of the right-handed gauge boson can be obtained by looking for $V + A$ current effects in various low-energy processes. Three classes of probes are available:

(i) purely leptonic processes such as $\mu \to e\bar{\nu}_e\nu_\mu$;

(ii) semileptonic decays such as $n \to pe^-\bar{\nu}_e$, $K \to \pi\mu\bar{\nu}_\mu$, $\pi e\bar{\nu}_e$, etc.; and

(iii) nonleptonic processes such as $K \to 2\pi$, $K \to 3\pi$, hyperon decays, and $K_L - K_S$ masses.

We will analyze all these various processes to put bounds on the two parameters that characterize the presence of the right-handed currents, i.e., m_{W_R} and ζ. It turns out that in the processes involving neutrinos the bound on m_{W_R} depends on the nature of the right-handed component of the neutrino. The point is that, since the neutrino is an electrically neutral fermion, it can admit two kinds of mass terms:

(i) Dirac mass for which

$$L_D = m_D\bar{\nu}_L\nu_R + \text{h.c.}; \quad \text{or} \tag{6.3.1}$$

(ii) Majorana masses

$$L_M = m_L\nu_L^T C^{-1}\nu_L + m_R\nu_R^T C^{-1}\nu_R + \text{h.c.} \tag{6.3.2}$$

or a combination of both [10]. In the Dirac neutrino case both helicity components have the same mass: therefore, there is no kinematical obstacle to the manifestation of $V + A$ currents in lepton involved processes. However, in the case of the Majorana neutrinos, m_R is an arbitrary parameter and therefore there exists the possibility that $V + A$ current-mediated processes involving neutrinos may be kinematically forbidden if m_R is large. In fact, an interesting theoretical possibility known as the seesaw mechanism [11] (see later) is that $m_R \approx m_{W_R}$ in which case most lepton-involved processes will arise either from the pure left-handed current terms [i.e., first term in eq. (6.2.22)] or the left–right mixing term [third term in eq. (6.2.22)]. If, however, m_R is chosen in the MeV range [12], by adjustment of parameters, some semileptonic processes may provide a useful bound on m_{W_R}. Let us proceed to discuss these bounds for each case.

Case (i). Dirac Neutrino Case

Bounds on m_{W_R} for the case of the Dirac neutrino were first obtained [13] in 1977. An analysis of various leptonic and strangeness-conserving semileptonic decays indicated that the most severe bounds on η and ζ came from the measurement of electron polarization in ^{14}O-decay and the measurement of the Michael ρ parameter in μ decay. The two best experimental values [14, 15] are

$$\rho_{\text{expt}} = 0.7518 \pm 0.0026$$

and $P/P_{V-A} = 1.001 \pm 0.008$ in Gamow–Teller transition.

Pure $V - A$ theory implies $\rho = \frac{3}{4}$ and $P/P_{V-A} = 1$. Thus we obtain

$$\sqrt{\eta} \leq 0.35 \qquad\qquad (6.3.3)$$

and

$$\tan \zeta < 0.05.$$

This corresponds to $m_{W_R} \geq 200$ GeV.

It was noted [13] that a crucial parameter in the probe of $V + A$ currents is the angular asymmetry parameter ζ in polarized muon decay. This parameter is expected to be $+1$ in pure $V - A$ theories. In 1977, the best value was [15]

$$\zeta P_\mu = 0.972 \pm 0.013.$$

Recent experimental searches for $V + A$ current effects by Carr et al. [16] in μ-decay has led to much better measurement of the value of ζP_μ

$$\zeta P_\mu > 0.9975$$

leading to bounds on $m_{W_R} \geq 432$ GeV for $-0.05 < \zeta < 0.035$. At such precision levels it is not clear whether the fourth-order radiative corrections play an important role or not.

Several other searches for right-handed currents are Ref. 17, which measured the ζ-parameter

$$\zeta P_\mu = 1.010 \pm 0.064.$$

There is also an ongoing experiment by Vanklinken et al. of the parameter P_F/P_{G_T} where a precision at the 10^{-3} level is expected. Finally, Yamazaki et al. [18] have looked for muon polarization in K_{μ_2} decay where for $V - A$ currents $P_\mu = -1$. Their result is

$$P_\mu = -0.967 \pm 0.047,$$

which allows a 4% admixture of $V + A$ currents. The most stringent bound that remains, however, is the one from Ref. 16.

Finally, analysis of the asymmetry [19] in ^{19}Ne β decay puts a more stringent bound on ζ,

$$0.005 \leq \zeta \leq 0.015,$$

although this involves assumptions that are more model-dependent than Ref. 13.

Case (ii). Light-Majorana Neutrinos ($m_{\nu_R} \leq 10\text{--}100$ MeV)

In this case, the right-handed neutrinos cannot be emitted in β-decay. It was pointed out [12] that the most stringent bounds in this case come from peak searches in π_{l_2} decay. There is now a new channel for the decay of the pion: $\pi \to e_R^- \nu_R$ with width given by

$$R = \frac{\Gamma(\pi \to e_R^- \nu_R)}{\Gamma(\pi \to e_L^- \nu_L)} = (|\eta|^2 + |\zeta|^2) \left(\frac{m_{\nu_R}}{m_e}\right)^2 \left(1 - \frac{m_{\nu_R}^2}{m_\pi^2}\right)^2, \qquad (6.3.4)$$

where $\eta = (m_{W_L}/m_{W_R})^2$ as defined before; ζ denotes the mixing between the light and heavy neutrino. Experiment searches for secondary peaks at these mass ranges for the heavy neutrino have been conducted. Lack of evidence for any secondary peak in π decay at TRIUMF [20] and KEK [21] translates into an upper bound on $R < 4 \times 10^{-3}$ for 45 MeV $< m_{\nu_R} < 74$ MeV, which implies

$$|\eta| \leq 10^{-3}. \qquad (6.3.5)$$

One can also derive bounds on the W_R mass from considerations of unitarity of the quark mixing matrix [22]. We will see in the next section that similar bounds appear from considerations of the ν_R contributions to neutrinoless double β decay.

Case (iii). Bound on m_{W_R} from $K_L - K_S$ Mass Difference

One of the most stringent constraints on M_{W_R} comes from the short-distance contributions to $K_L - K_S$ mass difference [23] due to the new contribution from the $W_L - W_R$ box graph (Fig. 6.1) in the case of the

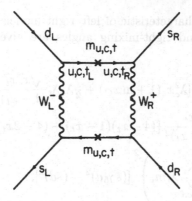

Figure 6.1.

two-generation left–right symmetric model. Their result was that

$$\Delta m_K \simeq \frac{G_F^2 m_{W_L}}{4\pi^2} (\cos^2\theta_C \sin^2\theta_C) \left(\frac{m_c^2}{m_{W_L}^2}\right) \langle K^0|O^{(1)}|\bar{K}^0\rangle$$

$$\times \left[1 + 8\eta \ln\left(\frac{m_c^2}{m_{W_L}^2}\right) \frac{\langle K^0|O^{(2)}|\bar{K}^0\rangle}{\langle K^0|O^{(1)}|\bar{K}^0\rangle}\right], \tag{6.3.6}$$

where

$$O^{(1)} = \bar{s}_L\gamma_\mu d_L \bar{s}_L\gamma^\mu d_L,$$

$$O^{(2)} = \bar{s}_L d_R \bar{s}_R d_L. \tag{6.3.7}$$

Evaluating all matrix elements in the vacuum saturation approximation [24] we find that the second term within the bracket becomes about -430η. Since the pure left–left contribution [25] [the expression outside the bracket in eq. (6.3.6)] explains the bulk of the $K_L - K_s^0$ mass difference at least in magnitude (presumably also in sign [26]), the second term must be small compared to 1. This implies that

$$\eta \leq \frac{1}{430} \tag{6.3.8}$$

or

$$m_{W_R} \geq 1.6 \text{ TeV}.$$

Because this bound is significantly higher than any obtained before, it needs to be examined critically as has been done in several subsequent papers [27].

There exist several corrections to the expression in (6.3.6): (i) contribution due to t quarks, and (ii) new contributions from tree graphs involving neutral Higgs exchanges. In the presence of these two contributions we can write

$$\Delta M_K = M_{LL} + M_{LR} + M_{H^0}, \tag{6.3.9}$$

where contributions characteristic of left–right models under the assumption of equal left- and right-mixing angles are given by [define $x_i = (m_i/m_W)^2$]

$$M_{LR} = \frac{G_F^2 m_{W_L}^2}{4\pi^2} 8\eta[\lambda_c^2 x_c (1 + \ln x_C) + \frac{1}{2}\lambda_c \lambda_t \frac{\sqrt{x_c x_t}}{(1-x_t)}(4-x_t)\ln X_t$$

$$+ \frac{1}{4}\lambda_t^2 \frac{x_t}{(1-x_t)^2}\{(4-x_t)(1-x_t) + (4-2x_t)\}\ln x_t]O^{(2)}, (6.3.10)$$

$$M_{H^0} = \frac{2G_F}{m_H^2}\left(\sum_{i=u,c,t}\lambda_i m_i\right)^2 [(\bar{s}\gamma_5 d)^2 - (\bar{s}d)^2], \tag{6.3.11}$$

$$\lambda_u = s_1 c_1 c_3,$$

$$\lambda_c = -s_1 c_2 (c_1 c_2 c_3 - s_2 s_3 e^{-i\delta}),$$

with

$$\lambda_t = -s_1 s_2 (c_1 s_2 c_3 + c_2 s_3 e^{-i\delta}).$$

Using the vacuum saturation assumption we obtain

$$\langle(\bar{s}\gamma_5 d)^2 - (\bar{s}d)^2\rangle_{K^0 \bar{K}^0} = \frac{4}{3}B(7.7)f_K^2 M_K. \tag{6.3.12}$$

These contributions, however, do not provide any clean way to avoid the constraint in Eq. (6.3.10).

Case (iv) Astrophysical bounds:

It has been pointed out [27] that observations of neutrinos from SN1987a by the IMB and Kamiokande collaboration [28] imply that $m_{W_R} \geq 22$ TeV for the case of Dirac neutrinos. The reason for this bound is that for lower values of M_{W_R}, the right-handed charged-current interactions are so "strong" that electron–nucleon interaction inside the supernova core can produce large number of light ν_R's, which then escape the supernova thereby draining its energy away from the observed left-handed neutrinos, thereby contradicting observations of the energetic ν_Ls. This is the most stringent restriction on m_{W_R}. However, if the right-handed neutrino had new interactions of strength $\sim G_F$, such bounds could be avoided; the same interactions would however imply that the right-handed neutrino is in equilibrium at the epoch of nucleosynthesis and must count as one extra species of neutrino, which is barely allowed by present analysis. Therefore, these bounds on m_{W_R} are on a somewhat different footing from those obtained from laboratory experiments.

There are also limits from neutrinoless double beta decay on M_{W_R}, for the case of Majorana ν_R, which imply $M_{W_R} \geq 1000$ GeV or so [29].

The results on the bounds on m_{W_R} are summarized in Table 6.1.

The question that could be asked is that should the right-handed gauge bosons be discovered at lower mass, how does one avoid these bounds? A

Table 6.1.

Process	(m_{W_R}) min
Dirac ν	380 GeV
$P_\mu \zeta$	
Majorana ν, $m_{\nu_R} \leq 100$ MeV	2 TeV
$K_L - K_S$ mass difference + manifest $L - R$	1.6 TeV (3 TeV)
Sym (with QCD effects included)	
Dirac ν: SN1987A	22 Tev
Majorana ν and Arbitrary m_{ν_R}	1000 Gev

simple way out of the above bounds [31], is to assume that the left- and right-mixing angles themselves are different from each other, in which case the constraint on m_{W_R} gets weaker.

6.3.1   Limits on the $W_R - W_L$ Mixing Parameter ζ

We see from Eq. (6.2.20) that the presence of right-handed gauge bosons brings another parameter ζ into the theory, i.e., the mixing between W_R and W_L. This parameter is also severely constrained by low-energy phenomenology.

- The nonleptonic decays of kaon such as $K \to 3\pi$ and $K \to 2\pi$ receive new contributions in the presence of ζ regardless of whether the neutrino is a Majorana or Dirac particle. The essential method is to assume that there is at most a 10% deviation from the equality $[F_i^5, H_{wk}] = [F_i, H_{wk}]$, which is known to hold in the absence of right-handed currents. This implies $\zeta < 4 \times 10^{-3}$ [31a].

- The process $b \to s\gamma$ amplitude has an $\frac{m_t}{m_b}$ enhancement in the presence of ζ relative to the standard model. Using the present experimental number for this process, one gets a limit $-0.01 \leq \zeta \leq 0.003$ [31b].

- For heavy right-handed neutrinos, there will be new contributions to β decay and K decay but not to muon decay, there altering the universality properties. From these considerations, one can get a limit $\zeta \leq 10^{-3}$ [31c].

Decay Properties and Possible High-Energy Signature for W_R and Z_R Bosons

If the W_R- and Z_R-boson masses are in the 300–400 GeV range, they can be produced in the high-energy machines of the coming generation in $p\bar{p}$ colliders. It may therefore be useful to list their decay modes and properties.

Assuming that $2m_{\nu_R} \ll m_{Z_R}$, we obtain [32]

$$\Gamma_{Z_2} \simeq 12 \text{ GeV},$$
$$B(Z_R \to 2\nu) \simeq 1\%,$$
$$B(Z_R \to \mu^+\mu^-) \simeq 2\%,$$
$$B(Z_R \to u\bar{u}) \simeq 7\%,$$
$$B(Z_R \to d\bar{d}) \simeq 14\%,$$
$$B(Z_R \to 2N_R) \simeq 9\%. \tag{6.3.13}$$

For the W_R^+ the leptonic branching ratio is $B(W_R^+ \to N_\mu\mu^+) \simeq 8\%$.

It has been pointed out [32] that, if neutrinos are Majorana particles, then, in $p\bar{p}$ collision, we obtain $p\bar{p} \to \mu^+\mu^+ + X$ without any missing energy. This should provide a clean signature both for the W_R boson as well as for the Majorana character of the neutrino.

In high-energy ep colliding machines (i.e., HERA) possible effects of m_{W_R} as heavy as 600 GeV can be detected through its effect on the y distribution in $ep \to eX$ processes. To detect a right-handed gauge boson in the mass range of TeV's, we need higher-energy machines such as the Tevatron at Fermilab or the LHC at CERN.

Finally, another good way to look for right-handed charged currents is to observe the absorption cross section for μ_L^+ at high energies, i.e., $\mu_L^+ N \to \bar{\nu}_R X$, i.e.,

$$\frac{\sigma(\mu_L^+)}{\sigma(\mu_R^+)} = \left(\frac{m_{W_L}}{m_{W_R}}\right)^4. \tag{6.3.14}$$

The main problem here is to isolate a pure μ_L^+ beam from the decay of π^+ in flight. From simple kinematic considerations it follows that for pion momenta p_π^+, the decay μ^+'s with $p_\mu \simeq p_\pi$ have left-handed helicity, which is of interest in the detection of right-handed currents. It must, however, be noted for $m_{W_R} \simeq 800$ GeV, the ratio of cross sections in (6.3.14) is 10^{-4} and this means that the momentum spread of π's and μ's must be less than 10^{-4}, which may not be so easy to attain.

§6.4   Properties of Neutrinos and Lepton-Number-Violating Processes

(a) Searches for neutrino mass:

An important prediction of the left–right symmetric models that distinguishes it from the standard model is the nonzero mass of neutrinos. Recent evidence for neutrino masses have therefore provided a major boost to the credibility of these models. Let us therefore begin this section with a brief overview of the evidences for neutrino masses [33].

The neutrino mass searches can be divided into two categories: direct laboratory searches and indirect searches involving neutrino oscillations. The simplest direct experiment involves searching for anomalous end point behavior in the beta spectrum of beta-decaying nuclei such as tritium. The basic principle here is that the differential beta spectrum depends on phase-space factors that involve factors such as $p_{\nu_e} E_{\nu_e}$, which denote the momentum and energy, respectively, of the neutrino. Since $p_{\nu_e} = \sqrt{E_{\nu_e}^2 - m_{\nu_e}^2}$, a nonvanishing neutrino mass will effect the electron count E_e is large (which means from energy momentum conservation that E_{ν_e} is close to m_{ν_e}). From these experiments, one only has an upper limit at the level of 2.5 eV [34]. Direct limit on the ν_μ is derived from the decay of $\pi^+ \rightarrow \mu^+ + \nu_\mu$ using energy momentum conservation and measurement of the muon momentum. This yields a limit of $m_{\nu_\mu} \leq 160$ KeV. The limit on the ν_τ is derived from the multipion decay of the τ lepton. The most recent limit is $m_{\nu_\tau} \leq 18.2$ MeV [34].

Another class of direct search experiments that provides information about neutrino mass is the process of neutrinoless double beta decay [35]. In this process a nucleus (A,Z) decays to a daughter nucleus $(A, Z+2)+2e^-$ without any accompanying neutrinos and can go only if the neutrino is its own anti-particle (which is same as it being a Majorana fermion, as we explain shortly). The most stringent limit on a parameter, which we call effective electron neutrino mass $\rangle m_{\nu_e} \langle$ (defined later), is 0.2 eV [36] and has been found by the Heidelberg–Moscow collaboration using enriched Germanium ^{76}Ge in an underground experiment at Gran Sasso, Italy.

Another indirect way to look for neutrino masses is to realize that a neutrino produced in a weak-interaction process is really a coherent superposition of several states if its mass is nonzero. We will call the state produced in a weak process a weak eigenstate, ν_ℓ with $\ell = e, \mu, \tau$. If we denote a mass eigenstate as ν_i with mass m_i $(i = 1, 2, 3)$, then the state ν_α produced in the weak process at $t = 0$ can be written as

$$|\nu_\ell\rangle = \sum_i U_{\ell i} |\nu_i\rangle , \qquad (6.4.1)$$

where U is a unitary matrix similar to the quark mixing matrices discussed in Chapter 3.

For a simple-minded approach to the propagation of this state, we assume that the 3-momentum $\mathbf{p}$ of the different components in the beam are the same. However, since their masses are different, the energies of all these components cannot be equal. Rather, for the component ν_i, the energy is given by the relativistic energy–momentum relation

$$E_i = \sqrt{p^2 + m_i^2}. \qquad (6.4.2)$$

After a time t, the evolution of the initial beam of neutrinos gives

$$|\nu_\ell(t)\rangle = \sum_i e^{-iE_it}U_{\ell i}|\nu_\alpha\rangle .$$ (6.4.3)

In writing this, we assume that the neutrinos ν_i are stable particles; otherwise the analysis below will need to be modified.

Since all E_i's are not equal if the masses are not, eq. (6.4.3) represents a different superposition of the physical eigenstates ν_i compared to eq. (6.4.1). In general, this state has not only the properties of a ν_ℓ, but also of other flavor states. The amplitude of finding a $\nu_{\ell'}$ in the original ν_ℓ beam is

$$\langle \nu_{\ell'} \mid \nu_\ell(t)\rangle = \sum_{i,j}\langle \nu_j \mid U_{j\ell'}^\dagger e^{-iE_it}U_{\ell i} \mid \nu_i\rangle$$

$$= \sum_i e^{-iE_it}U_{\ell i}U_{\ell' i}^*$$ (6.4.4)

using the fact that the mass eigenstates are orthonormal. The probability of finding a $\nu_{\ell'}$ in an originally ν_ℓ beam is

$$P_{\nu_\ell\nu_{\ell'}}(t) = |\langle \nu_{\ell'}|\nu_\ell(t)\rangle|^2$$

$$= \sum_{i,j}|U_{\ell i}U_{\ell' i}^*U_{\ell j}^*U_{\ell' j}|\cos[(E_i - E_j)t - \varphi_{\ell\ell' ij}] ,$$ (6.4.5)

where

$$\varphi_{\ell\ell' ij} = \text{Arg}\,(U_{\ell i}U_{\ell' i}^*U_{\ell j}^*U_{\ell' j}) .$$ (6.4.6)

In all practical situations, neutrinos are extremely relativistic, so that we can approximate the energy–momentum relation as

$$E_i \simeq |p| + \frac{m_i^2}{2p} ,$$ (6.4.7)

and can also replace t by the distance x traveled by the beam. Thus we obtain

$$P_{\nu_\ell\nu_{\ell'}}(x) = \sum_{i,j}|U_{\ell i}U_{\ell' i}^*U_{\ell j}^*U_{\ell' j}|\cos\left(\frac{2\pi x}{L_{ij}} - \varphi_{\ell\ell' ij}\right) ,$$ (6.4.8)

where, writing $p = E$ for the sake of brevity, we defined

$$L_{ij} \equiv \frac{4\pi E}{\Delta_{ij}} ,$$ (6.4.9)

with

$$\Delta_{ij} \equiv m_i^2 - m_j^2 .$$

The quantities L_{ij} are called the *oscillation lengths*, which give a distance scale over which the oscillation effects can be appreciable.

Note that if the distance x is an integral multiple of all L_{ij}, we obtain $P_{\nu_\ell \nu_{\ell'}} = \delta_{\ell\ell'}$, as in the original beam. But at distances where that condition is not satisfied, we can see nontrivial effects, which are sought for in the experiments. An important point to note is that these experiments provide information only about mass difference squares and not about the absolute masses. One therefore still needs the direct search experiments to supplement the oscillation experiments

A simplified version of the formula in Eq. (6.4.6) for the two-neutrino case is given below:

$$P_{\nu_\ell, \nu'_{e ll}} = \sin^2 2\theta \sin^2 \frac{1.27\Delta m^2 (eV^2) L(\text{meter})}{E(\text{MeV})}. \qquad (6.4.10)$$

This formula provides a simple way to read off the mass differences squared's probed in a given experiment.

There exist three classes of oscillation experiments that have provided positive signals and therefore have indicated nonzero neutrino masses. The first two are "natural" sources of neutrinos so that one only has to build a detector to detect the final neutrinos.

The first and the oldest such experiment is the detector looking for neutrinos produced in the core of the Sun in its thermonuclear fusion responsible for sunlight that sustains life on Earth. For this case $L \sim 10^{11}$ meters and $E \sim 1 - 10$ MeV's implying that the probed $\Delta m^2 \approx 10^{-10}$ eV². There is now solid evidence for this oscillation. There is the possibility of matter effect from the dense core of the Sun, known as the Mikheyev–Smirnov–Wolfenstein effect [41], which can reduce this mass difference to 10^{-6} eV² level. We do not discuss this here and instead refer to books on the subject. It is becoming clear after the results from the SNO experiment that the final states to which ν_e oscillates consists predominantly of $\nu_{\mu,\tau}$.

The second one is the evidence for the oscillation of the atmospheric muon neutrinos produced at the top of the atmosphere in cosmic ray collisions. The longest distance in this case is 10,000 km and energies are of order 1000 MeV leading to the probed mass range to be $\Delta m^2 \sim 10^{-3}$ eV². Here it is almost certain that the oscillation is between a muon to a tau neutrino.

Finally there are indications of oscillation of accelerator muon neutrinos in an experiment at Los Alamos (the LSND experiment [37]). Here $E \sim 30$ MeV and $L \sim 20$ meters leading to $\Delta m^2 \sim eV^2$. This result will be checked in an experiment under way at Fermilab known as Mini Boone experiment.

(b) Majorana Mass for Neutrinos: Model-Independent Considerations

Two kinds of mass terms are allowed by proper Lorentz transformations for a neutral-spin-1/2 fermions such as the neutrino ν. Let us take ν as a four-component spinor and $\nu_{L,R}$ as its left and right chiral projections. We can write a Dirac mass $\mathcal{L}_D$ and a Majorana mass $\mathcal{L}_M$ as follows (we consider only one neutrino species):

$$\mathcal{L}_D = m_D \bar{\nu}_L \nu_R + \text{h.c.} \tag{6.4.11}$$

and

$$\mathcal{L}_M = m_L \nu_L^T C^{-1} \nu_L + m_R \nu_R^T C^{-1} \nu_R \tag{6.4.12}$$

(where m_D and $m_{L,R}$ are arbitrary complex numbers).

The first point to note is that $\mathcal{L}_D$ is invariant under a global U(1) symmetry under which $\nu \to e^{i\theta}\nu$, where $\mathcal{L}_M$ is not. This U(1) symmetry may be identified with the lepton number L, with $L(\nu) = -L(\bar{\nu}) = 1$. The same is true of the standard model where $m_D = 0$ as are $M_{L,R}$.

Once one includes $\mathcal{L}_M$ in the Lagrangian, it breaks the lepton number by two units ($\Delta L = 2$). Therefore, in the presence of $\mathcal{L}_M$, the $\Delta L = 2$ type lepton-number-violating processes such as neutrinoless double β decay, $K^+ \to \pi^- e^+ e^+$, etc., will take place. Thus, observation of any such process will constitute strong evidence for the Majorana character of the neutrino mass. Furthermore, both types of mass terms involve as yet unobserved physics (i.e., right-handed neutrino or lepton number violation). This means that discovery of the neutrino mass is an important step in uncovering new physics beyond the standard model.

Let us now give a brief discussion of the Majorana neutrino and its wave function. We start with the Lagrangian neutrino:

$$\mathcal{L} = i\bar{\nu}\gamma^\lambda \partial_\lambda \nu + \mathcal{L}_{\text{mass}} + \mathcal{L}_{\text{wk}}. \tag{6.4.13}$$

It is convenient for this discussion to choose Weyl basis with the following choice of γ-matrices ($k = 1, 2, 3$):

$$\gamma^k = \begin{pmatrix} 0 & \sigma_k \\ -\sigma_k & 0 \end{pmatrix}, \quad \gamma^4 = \begin{pmatrix} 0 & 1 \\ 1 & 0 \end{pmatrix}, \quad \gamma^5 = \begin{pmatrix} 1 & 0 \\ 0 & -1 \end{pmatrix}. \tag{6.4.14}$$

We can write a four-component Dirac spinor ν in the basis as

$$\nu = \begin{pmatrix} u \\ i\sigma_2 v^* \end{pmatrix}, \tag{6.4.15}$$

where u and $i\sigma_2 v^*$ denote the left $[(1 + \gamma_5)/2]$ and right $[(1 - \gamma_5)/2]$ chiral projections of ν. Ignoring $\mathcal{L}_{\text{wk}}$ we can write the field equations following from eqs. (6.4.11), (6.4.12), (6.4.13) in terms of u and v as follows:

$$-i(\sigma\nabla - \partial_t)\begin{pmatrix} u \\ v \end{pmatrix} + \sigma_2 \begin{pmatrix} m_L^* & m_D \\ m_D & m_R \end{pmatrix} \begin{pmatrix} u^* \\ v^* \end{pmatrix} = 0. \tag{6.4.16}$$

(In this case, m_D can be chosen real by redefining the phase of either u or v.) This matrix is a complex symmetric matrix (to be denoted henceforth by M) connecting the two-component spinors. In a more general situation this becomes a more complicated matrix. The various cases known as Dirac, Majorana, or pseudo-Dirac neutrinos correspond to various forms of the matrix M:

$$M \equiv \begin{pmatrix} m_L^* & m_D \\ m_D & m_R \end{pmatrix}. \tag{6.4.17}$$

(i) Dirac: $m_L = m_R = 0$; $m_D \neq 0$.

(ii) Majorana: either m_L or m_R or both nonzero; m_D arbitrary.

(iii) Pseudo-Dirac [38]: $m_D \neq 0$ and $m_L = m_R \ll m_D$.

As is evident, case (iii) is a special case of the Majorana mass matrix. In fact, even the Dirac case is a special case, where an additional U(1) symmetry appears in the Lagrangian in the limit $m_L = m_R = 0$. In summary, a Dirac neutrino consists of two Majorana neutrinos with equal masses and opposite CP properties. To see this note that in case (i) one mass eigenvalue is negative, which can be made positive by defining $\psi = -C\bar{\psi}^T$, i.e., a state odd under CP; a pseudo-Dirac neutrino consists of two Majorana neutrinos with opposite CP properties but slightly different masses $(m_M \pm M_D)$. This case will become of interest if the neutrino mass measured in tritium decay and that obtained from $(\beta\beta)_{0\nu}$ decay happen to be different, as is sometimes thought.

(This situation is similar to the $K^0 - \bar{K}^0$ mixing case and will become almost identical to it if the tiny Majorana mass is induced by weak interactions instead of being present from the beginning.)

To proceed further, we can diagonalize the complex and symmetric mass matrix M, which can be done, in general, as follows:

$$UMU^T\Lambda = M_D, \tag{6.4.18}$$

where M_D is diagonal with positive, real eigenvalues and Λ is a diagonal unitary matrix. In terms of the eigenstates

$$\chi \equiv \begin{pmatrix} \chi_1 \\ \chi_2 \end{pmatrix} \equiv \Lambda U \begin{pmatrix} u \\ v \end{pmatrix} \tag{6.4.19}$$

the field equations (6.4.16) factorize

$$(\sigma\nabla - \partial_t)\begin{pmatrix} \chi_1 \\ \chi_2 \end{pmatrix} + \begin{pmatrix} m_1 e^{i\alpha_1}\chi_1^* \\ m_2 e^{i\alpha_2}\chi_2^* \end{pmatrix} = 0. \tag{6.4.20}$$

It therefore suffices to study one of the eqs. to learn about the nature of the Majorana wave function. Using methods already given earlier [39] we can expand χ as follows:

$$\chi(x) = \frac{1}{\sqrt{V}} \sum_{\lambda=+,-} (a_\lambda(k)u_\lambda(k)e^{ik\cdot x} + a_\lambda^\dagger(k)v_\lambda(k)e^{-ik\cdot x}), \tag{6.4.21}$$

and get a detailed solution of the form

$$\chi(x,t) = \sum_p [a_{p,+}e^{-ip.x} - a_{p,-}^\dagger e^{ip.x}]\alpha\sqrt{E+p} \tag{6.4.22}$$

$$+ \sum_{p} [a_{\mathbf{p},-} e^{-p.x} + a_{\mathbf{p},+}^{\dagger} e^{ip.x}] \beta \sqrt{E - p},$$

where choosing the momentum along the z direction leads to $\alpha = \begin{pmatrix} 1 \\ 0 \end{pmatrix}$
and $\beta = \begin{pmatrix} 0 \\ 1 \end{pmatrix}$. In the presence of small masses, only the first terms are
important; i.e., the field χ creates the "down" helicity state and destroys the
"up" one to zeroth order in m/E, and the $\chi^{\dagger}$ field will dominantly create
the "up" and destroy the "down" ones, respectively. The two component
field behaves like a Weyl fermion.

At this point we wish to note the way to understand the smallness of the
neutrino masses in unified gauge theories in the case of Majorana neutrinos. Since we work with chiral spinors, we chose appropriate Higgs boson
couplings so that the mass matrix in eq. (6.4.17) has the following form:

$$M = \begin{pmatrix} 0 & m_D \\ m_D & m_R \end{pmatrix}. \tag{6.4.23}$$

The eigenvalues of this mass matrix are (for $m_R \gg m_L$)

$$m_N \simeq m_R \quad \text{(heavy neutrino)}$$

and

$$m_\nu \simeq -\frac{m_D^2}{m_R} \quad \text{(light neutrino)}. \tag{6.4.24}$$

Thus, the physical Majorana neutrino mass can be small without any
unnatural fine tuning or parameters if there is a heavy scale m_R in the
theory. This is known as the seesaw mechanism [11].

(c) Neutrino Masses and Left–Right Symmetry

Our knowledge of weak interactions since the time of Fermi and Pauli has
been intimately connected with our understanding of the nature of the
neutrino. When $V - A$ theory was proposed by Sudarshan and Marshak,
Feynmann and Gell-Mann, they based their argument on the assumption
that neutrinos are massless. If $m_\nu = 0$, then the neutrino spinor obeys
the Weyl equation, which is invariant under γ_5 transformations, i.e., $\nu \to$
$\gamma_5 \nu$. Marshak and Sundarshan argued that, since neutrinos participate only
in weak interactions, the weak Hamiltonian H_{wk} ought to be invariant
under separate γ_5 transformations of the various fermions participating in
it, and this leads to the successful $V - A$ theory of charged-current weak
processes. From this point of view, the existence of $V + A$ currents ought
to be connected with a nonvanishing neutrino mass and, the small neutrino
mass and the suppression of $V + A$ currents should be connected. We will
show in this section that in the framework of left–right symmetric theories

this connection comes out very naturally, i.e., we show that

$$m_\nu = \gamma \frac{m_e^2}{m_{W_R}}. \tag{6.4.25}$$

Furthermore, we analyze the implications of the left–right symmetric models for lepton number nonconservation.

The basic equation that leads to the connection between neutrino mass and the magnitude of $V + A$ currents (or parity violation) is the formula for the electric charge given in eq. (6.1.1)

$$Q = I_{3_L} + I_{3_R} + \frac{B - L}{2}. \tag{6.4.26}$$

If we work at a distance scale where weak left-handed symmetry is a good symmetry, then from eq. (6.1.1) we obtain the relation

$$\Delta I_{3_R} = -\tfrac{1}{2}\Delta(B - L) \tag{6.4.27}$$

for processes involving leptons; this implies that $\Delta L = 2\Delta I_{3R}$. Because in our case $|\Delta I_{3R}| = 1$, this means that neutrinos must be Majorana particles and their Majorana mass must be connected to the strength of parity violation. We now show that the symmetry-breaking mechanism discussed in Section 6.2 provides a realization of this idea.

Below we present an explicit Higgs model for this. As discussed in Section 6.2, we will use only those Higgs multiplets that are bilinears in the fundamental fermion fields of the theory, i.e.,

$$\Delta_L(1, 0, +2) + \Delta_R(0, 1, +2)$$

and

$$\phi = (\tfrac{1}{2}, \tfrac{1}{2}, 0).$$

The gauge symmetry is then broken in stages as follows (we also show the neutrino mass matrices for each stage):

$$SU(2)_L \times SU(2)_R \times U(1)_{B-L}$$

$$\downarrow$$

$$\langle \Delta_L^0 \rangle \simeq 0, \qquad \langle \Delta_R^0 \rangle = v_R \neq 0, \qquad M_\nu = \begin{pmatrix} 0 & 0 \\ 0 & v_R \end{pmatrix}$$

$$\downarrow$$

$$SU(2)_L \times U(1)$$

$$\downarrow$$

$$\langle \phi \rangle = \begin{pmatrix} k & 0 \\ 0 & k' \end{pmatrix}, \qquad M_\nu = \begin{pmatrix} 0 & \tfrac{1}{2}hk \\ \tfrac{1}{2}hk & fv_R \end{pmatrix}$$

$$\downarrow$$

$$U(1)_{\text{em}}.$$

Here M_ν denotes the Majorana–Dirac mass matrix for the neutrino at the different stages of symmetry breakdown, which arises from the following leptonic Higgs couplings (eq. (6.2.29)):

$$L_Y = \psi_L(h\phi + \tilde{h}\tilde{\phi})\psi_R + if(\psi_L^T C^{-1}\tau_2\psi_L\Delta_L + \psi_R^T C^{-1}\tau_2\psi_R\Delta_R) + \text{h.c.}$$
$$(6.4.28)$$

By diagonalizing the above M_ν we obtain the following eigenstates and masses:[1]

$$\nu = \nu_L\cos\xi + \nu_R\sin\xi, \qquad m_\nu \simeq \frac{h^2 k^2}{2fv_R},$$

$$N = -\nu_L\sin\xi + \nu_R\cos\xi, \qquad m_N \simeq 2fv_R, \qquad (6.4.29)$$

where $\tan\xi(m_\nu/m_N)^{1/2}$. As claimed before

$$v_R \to \infty, \qquad m_\nu \to 0.$$

Leptonic charged currents now look as follows:

$$\begin{pmatrix} \nu\cos\xi + N\sin\xi \\ e^- \end{pmatrix}_L \quad \text{and} \quad \begin{pmatrix} -\nu\sin\xi + N\cos\xi \\ e^- \end{pmatrix}_R. \qquad (6.4.30)$$

For convenience let us reparametrize the m_ν and m_N in terms of m_e and m_{W_R}:

$$m_\nu \simeq \frac{r^2}{\beta}\frac{m_e^2}{m_{W_R}},$$

$$m_N \simeq \beta m_{W_R}. \qquad (6.4.31)$$

r and β are free dimensionless parameters and we can obtain different values for m_N and m_ν in the electron volt range using $m_{W_R} \geq 250$ GeV– 2.5 TeV for different choices of these parameters. On the other hand, if the neutrino masses are in the range of 10^{-2} eV for ν_τ, as is suggested by the atmospheric neutrino data, then one must take $M_{W_R} \geq 10^{11}$ GeV.

An interesting feature of the seesaw formula is that, if we assume r and β to be independent of generations, the neutrino masses scale as square of

---

[1]As noted in Section 6.2, as the $SU(2)_L \times U(1)$ symmetry gets broken by $\kappa, \kappa' \neq 0$, the left-handed triplet field Δ_L acquires a nonzero vacuum expectation value, i.e., $\langle\Delta_L\rangle = v_L \neq 0$. It has, however, been shown by a detailed analysis of the potential that it is proportional to

$$v_L \simeq \gamma\kappa^2/4v_R.$$

It also turns out that γ is nonzero only when the Yukawa couplings h and f are nonzero and $\gamma \approx h^2 f^2$. Because h and f are small numbers ($\approx 10^{-2}$), γ is expected to be of order 10^{-8}. In the presence of nonzero v_L we find

$$\tfrac{1}{2}m_\nu \approx \gamma f\kappa^2/v_R - \tfrac{1}{4}h^2\kappa^2/fv_R, \qquad (6.4.26a)$$

which reduces to eq. (6.4.28) for $\gamma \approx 10^{-8}$. (See Ref. 40 for a natural way to obtain such a small number.)

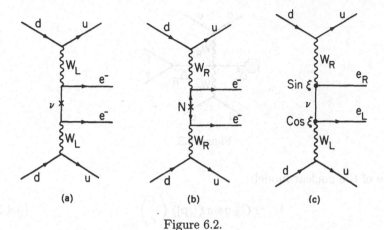

Figure 6.2.

the mass of the charged lepton of the corresponding generation leading to

$$m_{\nu_e} : m_{\nu_\mu} : m_{\nu_\tau} = m_e^2 : m_\mu^2 : m_\tau^2 \ldots \qquad (6.4.32)$$

(d) Double β Decay

Now we proceed to discuss lepton-number-violating processes, $\Delta L \neq 0$. The most important among them is neutrinoless double β decay: $(A, Z) \rightarrow (A, Z + 2) + e^- + e^-$. We will now discuss the implications of our model for the lifetime for this process. Neutrinoless double β-decay has been the subject of a great deal of discussion [35] in recent years because it provides a sensitive test of lepton number conservation. Owing to more available phase space, this decay rate is enhanced over the lepton-number-conserving process $(A, Z) \rightarrow (A, Z + 2) + e^- + e^- + \bar{\nu}_e + \bar{\nu}_e$. In order to see the general orders of magnitude, if we assume the strength of $(\beta\beta)_{0\nu}$ amplitude to be $G_F^2 \eta_0$ compared to that of $(\beta\beta)_{2\nu}$ as G_F^2, $T_{1/2}(2\nu) \simeq 10^{20}$ yr where as $T_{1/2}(0\nu) \simeq 10^{14}|\eta_0|^{-2}$. Thus, nonobservation of the $(\beta\beta)_{0\nu}$ process implies that $\eta_0 \leq 10^{-3}$. The present experimental bounds are however much more stringent and imply $\eta_0 < 10^{-5}$ [55]. In the left–right symmetric model there are three distinct contributions to $(\beta\beta)_{0\nu}$ [Figs. 6.2(a), (b), (c)]: ν-mass, N-mass, and left–right mixing contribution. All these contributions are incoherent. The strengths of the three amplitudes are roughly given by

$$M_a \simeq G_F^2 m_\nu \left\langle \frac{e^{-m_\nu r}}{r} \right\rangle_{\text{Nuc}}. \qquad (6.4.33)$$

This is a $0^+ \rightarrow 0^+$ transition

$$M_b \simeq G_F^2 \eta^2 \frac{1}{m_N} \left\langle \delta^3(r) \right\rangle_{\text{Nuc}}. \qquad (6.4.34)$$

This is also a $0^+ \rightarrow 0^+$ transition and is likely to be suppressed by a hard-core piece of the nuclear potential and is quite dependent on the

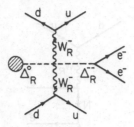

Figure 6.3.

nature of the nuclear model:

$$M_c \simeq G_F^2 \eta \sin \xi \langle |\mathbf{p}| \rangle \left\langle \frac{1}{r} \right\rangle_{\text{Nuc}}. \tag{6.4.35}$$

This is, however, a $0^+ \to 2^+$ transition and is a characteristic signature of right-handed currents. A nucleus well suited for study of this contribution is $^{76}\text{Ge} \to {}^{76}\text{Se}$.

To discuss the predictions for $T_{1/2}(0\nu)$ of left–right models, we will consider the equivalent of η_0 in the above three cases.

Case C. $\eta_0 = \eta \sin \xi$

Phenomenologically, $\eta < \frac{1}{400}$ and for the case $m_N \approx 100$ GeV, $\sin \xi (m_\nu/m_N)^{1/2} \approx 10^{-5}$, which gives $\eta_0 \lesssim 2(10^{-8})$. However, phenomenologically, m_N can be as low as 100 MeV, in which case $\eta_0 \simeq 10^{-6}$. The present experimental limits on η_0 are of order 10^{-4}.

Case B.

The equivalent η_0 parameter in this case is

$$\eta_0 \simeq \eta^2 \frac{1}{m_N} \langle \delta^3(r) \rangle_{\text{Nuc}} \frac{1}{|\mathbf{p}| \langle 1/r \rangle}. \tag{6.4.36}$$

For a 100 GeV right-handed neutrino and $\eta < 10^{-3}$, a very crude estimate for this parameter is $\eta_0 \approx 10^{-7}$, which is beyond the reach of any planned experiment. On the other hand, if η is left free, one gets correlated bounds between m_N and m_{W_R}.

Case A. $\eta_0 = m_\nu/|\mathbf{p}| \approx 10^{-5}$

For p (the momentum difference of the electrons) of order 2 MeV and $m_\nu \approx 10$ eV.

In this model there exist additional contributions to $(\beta\beta)_{0\nu}$-decay that do not involve neutrinos but the exchange of doubly charged Higgs bosons (Fig. 6.3), which are quite important and must be included in a detailed analysis of data. Thus, study of $(\beta\beta)_{0\nu}$ decay can provide important information about the left–right symmetry of weak interactions.

There are a variety of other processes such as $\mu \to e\gamma$, $\mu^- + (A, Z) \to e^+ +$ $(A, Z - 2)$, and muonium–antimuonium conversion that are characteristic of the left–right symmetric model.

§6.5 Baryon Number Nonconservation and Higher Unification

(a) Selection Rules for Baryon Nonconservation

This section is devoted to the study of baryon number nonconserving processes that arise within the framework of left–right symmetric theories. The reason that these models lead to breakdown of the baryon number is because the U(1) generator corresponds to the $B - L$ quantum number and that this U(1) symmetry is spontaneously broken. To study the selection rules for baryon number violation, we note the formula for electric charge:

$$Q = I_{3L} + I_{3R} + \frac{B - L}{2}. \tag{6.5.1}$$

Restricting ourselves to distance scales where $\Delta I_{3L} = 0$, we find that electric charge conservation implies

$$\Delta I_{3R} = \tfrac{1}{2}\Delta(B - L). \tag{6.5.2}$$

Because in our case $(\Delta I_{3R}) = 1$, this implies the selection rule $|\Delta(B-L)| = 2$. This has the following implications:

(i) $\Delta L = 2$, $\Delta B = 0$.   (see Section 6.4)

(ii) $\Delta L = 0$, $\Delta B = 2$.

(iii) $\Delta L = -1$, $\Delta B = 1$ [or $\Delta(B + L) = 0$].

In this section we will give the explicit Higgs model realizations of these selection rules for baryon nonconservation. First, we note that the minimal model described in the previous chapters [i.e., Higgs multiplets ϕ, $\Delta_L(3, 1, +2) + \Delta_R(1, 3, +2)$] has an extra symmetry under which all fields transform as $\phi \to e^{i\pi B}\phi$, where B is the baryon number of ϕ. This symmetry is not unbroken by vacuum. Thus, even though $\langle \Delta_R \rangle \neq 0$ breaks the local $B - L$ symmetry, a final global symmetry survives in the end that can be identified with the baryon number. So, to realize the full potential of left–right models, the Higgs sector has to be appropriately chosen. A more convenient framework for this is the partial unified gauge theory based on the group $SU(2)_L \times SU(2)_R \times SU(4)_C$ and the necessary Higgs multiplets to break $B - L$ symmetry [9].

(b) Partial Unification Model Based on $G_1 \equiv SU(2)_L \times SU(2)_R \times SU(4)_C$

This group was suggested in [1] and the symmetry breaking pattern of interest for baryon nonconservation was first discussed in [7] using the following

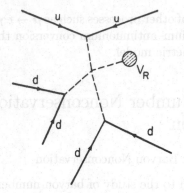

Figure 6.4.

multiplets (extending the work of [6]):

$$\phi(2,2,0),$$
$$\Delta_L(3,1,10) + \Delta_R(1,3,10).$$

Under $SU(2)_L \times SU(2)_R \times U(1)_{B-L} \times SU(3)_C$ the Δ's decompose as follows:

$$\Delta_L(3,1,10) = \{\Delta_{ll}(3,1,-2,1) + \Delta_{lq}(3,1,-2/3,3) + \Delta_{qq}(3,1,2/3,6)\}_L \tag{6.5.3}$$

and similarly for Δ_R. Note that all the submultiplets of Δ and ϕ are bilinears of the fermion fields, i.e., two leptons, two quarks, or quarks and leptons. Let us denote the fermions as follows:

$$\psi \rightarrow \begin{pmatrix} u_1 & u_2 & u_3 & \nu \\ d_1 & d_2 & d_3 & e \end{pmatrix}. \tag{6.5.4}$$

The G_1-invariant Yukawa couplings can be written as

$$\mathcal{L}_Y = f(\psi_{La}^T C^{-1} \tau_2 \tau \psi_{Lb} \cdot \mathbf{\Delta}_{Lab}^\dagger + L \rightarrow R) + \text{h.c.}, \tag{6.5.5}$$

where $a, b = 1,\ldots,4$ are the $SU(4)_C$ indices. In this notation the $SU(2)_R$ breaking occurs due to

$$\langle \Delta_{R^{44}}^{1+i2} \rangle = v_R \neq 0. \tag{6.5.6}$$

To obtain baryon-violating processes, note that the most general Higgs potential contains a term of the form

$$V' = \lambda \varepsilon^{a_1 a_2 a_3 a_4} \varepsilon^{b_1 b_2 b_3 b_4} \varepsilon^{pq} \varepsilon^{p'q'} \varepsilon^{rs} \varepsilon^{r's'}$$
$$\times \Delta_{a_1 b_1}^{pp'} \Delta_{a_2 b_2}^{qq'} \Delta_{a_3 b_3}^{rr'} \Delta_{a_4 b_4}^{ss'} + \text{all permutations.} \tag{6.5.7}$$

Equations (6.5.5), (6.5.6), and (6.5.7) lead to the Feynman diagram in Fig. 6.4, which causes $\Delta B = 2$ transitions such as $N - \bar{N}$ oscillation and $p + n \rightarrow \pi's$, etc. [42]. Let us now proceed to estimate the strength [7] of these processes $G_{\Delta B=2}$:

$$G_{\Delta B=2} \simeq \frac{f^3 \lambda v_R}{m_{\Delta_{qq}}^6}. \tag{6.5.8}$$

This is the strength of the free-quark Hamiltonian defined at mass scale $m_{\Delta qq}$. The actual low-energy $\Delta B = 2$ transition strength can be obtained from this; first, by doing renormalization group corrections to its strength [43]; and then taking account of the hadronic wave function effects [44] and nuclear effects [45,46,50,51]. All these corrections have the effect of reducing the $N - \bar{N}$ transition strength from $G_{\Delta B=2}$ (defined above) to $G_{\Delta B=2} \times 10^{-4}$. Thus, it is reasonable to say that, the $N - \bar{N}$ mixing strength δm is given by

$$\delta m \approx G_{\Delta B=2} \times 10^{-4} \text{ GeV}. \tag{6.5.9}$$

To present an estimate for δm in our model we note that v_R breaks both $SU(4)_C$ as well as $SU(2)_R$. Therefore we expect $m_{\Delta qq} \approx v_R$ to be consistent with the minimal fine tuning hypothesis. It can be argued from considerations of CP violation that $m_{W_R} \leq 36$ TeV, which implies 3 TeV $\leq$ $v_R \leq 100$ TeV. If we choose $m_{\Delta qq} \simeq v_R \approx 30$ TeV as a typical value and $f \approx \lambda \approx 10^{-1}$, we obtain

$$\delta m \simeq 10^{-31} \text{ GeV}. \tag{6.5.10}$$

This corresponds to $\tau_{N-\bar{N}} \simeq h/\delta m \simeq 10^7$ s.

The phenomenon of $N - \bar{N}$ oscillation can be connected to nuclear instability from which we can get experimental information on $\delta m_{N-\bar{N}}$. We present below the phenomenology of free and bound neutron oscillation to see whether these theoretical ideas can be experimentally tested. (We will define $\tau_{N-\bar{N}} = \hbar/\delta m_{N-\bar{N}}$ in what follows.)

Bounds on $\tau_{N-\bar{N}}$ from Stability of Matter

A question of crucial importance in planning experiments to detect $N - \bar{N}$ oscillation is the theoretical lower bound on $\tau_{N-\bar{N}}$ that is consistent with the known limits on matter stability. The main point is that the interaction Hamiltonian, which gives rise to $N - \bar{N}$ oscillation, can also lead to $N_1 + N_2 \to \pi's$ where $N_{1,2}$ can be protons and/or neutrons and thus lead to nuclear decay. Similarly, a neutron in the nucleus can convert into an antineutron which subsequently annihilates with a nucleon from the surrounding nuclear matter leading also to the same effect, i.e., nuclear instability. However, there already exist limits on nuclear instability from the various experiments that have looked for proton decay. This puts an upper bound on the strength of neutron–antineutron mixing, as we discuss below. From a preliminary discussion of the connection between $\delta m_{N-\bar{N}}$ and the amplitude for $N + P \to \pi's$, $A_{N+P} \to \pi's$, i.e.,

$$\delta m_{N-\bar{N}} \approx \sqrt{M/\tau_{NP \to \pi's}} \tag{6.5.11}$$

it was suggested earlier that $\tau_{N-\bar{N}} \geq 10^5$ s. Subsequently, more refined analysis has been carried out [60], and in this section we will discuss them.

For this purpose we study $N - \bar{N}$ oscillation inside the nucleus. The N–$\bar{N}$ mass matrix in the effective nuclear potential can be written as

$$M = \begin{array}{c} N \\ \bar{N} \end{array} \begin{array}{c} N \quad\quad \bar{N} \\ \left(\begin{array}{cc} V & \delta m \\ \delta m & V - iW \end{array} \right) \end{array} \quad\quad (6.5.12)$$

where δm represents the $N - \bar{N}$ transition mass; and V and $V - iW$ denote, respectively, the nuclear potentials seen by the neutron and the antineutron. The imaginary part of the $\bar{N}$-term represents the absorption of antineutrons by the nucleus. Diagonalization of the matrix in eq. (6.5.12) leads to two states $|N_{\pm}\rangle$ with mass eigenvalues $m_{\pm}$ given, respectively, by

$$m_+ \simeq V + \frac{(\delta m)^2(V - V)}{(V - \bar{V})^2 + W^2} - \frac{iW(\delta m^2)}{(V - \bar{V})^2 + W^2} \quad\quad (6.5.13)$$

and

$$m_- \simeq V - iW. \quad\quad (6.5.14)$$

The $|N_+\rangle$ eigenstate represents the conversion of N and $\bar{N}$ and subsequent nuclear decay with $\Delta B = 2$ with a lifetime τ_+ given by

$$\tau_+^{-1} \simeq \frac{W(\delta m)^2}{(V - \bar{V})^2 + W^2}. \quad\quad (6.5.15)$$

Since τ_+ is known from present limits on nuclear instability and V, $\bar{V}$, and W can be obtained from low-energy nuclear physics, eq. (6.5.14) can be translated to give an upper bound on δm or lower bound on $\tau_{N-\bar{N}} \geq h/\delta m$, i.e.,

$$\tau_{N-\bar{N}} \simeq \left\{ \tau_+ h \frac{W}{|V - \bar{V}|^2 + W^2} \right\}^{1/2}. \quad\quad (6.5.16)$$

There are lower bounds on τ_+ from existing experiments on proton decay [52]. However, we need information on $(V - \bar{V})$ and W from nuclear physics. There exist considerable uncertainties in their values, as inferred from different types of experiments.

One source of information on the $N\bar{N}$ potential is the $\bar{p}$-atom scattering experiments [62], which lead to the following values:

$$\bar{V} + iW = \text{(a)} \quad 240 - i120 \text{ MeV} \quad \text{(Barnes et al. [62])},$$

$$\text{(b)} \quad 165 - i165 \text{ MeV} \quad \text{(Poth et al. [47])},$$

$$\text{(c)} \quad 70 - i210 \text{ MeV} \quad \text{(Robertson et al. [47])}. \quad (6.5.17)$$

The variations could be attributed to the fact that $\bar{p}$-atom scattering is a surface effect where nuclear density is not very well known.

Another source of information on $\bar{V} - iW$ is from $\bar{p}p$ scattering data at low energies [48]. This leads to

$$W = -850 \text{ MeV} \quad \text{for } I = 1$$

$$= -659 \text{ MeV} \quad \text{for } I = 0.$$

Auerbach et al. [49] writes a potential with comparable values for the parameters

$$\text{(d)} \quad \bar{V} + iW = 1000 + i700 \text{ MeV}.$$

If we choose $\tau_+ \geq 1 \times 10^{31}$ yr as is indicated by recent data [56] and use $V = 0\text{--}50$ MeV we obtain for various cases

$$
\begin{aligned}
\tau_{N-\bar{N}} &\geq 1.5 \times 10^7 \text{s} && [\text{Case (a)}], \\
&\geq 2 \times 10^7 \text{s} && [\text{Case (b)}], \\
&\geq 2.6 \times 10^7 \text{s} && [\text{Case (c)}], \\
&\geq 0.9 \times 10^7 \text{s} && [\text{Case (d)}].
\end{aligned}
\tag{6.5.18}
$$

This corresponds to a value of $\delta m \approx 10^{-23}$ eV. Another way to obtain a lower limit on $\tau_{N-\bar{N}}$ has been discussed in [50]. In this method we consider the annihilation of $\bar{N}$ in nucleus via the ρ' virtual state, ρ' being the only resonance available with $J^P = 1^-$ and $I - 1$ in the 2 GeV mass region. Taking $g_{\rho'NN} \approx g_\rho$ and $g_{\rho'\bar{u}d}$ as the value of the quark–antiquark bound state wave function at the origin, we obtain

$$\tau_{N-\bar{N}} \geq 3 \times 10^6 \text{ s.} \tag{6.5.19}$$

We caution that there are uncertainties in the value of $g_{\rho'NN}$.

Finally, there are two recent analyses [51] where the nuclear effects have been studied with greater care and their conclusion is that the present lower limit on $\tau_{N-\bar{N}}$ is between $(2\text{--}8)\times 10^7$ s [52].

We point out, however, that there is an additional contribution to nuclear instability that arises from the same $\Delta B = 2$ interaction that gives rise to the $N - \bar{N}$ oscillation. This has to do with the direct decay of two nucleons in contact with pions in the presence of $\Delta B = 2$ interactions. Of course, the hard-core nature of the nuclear potential is likely to suppress this contribution. In any case, if we denote this contribution by $\Gamma_{\Delta B=2}$, the total width for nuclear decay is given by

$$\Gamma = (\Gamma_+)_{\text{ocs}} + \Gamma_{\Delta B=2}, \tag{6.5.20}$$

where $\tau_+^{-1} \equiv (\Gamma_+)_{\text{ocs}}$ and $\tau_{\text{Nuc}} \simeq \hbar/\Gamma$, and this additional contribution has the effect of increasing $\tau_{N-\bar{N}}$ somewhat. It has also been pointed out that the parameter $\delta m_{N-\bar{N}}$ may receive additional unknown nuclear renormalization that could effect the connection between free and bound $N - \bar{N}$ oscillation [53]. Let us now proceed to discuss ways of measuring $N - \bar{N}$ oscillation.

Free Neutron–Antineutron Oscillation

It was pointed out by Glashow [54] that the existence of the earth's magnetic field splits the $N - \bar{N}$ states by an amount $\Delta E = 2 \mu B$, where μ is the magnetic moment of the neutron and leads to $\Delta E \approx 10^{-11}$ eV $\gg \delta m$. This

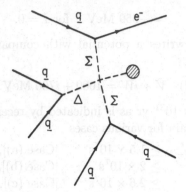

Figure 6.5.

would pose an immediate difficulty in searching for $N - \bar{N}$ mixing. However, it was subsequently demonstrated [55] that reducing $\Delta E \approx 10^{-14}$ eV is sufficient to simulate the conditions of a free neutron beam $[(\Delta E t/\hbar) \ll 1]$ thus making the experiment feasible. This would require degaussing the earth's magnetic field by a factor of 1000. For a free neutron beam the probability for detecting an $\bar{N}$ starting with a beam of neutrons is given by [as can be inferred from the analyzing equation (6.5.11)] $W = 0$ and $V = \bar{V} = m_N$. For thermal neutrons we can expect to have a flight time of about $t \approx 10^{-2}$ s. If we have a neutron beam intensity of about 10^{12} s^{-1}, then we can expect about 10 antineutrons per year making experimental detection of $N - \bar{N}$ oscillation feasible [56]. Observation of $N - \bar{N}$ oscillation will provide a clear vindication of the local $B - L$ symmetry models [57].

It has also been pointed out [58] that by extending this model to include a $(2, 2, 15) \equiv \Sigma$ Higgs multiplet we can realize the third selection rule, which follows from electric charge conservation, i.e., $\Delta(B + L) = 0$. The argument is that in the presence of Σ. there is a term in the Higgs potential of the form

$$V'' = \lambda \Delta_{aa'} \Delta_{bb'} \Sigma_c^{a'} \Sigma_d^{b'} \varepsilon^{abcd}, \qquad (6.5.21)$$

where for simplicity we have dropped the SU(2) indices. This gives rise to the Feynman diagram in Fig. 6.5, which obeys the $\Delta(B + L) = 0$ selection rule. The strength of this process is

$$G_{B+L=0} \approx \frac{f\lambda'h^2 V_R}{M_{\Delta qq}^2 m^4 \Sigma}. \qquad (6.5.22)$$

This can lead to decays of proton of type $p \rightarrow e^- \pi^+ \pi^+$, $n \rightarrow e^- \pi^+$, etc., with a lifetime of the order 10^{31}–10^{32} yr. However, in general grand unified models, this process becomes highly suppressed due to the survival hypothesis. Composite models can also lead to $\Delta B = -\Delta L$ type processes [59].

Experimental Search for Free Neutron–Antineutron Oscillation

A dedicated search for free $N - \bar{N}$ oscillation has been carried out at Grenoble. The first stage of the experiment yielded a limit $\tau_{N-\bar{N}} \geq 10^6$ s [60]. This experiment has been subsequently refined to yield the limit $\tau_{N-\bar{N}} \geq 8.6 \times 10^7$ s [61]. Another experiment carried out at Pavia has also reported a limit $\tau_{N-\bar{N}} \geq 0.5 \times 10^6$ s [62].

§6.6   Sin²θ_W and the Scale of Partial Unification

In this section we would like to consider the possibility that $G_1 \equiv \mathrm{SU}(2)_L \times \mathrm{SU}(2)_R \times \mathrm{SU}(4)_C$ is a partial unification group of the left–right symmetric model. G_1 has two couplings $g_2 = g_{2L} = g_{2R}$ and g_4 as against g_2, g_{B-L} and g_3 of the $\mathrm{SU}(2)_L \times \mathrm{SU}(2)_R \times \mathrm{U}(1)_{B-L} \times \mathrm{SU}(3)_C$ group. Owing to this unification we can obtain, at the unification scale M_c, a relation between $\sin^2 \theta_W$ and α_s. To obtain this relation we note that

$$\frac{1}{e^2} = \frac{2}{g_2^2} + \frac{1}{g_{BL}^2}. \tag{6.6.1}$$

But at M_c, $g_{BL}(M_c) = \frac{3}{2} g_4(M_c) = \frac{3}{2} g_3(M_c)$; from this it follows that

$$\sin^2 \theta_W(M_c) = \frac{1}{2} - \frac{1}{3} \frac{\alpha(M_c)}{\alpha_s(M_c)}. \tag{6.6.2}$$

In order to extrapolate this relation to $\mu = m_W$ we have to know the pattern of symmetry breaking from G_1 to $G_{123} \equiv \mathrm{U}(1)_Y \times \mathrm{SU}(2) \times \mathrm{SU}(3)_c$. Let us assume the pattern

$$\mathrm{SU}(2)_L \times \mathrm{SU}(2)_R \times \mathrm{SU}(4)c$$
$$\downarrow M_c$$
$$\mathrm{SU}(2)_L \times \mathrm{SU}(2)_R \times \mathrm{U}(1)_{B-L} \times \mathrm{SU}(3)_c$$
$$\downarrow M_R$$
$$\mathrm{U}(1)_Y \times \mathrm{SU}(2)_L \times \mathrm{SU}(3)_C.$$

The Higgs multiplets used to implement this pattern of symmetry breaking are those given in eq. (6.5.2). For this case we obtain, using the techniques of Georgi, Quinn, and Weinberg, that

$$\sin^2 \theta_W = \frac{1}{2} - \frac{1}{3} \frac{\alpha(m_W)}{\alpha_s(m_W)} - \frac{\alpha(m_W)}{4\pi} \left[\{(A_{2R} - A'_{2L}) + \frac{2}{3}(A_{BL} - A_3)\} \right.$$
$$\left. \times \ln \frac{M_c}{m_W} + \{\frac{5}{3} A_Y - A_{2R} - \frac{2}{3} A_{BL} + A_{2L}\} \ln \frac{M_R}{m_w} \right], \tag{6.6.3}$$

where the A_i's are the coefficients in the β function given for the $\mathrm{SU}(N)$ group by

$$A = -\frac{11}{3} N + \frac{4}{3} N_G + \frac{1}{6} T(R), \tag{6.6.4}$$

where N_G, is the number of generations and $T(R)$ is defined in terms of the generators θ_i of the group on the space of the Higgs multiplets

$$\text{Tr}(\theta_i \theta_j) = T(R)\delta ij. \tag{6.6.5}$$

If we keep only the fermion contribution, then this equation becomes

$$\text{Sin}^2 \theta_W = \frac{1}{2} - \frac{1}{3}\frac{\alpha(m_W)}{\alpha_s(m_W)} - \frac{11\alpha(m_W)}{6\pi}\left(\ln\frac{M_c}{m_W} + \ln\frac{M_R}{m_W}\right). \tag{6.6.6}$$

This implies that for $\sin^2 \theta_W \simeq 0.23$, $M_c M_R \simeq m_W^2 \times 10^{29}$ and for $\sin^2 \theta_W \simeq 0.25$, $M_c M_R \simeq m_W^2 \times 10^{27}$. If we choose $M_c = M_R$, this implies a partial unification scale of $10^{15} - 10^{16}$ GeV. On the other hand, if we assume $M_c \simeq M_{P_l} \simeq 10^{19}$ GeV, this implies $M_R \simeq 10^{12} - 10^{14}$ GeV. In either case, partial unification implies a very high scale for M_R; thus the $N - \bar{N}$ oscillation is highly suppressed and so are the $\Delta(B+L) = 0$ processes. It has recently been pointed out [63] that if the parity and $\text{SU}(2)_R$ breaking scales are decoupled, then at $\mu \approx m_{W_R}$, $g_L \neq g_R$ and in that case both M_c, and M_{W_R} can be lowered to the level of 100 TeV or so without conflicting with the observed value of $\sin^2 \theta_W$.

§6.7 Left–Right Symmetry—An Alternative Formulation

In recent years, an alternative formulation of the left–right symmetric models has been discussed, which ameliorates some of the naturalness problems of the standard left–right symmetric models [64–67], at the price of including isosinglet vectorlike heavy fermions with charge $+\frac{2}{3}$, color triplet (P), charge $-\frac{1}{3}$, color triplet (N), and charge -1, color singlet (E). This model has the following advantages:

(1) It generates charged fermion masses through a seesaw-type mechanism so that Yukawa coupling to light fermions need not be too small [64].

(2) It provides a natural mechanism for generating an ultralight Dirac neutrino for all three generations without requiring the scale of right-handed interactions to be beyond the TeV scale [65].

(3) It provides a natural setting for solving the problem of weak and strong CP-violation without invoking any new discrete symmetry [66].

(4) By taking one set of heavy fermions, one can generate a hierarchy of quark and lepton masses as well as their mixings [67].

(5) The model has an extremely simple Higgs sector consisting of two isodoublets, and in the unitary gauge only two neutral Higgs scalars are left over.

Let us now give some details: The gauge group for the model is of course $SU(2)_L \times SU(2)_R \times U(1)_{B-L}$ with quark doublet Q and lepton doublet $\Psi \equiv (\nu, e)$, transforming under the gauge group as in eq. (6.2.1). In addition, the model includes vectorlike isosinglet heavy quarks P, N, and lepton E as stated earlier. We will consider one heavy fermion set per generation. The $B - L$ quantum numbers of P, N, and E are $\frac{4}{3}$, $-\frac{2}{3}$, and -2, respectively.

The Higgs sector consists of only a pair of left–right symmetric doublets $\chi_L(2, 1, 1) \oplus \chi_R(1, 2, 1)$ with the following Higgs potential:

$$V = -(\mu_L^2 \chi_L^+ + \chi_L + \mu_R^2 \chi_R^+ \chi_R) + \frac{\lambda_1}{2}[(\chi_L^+ \chi_L)^2 + (\chi_R^+ \chi_R)^2] + \lambda^2(\chi_L^+ \chi_L)(\chi_R^+ \chi_R).$$
(6.7.1)

Note that the χ masses break parity invariance softly. The minimum of the potential corresponds to

$$\langle \chi_L^0 \rangle = \frac{v_L}{\sqrt{2}}, \qquad \langle \chi_R^0 \rangle = \frac{v_r}{\sqrt{2}}, \qquad (6.7.2)$$

where

$$\frac{v_L^2}{2} = \frac{\lambda_2 \mu_R^2 - \lambda_1 \mu_L^2}{\lambda_2^2 - \lambda_1^2}; \qquad \frac{v_R^2}{2} = \frac{\lambda_2 \mu_L^2 - \lambda_1 \mu_R^2}{\lambda_2^2 - \lambda_1^2}. \qquad (6.7.3)$$

Choosing $\mu_R \geq \mu_L$ guarantees $v_R \geq v_L$, which in turn implies that the right-handed charged-current effects are suppressed. In the unitary gauge there are only two physical Higgs bosons: $\sigma_L \equiv \sqrt{2}\, \text{Re}\, \chi_L^0$ and $\sigma_R \equiv \sqrt{2}\, \text{Re}\, \chi_R^0$; these two states mix at the tree level with mixing angle $\xi \simeq (\lambda_2/\lambda_1)(v_L/v_R)$ for $v_R \gg v_L$. Their masses are given by

$$M_{\sigma_R}^2 \simeq \lambda_1 v_R^2; \qquad M_{\sigma_L}^2 \simeq \lambda_1 \left(1 - \frac{\lambda_2^2}{\lambda_1^2}\right) v_R^2. \qquad (6.7.4)$$

In the gauge sector, in contrast to the model described previously in this chapter, there is $W_L - W_R$ mixing at the tree level. It arises at the one-loop level and is finite (see below). The masses of W_L and W_R are given by

$$M_{W_L} = \frac{g v_L}{2}; \qquad M_{W_R} = \frac{g v_R}{2}. \qquad (6.7.5)$$

In the neutral-current sector, the photon field is given as in eq. (6.2.18), the two massive neutral gauge bosoms are

$$Z_1 = \cos \alpha Z_L + \sin \alpha Z_R,$$
$$Z_2 = -\sin \alpha Z_L + \cos \alpha Z_R, \qquad (6.7.6)$$

where Z_L and Z_R are defined in eq. (6.2.18) with

$$\tan 2\alpha = \frac{2 v_L^2 \sec^2 \theta_W \sqrt{\sec^2 \theta_W}}{v_R^2 \cot^2 \theta_W \sec 2\theta_W + v_L^2 (\tan^2 \theta_W \sec 2\theta_W - \text{cosec}^2 \theta_W \sec^2 \theta_W)}. \qquad (6.7.7)$$

Obviously $\alpha \simeq v_L^2/v_R^2 \ll 1$.

In order to discuss the fermion masses in this model, let us write down the most general Yukawa interaction

$$
\begin{aligned}
L_Y = & \bar{Q}_L h_u \chi_L P_R + \bar{Q}_L h_d \chi_L N_R \\
& + \bar{\psi}_L h_2 \chi_L E_R + L \leftrightarrow R \\
& + \bar{P}_L M_P P_R + \bar{N}_L M_N N_R + \bar{E}_L M_E E_R + \text{ h.c.}
\end{aligned}
\tag{6.7.8}
$$

The h's and M_P's are, in general, complex. The scale of the masses $M_{P,N,E}$ is larger than the V_R scale. Subsequent to spontaneous symmetry breaking, the fermion mass matrices connecting the light and heavy fermions are given by

$$
M_\alpha = \begin{pmatrix} 0 & h_a v_L/\sqrt{2} \\ h_a^+ v_R/\sqrt{2} & M_a \end{pmatrix},
\tag{6.7.9}
$$

where a goes over u, d, and e, respectively. On diagonalizing this matrix, one typically gets for the light fermion mass

$$
M_{f_a} \approx \frac{h_a^2 v_l v_r}{2 M_a}.
\tag{6.7.10}
$$

We note that, even if $M_a \approx v_R$, to get the correct pattern of fermion masses, the smallest value required for h_a is of the order $\approx 10^{-2}$. Thus, this somewhat alleviates the naturalness problem of standard models. Equation (6.7.10) is a seesaw-type formula, now applied to the quark sector. Secondly, note that $m_{v_i} = 0$ at the tree levels and is therefore likely to be small since it comes from the loop effects. Finally, regardless of the nature of CP violation,

$$
\text{Arg Det } M = 0.
\tag{6.7.11}
$$

Therefore, if we set the QCD $\theta = 0$ using softly broken parity invariance, the model solves the strong CP-problem [66].

§6.8   Higher Order Effects

There are a number of interesting higher order effects for the seesaw type models.

(i) Loop-Induced $W_L - W_R$ Mixing

At the tree level, $W_L - W_R$ mixing vanishes as we saw, and it arises at the one-loop level via the diagram in Fig. 6.6. It is given by

$$
\zeta_{L-R} \simeq \frac{\alpha}{4\pi \sin^2 \theta_W} \cdot \frac{m_b m_t}{M_{W_R}^2}.
\tag{6.8.1}
$$

Therefore, for $m_b \simeq 5$ GeV and $m_t \simeq 100$ GeV, we get $\zeta_{L-R} \approx 10^{-7}$ for $M_{W_R} \approx 5$ TeV and $\approx 10^{-8}$ for $M_{W_R} \approx 15$ TeV. This is smaller than the

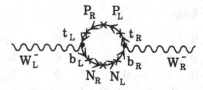

Figure 6.6.

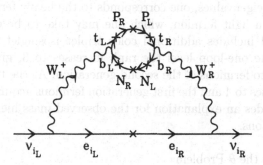

Figure 6.7.

most stringent bounds on ζ that arise from SN1987A neutrino observations [31a].

(ii) Loop Induced Neutrino Masses

An interesting feature of this model is that neutrino masses arise at the two-loop level from the diagram in Fig. 6.7. Therefore, its magnitude is automatically suppressed to become

$$m_{\nu_i} \simeq \frac{\alpha m_{e_i}}{4\pi \sin^2 \theta_W} \zeta_{L-R} I, \qquad (6.8.2)$$

where I denotes the effects of integration (and is of order 1–10). It turns out that the dominant contribution comes from the longitudinally Higgs boson exchanges, and one finds $m_{\nu_i} \simeq (10^{-8} - 10^{-7}) m_{e_i}$ for $m_{W_R} \simeq 5$ TeV where i denotes the generation index. The smallness of the neutrino mass is therefore automatic in the quark seesaw-type model (e.g., $m_{\nu_e} \simeq 10^{-2}$ eV, $m_{\nu_\mu} \simeq 1$ eV, and $m_{\nu_t} \simeq 20$ eV). Furthermore, the neutrino is a Dirac particle. Thus, lepton number violating processes such as double beta decay are forbidden.

(iii) Fermion Mass and Mixing Hierarchy

If we construct the seesaw models in such a way as to allow only one set of heavy quarks and leptons (P, N, E) (instead of three), the seesaw matrix in eq. (6.7.9) has the form

$$M_a = \begin{pmatrix} 0 & 0 & 0 & h_a v_L/\sqrt{2} \\ 0 & 0 & 0 & h_a v_L/\sqrt{2} \\ 0 & 0 & 0 & h_a v_L/\sqrt{2} \\ h_a v_R/\sqrt{2} & h_a v_R/\sqrt{2} & h_a v_R/\sqrt{2} & M_a \end{pmatrix}. \quad (6.8.3)$$

This matrix has rank 2 and therefore two zero eigenvalues at the tree level. Of the nonzero eigenvalues, one corresponds to the heavy fermion and one corresponds to a light fermion, which one may take to be either t, b, or τ. If the model includes additional color triplet–isosinglet Higgs bosons [67], then at the one-loop level the rank increases to 3, giving one-loop radiative mass to fermions of the second generation. At the two-loop level, the rank increases to 4 and the first generation fermions acquire mass. This, therefore, provides an explanation for the observed mass hierarchy among quarks and leptons.

(iv) Solution to the θ Problem

As mentioned in Section 6.7, the $\bar{\theta}$ at the tree level vanishes automatically, due to a combination parity invariance and the seesaw nature of quark mass matrices. It has been shown in [66] that $\bar{\theta}$ also vanishes at the one-loop level. It arises only at the two-loop level thus explaining its smallness. Thus, without any new discrete symmetry, we have a solution to the strong CP problem.

The seesaw model therefore has a number of attractive features. The price we pay is the inclusion of the heavy fermions P, N, E. Their masses are expected to be in the TeV range and, therefore, most of their radiative effects are suppressed by this heavy mass factor.

§6.9 Conclusions

In summary, we have noted some compelling intuitive reasons for suspecting the existence of a next level of unification symmetry, which is left–right symmetric. It not only restores parity to the status of a conserved quantum number of all fundamental interactions, but it also brings $B - L$ to the level of a local symmetry like electric charge. This is very appealing, since it extends the Gell-Mann–Nishijima electric charge formula from the domain of strong interactions to that of weak interactions. Extending this analogy a little further leads us to the suggestion that, just as the quark picture provides an underlying dynamical basis for the Gell-Mann–Nishijima formula for strong interactions, an underlying preonic substructure of quarks and leptons may form the basis of $B - L$ as the gauge generator of weak interactions. Indeed, most attractive preon models bear out this conjecture. Thus, left–right symmetric models may provide the big leap forward from a successful geometrical picture of weak interactions based on the $SU(2)_L \times U(1) \times SU(3)_c$ model to a dynamical model that may provide a natural basis for understanding many of the puzzles of weak interactions.

Clearly, further experimental and theoretical work is needed before this appealing scenario receives its rightful place in the annals of particle physics. At the experimental level we believe that any or all of the following pieces of evidence would constitute manifestations of left–right symmetry:

(i) nonvanishing neutrino mass;

(ii) $\Delta B = 2$ transitions such as nucleon + nucleon $\rightarrow$ pions and $N - \bar{N}$ oscillations; and

(iii) lepton number violating and lepton flavor changing processes such as $(\beta\beta)_{0\nu}$-decay, $\mu^- A \rightarrow e^+ A$, $\mu^- e^+ \rightarrow \mu^+ e^- \mu \rightarrow 3e$.

Exercises

6.1. Consider the most general $U(1)_L \times U(1)_R$-gauge and parity invariant potential involving a pair of Higgs bosons ($\phi_L \xrightarrow{P} \phi_R$), and show that for two ranges of parameters in the Higgs potential there are two possible minima: (a) $\langle \phi_L \rangle = \langle \phi_R \rangle = 0$ or (b) $\langle \phi_R \rangle \neq 0$ and $\langle \phi_L \rangle = 0$, or vice versa. Indicate the ranges of parameters for which these results obtain. Include the one-loop corrections into the result and show how this affects the results presented above.

6.2. Prove, by a detailed analysis of the Higgs potential of the left–right symmetric model with ϕ, $\Delta_L \oplus \Delta_R$ Higgs bosons, that $v_L \equiv \langle \Delta_L^0 \rangle$ satisfies the relation

$$v_L \simeq \gamma \frac{\kappa^2}{V_R} + \gamma' \frac{\kappa\kappa'}{V_R} + \gamma'' \frac{\kappa'^2}{V_R}.$$

Show that if we include in the model a parity odd real scalar field σ with v.e.v.

$$\langle \sigma \rangle \equiv M_P \gg V_R,$$

then

$$v_L \simeq \gamma \frac{\kappa^2 V_R}{M_P^2}.$$

6.3. Discuss the possible transformations of the field $\phi(2, 2, 0)$ under parity operation and show that:

(a) for the simplest possibility of $\phi \xrightarrow{P} \phi^\dagger$, the quark mixing angles in the left- and right-handed sectors are the same;

(b) if $\phi \xrightarrow{P} \phi^T$ the left-handed quark mixing angles vanish;

(c) the mixing angles also vanish when $\kappa = \kappa'$;

(d) if we want the quark mixings in the left- and right-handed weak interactions to be different, we need at least two ϕ's. Give the transformation property of ϕ under parity transformation when this happens.

6.4. The left–right symmetric theories could be carried one step further if we chose the gauge group $SU(2)_L \times SU(2)_R, \times U(1)_L \times U(1)_R$. Discuss the choice of Higgs bosons, symmetry breaking, and anomaly freedom in these theories. Discuss the properties of the extra Z-boson in these theories. What would be a natural grand unifying group for this model? Discuss some salient features of this model.

6.5. Show that in the $SU(2)_L \times SU(2)_R \times SU(4)_C$, partial unification model, with the Higgs multiplets $\phi(2,2,0), \Delta_L(3,1,10) \oplus \Delta_R(1,3,10)$, proton is stable to all orders. How will the stability of the proton be affected if we included a (1, 1, 6) Higgs multiplet in the model. Show that the model without (1, 1, 6) Higgs multiplet in the model. Show that the model without (1, 1, 6) Higgs also leads to hydrogen–antihydrogen oscillation $H \leftrightarrow \bar{H}$. Estimate its strength in this theory. Can you think of any possible bounds that may arise on the $H - \bar{H}$ transition time from existing observations?

6.6. Describe an $SU(2)_L \times SU(2)_R \times SU(4)_C$, generalization of the quark seesaw model of Section 6.7. What kind of baryon-violating signatures are present in the model, if any? Support your answer with explicit diagrams.

References

[1] J. C. Pati and A. Salam, *Phys. Rev.* **D1O**, 275 (1974);
    R. N. Mohapatra and J. C. Pati, *Phys. Rev.* **D11**, 566, 2558 (1975);
    G. Senjanovic and R. N. Mohapatra, *Phys. Rev.* **D12**, 1502 (1975).

[2] J. F. Wilkerson et al., *Phys. Rev. Lett.* **58**, 2023 (1987);
    H. Kawakami et al., *Phys. Lett.* **187B**, 198 (1987).

[2] Recent books on massive neutrinos are:
    R. N. Mohapatra and P. B. Pal, *Massive Neutrinos in Physics and Astrophysics*, World Scientific, Singapore, 1991;
    B. Kayser, F. Gibradebu, and F. Perrier, *Massive Neutrinos*, World Scientific Singapore, 1989.

[3] R. Davis, Jr., *Neutrino '88* (edited by J. Schnops et al.), World Scientific, Singapore, 1988;
    K. K. Hirata, et al., *Phys. Rev. Lett.* **63**, 16 (1989); **65**, 1297, 1301 (1990);
    Y. Fukuda et al., *Phys. Rev. lett.* **81**, 1562 (1998); *ibid* **82**, 2644 (1999);
    W. Hampel et al., *Phys. Lett.* **B 388**, 384 (1996);
    J. N. Abdurashitov et al., *Phys. Rev. Lett.* **77**, 4708 (1996);

[4] R. N. Mohapatra and R. E. Marshak, *Phys. Lett.* **91B**, 222 (1980);
A. Davidson, *Phys. Rev.* **D20**, 776 (1979).

[5] D. Chang, R. N. Mohapatra, and M. K. Parida, *Phys. Rev. Lett.* **50**, 1072 (1984);
Phys. Rev. **D30**, 1052 (1984).

[6] R. N. Mohapatra and G. Senjanović, *Phys. Rev. Lett.* **44**, 912 (1980);
Phys. Rev. **D23**, 165 (1981).

[7] R. N. Mohapatra and R. E. Marshak, *Phys. Rev. Lett.* **44**, 1316 (1980).

[8] R. N. Mohapatra, F. E. Paige, and D. P. Sidhu, *Phys. Rev.* **D17**, 2642 (1978).

[9] V. Barger, E. Ma, and K. Whisnant, *Phys. Rev.* **D26**, 2378 (1982);
I. Liede, J. Malampi, and M. Roos, *Nucl. Phys.* **B146**, 157 (1978);
T. Rizzo and G. Senjanovic, *Phys. Rev.* **D24**, 704 (1981);
V. Barger, J. Hewett, and T. Rizzo, *Phys. Rev.* **D42**, 152 (1990).

[10] For a review, see B. Kayser and R. N. Mohapatra, in *Perspectives in Neutrino Physics*, ed. D. O. Caldwell, Springer-Verlag (2001).

[11] M. Gell-Mann, P. Ramand, and R. Slansky, in *Supergravity* (edited by D. Freedman et al.), North-Holland, Amsterdam, 1979;
T. Yanagida, KEK lectures, 1979;
R. N. Mohapatra and G. Senjanovic, *Phys. Rev. Lett.* **44**, 912 (1980).

[12] M. Gronau and S. Nussinov, Fermilab preprint, 1982;
M. Gronau and R. Yahalom, *Nucl. Phys.* **B236**, 233 (184).

[13] M. A. B. Beg, R. Budny, R. N. Mohapatra, and A. Sirlin, *Phys. Rev. Lett.* **38**, 1252 (1977);
For a subsequent extensive analysis see
J. Maalampi, K. Mursula, and M. Roos, *Nucl. Phys.* **B207**, 233 (1982).

[14] M. Roos et al., *Phys. Lett.* **111B**, 1 (1982).

[15] F. W. Koks and J. Vanklinken, *Nucl. Phys.* **A272**, 61 (1976).

[16] J. Carr et al., *Phys. Rev. Lett.* **51**, 627 (1983);
A. Jodidio et al., *Phys. Rev.* **D34**, 1967 (1986).

[17] F. Corriveau et al., *Phys. Rev.* **D24**, 2004 (1981); *Phys. Lett.* **129B**, 260 (1983).

[18] T. Yamazaki et al., KEK preprint, 1983.

[19] B. Holstein and S. Treiman, *Phys. Rev.* **D16**, 2369 (1977).

[20] D. Bryman, Talk at Mini-Conference on Low-Energy Tests of Conservation Law, 1983.

[21] T. Yamazaki et al., KEK preprint, 1983.

[22] J. Deutsch, Proceedings of the workshop on *Breaking of Fundamental Symmetries in Nuclei*, ed. J. Ginochio and S. P. Rosen (World Scientific, 1989).

[23] G. Beall, M. Bender, and A. Soni, *Phys. Rev. Lett.* **48**, 848 (1982).

[24] Earliest use of vacuum saturation of short distance contribution to $K_L - K$, mass difference was by

R. N. Mohapatra, J. S. Rao, and R. E. Marshak, *Phys. Rev.* **171**, 1502 (1968);

B. L. Ioffe and E. Shabalin, *Sov. J. Nucl. Phys.* **6**, 328 (1967).

[25] M. K. Gaillard and B. W. Lee, *Phys. Rev.* **D10**, 897 (1974).

[26] See J. Trampetic, *Phys. Rev.* **D27**, 1565 (183) for a discussion of this point.

[27] R. N. Mohapatra, G. Senjanovic, and M. Tran, *Phys. Rev.* **D28**, 546 (1983);
G. Ecker, W. Grimus, and H. Neufeld, *Phys. Lett.* **127B**, 365 (1983);
H. Harari and M. Leurer, *Nucl. Phys.* **B223**, 221 (1983);
F. Gilman and M. Reno, *Phys. Rev.* **D29**, 937 (1974).

[28] R. Barbieri and R. N. Mohapatra, *Phys. Rev.* **D39**, 1229 (1989);
G. Raffelt, and D. Seckel, *Phys. Rev. Lett.* **60**, 1793 (1988).

[29] R. Bionta et al., *Phys. Rev. Lett.* **58**, 1494 (1987);
K. Hirata et al., *Phys. Rev. Lett.* **58**, 1490 (1987).

[30] R. N. Mohapatra, *Phys. Rev.* **D34**, 909 (1986).

[31] A. Datta and A. Raychaudhuri, *Phys. Rev.* **D28**, 1170 (1983);
F. Olness and M. E. Ebel, *Phys. Rev.* **D30**, 1034 (1984);
P. Langacker and S. Uma Sankar, *Phys. Rev.* **D40**, 1569 (1989).

[31a] J. Donoghue and B. Holstein, *Phys. Lett.* **113 B**, 383 (1982).

[31b] K. S. Babu, K. Fujikawa and A. Yamada, *Phys. Lett.* **B 333**, 196 (1994);
P. Cho and M. Misiak, *Phys. Rev.* **D49**, 5894 (1994);
T. Rizzo, *Phys. Rev.* **D 50**, 3303 (1994).

[31c] L. Wolfenstein, *Phys. Rev.* **D 29**, 2130 (1984).

[32] J. Gunion and B. Kayser, *Proceedings of the 1984 Snowmass Meeting* (edited by R. Donaldson et al.), p. 153;
G. Altarelli, B. Mele, and M. Ruiz Altaba, CERN preprint (1989);
F. Feruglio, L. Maiani, and A. Masiero, Padova preprint (1989);
W. Keung and G. Senjanovic, *Phys. Rev. Lett.* **50**, 1427 (1983).

[33] For a review, see B. Kayser, K. Macferland, and P. Fisher, *Ann. Rev. Nucl. Part. Sc.* **49**, 481 (1999).

[34] Particle Data Group, *Euro. Phys. Journ.* **C 15**, 1 (2000).

[35] For an extensive discussion of this topic, see, *Neutrinoless Double Beta Decay and Related Processes*, ed. H. Klapdor-Klein-Gorthaus and S. Stoica (World Scientific, 1995); for a recent review, see H. Klapdor-Kleingrothaus et al. hep-ex/9910205; H. V. Klapdor-Kleingrothaus, J. Hellmig and M. Hirsch, *J. Phys.* **G 24**, 483 (1998); for earlier references, see
H. Primakoff and S. P. Rosen, *Rep. Progr. Phys.* **22**, 121 (1959); *Proc. Phys. Soc. (London)* **78**, 464 (1961);
A. Halprin, P. Minkowski, H. Primakoff, and S. P. Rosen, *Phys. Rev.* **D13**, 2567 (1976);
M. Doi, T. Kotani, H. Nishiura, K. Okuda, and E. Takasugi, *Prog. Theor. Phys.* **66**, 1765 (1981); **68**, 348 (1982) (E);
W. Haxton, G. J. Stephenson, Jr., and D. Strottman, *Phys. Rev. Lett.* **47**, 153 (1981);

Phys. Rev. **D25**, 2360 (1982);

J. D. Vergados, *Phys. Rev.* **C24**, 640 (1981).

D. Caldwell et al., *Phys. Rev. Lett.* **59**, 419 (1987);

H. Ejiri et al., *J. Phys.* **G13**, 839 (1987).

[36] L. Baudis et al. hep-ex/9902014.

[37] C. Athanasopoulos et al. *Phys. Rev. Lett.* **77**, 3406 (1996).

[38] L. Wolfenstein, *Phys. Lett.* **107B**, 77 (1981).

J. Valle, *Phys. Rev.* **D27**, 1672 (1983);

S. Petcov, *Phys. Lett.* **110B**, 245 (1982);

M. Doi, M. Kenmoku, T. Kotani, H. Nishiura, and E. Taskasugi, Osaka preprint OS-GE-83-48, 1983.

[39] K. M. Case, *Phys. Rev.* **107**, 307 (1957); M. Zralek, hep-ph/99 ; B. Kayser and R. N. Mohapatra, Ref. [10].

[40] D. Chang and R. N. Mohapatra, *Phys. Rev.* **D32**, 1248 (1985).

[41] S. Mikheyev and A. Y. Smirnov, *Nuovo Cimento* **9C**, 17 (1986);

L. Wolfenstein, *Phys. Rev.* **D17**, 2369 (1978).

[42] R. N. Mohapatra, *Nucl. Instr. Methods* **A284**, 1 (1989), for a review.

[43] W. Caswell, J. Milutinovic, and G. Senjanović, *Phys. Rev.* **D26**, 161 (1982);

S. Rao and R. Shrock, *Phys. Lett.* **116B**, 238 (1982).

[44] J. Pasupathy, *Phys. Lett* **114B**, 172 (1982);

S. Rao and R. Shrock, Phys. Lett. 116B, 238 (1982);

U. Sarkar and S. P. Misra, *Phys. Rev.* **D28**, 249 (1983).

[45] K. Chetyrkin et. al., *Phys. Lett.* **99B**, 358 (1981);

P. G. Sandars, *J. Phys.* **G6**, L161 (1980);

[46] W. Alberico et al., *Phys. Lett.* **114B**, 266 (1982);

For a review see R. N. Mohapatra, *Proceedings of the Harvard Workshop on N − N̄ Oscillation*, 1982.

[47] Barnes et al., *Phys. Rev. Lett.* **29**, 1132 (1972);

Poth et al., *Nucl. Phys.* **A294**, 435 (1977);

Roberson et al., *Phys. Rev.* **C16**, 1945 (1977);

For a review see

C. J. Batty, Rutherford Laboratory preprint, 1981.

[48] J. Cote et al., *Phys. Rev. Lett.* **48**, 13198 (1982).

[49] R. Auerbach et al., *Phys. Rev. Lett.* **46**, 702 (1980).

[50] Riazzuddin, *Phys. Rev.* **D25**, 885 (1982).

[51] C. Dover, A. Gal, and J. Richards, *Phys. Rev.* **D27**, 1090 (1983);

A. Kerman et al., MIT preprint, 1983.

[52] L. Jones et al., *Phys. Rev. Lett.* **52**, 720 (1984);

M. L. Cherry et al., *Phys. Rev. Lett.* **50**, 1354 (1983);

J. Chung et al., hep-ex/020593;

T. Takita et al., *Phys. Rev.* **D34**, 902 (1986);

M. Beger et al., *Phys. Lett.* **B240**, 237 (1990).

[53] P. K. Kabir, *Phys. Rev. Lett.* **51**, 231 (1983).

[54] S. L. Glashow, Cargese lectures, 1979.

[55] R. N. Mohapatra and R. E. Marshak, *Phys. Lett.* **94B**, 183 (1980).

[56] M. Baldoceolin et al., CERN preprint (1983).

[57] For a recent proposal to measure neutron-anti-enutron oscillation at the Oak-Ridge National Laboratory, see Y. Kamyshkov, in *Proceedings of the Workshop on Nucleon Stability*, ed. Y. Kamyshkov et al. (1996):

[58] J. C. Pati, A. Salam, and U. Sarkar, *Phys. Lett.* **B133**, 330 (1983).

[59] H. Harari, R. N. Mohapatra, and N. Seiberg, *Nucl. Phys.* **B209**, 174 (1982).

[60] G. Fidecaro et al., *Phys. Lett.* **B156**, 122 (1985).

[61] M. Baldoceolin et al., *Phys. Lett.* **B236**, 95 (1990).

[62] S. Ratti et al., *Z. Phys.* **C43**, 175 (1989).

[63] D. Chang, R. N. Mohapatra, J. Gipson, R. E. Marshak, and M. K. Parida, *Phys. Rev.* **D31**, 1718 (1985).

[64] A. Davidson and K. C. Wali, *Phys. Rev. Lett.* **59**, 393 (1987);
S. Rajpoot, *Phys. Lett.* **B191**, 122 (1987).
D. Chang and R. N. Mohapatra, *Phys. Rev. Lett.* **58**, 1600 (1987).

[65] R. N. Mohapatra, *Phys. Lett.* **210B**, 517 (1988);
K. S. Babu and X. He, *Mod. Phys. Lett.* **A4**, 61 (1989).

[66] K. S. Babu and R. N. Mohapatra, *Phys. Rev. Lett.* **62**, 1079 (1989); *Phys. Rev.* **D41**, 1286 (1990).

[67] B. S. Balakrishna, *Phys. Rev. Lett.* **60**, 1602 (1988);
B. S. Balakrishna, A. Kagan and R. N. Mohapatra, *Phys. Lett.* **205B**, 345 (1988);
B. S. Balakrishna and R. N. Mohapatra, *Phys. Lett.* **216B**, 349 (1989).

7

SO(10) Grand Unification

§7.1   Introduction

The possibility of SO(10) as a grand unification group of the standard $SU(2)_L \times U(1)_Y \times SU(3)_c$ was first noted by Georgi [1] and Fritzsch and Minkowski [1]. Unlike SU(5), SO(10) is a group of rank 5 with the extra diagonal generator of SO(10) being $B - L$ as in the left–right symmetric groups. The advantages of SO(10) over SU(5) grand unification are that:

(a) only one 16-dimensional spinor representation of SO(10) has the correct quantum numbers to accommodate all fermions (including the right-handed neutrino) of one generation;

(b) the gauge interactions of SO(10) conserve parity thus making parity a part of a continuous symmetry: this has the advantage that it avoids the cosmological domain wall problem associated with parity symmetry breakdown; and

(c) it is the minimal left–right symmetric grand unified model that gauges the $B - L$ symmetry and is the only other simple grand unification group that does not need mirror fermions [2]. The model does not have any global symmetries.

Before proceeding to a discussion of the model, we first spell out the various maximal subgroups of SO(10) as well as the decomposition of the small-dimensional representations under the interesting subgroups

$$SO(10) \underset{\longrightarrow}{\overset{\longrightarrow}{}} \begin{array}{l} SU(5) \times U(1), \\ SO(6) \times SO(4), \end{array} \quad \text{or} \quad SU(4) \times SU(2)_L \times SU(2)_R \times D.$$
$$(7.1.1)$$

Decomposition of some of the smaller SO(10) irreducible multiplets under (a) $SU(2)_L \times SU(2)_R \times SU(4)_C$ and under (b) $SU(5) \times U(1)$ are

$\{10\}$ (a) $(1,1,6) + (2,2,1)$;

(b) $\{5\} + \{\bar{5}\}$.

$\{16\}$ (a) $(2,1,4) + (1,2,\bar{4})$;

(b) $\{10\} + \{\bar{5}\} + \{1\}$.

$\{45\}$ (a) $(3,1,1) + (2,2,6) + (1,1,15) + (1,3,1)$;

(b) $\{24\} + \{10\} + \{\overline{10}\} + \{1\}$.

$\{54\}$ (a) $(1,1,1) + (2,2,6) + (1,1,20) + (3,3,1)$;

(b) $\{15\} + \{\overline{15}\} + \{24\}$.

$\{120\}$ (a) $(2,2,15) + (3,1,6) + (1,3,\bar{6}) + (2,2,1) + (1,1,20)$;

(b) $\{5\} + \{\bar{5}\} + \{10\} + \{\overline{10}\} + \{45\} + \{\overline{45}\}$.

$\{126\}$ (a) $(3,1,10) + (1,3,\overline{10}) + (2,2,15) + (1,1,\bar{6})$;

(b) $\{1\} + \{\bar{5}\} + \{10\} + \{\overline{15}\} + \{45\} + \{\overline{50}\}$.

$\{210\}$ (a) $(1,1,1) + (1,1,15) + (2,2,20) + (3,1,15) + (1,3,15) + (2,2,6)$;

(b) $\{1\} + \{5\} + \{\bar{5}\} + \{10\} + \{\overline{10}\} + \{24\} + \{40\} + \{\overline{40}\} + \{75\}$.

In order to discuss the SO(10) model it is convenient to give the representations in a simple algebraic formulation, which we do in the next section, and identify the fermions in the spinor representation.

§7.2   SO(2N) in an SU(N) Basis [3]

There exist different ways [3, 4] to discuss the algebra of SO(10). A particularly convenient way is to use the spinor $SU(N)$ basis [3], which we present below. Consider a set of N operators $\chi_i (i = 1, \ldots, N)$ and their Hermitian conjugate χ_i^+ satisfying the following anticommutation relations:

$$\{\chi_i, \chi_j^+\} = \delta_{ij},$$
$$\{\chi_i, \chi_j\} = 0. \qquad (7.2.1)$$

We use the symnol $\{\,,\,\}$ to denote anticommutation and $[\,,\,]$ to denote the commutation operation. It is well known that the operators T_j^i defined as

$$T_j^i = \chi_i^+ \chi_j \qquad (7.2.2)$$

satisfy the algebra of the U(N) group, i.e.,

$$[T_j^i, T_l^k] = \delta_j^k T_l^i - \delta_l^i T_j^k. \tag{7.2.3}$$

Now let us define the following $2N$ operators, $\Gamma_\mu (\mu = 1, \ldots, 2N)$:

$$\Gamma_{2j-1} = -i(\chi_j - \chi_j^+)$$

and

$$\Gamma_{2j} = (\chi_j + \chi_j^+), \qquad j = 1, \ldots, N. \tag{7.2.4}$$

It is easy to verify, using eqs. (7.2.1), that

$$\{\Gamma_\mu, \Gamma_\nu\} = 2\delta_{\mu\nu}. \tag{7.2.5}$$

Thus, the Γ_μ's form a Clifford algebra of rank $2N$ (of course, $\Gamma_\mu = \Gamma_\mu^+$). Using the Γ_μ's we can construct the generators of the SO(2N) group as follows:

$$\Sigma_{\mu\nu} = \frac{1}{2i}[\Gamma_\mu, \Gamma_\nu]. \tag{7.2.6}$$

The $\Sigma_{\mu\nu}$ can be written down in terms of χ_j and χ_j^+ as follows:

$$\Sigma_{2j-1,2k-1} = \frac{1}{2i}[\chi_j, \chi_k^+] - \frac{1}{2i}[\chi_k, \chi_j^+] + i(\chi_j\chi_k + \chi_j^+\chi_k^+),$$

$$\Sigma_{2j,2k-1} = \tfrac{1}{2}[\chi_j, \chi_k^+] + \tfrac{1}{2}[\chi_k, \chi_j^+] - (\chi_j\chi_k + \chi_j^+\chi_k^+),$$

$$\Sigma_{2j,2k} = \frac{1}{2i}[\chi_j, \chi_k^+] - \frac{1}{2i}[\chi_k, \chi_j^+] - i(\chi_j\chi_k + \chi_j^+\chi_k^+). \tag{7.2.7}$$

It is well known that the spinor representation of SO(2N) is 2^N dimensional. To write it in terms of the SU(N) basis, let us define a "vacuum" state $|0\rangle$ that is SU(N) invariant. The 2^N-dimensional spinor representation is then given in Table 7.1.

This representation can be split into the 2^{N-1}-dimensional representations under a chiral projection operator. We now proceed to construct this operator. Define

$$\Gamma_0 = i^N \Gamma_1 \Gamma_2 \ldots \Gamma_{2N}. \tag{7.2.8}$$

Also define a number operator $n_j = \chi_j^+ \chi_j$.

Using eq. (7.2.8) Γ_0 can be written as follows:

$$\Gamma_0 = [\chi_1, \chi_1^+][\chi_2, \chi_2^+] \cdots [\chi_N, \chi_N^+]$$

$$= \prod_{j-1}^{N}(1 - 2n_j). \tag{7.2.9}$$

Using the property of the number operator $n_j^2 = n_j$, we can show that $1 - 2n_j = (-1)^{n_j}$ and so we get

$$\Gamma_0 = (-1)^n, \qquad n = \sum_j n_j. \tag{7.2.10}$$

It is then easily checked that

$$[\Sigma_{\mu\nu}, (-1)^n] = 0. \tag{7.2.11}$$

The "chirality" projection operator is therefore given by $\frac{1}{2}(1 \pm \Gamma_0)$. Each irreducible "chiral" subspace is therefore characterized by odd or even numbers of χ particles. To make it more explicit let us consider $N = 5$ and define a column vector $|\psi\rangle$ as follows:

$$|\psi\rangle = |0\rangle\psi_0 + \chi_j^+|0\rangle\psi_j + \frac{1}{2}\chi_j^+\chi_k^+|0\rangle\psi_{ik} + \frac{1}{12}\varepsilon^{ijklm}\chi_k^+\chi_l^+\chi_m^+|0\rangle\bar{\psi}_{jl}$$
$$+ \frac{1}{24}\varepsilon^{jklmn}\chi_k^+\chi_l^+\chi_m^+\chi_n^+|0\rangle\bar{\psi}_j + \chi_1^+\chi_2^+\chi_3^+\chi_4^+\chi_5^+|0\rangle\bar{\psi}_0, \tag{7.2.12}$$

where $\bar{\psi}$ is not the complex conjugate of ψ but an independent vector. We will denote complex conjugate by $*$. The column vector ψ is then given by:

$$\psi = \begin{pmatrix} \psi_0 \\ \psi_j \\ \psi_{jk} \\ \bar{\psi}_{jk} \\ \bar{\psi}_j \\ \bar{\psi}_0 \end{pmatrix}. \tag{7.2.13}$$

Under chirality

$$\psi = \begin{pmatrix} \psi_+ \\ \psi_- \end{pmatrix},$$

where

$$\psi_\pm = \frac{1}{2}(1 \pm \Gamma_0)\psi$$

and

$$\psi_+ = \begin{pmatrix} \psi_0 \\ \psi_{ij} \\ \bar{\psi}_j \end{pmatrix}, \qquad \psi_- = \begin{pmatrix} \bar{\psi}_0 \\ \bar{\psi}_{ij} \\ \psi_i \end{pmatrix}. \tag{7.2.14}$$

For the $N = 5$ case $\bar{\psi}_i$ and ψ_{ij} represent the $\bar{5}$- and 10-dimensional representation of SU(5) and ψ_0 is the singlet. All the fermions are assigned to ψ_+. It is then easy to write down the gauge interaction of the fermions. We further note that for the case of SO(10) the formula for electric charged Q is given by

$$Q = \frac{1}{2}\Sigma_{78} - \frac{1}{6}(\Sigma_{12} + \Sigma_{34} + \Sigma_{56}). \tag{7.2.15}$$

We next tackle the problem of spontaneous generation fermion masses in the SO(10) model.[1]

---

[1]It is also worth pointing out that to discuss SO(2N + 1) groups we have to adjoin Γ_0 to the Clifford algebra generated by $\Gamma_1 \ldots \Gamma_{2N}$. There is no chirality operator in this model. So the irreducible representation of SO(2N + 1) is 2^{2n} dimensional. The irreducible spinor representation can also be constructed in the same manner.

§7.3 Fermion Masses and the "Charge Conjugation" Operator

As is well known, in the framework of gauge theories, at the present state of art, the fermion masses arise from Yukawa couplings of fermions to Higgs bosons and subsequent breakdown of the gauge symmetry by nonzero vacuum expectation values of the Higgs mesons. In general, in grand unified theories, both particles and antiparticles belong to the same irreducible representation of the gauge group. So, to generate all possible mass terms, we must write down gauge-invariant Yukawa couplings of the form

$$\tilde{\psi} B C^{-1} \Gamma_\mu \psi \phi_\mu, \qquad \tilde{\psi} B C^{-1} \Gamma_\mu \Gamma_\nu \Gamma_\lambda \psi \phi_{\mu\nu\lambda}, \qquad (7.3.1)$$

where $\tilde{\psi}$ stands for the transpose of ψ, B is the equivalent of the charge conjugation matrix for $SO(10)$, and C is the Dirac charge conjugation matrix. The ϕ_μ, $\phi_{\mu\nu\lambda}$, etc., are the Higgs mesons belonging to totally irreducible antisymmetric representations of appropriate dimensions of $SO(2N)$, i.e., ϕ_μ is 2^N dimensional; $\phi_{\mu\nu\lambda}$ is $[2N(2N-1)(2N-2)/6]$ dimensional, etc. (for $N = 5$, ϕ_μ, $\phi_{\mu\nu\lambda}$, are, respectively, 10 and 120 dimensional). To see the need for inserting B we note that under the group transformation

$$\delta\psi = i\varepsilon_{\mu\nu}\Sigma_{\mu\nu}\psi,$$
$$\delta\psi^+ = -i\varepsilon_{\mu\nu}(\Sigma_{\mu\nu}\psi)^+ = -i\varepsilon_{\mu\nu}\psi^+\Sigma_{\mu\nu}, \qquad (7.3.2)$$
$$\delta\tilde{\psi} = i\varepsilon_{\mu\nu}\tilde{\psi}\tilde{\Sigma}_{\mu\nu}.$$

Thus, $\tilde{\psi}$ does not transform like a conjugate spinor representation of $SO(2N)$. However, if we introduce a $2^N \times 2^N$ matrix

$$B^{-1}\tilde{\Sigma}_{\mu\nu}B = -\Sigma_{\mu\nu}. \qquad (7.3.3)$$

Then

$$\delta(\tilde{\psi}B) = -i\varepsilon_{\mu\nu}\tilde{\psi}B\Sigma_{\mu\nu}. \qquad (7.3.4)$$

Thus, B has the correct transformation property under $SO(2N)$. It is easy to see that eq. (7.3.3) requires that

$$B^{-1}\tilde{\Gamma}_\mu B = \pm\Gamma_\mu. \qquad (7.3.5)$$

We will choose the negative sign on the right-hand side. Since the Γ_μ's are represented by symmetric matrices for even μ in the spinor basis of Table 7.1, one obvious representation of B in the spinor space is (for odd N)

$$B = \prod_{\mu=\text{odd}} \Gamma_\mu. \qquad (7.3.6)$$

Using eqs. (7.2.12) and (7.3.6) we conclude that

$$
B \begin{pmatrix} \psi_0 \\ \psi_{ij} \\ \bar{\psi}_i \\ \psi_i \\ \bar{\psi}_{ij} \\ \bar{\psi}_0 \end{pmatrix} = \begin{pmatrix} \bar{\psi}_0 \\ -\bar{\psi}_{ij} \\ \psi_i \\ -\psi_i \\ \psi_{ij} \\ \psi_0 \end{pmatrix}. \tag{7.3.7}
$$

Since

$$
\tilde{\psi} B C^{-1} \Gamma_\mu \psi = \langle \psi^* | B C^{-1} \Gamma_\mu | \psi \rangle.
$$

The Yukawa coupling (for $N = 5$) of f

$$
\phi_\mu \langle \psi^* | B C^{-1} \Gamma_\mu | \psi \rangle = \sum_{j=\pm} \langle \psi_j^* | B C^{-1} \Gamma_\mu | \psi_\mu \rangle \phi_\mu, \tag{7.3.8}
$$

where all quantities are listed in this and the previous sections. Note that in writing eq. (7.3.8) we used the fact that $[\Gamma_0, B\Gamma_\mu]$. To get fermion masses all we have to do is set $\langle \phi_\mu \rangle \neq 0$ for appropriate μ and evaluate $\langle \psi_+^* | B C^{-1} \Gamma_\mu | \psi_+ \rangle$ using the anticommutation relations of χ_j's and the fact that $\chi_j | 0 \rangle = 0$. In the next section we give explicit examples for the case of the SO(10) grand unified group.

Fermion Masses in SO(10); an Application

As an explicit application of our techniques we will calculate the fermion masses for SO(10) theory with Higgs mesons belonging to both 10-dimensional (ϕ_μ) and 120-dimensional ($\phi_{\mu\nu\lambda}$) representations. Before doing that we would like to identify the various particle states belonging to the 16-dimensional spinor representation of SO(10). We identify

$$
\psi_0 = \nu_L^c, \qquad \bar{\psi}_i = \begin{pmatrix} d_1^c \\ d_2^c \\ d_3^c \\ e^- \\ \nu \end{pmatrix}, \qquad \psi_{ij} = \begin{pmatrix} 0 & u_3^c & -u_2^c & u_1 & d_1 \\ & 0 & u_1^c & u_2 & d_2 \\ & & 0 & u_3 & d_3 \\ & & & 0 & e^+ \end{pmatrix}.
$$

$$
\tag{7.3.9}
$$

We remind the reader that $\bar{\psi}_i$ and ψ_{ij} are the usual SU(5) representations of Georgi and Glashow (see Section 5.4).

(a) *10-Dimensional Higgs*

Since we want color symmetry unbroken, the components of ϕ_μ, which can acquire v.e.v.'s, are ϕ_9 and ϕ_{10}. Let us set

$$
\langle \phi_9 \rangle = v_1, \qquad \langle \phi_{10} \rangle = v_2. \tag{7.3.10}
$$

We have to evaluate

$$
\mathcal{L}_{\text{mass}} = -i\kappa v_1, \langle \psi^* | B(\chi_5 - \chi_5^+) | \psi_+ \rangle + \kappa_2 \langle \psi^* | B(\chi_5 + \chi_5^\dagger) | \psi_+ \rangle. \tag{7.3.11}
$$

Using eqs. (7.2.12) and (7.3.6), and after some algebra, we obtain

$$\mathcal{L}_{\text{mass}} = \kappa(v_2 - v_1)(\bar{d}_L d_R + \bar{e}_L e_R) + \kappa(v_2 + v_1)(\bar{u}_L u_R + \bar{\nu}_L \nu_R). \quad (7.3.12)$$

We thus see that

$$m_d = m_e, \qquad m_u = m_\nu. \quad (7.3.13)$$

It is also easily seen that, if there are more than one-spinor multiplets of fermions corresponding to different families of particles, the mass matrix is symmetric. Also, we remind the reader that the relevant neutrino mass here is the Dirac mass.

(b) 120-Dimensional Case

The invariant Yukawa coupling of O(10) spinor ferminos ψ_+ to the 120-dimensional Higgs field $\phi_{\mu\nu\lambda}$ can be written down as follows:

$$\mathcal{L}_Y^{ab} = \kappa_{ab}\tilde{\psi}_a BC^{-1}\Gamma_\mu \Gamma_\nu \Gamma_\lambda \psi_b \phi_{\mu\nu\lambda}, \quad (7.3.14)$$

where a and b stand for the different generations of fermions. Using the fact that

$$\tilde{B} = -B \quad \text{and} \quad \tilde{C} = -C,$$

we find that

$$\mathcal{L}_Y^{ab} = -\mathcal{L}_Y^{ab}. \quad (7.3.15)$$

So, if we restrict ourselves to only one generation, it does not contribute to the fermion masses. It will, however, contribute to mixings between various generations. To analyze the kind of mixing pattern that this representation generates, we note that under SU(5), $\phi_{\mu\nu\lambda}$ breaks up as follows:

$$\{120\} = \{45\} + \{45^*\} + \{10\} + \{10^*\} + \{5\} + \{5^*\}. \quad (7.3.16)$$

Thus, we can choose either of the following patterns of vacuum expectation values:

(i) A linear combination of $\{45\}$ and $\{5\}$ acquires v.e.v.
This means that we have

$$\langle \phi_{789} \rangle \neq 0, \qquad \langle \phi_{7810} \rangle \neq 0. \quad (7.3.17)$$

Inserting this into the Yukawa couplings and proceeding with the calculation as in the case of $\{10\}$-dimensional Higgs, we find that the mixing between the various generations of quarks and leptons are related as follows:

$$m_{d_a} = 3m_{E_a E_b}, \quad (7.3.18)$$
$$m_{u_a u_b} = 3m_{\nu_a \nu_b}, \quad (7.3.19)$$

where $m_{\nu_a \nu_b}$ stands for mixing terms in the mass matrix between generations a and b; d_a means the $-1/3$ charged quark of an ath generation and similarly for u, E^-, ν.

(ii) Only the SU(5) {45}-dimensional Higgs acquires v.e.v.
This means the following fields acquire v.e.v.'s:

$$\langle \phi_{789} \rangle = -3\langle \phi_{2k-1,2k,9} \rangle, \qquad k = 1, 2, 3,$$
$$\langle \phi_{7890} \rangle = -3\langle \phi_{2k-01,2k,10} \rangle, \qquad k = 1, 2, 3. \qquad (7.3.20)$$

In this case the mixing pattern is very different from case (i); we get

$$m_{E_a E_b} = 3m_{d_a d_b},$$
$$m_{\nu_a \nu_b} = 0,$$
$$m_{u_a u_b} \neq 0. \qquad (7.3.21)$$

(c) *126-Dimensional Case*

The invariant Yukawa coupling in this case involves five Γ matrices

$$\mathcal{L}_Y^{ab} = \kappa_{ab} \tilde{\psi}_a BC^{-1} \Gamma_\mu \Gamma_\nu \Gamma_\lambda \Gamma_\sigma \Gamma_\alpha \psi_b \phi_{\mu\nu\lambda\sigma\alpha}. \qquad (7.3.22)$$

We first note that $\mathcal{L}^{ab} = \mathcal{L}^{ba}$. Thus, this makes a symmetric contribution to various masses. We may choose the following vacuum expectation values for the Higgs field to be consistent with the local color SU(3) symmetry remaining exact:

$$\langle \phi_{1278\mu} \rangle = \langle \phi_{3478\mu} \rangle = \langle \phi_{5678\mu} \rangle \neq 0, \qquad (7.3.23)$$

where $\mu = 9$ or 10. Substituting this into the Yukawa couplings we get for one generation ($a = b = 1$) the following mass relations:

$$m_e = 3m_d, \qquad m_\mu = 3m_\nu. \qquad (7.3.24)$$

Neutrino Mass in the SO(10) Model

As we saw from eq. (7.3.13) the smallness of the neutrino mass is not easy to understand. One idea is to give nonzero vacuum expectation value to the SU(5) singlet part of {126} [note the decomposition of {126} under SU(5)]:

$$\{126\} = \{1\} + \{5\} + \{10\} + \{10^*\} + \{50\} + \{50_*\}. \qquad (7.3.25)$$

Under $SU(2)_L, \times SU(2)_R \times SU(4)$ its decomposition can be written as

$$\{126\} = (3, 1, 10) + (1, 3, 10) + (2, 2, 15) + (1, 1, 6). \qquad (7.3.26)$$

The Higgs representation of eq. (7.3.26), which acquires vacuum expectation values in this case, is the right-handed triplet (1, 3, 10) as in the left–right symmetric models. Thus, as in this case, the smallness of the neutrino mass can be understood as a consequence of suppression of $V + A$ currents.

In our notation the SU(5) singlet component of $\phi_{\mu\nu\lambda\sigma\alpha}$ has the form $\chi_1^+ \chi_2^+ \chi_3^+ \chi_4^+ \chi_5^+$ or $\chi_1 \chi_2 \chi_3 \chi_4 \chi_5$. Substituting this into eq. (7.3.22) we note that this gives a Majorana mass only to the right-handed neutrino

Table 7.1. Construction of the states belonging to the spinor representation of $SO(N)$ dimensionality.

$SO(2N)$ spinor state	$SU(N)$ dimension
$|0\rangle$	1
$\chi_j^+|0\rangle$	N
$\chi_j^+\chi_k^+|0\rangle$	$\frac{N(N-1)}{2}$
$\chi_j^+\chi_k^+\chi_l^+|0\rangle$	$\frac{N(N-1)(N-2)}{6}$
$\vdots$	
$\chi_1^+\chi_2^+\cdots\chi_N^+|0\rangle$	1
Total	2^N

N_R^T, CN_R, whereas Dirac masses arise from the introduction of {10}-dimensional Higgs representations as in eq. (7.3.13). Thus, in the $SO(10)$ model, we predict (as in eq. 6.4.20)

$$m_\nu = \frac{m_\mu^2}{m_{N_R}}. \tag{7.3.27}$$

An important point worth noting here is that, in deriving eq. (7.3.27), the mass matrix for neutrinos ν, N is assumed to be of the form in eq. (6.4.19). But in mass realistic models, minimization of the potential leads to a mass matrix of the form:

$$M = \begin{pmatrix} f\frac{k^2}{V_R} & m_D \\ m_D & fV_R \end{pmatrix} \tag{7.3.28}$$

This leads to a new type of seesaw formula which we will call type II seesaw formula:

$$m_\nu \simeq f\frac{k^2}{v_R} - m_D(fv_R)^{-1}m_D.$$

If D-parity symmetry present in $SO(10)$ model is broken at the GUT scale separately from $SU(2)_R$ symmetry by introducing a {210}- or {45}-dimensional Higgs multiplet [5a] fk^2/V_R in eq. (7.3.28) is replaced by $fk^2V_R/M_{\mathrm{GUT}}^2$, which become small if $V_R \ll M_{\mathrm{GUT}}$.

§7.4 Symmetry-Breaking Patterns and Intermediate Mass Scales

The $SO(10)$ group has rank 5 where the low-energy standard electroweak group has rank 4. Because of this there exist several intermediate symmetries through which SO(10) can descend to the $SU(3)_c \times SU(2)_L \times U(1)_Y$ group (G_{321}). To list these various possibilities let us recall that SO(10) has two maximal subgroups that contain the G_{321} group: (i) $SU(5) \times U(1)$; and (ii) $SU(2)_L \times SU(2)_R \times SU(4)_c \times D$; where D is a discrete symmetry which transforms [6, 7]

$$q_L \xrightarrow{D} q_L^c. \tag{7.4.1}$$

In terms of the SO(10) generators D is given by

$$D = \Sigma_{23}\Sigma_{67}. \tag{7.4.2}$$

The breaking of D parity has important cosmological, as well as physics [8], implications.

The subsequent stages of symmetry breaking depend on whether we are in Case (i) or Case (ii).

Case (i)

$$SO(10) \rightarrow SU(5) \times U(1) \rightarrow SU(5)$$
$$\downarrow M_x$$
$$SU(3)_c \times SU(2)_L \times U(1)_Y.$$

For Case(ii) there exists a large variety of possibilities since $SU(2)_R \times SU(4)_c \times D$ can break down to $U(1)_Y \times SU(3)_c$ in many ways. We see that, unlike the SU(5) grand unified model, the SO(10) model has many ways in which it can appear. So, how do we experimentally test SO(10) grand unification? This question has been addressed in Refs. 7–10 where the low-energy constraints on $\sin^2 \theta_W$ and α_s are imposed to isolate the values of the intermediate mass scales corresponding to various symmetry-breaking chains. There exist two distinct possiblities in Case (ii) depending on whether D parity and $SU(2)_R$ local symmetry are broken together or at separate scales.

Let us first deal with Case (i). It has been shown in Case (i) that the proton lifetime is predicted to be less than that of the SU(5) model. Thus, the $SU(5) \times U(1)$ intermediate symmetry is ruled out by experiment.

Case (ii) leads to two distinct symmetry-breaking patterns if we assume the D parity and $SU(2)_R$ break down simultaneously:

$$\text{(iiA)} \quad SO(10) \xrightarrow{M_U} G_{224D} \xrightarrow{M_c} G_{2213D} \xrightarrow{M_{R+}} G_{2113} \xrightarrow{M_{R^0}} G_{213};$$

$$\text{(iiB)} \quad SO(10) \xrightarrow{M_U} G_{224D} \xrightarrow{M_{R+}} G_{214D} \xrightarrow{M_c} G_{2113} \xrightarrow{M_{R^0}} G_{213}.$$

In both these cases we can write down the formulas for $\sin^2 \theta_W(m_W)$ and $\alpha_s(m_W)$ following the same procedure as in Chapter 5.

Let us illustrate the method of derivation for Case (iiA), and one can use the same procedure in all cases. First, we normalize all generators in the same way. To see what this means, consider a particular representation of SO(10), say the spinor representation to which the various fermions are assigned. Consider the $SU(2)_L$ quantum numbers for all members of the [16]-dimensional representation, we find

$$T_r I_{3L}^2 = 3I_{3L}^2 (\text{quarks}) + I_{3L}^2 (\text{leptons})$$
$$= \tfrac{3}{2} + \tfrac{1}{2} = 2. \tag{7.4.3}$$

Now consider the $U(1)_Y$, generator Y (since $Y_{\nu_R} = 0$):

$$T_r \frac{Y^2}{4} = \tfrac{3}{4}(Y_{U_L}^2 + Y_{d_L}^2 + Y_{U_R}^2 + Y_{d_R}^2)$$
$$+ \tfrac{1}{4}(Y_{\nu_L}^2 + Y_{e_L}^2 + Y_{e_R}^2) = \tfrac{10}{3}. \tag{7.4.4}$$

We can now define a generator $I_Y = \sqrt{\tfrac{3}{5}}(Y/2)$, which is normalized the same way as I_{3L}. Finally,

$$Tr\left(\frac{B-L}{2}\right)^2 = \frac{4}{3}. \tag{7.4.5}$$

So the properly normalized generator $I_{BL} = \sqrt{3}2(B - L/2)$. The next ingredient we need is that, if at a mass scale μ, the gauge symmetry $G_1 \times G_2$ breaks down to G, such that the corresponding generators T_1, T_2, and T_0 are related as

$$T = p_1 T_1 + p_2 T_2, \tag{7.4.6}$$

then the respective gauge couplings g_1, g_2, and g satisfy the following relation at μ:

$$\frac{1}{g^2(\mu)} = \frac{p_1^2}{g_1^2(\mu)} + \frac{p_3^2}{g_2^2\mu}. \tag{7.4.7}$$

The next piece of information we need is that between two mass scales where a given local symmetry is exact, its fine-structure constant $\alpha_i = g_i^2/4\pi$ obeys the following evolution equation:

$$\frac{1}{\alpha_i(\mu_1)} = \frac{1}{\alpha_i(\mu_2)} - \frac{b_i}{2\pi} \ln\left(\frac{\mu_2}{\mu_1}\right). \tag{7.4.8}$$

Now we are ready to apply this to chain (iiA). First note that, in this case, the $SU(2)_L$ symmetry remains exact from $\mu = M_{W_L}$ to $\mu - M_U$ and evolves independently. Therefore, we get the corresponding α_{2L} to obey

$$\frac{1}{\alpha_{2L}(M_W)} = \frac{1}{\alpha_{2L}(M_U)} - \frac{b_{2L}}{2\pi} \ln\left(\frac{M_U}{M_W}\right). \tag{7.4.9}$$

Constraint of grand unification of course implies that $\alpha_{2L}(M_U) = \alpha_U$.

Turning now to $SU(3)_C$, it also remains exact up to M_C, above which it becomes part of the $SU(4)_C$ group. Therefore, we get

$$\frac{1}{\alpha_{2L}(M_W)} = \frac{1}{\alpha_3(M_C)} - \frac{b_3}{2\pi} \ln\left(\frac{M_c}{M_W}\right)$$

$$\times \frac{1}{\alpha_U(M_U)} - \frac{b_4}{2\pi} \ln\left(\frac{M_U}{M_C}\right) - \frac{b_3}{2\pi} \ln\left(\frac{M_C}{M_W}\right), \quad (7.4.10)$$

where we have used $\alpha_4(M_U) = \alpha_U$.

Let us now turn to $U(1)_Y$-gauge coupling α_Y. This evolves between M_W and M_{R^0} where it gets merged into $U(1)_{I_{3R}} \times U(1)_{B-L}$ and the relation is

$$\frac{Y}{2} = I_{3R} + \frac{B-L}{2},$$

or

$$I_Y = \sqrt{\tfrac{3}{5}} I_{3R} + \sqrt{\tfrac{2}{5}} I_{BL}. \quad (7.4.11)$$

Therefore, using eq. (7.4.7), we get

$$\frac{1}{\alpha_Y(M_{R^0})} = \frac{3}{5} \frac{1}{\alpha_{I_{3R}}(M_{R^0})} + \frac{2}{5} \frac{1}{\alpha_{I_{BL}}(M_{R^0})}. \quad (7.4.12)$$

At M_{R^+}, I_{3R} becomes part of $SU(2)_R$ leading to the boundary conditions $\alpha_{I_{3R}}(M_{R^+}) = \alpha_{2R}(M_{R^+})$. Similarly, I_{BL} becomes part of $SU(4)_C$ at M_C. Let us now start evolving α_Y

$$\frac{1}{\alpha_Y(M_W)} = \frac{1}{\alpha_Y(M_{R^0})} - \frac{b_Y}{2\pi} \ln\left(\frac{M_{R^0}}{M_W}\right). \quad (7.4.13a)$$

Using eq. (7.4.12), we get

$$\frac{1}{\alpha_Y(M_W)} = \frac{3}{5} \frac{1}{\alpha_{I_{3R}}(M_{R^0})} + \frac{2}{5} \frac{1}{\alpha_{I_{BL}}(M_{R^0})} - \frac{b_Y}{2\pi} \ln\left(\frac{M_{R^0}}{M_W}\right)$$

$$= \frac{3}{5} \left[\frac{1}{\alpha_{2R}(M_{R^+})} - \frac{b_{I_{3R}}}{2\pi} \ln\left(\frac{M_{R^+}}{M_{R^0}}\right) \right]$$

$$+ \frac{2}{5} \left[\frac{1}{\alpha_4(M_C)} - \frac{b_{I_{BL}}}{2\pi} \ln\left(\frac{M_C}{M_{R^0}}\right) \right] - \frac{b_Y}{2\pi} \ln\left(\frac{M_{R^0}}{MW}\right) \quad (7.4.13b)$$

Now, using the evolution of $SU(4)_C$ and $SU(3)_R$ all the way to M_U, we get [using $\alpha_{2R}(M_U) - \alpha_4(M_U) - \alpha_U$]

$$\frac{1}{\alpha_Y(M_W)} = \frac{1}{\alpha_U} - \frac{3}{5} \frac{b_{2R}}{2\pi} \ln\left(\frac{M_U}{M_{R^+}}\right) - \frac{3}{5} \frac{b_{I_{3R}}}{2\pi} \ln\frac{M_{R^+}}{M_{R^0}}$$

$$- \frac{2}{5} \frac{b_4}{2\pi} \ln\left(\frac{M_U}{M_C}\right) - \frac{2}{5} \frac{b_{I_{BL}}}{2\pi} \ln\left(\frac{M_C}{M_{R^0}}\right)$$

$$- \frac{b_Y}{2\pi} \ln\left(\frac{M_{R^0}}{M_W}\right). \quad (7.4.13c)$$

Noting that the electric charge

$$Q = I_{3L} + \frac{Y}{2}$$

$$= I_{3L} + \sqrt{\tfrac{5}{3}} I_Y, \quad (7.4.14)$$

we get

$$\frac{1}{\alpha_{em}(M_N)} = \frac{1}{\alpha_{2L}(M_W)} + \frac{5}{3}\frac{1}{\alpha_Y(M_W)}, \tag{7.4.15}$$

since $\sin^2\theta_W = \alpha_{em}/\alpha_{2L}$, using eq. (7.4.15), (7.4.13a) and (7.4.9), we get

$$\sin^2\theta_W = \frac{3}{5} - \frac{5}{8}\frac{\alpha_{em}}{2\pi}\left[(b_2 - b_Y)\ln\left(\frac{M_{R^0}}{M_W}\right)\right.$$

$$+ (b_2 - \tfrac{3}{5}b_{I_{3R}} - \tfrac{2}{5}b_{I_{BL}})\ln\left(\frac{M_{R^+}}{M_{R^0}}\right)$$

$$+ (b_2 - \tfrac{3}{5}b_{2R} - \tfrac{2}{5}b_{I_{BL}})\ln\left(\frac{M_C}{M_{R^+}}\right)$$

$$\left. + (b_2 - \tfrac{3}{5}b_{2R} - \tfrac{2}{5}b_4)\ln\left(\frac{M_U}{M_C}\right)\right]. \tag{7.4.16}$$

Note that

$$b_4 = \tfrac{44}{3} - \tfrac{4}{3}N_g - T_4,$$
$$b_2 = \tfrac{22}{3} - \tfrac{4}{3}N_g - T_L,$$
$$b_2 = -\tfrac{4}{3}N_g - T_i \quad \text{for} \quad i = T_Y, T_{3R}, \text{ and } I_{BL}, \tag{7.4.17}$$

where N_g is the number of generations and T's are Higgs boson contributions to the beta functions. Substituting eq. (7.4.17) into eq. (7.4.16) we get, for Case (iiA)

$$\sin^2\theta_W(m_W) = \frac{3}{8} - \frac{\alpha(m_W)}{48\pi}\left[(110 + 3T_Y - 5T_L)\ln\frac{M_{R^0}}{m_W}\right.$$

$$+ (110 + 3T_{R^0} + 2T_{BL} - 5T_L)\ln\left(\frac{M_{R^+}}{M_{R^0}}\right)$$

$$+ (44 + 3T_{R^0} + 2T_{B-L} - 5T_L)\ln\left(\frac{M_c}{M_{R^+}}\right)$$

$$\left. + (-44 + 3T_R + 2T_4 - 5T_L)\ln\left(\frac{M_U}{M_c}\right)\right], \tag{7.4.18}$$

$$\frac{\alpha(m_W)}{\alpha_s(m_W)} = \frac{3}{8} - \frac{a(m_W)}{16\pi}\left[(66 + T_L + T_Y - \tfrac{8}{3}T_s)\ln\left(\frac{M_{R^0}}{m_W}\right)\right.$$

$$+ (66 + T_L + T_{R^0} + \tfrac{2}{3}T_{BL} - \tfrac{8}{3}T_s)\ln\left(\frac{M_{R^+}}{M_{R^0}}\right)$$

$$+ (44 + T_L + T_R + \tfrac{2}{3}T_{BL} - \tfrac{8}{3}T_s)\ln\left(\frac{M_C}{M_{R^+}}\right)$$

$$\left. + (44 + T_L + T_R - 2T_4)\ln\left(\frac{M_U}{M_C}\right)\right]. \tag{7.4.19}$$

The T_i's in eqs. (7.4.18) and (7.4.19) denote the contributions of the Higgs boson multiplets to the β functions for the ith symmetry group. Their contributions have to be included in accordance with the minimal find tuning and extended survival hypotheses [11]. Already a very important conclusion can be drawn by looking at eqs. (7.4.18) and (7.4.19). Note that if we ignore the Higgs contributions to these equations in the range $M_{R^0} < \mu < M_R^+$, then M_{R^0} drops out of both equations. (In fact, all the T_i's are small, of order unity, in this mass range.) Thus, an important feature of the SO(10) grand unified models is that M_{R^0} can be as light as possible without affecting the values of $\sin^2 \theta_W(m_W)$ and $\alpha_s(m_W)$. This point has been emphasized before in Refs. 9 and 10. In fact, as was noted in Ref. 10, if all Higgs contributions are ignored, then we find

$$M_U = M_5 \left(\frac{M_5}{M_{R^+}} \right)^{1/2}, \tag{7.4.20}$$

which is even independent of the scale M_c [where M_5 is the unification scale in the SU(5) model]. From this we can conclude that if M_{R^+} is lower than the SU(5) scale, M_u for SO(10) becomes higher, making the proton lifetime longer, i.e.,

$$\tau_{10} = \tau_5 \left(\frac{M_5}{M_{R^+}} \right)^2. \tag{7.4.21}$$

Thus, a longer proton lifetime in the context of SO(10) grand unification implies that there must exist an intermediate left–right symmetric scale. Furthermore, we can relate the prediction for $\sin^2 \theta_W$ in the SO(10) model to that of the SU(5) model to get an idea of how small M_{R^+} can be. Ignoring the Higgs contributions we find

$$\Delta(\sin^2 \theta_W) = \sin^2 \theta_{10} - \sin^2 \theta_5 = \frac{11\alpha(m_W)}{24\pi} \left[\ln \frac{M_U^2 M_5^5}{M_C^4 M_{R^+}^3} \right]. \tag{7.4.22}$$

We conclude that if $M_C = M_U$ and $M_{R^+} \simeq 1$ TeV, then $\Delta(\sin^2 \theta_W) \simeq 0.06$, which, on using the SU(5) result for $\sin^2 \theta_W$, implies $\sin^2 \theta_{10} \simeq 0.27$ [12] which is in conflict with experimental data.

Let us now look at the symmetry-breaking pattern (iiB) where $M_{R^+} > M_C$. The equations for $\sin^2 \theta_W$ and α_s can be written as follows:

$$\sin^2 \theta_W(m_W) = \frac{3}{8} - \frac{\alpha(m_W)}{48\pi} \left[(110 + 3T_Y - 5T_L) \ln \left(\frac{M_{R^0}}{M_W} \right) \right.$$

$$+ (110 + 3T_{R^0} + 2T_{BL} - 5T_L) \ln \left(\frac{M_C}{M_{R^0}} \right)$$

$$+ (22 + 3T_{R^0} + 2T_4 - 5T_L) \ln \left(\frac{M_{R^+}}{M_C} \right)$$

$$\left. + (-44 + 3T_R + 2T_4 - 5T_L) \ln \left(\frac{M_U}{M_{R^+}} \right) \right], \tag{7.4.23}$$

$$\frac{\alpha(m_W)}{\alpha_s(m_W)} = \frac{3}{8} - \frac{\alpha(m_W)}{16\pi} \left[(66 + T_L + T_Y - \frac{8}{3}T_s) \ln \left(\frac{M_C}{M_{R^0}} \right) \right.$$

$$+ (66 + T_L + T_{R^0} + \tfrac{2}{3}T_{BL} - \tfrac{8}{3}T_s) \ln \left(\frac{M_C}{M_{R^0}} \right)$$

$$+ (66 + T_L + T_{R^0} - 2T_4) \ln \left(\frac{M_{R^+}}{M_C} \right)$$

$$+ (44 + T_L + T_R - 2T_4) \ln \left(\frac{M_U}{M_{R^+}} \right) \Bigg] . \tag{7.4.24}$$

Again, as in chain (iiA), we note that, ignoring the Higgs contributions, the equations become independent of M_{R^0} and, therefore, a low M_{R^0} is allowed by the second symmetry breaking chain without conflicting with low-energy data. However, as far as the proton lifetime is concerned, the result is sensitive to the Higgs contributions and no unambiguous prediction can be made; nevertheless, no low values of M_C or M_{R^+} can be obtained.

We conclude that, within the framework of conventional SO(10) grand unification, the only interesting physics—beyond the predictions of the standard subgroup G_{123}—is a low-mass ($\approx$ 300 GeV to 1 TeV) right-handed neutral Z_R boson [9]. The charged right-handed W boson W_{R^+} has to be extremely heavy ($\simeq 10^{11}$ GeV) due to the constraints of grand unification.

§7.5 Decoupling Parity and SU(2)$_R$ Breaking Scales

So far, in our discussion of symmetry breaking in the left–right symmetric models, we have broken parity and SU(2)$_R$ symmetry at the same scale. But, in a recent paper, Chang, Mohapatra, and Parida have suggested [7] an alternative approach where, by including a real parity odd SU(2)$_R$ singlet σ-field in the theory, we can decouple the parity and SU(2)$_R$ breaking scales.

The method can be illustrated in the case of the left–right symmetric model by choosing the following set of Higgs multiplets:

$$\Delta_L(3,1,+2) + \Delta_R(1,3,+2), \tag{7.5.1}$$

$$\sigma(2,2,0),$$

$$\sigma(1,2,0).$$

The relevant part of the Higgs potential can be written as

$$V' = - \mu^2 \sigma^2 + \lambda \sigma^4 + m\sigma(\Delta_L^\dagger \Delta_L - \Delta_R^\dagger \Delta_R) + \mu^2 (\Delta_L^\dagger \Delta_L + \Delta_R^\dagger \Delta_R)$$

$$+ \text{quartic terms.} \tag{7.5.2}$$

From eq. (7.5.2) we observe that parity symmetry is broken by $\langle \sigma \rangle = \mu/\sqrt{2\lambda}$, which then makes the Δ_L, Δ_R mass terms asymmetric. By choosing $m > 0$, we see that Δ_R mass can be negative if $\mu_4^2 < m\mu/\sqrt{2\lambda_2}$ and this, then, triggers SU(2)$_R$ breaking at a different (lower) scale $M_R = $

$(\mu_4^2 - m\mu/\sqrt{2\lambda})/\gamma$, where γ is a function of scalar couplings. Thus, parity and $SU(2)_R$ scales are decoupled.

This has the following implications: (i) the Higgs masses are parity asymmetric even above the $SU(2)_R$ breaking scale; and (ii) this asymmetry leads to $g_L \neq g_R$ at the scale M_R, which, therefore, has important implications for low-energy phenomenology.

It is interesting that this idea can be embedded in an $SO(10)$ grand unified model. To do this, we have to search for an irreducible representation of $SO(10)$ that has an $SU(2)_R$ gauge singlet parity odd field. Such a representation turns out to be the $\{210\}$-dimensional representation of $SO(10)$. The $\{210\}$-dimensional representation is denoted by the totally antisymmetric fourth-rank tensor $\phi_{\mu\nu\lambda\sigma}$. The relevant D-parity odd component is

$$\sigma = \phi_{78910}. \tag{7.5.3}$$

Thus, if we break $SO(10) \xrightarrow[M_u]{\{210\}} SU(2)_L \times SU(2)_R \times SU(4)_C$, we break parity symmetry, which changes the pattern of gauge symmetry breaking. These have been studied in detail in Refs. 7 and 8. A major outcome of detailed analysis of various $SO(10)$ breaking chains is that we can now lower both the $SU(4)_C$ as well as W_R, Z_R scales, which then imply observable right-handed current effects at low energies. For detailed discussion of these ideas we refer the reader to Refs. 7 and 8; but we simply note the symmetry-breaking chain favorable for low-energy phenomenology

$$SO(10) \xrightarrow[\{54\}]{M_u} > SU(2)_L \times SU(2)_R \times SU(4)_C \times P$$

$$M_P \downarrow \{210\}$$

$$SU(2)_L \times SU(2)_R \times SU(4)_C$$

$$M_C \downarrow \{210\}$$

$$SU(2)_L \times U(1)_R \times U(1)_{B-L} \times SU(3)_C$$

$$M_{R^0} \downarrow \{126\}$$

$$SU(2)_L \times U(1)_Y \times SU(3)_C.$$

This leads to $M_{R^+} \simeq M_C \simeq 10^5$ GeV, $M_{R^0} \simeq 1$ TeV for $M_U \simeq 10^{16}$ GeV and $\sin^2 \theta_W \simeq 0.227$ for a D-parity breaking scale of 10^{14} GeV. In this analysis [8] the two-loop β-function effects have been taken into account. This leads to testable experimental predictions such as $K_L^0 \to \mu\bar{e}$ decay with $B(K_L^0 \to \mu\bar{e}) \simeq 10^{-9}$ and neutron–antineutron oscillation with mixing time $\tau_{N-\bar{N}} \simeq 10^7 - 10^8$ s. The proton lifetime is now predicted to be $\simeq 10^{35\pm2}$ yr which may barely be within reach of the on-going experiments. If proton decay can be measured, a further check of this symmetry-breaking pattern comes from measurement of the branching ratios, i.e.,

$$\Gamma(p \to e^+\pi^0) = \Gamma(p \to \bar{\nu}\pi^+), \tag{7.5.4}$$

which is different from the SU(5) prediction of

$$\tfrac{2}{5}\Gamma(p \to e^+\pi^0) = \Gamma(p \to \bar{\nu}\pi^+).\tag{7.5.5}$$

§7.6 Second Z' Boson

Since SO(10) contains the left–right symmetric group, it has an extra neutral gauge boson. However, unlike the left–right symmetric model, the constraints of grand unification relate both g and g' at the grand unification scale. Therefore this gives a specific symmetry-breaking scheme and the spectrum of the Higgs boson, all neutral-current couplings at low energies are predicted in terms of the $SU(2)_L$-gauge coupling. In limit of exact SO(10) symmetry, we can write the neutral-current Lagrangian (including electromagnetism) symbolically as

$$\mathcal{L}_{\mathrm{N.C.}} = eQ_{em}A + g_u Q_z Z + g_u Q_x Z_x,\tag{7.6.1}$$

where in terms of SO(10) generators we have

$$Q = I_{eL} + I_{3R} + \sqrt{\tfrac{2}{3}}I_{BL},$$

$$Q_z = \sqrt{\tfrac{5}{9}}(I_{3L} - \tfrac{3}{5}I_{3R} - \sqrt{\tfrac{6}{25}}I_{BL}),$$

and

$$Q_x = \frac{1}{\sqrt{10}}\{2I_{3R} - \sqrt{6}I_{BL}\},\tag{7.6.2}$$

where A is the photon field with $e = \sqrt{\tfrac{3}{8}}g_u$,

$$Z = \sqrt{\tfrac{5}{8}}W_{3L} - \frac{3}{\sqrt{40}}W_{3R} - \sqrt{\tfrac{3}{20}}B,$$

$$Z_R = \sqrt{\tfrac{2}{5}}W_{3R} - \sqrt{\tfrac{3}{5}}B.\tag{7.6.3}$$

Noting that at the grand unification scale $\sin\theta_W = \sqrt{\tfrac{3}{8}}$, we can obtain eq. (7.6.3) from eq. (6.2.18). The experimental limits on the extra $Z_R(Z_R = Z')$ boson has been very extensively studied recently.[2]

Exercises

7.1. Derive the D-parity operator and show that the ϕ_{78910} component of the $\{210\}$-dimensional representation and $H_{12} = H_{34} = H_{56}$ compo-

[2]G. Altarelli, R. Casalbuoni, F. Feruglio, and R. Gatto, *Mod. Phys. Lett.* **A5**, 495 (1990); *Nucl. Phys.* **B342**, 15 (1990).

nents are odd under D parity. Give an explicit example where D-parity breaking splits the $SU(2)_L$-and $SU(2)_R$-gauge couplings.

7.2. Obtain the relation between quark and lepton masses in the SO(10) model, if the {10}-dimensional Higgs boson is complex. Also show that for a complex {10}-dimensional representation, the SO(10) model has a $U(1)_{PQ}$ symmetry. What additional Higgs bosons will be needed so that the $B - L$ and $U(1)_{PQ}$ symmetry break at the same scale? Obtain the GUT scale, M_{B-L}, and the neutrino masses using $\sin^2 \theta_W$ and α_s constraints on the low-energy structure of the model. [See R. N. Mohapatra and G. Senjanović, *Z. Phys.* **C17**, 53 (1983).]

7.3. Using the spinor formalism outlined in Chapter 7, analyze the spinor representation of the SO(14) and SO(18) groups. Examine their viability as models for generations. What kind of Higgs bosons would you choose to make the model realistic? Is it possible to use SO(13) and SO(11) as grand unification groups?

7.4. Prove eq. (7.5.4).

References

[1] H. Georgi, in *Particles and Fields* (edited by C. E. Carlson), A.I.P., 1975; H. Fritzsch and P. Minkowski, *Ann. Phys.* **93**, 193 (1975).

[2] Y. Tosa and S. Okubo, *Phys. Rev.* **D23**, 2486 (1981).

[3] R. N. Mohapatra and B. Sakita, *Phys. Rev.* **D21**, 1062 (1980).

[4] F. Wilczek and A. Zee, *Phys. Rev.* **D25**, 553 (1982).

[5] M. Gell-Mann, P. Ramond, and R. Slansky, in *Supergravity* (edited by D. Freedman et al. North-Holland, Amsterdam, 1980; See also T. Yanagida, K.E.K. preprint, 1979; R. N. Mohapatra and G. Senjanović, *Phys. Rev. Lett.* **44**, 912 (1980).

[5a] D. Chang and R. N. Mohapatra, *Phys. Rev.* **D32**, 1248 (1985).

[6] T. W. B. Kibble, G. Lazaridis, and Q. Shafi, *Phys. Rev.* **D26**, 435 (1982).

[7] D. Chang, R. N. Mohapatra, and M. K. Parida, *Phys. Rev. Lett.* **52**, 1072 (1984); *Phys. Rev.* **D30**, 1052 (1984).

[8] D. Chang, R. N. Mohapatra, J. Gipson, R. E. Marshak, and M. K. Parida, *Phys. Rev.* **D31**, 1718 (1985).

[9] R. Robinett and J. L. Rosner, *Phys. Rev.* **D25**, 3036 (1982).

[10] Y. Tosa, G. C. Branco, and R. E. Marshak. *Phys. Rev.* **D28**, 1731 (1983).

[11] R. N. Mohapatra and G. Senjanovic, *Phys. Rev.* **D27**, 1601 (1983); F. del Aguila and L. Ibanez, *Nucl. Phys.* **B177**, 60 (1981).

[12] T. Rizzo and G. Senjanovic, *Phys. Rev.* **D25**, 235 (1982).

8
Technicolor and Compositeness

§8.1 Why Compositeness?

In the previous chapters we have emphasized that while the success of the $SU(2)_L \times U(1)_Y \times SU(3)_C$ model has indicated that the unified gauge theories are perhaps the right theoretical framework for the study of quark–lepton interactions, it still leaves many questions unanswered. Some of the outstanding questions are:

(a) the nature of the Higgs bosons and the origin of electroweak symmetry breaking;

(b) the apparent superfluous replication of quarks and lepton (and even Higgs bosons if electroweak symmetry is higher); and

(c) the origin of fermion masses that are much smaller than the scale of electroweak symmetry breaking: for instance, $m_{e,u,d} \sim 10^{-5}\Lambda_W$.

Historically, this is similar to the situation that existed in the domain of strong interactions in the early 1960s before Gell-Mann and Zweig introduced the quarks as the substructure of baryons and mesons. Most of the observed hadronic spectrum received a simple and elegant explanation in terms of bound states of the quarks and antiquarks. Crucial to the success of the quark idea was the introduction of the concept of color and non-abelian gauge theory of color interaction, which provided a solid theoretical framework for quark interactions. We may extrapolate this idea to the domain of weak interactions and ask whether the Higgs bosons, quarks, and leptons (perhaps even the W, Z, etc.) are composites of a new set of fermions and bosons (which we will call preons). There must, then, exist a new kind of strong interaction that binds the preons to form the composites, i.e., quarks, leptons, and Higgs bosons. Since we do not know anything about this hypothetical interaction (which has been called by a variety of names in the literature such as metacolor, technicolor, hypercolor, etc.; we will use the word hypercolor), we will assume that its properties are very similar to that of QCD.

(a) It (QHCD) is described by an unbroken non-abelian gauge symmetry under which preons are nonsinglets.

(b) Like QCD, it confines nonsinglet composites and only asymptotic states correspond to a singlet of QHCD. Quarks, leptons, and Higgs bosons will be singlets under QHCD.

Following the existing literature we will present our discussion in two stages. First, we will discuss partial compositeness where only Higgs bosons are treated as composites of new kinds of fermions (to be called techni–quarks) and quarks and leptons are left elementary. This has generally gone under the name of technicolor. Second, we will introduce total compositeness when quarks, leptons, Higgs (techni-quarks), etc., will all be treated as composites of preons.

§8.2 Technicolor and Electroweak Symmetry Breaking

The goal of dynamical symmetry breaking is to remove all scalar bosons from electroweak physics and replace them by a new set of fermions. In this process we may learn about the deeper structure of weak interactions. The Higgs fields, which led to such a successful picture of electroweak symmetry breaking, are therefore to be thought of as bilinear composites of these new sets of fermions. For purposes of easy identification, and to distinguish them from ordinary quarks and leptons, we will call them techni-fermions. For the formation of composites, we need a strong binding force. We cannot use the known strong interactions for this purpose since the new effective composites must be present at an energy scale of 100 GeV or more, whereas the usual quantum changeable interactions become strong only at $\Lambda_{QCD} \sim (1 - 2)10^{-1}$ GeV. We, therefore, need to introduce a force (to be called technicolor) that becomes strong in a scale of 1 TeV or so. This fact can be represented by the following expressions:

$$\frac{g^2}{4\pi}(\Lambda_{QCD} \approx 100 \text{ MeV}) \geq 1, \tag{8.2.1}$$

where

$$\frac{g_{TC}^2}{4\pi}(\Lambda_{TC} \leq 1\,\text{TeV}) \geq 1. \tag{8.2.2}$$

This is the general picture of dynamical symmetry breaking introduced by Susskind [1] and Weinberg [1]. To realize these ideas in the building of electroweak models, we start by giving the gauge group, and the assignment of fermions under it, as follows (we consider one generation of fermions). The gauge group is

$$G = G_{WK} \times G_{St} \times G_{TC}. \tag{8.2.3}$$

Let us choose $G_{St} \times G_{WK} = \text{SU}(3)_C \times \text{SU}(2)_L \times \text{U}(1)_Y$. We keep $G_{TC} = \text{SU}(N_{TC}), N_{TC} > 3$. Let us denote by $\psi_L \equiv \binom{\nu}{e^-}_L$; e_R^-; $Q_L \equiv \binom{u}{d}_L$, u_R, d_R the observed quarks and leptons of first generation. Let Q_T denote the techniquarks. We can choose their transformation properties under $G \equiv \text{SU}(N_{TC}) \times \text{SU}(3)_C \times \text{SU}(2)_L \times \text{U}(1)_Y$, as follows (this model has anomalies and will therefore need to be modified; but this will not affect the essential points of the ensuing discussion):

$$\begin{aligned}
\psi_L &\equiv (1, 1, 2, -1), \\
e_R^- &\equiv (1, 1, 1, -2), \\
Q_L &\equiv (1, 3, 2, 1/3), \\
u_R &\equiv (1, 3, 1, 4/3), \\
d_R &\equiv (1, 3, 1, -2/3), \\
Q_{TL} &\equiv (N, 1, 2, 1/3), \\
U_{TR} &\equiv (N, 1, 1, 4/3), \\
D_{TR} &\equiv (N, 1, 1, -2/3).
\end{aligned} \tag{8.2.4}$$

We then assume that the technicolor interactions behave in a manner very similar to quantum chromodynamics.

First of all, because $\text{SU}(N)_{TC}$ is an unbroken non-abelian gauge symmetry, it is asymptotically free. Like QCD, it is expected to have the property of confinement, which means that all technicolor nonsinglet states are confined and do not appear in collisions or as decay products. Finally, like QCD, it will break chiral symmetry of techni-quarks by formation of condensates, i.e.,

$$\langle \bar{Q}_T Q_T \rangle = \Lambda_{TC}^3 \neq 0. \tag{8.2.5}$$

It is now clear that since Q_{TL} transforms as a nonsinglet under $\text{SU}(2)_L \times \text{U}(1)_Y$, it will break the electroweak gauge symmetry down to $\text{U}(1)_{\text{em}}$ at the scale Λ_{TC}. This will give rise to the masses for the W and Z bosons. Before we discuss how this explicitly comes about in a model without elementary scalars, it is worth pointing out that, since we would expect $M_W \simeq g\Lambda_{TC}$, observations indicate $\Lambda_{TC} \approx 300$ GeV. This implies that if there is grand unification of both QCD and QTCD (technicolor dynamics), i.e., $g_S = g_{TC}$

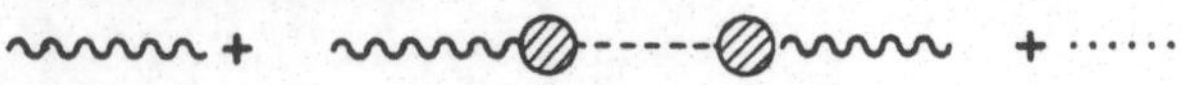

Figure 8.1.

at a scale $\mu \gg \Lambda_{TC}, \Lambda_{QCD}$, then the group $SU(N)_{TC}$ must be larger than $SU(3)$ and the number of techni-fermions must be smaller than the number of ordinary fermions.

Let us try to discuss the Higgs–Kibble mechanism for the case with no elementary Higgs bosons. The W bosons will couple to Q_{TC} as follows:

$$\mathcal{L}_{W,T} = \mathbf{W}_\mu \left(-\frac{ig}{2} \bar{Q}_{TL} \gamma_\mu \tau Q_{TL} \right)$$
$$- \frac{ig'}{6} B_\mu (\bar{Q}_{TL} \gamma_\mu Q_{TL} + 4\bar{U}_{TR} \gamma_\mu U_{TR} - 2\bar{D}_{TR} \gamma_\mu D_{TR}). \quad (8.2.6)$$

As a result of symmetry breakdown there will be composite techni-Goldstone bosons that will have odd parity and we can denote them by π_T. By using the ideas of PCAC (partially conserved axial current) they will dominate the axial techni-current, i.e.,

$$\tfrac{1}{2} \bar{Q}_T \tau \gamma_\mu \gamma_5 Q_T = F_\pi \partial_\mu \pi_T + \cdots \quad (8.2.7)$$

where $F\pi \approx \Lambda_{TC}$ up to some numerical factor of order 1. We can therefore write

$$\mathcal{L}_{W,T} = \mathbf{W}_\mu \left(-\frac{igF\pi}{2} \partial_\mu \pi_T \right) - \frac{ig'}{6} B_\mu (3F_\pi \partial_\mu \pi_T^3) + \cdots \quad (8.2.8)$$

$$\equiv -\frac{igF_\pi}{2} W_\mu^+ \partial_\mu \pi_T^- \text{ h.c.} - \frac{iF_\pi}{2} \partial_\mu \pi_T^3 (gW_\mu^3 + g'B_\mu) + \cdots, (8.2.9)$$

where ellipses indicate the rest of the interactions. The $W - \pi_T$ bilinears in eq. (8.2.9) will contribute to the self-energy of the fields W_μ^+ and $Z_\mu = (1/\sqrt{g^2 + g'^2})(gW_\mu^3 + g'B_\mu)$ through the diagram of Fig. 8.1. Let the massless propagator for the W^+ be chosen as

$$\Delta_{\nu,0}^{\mathbf{W}^+} \sim \frac{g_{\mu\nu} - q_\mu q_\nu/g^2}{q^2 + i\varepsilon}. \quad (8.2.10)$$

Since each vertex with $W - \pi_T$ contributes $(gF_\pi/2)q_\mu$ we can evaluate the modification to the propagator and find

$$\Delta_{\mu,\nu}^{W^+} = \frac{g_{\mu\nu} - q_\mu q_\nu/g^2}{q^2 - g^2 F_\pi^2/4}. \quad (8.2.11)$$

Thus, the W-boson propagator picks up a pole in q^2 at $q^2 = -g^2 F_\pi^2/4$. This gives a mass to the W boson:

$$M_W = \frac{gF_\pi}{2}. \quad (8.2.12)$$

Similarly, we find

$$M_Z = \sqrt{g^2 + g'^2} \frac{F_\pi}{3}.$$ (8.2.13)

If we identify $F_\pi = v$ in eq. (3.1.12), Chapter 3, these are precisely the gauge boson masses in the elementary Higgs picture. Thus, the electroweak symmetry breaking has been achieved without the need for elementary scalars. The correct mass relation between W and Z with $\rho_W = 1$ is also achieved. Thus, replacing the Higgs bosons by an underlying strongly interacting fermion leads to the same picture of electroweak symmetry breaking. The Goldstone bosons, which are now composite, will be absorbed by the gauge bosons. The underlying strong interaction will, however, give its own composite spectrum, which are technicolor singlets, but consist of the techni-quarks Q. We will call them techni-hadrons. Typically, their masses will be of the order of the scale of technicolor interactions, Λ_T, i.e., 1 TeV or so, unless they are protected by a symmetry to have zero or lighter mass. This will, in general, depend on the global symmetry of the technicolor theory. However, one model-independent statement can be made, i.e., because we expect the chiral symmetry of the techni-world to be spontaneously broken, there will be no light baryonic techni-composites (i.e., $M_B \approx 1\,\text{TeV}$). There will, however, be light mesonic techni-composites of which we give some examples in a subsequent section. Using similarity with QCD we may expect the techni-baryon (B_T) mass to be of order

$$M_{B_T} \sim \frac{N}{3} \times \frac{\Lambda_T}{\Lambda_{\text{QCD}}} M_{\text{proton}}.$$ (8.2.14)

§8.3 Techni-Composite Pseudo-Goldstone Bosons

In the previous section we chose only one doublet (U, D) of techni-quarks to break the electroweak symmetry. As a result, in the limit of vanishing electroweak coupling, we were left with the chiral symmetry $SU(2)_L \times SU(2)_R$, which broke down by techni-condensates [eq. (8.2.5)] to $SU(2)$ leading to three (π_L) Goldstone bosons. All of them were absorbed by the electroweak gauge bosons $W^\pm$, Z. Thus, no light scalar bosons remain in this theory. If, however, for some reason, the underlying technicolor theory has $N(N > 2)$ techni-quarks, then the chiral symmetry group of the techni-world will be $SU(N)_L \times SU(N)_R$. In analogy with QCD, this chiral symmetry will be expected to undergo spontaneous symmetry breakdown to $SU(N)_{L+R}$ leading to $N^2 - 1$ pseudo-scalar massless bosons. Of these, three will be absorbed by the $W_L^\pm$ and Z bosons leaving $N^2 - 4$ massless bosons. The question we would address now is the following: Are these $N^2 - 4$ pseudoscalar bosons massless? If not, what are their masses? All these $N^2 - 4$ bosons are actually pseudo-Goldstone bosons since the electroweak gauge interactions do not respect the entire chiral symmetry $SU(N)_L \times SU(N)_R$ prior to spontaneous symmetry breaking. Their masses arise at the one-loop level due to gauge interactions and are expected to be of order

$$M_P^2 \approx \frac{\alpha}{\pi}\Lambda_{TC}^2. \tag{8.3.1}$$

Naively, this implies $M_P \approx 30\,\text{GeV}$ for $\Lambda_{TC} \approx 1\,\text{TeV}$. These particles have the interesting property that they are both P and CP odd like the pion. If the Higgs bosons were elementary, then, in extended electroweak models, they do not have definite P and CP properties. In the standard $SU(2)_L \times U(1)$ model with one Higgs doublet, the physical Higgs boson has even parity and CP. Thus, experimental study of the P and CP properties of the scalar bosons, when they are discovered, can distinguish the technicolor picture from the elementary Higgs picture.

AN EXAMPLE. Let us consider an anomaly-free extension [2] of the toy technicolor model in eq. (8.2.4) by considering the techni-fermions [within the brackets we give their $SU(N)_H \times SU(3)_C \times SU(2)_L \times U(1)_Y$ quantum numbers]

$$Q_{CL} \equiv \begin{pmatrix} U_a \\ D_a \end{pmatrix}_L \equiv (N,3,2,1/3), \qquad a = 1,2,3,$$

$$\psi_L \equiv \begin{pmatrix} N \\ E_L \end{pmatrix} \equiv (N,3,2,-1),$$

$$U_{aR} \equiv (N,3,1,4/3), \qquad N_R \equiv (N,1,1,0),$$

$$D_{aR} \equiv (N,3,1,-2/3), \qquad E_R \equiv (N,1,1,-2). \tag{8.3.2}$$

In the limit of zero electroweak gauge couplings, the chiral symmetry of the techni-world is $SU(8)_L \times SU(8)_R$. This symmetry is spontaneously broken, by the following condensates, down to $SU(8)_{L+R}$:

$$\langle \bar{U}_a U_a \rangle = \langle \bar{D}_a D_a \rangle = \langle \bar{E}E \rangle =\rangle \bar{N}N \rangle \neq 0. \tag{8.3.3}$$

This leads to 63 massless pseudo-scalar particles of which 60 are the pseudo-Goldstone bosons according to our previous counting arguments. We can identify them by giving their color and isospin quantum numbers:

$$\theta_p^i \sim \bar{Q}\gamma_5 \lambda \tau^i Q, \quad \left\{ \begin{array}{ll} p = 1,\ldots,8, & \text{color indices,} \\ i = 1,2,3, & \text{isospin,} \end{array} \right.$$

$$\theta_p \sim \bar{Q}\gamma_5 \lambda_p Q,$$

$$T_a^i \sim \bar{Q}_a \gamma_5 \tau_i L, \qquad a = 1,\ldots,3, \quad \tau = 1,\ldots,3,$$

$$T_a \sim \bar{Q}_a \gamma_5 L,$$

$$P_i \sim (\bar{Q}\gamma_5 \tau^i Q - 3\bar{L}\gamma_5 \tau_i L),$$

$$P_0 \sim (\bar{Q}\gamma_5 Q - 3\bar{L}\gamma_5 L). \tag{8.3.4}$$

These add up to the 60 pseudo-Gold stone bosons predicted by the theory. We would expect their masses to be smaller than 1 TeV and presumably somewhere in the range of 100 GeV or so. For instance, we expect the masses of $P^{\pm}$ to be

$$M_{P^{\pm}}^2 \approx \frac{3\alpha}{4\pi} M_Z^2 \ln\left(\frac{\Lambda_{TC}^2}{M_Z^2}\right)$$

$$\approx (5-10)\,\text{GeV}^2. \tag{8.3.5}$$

This is among the lightest pseudo-scalar particles predicted by technicolor theories. The pseudo's with $SU(3)_C$ such as θ's will have higher mass due to the strong interaction corrections. Roughly, their masses will be

$$M_{\theta} \approx \frac{\sqrt{\alpha_{\text{strong}}(1\,\text{TeV})}}{\alpha_{\text{em}}(1\,\text{TeV})} M_p.$$

The existence of these lower mass states makes it possible to test these ideas in the current generation of accelerators. Their detailed decay properties and other implications [3] have been studied in the literature.

§8.4   Fermion Masses

In the standard electroweak model discussed in Chapter 3, the elementary Higgs boson serves two purposes: it breaks the electroweak symmetry and generates fermion masses. To achieve this purpose in technicolor theories we have to extend the scope of the technicolor models. The reason is that the analog of the Yukawa coupling in this case is a four-Fermi operator

$$\mathcal{L}_{\text{eff}} = \frac{1}{\Lambda_E^2} \bar{q}q\bar{Q}Q. \tag{8.4.1}$$

This can lead to fermion masses M_f, on substituting into eq. (8.2.5), i.e.,

$$M_f \approx \frac{\Lambda_T^3}{\Lambda_E^2}. \tag{8.4.2}$$

An inspection of the models discussed in the previous sections makes it clear that there is no scope for generating such interactions without further extensions. These extensions [3] are called extended technicolor models.

The idea in these models is to introduce new broken gauge interactions between the known fermions and techni-fermions, with the scale of breaking of the new interactions being of order $\Lambda_E > \Lambda_T$. To give an example, consider an $SU(8) \times SU(2)_L \times U(1)_Y$ gauge theory with techni- and known particles of one generation transforming under it as follows:

$$\left.\begin{pmatrix} U_1 & D_1 \\ U_2 & D_2 \\ U_3 & D_3 \\ N & E \end{pmatrix}_L\right\} \quad \text{technicolor} \quad \left\{ \begin{pmatrix} U_1 \\ U_2 \\ U_3 \\ N \end{pmatrix}_R \begin{pmatrix} D_1 \\ D_2 \\ D_3 \\ E \end{pmatrix}_R \right.$$

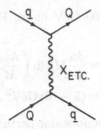

Figure 8.2.

$$\left.\begin{pmatrix} u_1 & d_1 \\ u_2 & d_2 \\ u_3 & d_3 \\ \nu & e \end{pmatrix}\right\}_L \quad \text{color} \quad \left\{ \begin{pmatrix} u_1 \\ u_2 \\ u_3 \\ \nu \end{pmatrix}_R \begin{pmatrix} d_1 \\ d_2 \\ d_3 \\ e^- \end{pmatrix}_R \right.$$
$$\mathrm{SU(2)}_L$$

Let us assume that SU(8) gauge theory is broken down to $\mathrm{SU(4)}_{TC} \times \mathrm{SU(3)}_C$ by some unknown mechanism at a scale Λ_{ETC}. Then the graphs of Fig. 8.2 will induce four-Fermi interactions

$$\mathcal{L}_{\mathrm{eff}} = \frac{1}{\Lambda_{\mathrm{ETC}}^2} \bar{q}\gamma_\mu Q \bar{Q}\gamma^\mu q. \tag{8.4.3}$$

On Fierz rearrangement this gives rise to eq. (8.4.1), which after techni-condensates form can lead to fermion masses. Clearly, this picture can be extended by including more light quarks by suitable extension of the extended technicolor group G_{ETC}. For instance, by choosing $G_{ETC} \equiv \mathrm{SU(12)}$, we can include two generations of fermions. By appropriate adjustment of the masses of the extended techni-gauge bosons that connect Q with the first and the second generations, we have enough free parameters to generate fermion mixings and masses. It has, however, been argued that these models predict large amounts of flavor violation in weak interactions, e.g., $K_1 - K_2$ mass differences receive too much contribution from the extended technicolor interactions [4].

Regardless of the detailed phenomenological implications, the technicolor models have introduced a new concept into physics, i.e., the existence of new strong interactions with a higher scale that can simulate many aspects of low-energy electroweak interactions without introducing elementary scalars. This picture of "partial compositeness" may be extended to a domain where not only the Higgs bosons but also the quarks, leptons, W bosons, and Higgs bosons all may be thought of as composites bound by a new strong-interaction force analogous to technicolor. This philosophy may be dubbed one of "total compositeness" and will be the subject of the subsequent sections.

§8.5 Composite Quarks and Leptons

Even though there are theoretical reasons [5] to suspect that the quarks and leptons may be composite, as yet, there is no experimental evidence for their compositeness. On the other hand, based on new effects predictable on the basis of compositeness, we can put an upper bound on the compositeness size (or a lower bound on the mass scale of compositeness, which we will call Λ_H). This is also of great interest to experimentalists as they build accelerators with higher and higher energies to look for new structures in particle interactions.

In order to identify effects due to compositeness, we must isolate the effects of known physics on the experimental parameters being studied to a certain level of accuracy. One such parameter where effects of known physics have been computed to a high degree of accuracy is the $(g-2)$ of the electron and the muon. Experimentally, $(g-2)$ is known to a great precision:

$$(g-2)_\mu = (23318406 \pm 30) \times 10^{10}. \tag{8.5.1}$$

Theoretical value, which takes into account QED effects to eighth order in the electric charge and other weak effects, for this parameter is given by [6]

$$(g-2)_{th} = (233183544 \pm 14) \times 10^{-10}. \tag{8.5.2}$$

The hadronic contribution to eq. (8.5.2) is $(140.4 \pm 3.8) \times 10^{-9}$, whereas the electroweak contribution is $(3.90 \pm 0.02) \times 10^{-9}$.

If there is any effect due to compositeness, it must therefore be less than

$$(\delta a)_\mu = [(g-2)_{expt} - (g-2)_{th}] < 0.5 \times 10^{-8}. \tag{8.5.3}$$

Attempts have been made [7] to estimate the contribution of compositeness to (δa). A crude estimate can be made on the basis of chiral symmetry and dimensional arguments: the latter says that, since the anomalous magnetic moment operator has dimension of inverse mass, we can write

$$\frac{1}{m_\mu}(\delta a) = f(m_\mu \Lambda_H) = 0. \tag{8.5.4}$$

Furthermore, since the $(g-2)$ vanishes in the limit of chiral symmetry, we conclude that

$$\lim_{m\mu \to 0} f(m_\mu, \Lambda_H) = 0. \tag{8.5.5}$$

This implies that the leading contribution to $(\delta a)_\mu$ from compositeness can be written as

$$(\delta a)_\mu \simeq \left(\frac{m\mu}{\Lambda_H}\right)^2 \leq 10^{-8}. \tag{8.5.6}$$

This implies that $\Lambda_H \geq$ TeV. The corresponding value $(\delta a)_e$ for the electron is known to be much less, i.e., $(\delta a)_e \leq 3.2 \times 10^{-10}$. However, since the analogous formula for $(\delta a)_e$ involves m_e^2, the bound on Λ_H is not as good.

Other sources of information on Λ_H are Bhabha scattering $e^+e^- \rightarrow e^+e^-$ off the forward direction. To study these contributions, we may note that in a composite theory, low-energy structure of the interaction Hamiltonian for quarks and leptons is given by the chiral symmetry of the theory. For one family of quarks and leptons this low-energy chiral symmetry is expected to be $SU(2)_L \times SU(2)_R \times SU(4)_C$ [8] or $SU(2)_R \times SU(2)_L \times SU(4)_L \times SU(4)_R$ [9]. Therefore, all four-Fermi operators consistent with these symmetries should appear at low energies. A more conservative approach may be simply to assume that the only electroweak symmetry below the compositeness scale is $SU(2) \times U(1) \times SU(3)_C$. It may be worth pointing out that at this stage there is no compelling argument for it. In other words, we may assume either $SU(2)_L \times U(1) \times SU(3)_C$ or $SU(2)_L \times SU(2)_R \times U(1)_{B-L} \times SU(3)_C$ or $SU(2)_L \times SU(2)_R \times SU(4)_C$ to constrain the nature of the four-Fermi interactions below the compositeness scale. Below we give a list of possible four-Fermi operators consistent with the $SU(2)_L \times SU(2)_R \times SU(4)_C$ symmetries for the case of two generations: the extensions to other cases are straightforward.

(i) $SU(2)_L \times SU(2)_R \times SU(4)_C$

Let the fermion multiplet F_a be denoted by (a denotes generation)

$$F_a = \begin{pmatrix} u_1 & u_2 & u_3 & \nu \\ d_1 & d_2 & d_3 & e \end{pmatrix}_a,$$

$$F_{a_L} \equiv (2,1,4) \quad \text{and} \quad F_{a_R} \equiv (1,2,4), \tag{8.5.7}$$

under the $SU(2)_L \times SU(2)_R \times SU(4)_C$ group. The possible invariant four-Fermi operators are then given by

$$O_{1L}^{ab,cd} = (\bar{F}_{a_L} \gamma_\mu \tau_i \times \lambda_p F_{b_L})(\bar{F}_{c_L} \gamma^\mu \tau_i \times \lambda_p F_{d_L}). \tag{8.5.8}$$

Similarly,

$$O_{1L}^{ab,cd} = O_{1(L \rightarrow R)}^{ab,cd}, \tag{8.5.9}$$

$$O_{e\varepsilon,\varepsilon'}^{ab,cd} = (\bar{F}_{a_\varepsilon} \gamma_\mu \lambda_p F_{b_\varepsilon})(\bar{F}_{c_{\varepsilon'}} \gamma^\mu \lambda_p F_{d_{\varepsilon'}}) \tag{8.5.10}$$

where $\varepsilon, \varepsilon' = L, R; p = 0, \ldots, 15$ and denotes the $SU(4)$ index; τ_i denotes the flavor $SU(2)$ matrices; and λ_p are $SU(4)$ matrices. The effective four-Fermi interaction at low energies can therefore be written as

$$\mathcal{L}_{\text{eff}} = \frac{1}{\Lambda_H^2} \sum_i C_i O_i, \tag{8.5.11}$$

where i goes over all the operators in eqs. (8.5.9) and (8.5.10). Of these, we can pick out the combination, involving electron fields only, that will contribute to Bhabha scattering off the forward angle. If all C_i are chosen to be of order one, then, for a particular value of momentum transfer Q, we can probe the compositeness scale $\Lambda_H \sim Q/\sqrt{\alpha_{\rm em}}$. This kind of analysis [10] leads to bounds on

$$\Lambda_H \geq 1.5\,{\rm TeV}. \tag{8.5.12}$$

The Hamiltonian in eq. (8.5.11), in general, also leads to flavor-changing effects. For instance, if we choose $a = c = 1$ and $b = d = 2$ in eq. (8.5.10), we can get an operator of the form (for $p = 0$)

$$\mathcal{L}_{\rm eff}^{\Delta S=2} \equiv \sum_{\varepsilon,\varepsilon'} \frac{C_{2\varepsilon\varepsilon'}}{\Lambda_H^2} \bar{d}_\varepsilon \gamma_\mu s_\varepsilon \bar{d}_{\varepsilon'} \gamma^\mu s_{\varepsilon'}. \tag{8.5.13}$$

This would contribute to the $K_1 - K_2$ mass difference with a dominant part coming from $\varepsilon = L$, $\varepsilon' = R$. A simple analysis following the literature on the subject [11] can be found

$$\Lambda_H^2 \geq \frac{32\pi^2}{G_F^2 \sin^2 \theta_C M_C^2}. \tag{8.5.14}$$

This implies

$$\Lambda_H \geq 5.5 \times 10^3\,{\rm TeV}. \tag{8.5.15}$$

This appears to be the strongest bound. Furthermore, from eq. (8.5.10), taking $p = 9, \ldots, 14$ and $a = b = 1$ and $c = d = 2$ we find an interaction of type

$$\mathcal{L}_{\rm eff} = \frac{\tilde{C}}{\Lambda_H^2} \bar{d}\gamma_\mu e \bar{\mu} \gamma^\mu s. \tag{8.5.16}$$

This would contribute to the process $K_L^0 \to \mu\bar{e}$ decay. The present experimental upper limit on the branching ratio for this decay model is [12]

$$B(K_L^0 \to \mu\bar{e}) \leq 10^{-8}. \tag{8.5.17}$$

This implies that $\Lambda_H \geq 100\,{\rm TeV}$. Again we see that these limits on the scale of compositeness are much more stringent than those derived previously.

It must, however, be pointed out that if there exists a family symmetry in the preon model, these limits will be considerably altered and will depend on unknown parameters corresponding to breaking of these symmetries. Thus, the two latter limits on Λ_H are to be taken with a degree of caution.

In summary, the present low-energy and high-energy scattering experiments indicate the scale of compositeness to be somewhere between 1 and 100–TeV.

§8.6   Light Quarks and Leptons and 't Hooft Anomaly Matching

As we saw in the last section, the scale of compositeness is at least of order
1–100 TeV. On the other hand, the quarks and leptons have masses ranging
from 10 MeV to 40 GeV. It is therefore to be expected that the fermion
masses owe their origin to a different mechanism than compositeness and,
strictly in the absence of those "unknown mechanisms," they should vanish. The disappearance of fermion masses is generally guaranteed by the
existence of chiral symmetries such as $\psi \to e^{i\gamma_5 \alpha}\psi$, etc., operating on the
preons as well as their composites. In other words, to understand the masslessness of quarks and leptons we must ensure that the chiral symmetries
of the preonic world must remain unbroken as the composites form. This
situation is in contrast to that existing in the case of QCD, where the chiral
symmetry of the quarks is dynamically broken by $\bar{q}q$ condensates, which
then leads to nucleon mass of order $k\Lambda_{\text{QCD}}$, where k is a number of order
1-10.

It was pointed out by 't Hooft in a classic paper [13] that the requirement that chiral symmetry remain unbroken at the composite level imposes
strong constraints on the possible models for compositeness. The origin of
the constraints can be understood as follows.

Let there be a global chiral symmetry $G_L \times G_R$ of the preon dynamics.
Let $J_{\mu,i_{L,R}}$ be the currents corresponding to this global symmetry group. It
is well known [14,15] that the three point functions involving these currents
have a singularity in the q^2 plane at $q^2 = 0$. To be more specific let us define

$$\Gamma_{\mu\nu\lambda}^{ijk}(q,k) = \int \exp(-ik \cdot x - iq \cdot y) d^4x \, d^4y \langle 0|T(V_\mu^i(x)V_\nu^j(y)A_\lambda^k(0))|0\rangle.$$

$$(8.6.1)$$

The internal indices i, j, k can be factored out

$$\Gamma_{\mu\nu\lambda}^{ijk} = d^{ijk}\Gamma_{\mu\nu\lambda}, \tag{8.6.2}$$

where

$$d^{ijk} = \text{Tr}\,[\lambda^i \{\lambda^j, \lambda^k\}_+], \tag{8.6.3}$$

λ's being the generators of the appropriate group. Lorentz invariance
implies that

$$\Gamma_{\mu\nu\lambda}(q,k) = A_1 q^\tau \varepsilon_{\tau\mu\nu\lambda} + A_2 k^\tau \varepsilon_{\tau\mu\nu\lambda} + A_3 q_\nu q^\sigma k^\beta \varepsilon_{\sigma\beta\mu\lambda} + A_4 k_\nu q^\sigma k^\beta \varepsilon_{\sigma\beta\mu\lambda}$$
$$+ A_5 q_\mu q^\sigma k^\beta \varepsilon_{\sigma\beta\nu\lambda} + A_6 k_\mu q^\sigma k^\beta \varepsilon_{\sigma\beta\nu\lambda}. \tag{8.6.4}$$

Following Ref. 15 let us restrict ourselves to the case when $k^2 = q^2$. The
invariant amplitude then depends on $(q+k)^2 = k_1^2$ and $k_2^2 = q^2 = k^2$. Bose
symmetry implies

$$A_1 = -A_2, \qquad A_3 = -A_6,$$

and

$$A_4 = -A_5. \tag{8.6.5}$$

Now let us impose the vector and axial vector Ward identities, i.e.,

$$\partial^\mu V_\mu = 0$$

and

$$\partial^\mu A_\mu = m\bar{\psi}\gamma_5\psi. \tag{8.6.6}$$

This implies

$$A_1 = \frac{k_1^2 - 2k_2^2}{2} A_4 + k_2^2 A_3,$$
$$A_1 - A_2 = 2mB, \tag{8.6.7}$$

where B is the Fourier transform similar to that in eq. (8.6.1) with A_μ replaced by $\bar{\psi}\gamma_5\psi$. If the amplitude is assumed to be regular at $k_1^2, k_2^2 \to 0$, then eq. (8.6.7) implies that for $m \to 0$, $A_1 = A_2 = 0$. Then keeping k_1^2 fixed and letting $k_2^2 \to 0$ we get $A_3(k_1^2, 0) = 0$, and taking $k_1^2 \to 0$ we get $A_3(0, k_2^2) = A_4(0, k_2^2)$. Thus at $k_1^2 = k_2^2 = 0$ the whole amplitude vanishes. This is not acceptable. This implies that either the vector or the axial vector current will not satisfy the Ward identities.

To study the nature of A_i further we assume unsubtracted dispersion relations k_1^2, i.e.,

$$A_i^{(u)}(k_1^2, k_2^2) = \frac{1}{\pi} \int \frac{\text{Im } A_i(s, k_2^2)}{s - k_1^2} \, ds. \tag{8.6.8}$$

It is then clear from eq. (8.6.7) that, in order to satisfy the Ward identities, A_1 must have a subtraction with subtraction constant

$$A_1 = A_1^{(u)} - \frac{1}{2\pi} \int \text{Im } A_4(s, k_2^2) \, ds. \tag{8.6.9}$$

Now going to the point $k_2^2 = 0$, eqs. (8.6.7) and (8.6.8) imply that

$$k_1^2 \text{ Im } A_4(k_1^2, 0) = 0$$

and

$$A_1(k_1^2, 0) = -\frac{1}{2\pi} \int \text{Im } A_4(s, k_2^2 = 0) \, ds. \tag{8.6.10}$$

Equation (8.6.10) requires that

$$\text{Im } A_4(k_1^2, 0) = c\delta(k_1^2). \tag{8.6.11}$$

This is the 't Hooft anomaly equation. The constant c is obtained by looking at the triangle graphs with internal fermion lines (Fig. 8.3) and picking out the Clebsch–Gordan coefficients. In order to apply it to constrain the

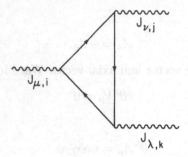

Figure 8.3.

confining models of weak interaction, we can choose a normalization such that we get c as follows:

$$c = \sum_f Tr[\lambda_a\{\lambda_b, \lambda_c\}]_f, \qquad (8.6.12)$$

where f denotes the number of fermions with some internal global chiral quantum numbers going around in Fig. 8.3

In our derivation of the 't Hooft anomaly equation (8.6.11) we did not assume anything about the underlying field theory. In other words, the same singularity structure should obtain, both in the component preonic field theory and in the composite theory. In the composite theory the singularity can come about in two ways:

(i) by spontaneous breakdown of chiral symmetry so that the resulting Goldstone bosons lead to the k_1^2 singularity of A_3;

(ii) by massless chiral fermions in which case c [defined in eq. (8.6.12)] must be the same as in the preonic theory, i.e.,

$$c_{\text{preons}} = c_{\text{composite}}. \qquad (8.6.13)$$

This is a strong constraint on the composite models. We give examples of this in the following sections.

§8.7   Examples of 't Hooft Anomaly Matching

In this section we give some examples of composite theories and show the impact of anomaly constraints.

(a) Two-Flavor QCD

Let us consider the usual quantum chromodynamic theory with two flavors of quarks u, d. In the limit of m_u, $m_d = 0$ the Lagrangian can be written as

$$\mathcal{L}_{\text{QCD}} = - \sum_{\psi = u,d} \bar{\psi}\gamma^{\mu}(\partial_{\mu} - \frac{ig_s}{2}\boldsymbol{\lambda}_c \cdot \mathbf{B}_{\mu})\psi - \frac{1}{4}\mathbf{f}_{\mu\nu} \cdot \mathbf{f}_{\mu\nu}. \qquad (8.7.1)$$

This theory has the global symmetry $SU(2)_L \times SU(2)_R \times U(1)_B \times U(1)_B^5$. Of these, the $U(1)_B^5$ that gives the axial baryonic symmetry is broken down to Z_2 by instanton effect. Thus, for our purpose, the continuous global symmetry is $SU(2)_L \times SU(2)_R \times U(1)_B$. Let us now calculate the anomaly coefficient c [eq. (8.6.12)] for this quark theory. The only three-point functions that are anomalous are $U(1)_B[SU(2)_L]^2$, $U(1)_B[SU(2)_R]^2$,

$$C_{qR} = C_{qL} = 3 \times B_q = 1, \qquad (8.7.2)$$

where B_q is the baryon number of the quarks and is 1/3.

The composites of this theory are the $SU(3)_c$ singlet three-quark states, i.e., proton, $p = (uud)$ and neutron, $n = (udd)$. They form representations of the chiral $SU(2)_L \times SU(2)_R \times U(1)_B$. Now there are two ways to satisfy the anomaly constraints:

(i) if the proton and the neutron are massless, then, since their baryon number is 1, they imply $C_{N_L} = C_{N_R} = 1$; or

(ii) if chiral symmetry is broken by quark–antiquark condensates. In this case the corresponding zero-mass Goldstone bosons, the pions, provide the singularity of the three-point function.

Thus, the 't Hooft anomaly does not prefer one possibility over the other in the case of two-flavor QCD. However, it rules out the possibility considered seriously in the early days whereby a massive parity-odd nucleon was supposed to provide the chiral partner of the nucleon in the exact symmetry limit.

It is easy to construct theories where 't Hooft anomaly matching will require that the component global chiral symmetry be spontaneously broken.

(b) Three-Flavor QCD

As an example of this kind consider three-flavor QCD with massless flavors u, d, s. The quark theory has the chiral symmetry $G_q = SU(3)_L \times SU(3)_R \times U(1)_B$. The anomalous three-point functions in this case are:

(i) $U(1)_B[SU(3)_L]^2$;

(ii) $U(1)_B[SU(3)_R]^2$;

(iii) $[SU(3)_L]^3$;

(iv) $[SU(3)_R]^3$.

The corresponding values of C at the quark level are

$$c_i = 1 = c_{ii},$$
$$c_{iii} = c_{iv} = d_{abc}. \qquad (8.7.3)$$

Let us now look at the composite theory. If the chiral symmetry is not spontaneously broken, the only source of singularities for the three-point functions (i)–(iv) is the massless spin-1/2 baryon octet (p, n, Σ^+, Σ^-, Σ^0, Λ, Ξ^0, Ξ^-). Obviously, their contributions to $c_{(i)-(iv)}$ are very different. For instance, $c_i = c_{ii} = 3$. Thus, the anomalies do not match unless the chiral symmetry is spontaneously broken down to $SU(2)_L \times SU(2)_R \times U(1)_B$, in which case in a manner analogous to Case (a) the anomalies match. It can therefore be argued that the 't Hooft anomaly implies that dynamics will require spontaneous breakdown of chiral symmetry in this case.

(c) Color-Flavor-Factorized Preon Model

We present a realistic preon model of quarks and leptons and show how for a particular kind of hypercolor binding force for the preons the 't Hooft anomaly constraints match, implying massless quarks and leptons.

The model is based on the observation of Pati and Salam [16] that the symmetry of the quark–lepton world is $SU(2)_L \times SU(2)_R \times SU(4)_C$ operating on them as follows:

$$F = \begin{pmatrix} u_1 & u_2 & u_3 & \nu \\ d_1 & d_2 & d_3 & e \end{pmatrix} \begin{matrix} \rightarrow SU(4)_C \\ \downarrow \\ SU(2)_L \times SU(2)_R. \end{matrix}$$

It then is suggestive to consider the quarks and leptons as bound states [17] of two-flavor fermions (F_u, F_d) and four-color bosons ($\phi_1 \phi_2 \phi_3 \phi_l$). Let the binding force be given by an $SU(N)_H$-hypercolor force analogous to QCD. Assume that, under $SU(N)_H \times SU(3)_C \times U(1)_{em}$,

$$\begin{pmatrix} F_u \\ F_d \end{pmatrix} \equiv \begin{pmatrix} N & 1 & 1/2 \\ N & 1 & -1/2 \end{pmatrix} \qquad (8.7.4)$$

and

$$\phi_i \equiv (\bar{N}, 3, 1/6),$$
$$\phi_l \equiv (\bar{N}, 1, -1/2). \qquad (8.7.5)$$

The anomaly constraints on this model were first analyzed by Barbieri, Mohapatra, and Masiero [8]. For this let us note that the preonic theory has the global symmetry

$$G_p^0 = SU(2)_L \times SU(2)_R \times \times U(1)_F \times U(4)_\phi. \qquad (8.7.6)$$

In order to satisfy the anomalies we must break the $U(1)_\phi$ either explicitly or spontaneously so that the surviving symmetry is

$$G_p = SU(2)_L \times SU(2)_R \times U(1)_F \times SU(4)_\phi. \qquad (8.7.7)$$

There are only two sets of preonic anomalies in this case corresponding to (i) $U(1)_F[SU(2)_L]^2$ and (ii) $U(1)_F[SU(2)_R]^2$, with the following values for C (choose F's to have fermion number 1):

$$C_i = C_{ii} = N_H. \qquad (8.7.8)$$

At the composite level we have the composites listed earlier as F. Their contribution to anomaly are (if only one set of composites appear)

$$C_i = 4 = C_{ii}. \qquad (8.7.9)$$

This implies that for one family of fermions $N_H = 4$.

It is then easy to note that we could get N_g families by choosing $N_H = 4$ N_g. Thus, in this case, masslessness of quarks and leptons is understood as a consequence of 't Hooft anomaly constraints.

§8.8   Some Dynamical Constraints on Composite Models

In this section we summarize some other dynamical constraints that have been discussed in connection with composite models with vectorlike non-abelian forces binding the constituents.

(a) Constraints Due to Heavy Constituent Decoupling

In his original paper on anomaly matching 't Hooft also proposed another kind of constraint on composite models. He argued that if one of the preonic constituents becomes heavier than the scale of compositeness, then it should decouple from the entire theory [18]. This would therefore change the global chiral symmetry from, say, $SU(N)_L \times SU(N)_R$ down to $SU(N-1)_L \times SU(N-1)_R$. Then again, in the resulting theory, the anomalies should match. This argument can be applied successively to move preons being made heavy and made to decouple. The anomalies for chiral groups at each stage should match. Thus, anomaly matching should basically be independent of N. Clearly, this is a much stronger constraint on the models and makes it extremely difficult to construct composite models. However, it has since been argued [19] that the nonperturbative effects may not be analytic in the preon mass so that for the large and small preon masses the theories may be quite difficult, thus avoiding this additional constraint.

(b) Mass Inequalities in Vectorlike Non-abelian Fermionic Constituent
Type Theories

Recently, Weingarten [20], Witten [20], and Nussinov [20] have derived
inequalities between composite fermions and scalar boson masses in vectorlike confining theories with purely fermionic constituents. Their results
can simply be stated as follows: if there is a vectorlike confining theory with
non-abelian $SU(N)_H$, gauge interaction providing binding of $SU(N)_H$, nonsinglet fermionic constituents to form singlet composites, then in this theory
the composite fermions and scalar bosons must satisfy the inequality

$$M_B \geq aM_s, \tag{8.8.1}$$

where B and S stand for baryons and scalar composites and a is a number
depending on N_H; for instance, for QCD, $a = 3/2$. Let us discuss its implications for composite models. Naively speaking, it implies that if quarks
and leptons are to be identified with the composite baryons, eq. (8.8.1)
implies that there must be spin-0 particles lighter than them in the composite spectrum. Of course, if such particles are neutral under the strong
and electroweak $U(1)_Y \times SU(2)_L \times SU(3)_C$ symmetry, then they are harm-
less provided some weak constraints are satisfied by their couplings [21].
However, in general, they carry color as well as electric charge. Therefore,
they are in obvious conflict with experiment. This would therefore rule
out most [22] fermionic-type vectorlike composite quark and lepton models. No such theorem exists for theories with scalar constituents. Therefore,
combining this with the philosophical elegance of the color-flavor factorized model would make the fermion–boson-type models of the previous
section quite interesting candidates for composite models of quarks and leptons. The presence of scalar bosons may, of course, indicate an underlying
supersymmetry in nature at the preonic level.

(c) Theorem of Weinberg and Witten [23]

A result of great significance for the composite models is a theorem due to
Weinberg and Witten, which says that the only possible massless states in
field theory carrying internal quantum numbers must have spin $\leq \frac{1}{2}$ unless
there are gauge symmetries in the theory. This is important for composite
model building since it says, for instance, that the composite spectrum
cannot have massless states with spin $J \geq 3/2$, which would, in principle,
be allowed by chiral symmetries needed to understand the smallness of
quark and lepton masses.

§8.9   Other Aspects of Composite Models

In this section we raise some general issues encountered in building
composite models:

(a) the existence of real Goldstone bosons;

(b) composite versus elementary gauge bosons;

(c) possibilities of baryon and lepton nonconservation;

(d) CP violation and baryon asymmetry of the universe.

(a) *Massless Goldstone Bosons.* We have discussed in the previous sections that global symmetries play a vital role in understanding the small masses of quarks and leptons. The nature of the global symmetry is dictated by the preonic structure of the theory. It may therefore happen (and it often does in actual model building) that there may be global symmetries that have to be broken spontaneously in order to satisfy 't Hooft anomaly constraints. In such a case there will appear real Goldstone bosons. It is important to point out that these Goldstone bosons are similar to the Majoron [21] discussed in Chapter 2. As pointed out there, since they lead to spin-dependent forces in the nonrelativistic limit, the phenomenological constraints on their couplings are relatively mild. To see this we note that the general form of the Goldstone boson coupling to light fermions is of the form

$$\mathcal{L}_{ffG} = \frac{M_f}{V} \bar{f} \gamma_5 f G, \tag{8.9.1}$$

where V is the scale of the symmetry breaking. The strength of the spin-dependent potential generated by it is $\approx (M_f/V)^2$ and is bounded phenomenologically to be less than one part in a million. For $M_f \approx 1$ GeV, this implies $V \geq$ TeV, since $V \approx \Lambda_H$, the compositeness scale; this is quite an acceptable constraint. The astrophysical constraint is, however, more stringent and may require [24] $\Lambda_H \approx 10^8$ GeV. In this case more careful analysis of the model is required to seek additional suppressions. This phenomenon is similar to the one discussed in Chapter 2 in connection with Majoron and is not repeated here.

(b) A second important aspect of composite models is the interpretation of the W, Z bosons, photon, color-gauge bosons, etc. There are two general points of view with regard to this: one [25] is to treat the massive gauge bosons W, Z as the composites of preons in much the same manner as the ρ, ω mesons are the quark–antiquark composites in QCD. The masses of the W and Z are then expected to be related to the compositeness scale, Λ_H. The photon and color-gauge bosons are associated with unbroken gauge theories $U(1)_{em} \times SU(3)_C$. In this picture there is no $SU(2)_L \times U(1)$ spontaneously broken gauge theory describing weak-interactions. The structure of the four-Fermi weak-interaction Lagrangian is given by the residual global $SU(2)_L$ symmetry of weak interaction together with neutrino charge radius interaction [26] as follows.

If $\mathbf{J}_{\mu_L}$ is denoted as the fermionic current associated with weak $SU(2)_L$, then the $SU(2)_L$, invariant current–current interaction is given by

$$\mathcal{L}_{\text{wk}}^{(0)} = 4\frac{G_F}{\sqrt{2}}\mathbf{J}_L^{\mu}\mathbf{J}_{\mu L} \tag{8.9.2}$$

and the neutrino charge–radius piece can be parametrized as follows:

$$\mathcal{L}_{\text{wk}}^{(\nu)} = 2\frac{G_F}{\sqrt{2}}\sin^2\theta_W\bar{\nu}\gamma^{\mu}(1+\gamma_5)\nu J_{\mu}^{\text{em}}. \tag{8.9.3}$$

The combination of these two terms are, of course, well known [26] to give the correct structure of low-energy weak interaction.

The low-energy effective Lagrangian can also be written in terms of $\mathbf{W}_{\mu_L}$ bosons that belong to the weak isospin-1 representation of $SU(2)_L$ along with an additional $\gamma - W^3$ mixing term. Hung and Sakurai [26] have demonstrated that this also gives the correct description of electroweak physics at low energy. In this picture the Weinberg angle is replaced by the combination $e\lambda/g$, where e is the electric charge, g is the coupling of the $\mathbf{W}$ boson to the weak isospin current $\mathbf{J}_{\mu_L}$, and λ is the $W^3 - \gamma$ mixing parameter.

Several questions have been raised concerning this very interesting alternative approach to weak interactions.

(i) λ is the analog of the $\rho - \gamma$ coupling in QCD, which is known to be a very small number ($\sim 10^{-2}$), whereas neutral-current data say that $e\lambda/g \approx 0.23$ implying $\lambda \approx 1$. The question then arises that if QHCD is so similar to QCD, how can we understand this discrepancy? It has been argued that this may be related to the question of chiral symmetry breaking of QCD and nonbreaking in the case of QHCD [27].

(ii) In this model M_W and M_Z are unpredictable. How then can we understand the discovery of W and Z bosons with mass values exactly predicted by the standard gauge model? A phenomenological answer to this question is to assume that, exactly like the ρ dominance of isospin current in strong interactions, the W boson dominates [28] the weak current. We can then prove that

$$\lambda \simeq \frac{e}{g}, \tag{8.9.4}$$

which leads to a prediction of the W and Z masses exactly as in the standard electroweak model. This also explains the universality of weak interactions.

This model is philosophically different from the standard gauge theory approach and is under a great deal of discussion [29] due to the new events from the CERN $p\bar{p}$ collider experiments such as anomalous $Z \to e^+e^-\gamma$, $\mu^+\mu^-\gamma$, etc.

The other approach to the W, Z boson, photon, color boson, etc., is to assume that the preons carry the spontaneously broken electroweak gauge theory, which is then transmitted to their composites such as quarks and leptons in much the same way that the electroweak interactions are transmitted from quarks to nucleons.

(c) An important conceptual triumph of grand unified theories (or a geometrical picture of unification) is that they provide a natural framework for baryon- and lepton-number violation. Combined with CP-violating interactions they meet all the basic requirements for understanding a great cosmological riddle–the origin of matter in the universe. If composite models are to provide a unified description of physics, not only at small distances but also on large scales such as the universe, they must provide a basis for understanding the origin of matter. This requires the introduction of baryon-violating interactions as well as CP violation into composite models. While there exist no satisfactory realistic models with the above features, some general intuitive inferences can be listed.

(i) There are two possible sources of CP violation in composite models: (a) the fermion condensates that break chiral symmetries; and (b) the CP-violating interactions arising out of hypercolor anomalies. For the former possibility an effective Lagrangian can be written down [30] that provides a realistic superweak model of CP violation and provides a mechanism for baryon generation [31].

(ii) Baryon-violating interactions could also arise through multifermion condensates along with CP violation. The other possibility is the grand unified composite models where hypercolor and color interactions unify at a superheavy scale. This may provide a mechanism for baryon generation of the universe.

§8.10 Symmetry Breaking via Top-Quark Condensate

In recent years, a new scheme for dynamical electroweak symmetry breaking has been proposed, based on the idea that due to some unknown interaction the top-quark condensates result [32–34], i.e.,

$$\langle \bar{t}_L t_R \rangle \neq 0. \tag{8.10.1}$$

To study these models it is assumed that at a superheavy scale, unknown physics at a heavier scale ($\mu \ll \Lambda$), produces a four-Fermi interaction involving the heavy-quark multiplets of the third generation, $Q_L \equiv (t_L, b_L)^T$, and t_R, b_R. There are, of course, several possible four-Fermi terms, but, for simplicity, the authors of Refs. 32–34 choose

$$H_{\text{eff}} = \frac{1}{M^2} \bar{Q}_L t_R \bar{t}_R Q_L. \tag{8.10.2}$$

This Hamiltonian is invariant under the $SU(2)_L \times U(1)_Y$-gauge symmetry. The question of the dynamical breaking of this symmetry by the formation of $\langle \bar{t}_L t_R \rangle$ condensate can be studied in a manner identical to that used by Nambu and Jona-Lasino in their classic paper on the application of the BCS model to particle physics. Working in the bubble approximation, one can convince oneself that, for

$$M^2 < \frac{N_c \Lambda^2}{8\pi^2} \qquad (8.10.3)$$

(N_c is the number of colors), $t\bar{t}$ condensate forms and electroweak symmetry is broken.

To get quantitative prediction from this we follow the procedure of Bardeen, Hill, and Lindner [34], who parametrize the four-Fermi interaction in (8.10.2) as follows:

$$\mathcal{L}_{\text{eff}} = L_{\text{kin}} - M^2 \phi_0^\dagger \phi_0 - \bar{Q}_L t_R \phi_0 - \bar{t}_R Q_L \phi_0^\dagger, \qquad (8.10.4)$$

where $\phi_0 = (\phi_0^0, \phi_0^-)^T$. This Lagrangian, when extrapolated to energies lower than Λ, generates kinetic energies and ϕ_0^4-type couplings for the Higgs double ϕ: i.e., for $\mu \ll \Lambda$,

$$\mathcal{L}_{\text{eff}}(\mu \ll \Lambda) = \mathcal{L}_{\text{kin}} - Z_H D_\mu \phi_0^\dagger D_\mu \phi_0 - h_0 \bar{Q}_L t_R \phi_0$$
$$- h_t \bar{t}_R Q_L \phi_0^\dagger - M'^2 \phi_0^\dagger \phi_0 - \lambda_H (\phi_0^\dagger \phi_0)^2. \qquad (8.10.5)$$

In the bubble approximation

$$Z_H = \frac{N_c}{16\pi} \ln \left(\frac{\Lambda^2}{\mu^2} \right),$$

$$\lambda_H = \frac{N_c}{8\pi^2} \ln \left(\frac{\Lambda^2}{\mu^2} \right),$$

$$M'^2 = M^2 + a\Lambda^2. \qquad (8.10.6)$$

It is clear from eq. (8.10.6) that the parameters Z_H and λ_H, satisfy the compositeness condition

$$Z_H(\mu = \Lambda) = 0,$$
$$\lambda_H(\mu = \Lambda) = 0. \qquad (8.10.7)$$

We can now redefine the fields to go to the canonical field theory basis where the kinetic energy term has coefficient 1. Denoting the redefined field by ϕ, we have

$$\phi_0 = \frac{1}{\sqrt{Z_H}} \phi,$$

$$\lambda = \frac{\lambda_H}{Z_H^2},$$

$$h = \frac{h_0}{\sqrt{Z_H}}. \qquad (8.10.8)$$

We see that the low-energy theory looks exactly like the standard model. In eq. (8.10.8), λ and h_t, are the couplings familiar from the standard model. The only difference now is that the compositeness condition eq. (8.10.7) imposes nontrivial boundary conditions on them:

$$\lambda(\mu = \Lambda) = \infty,$$
$$h_t(\mu = \Lambda) = \infty. \tag{8.10.9}$$

Equations (8.10.9) are extremely important because, given a value of Λ, one can use the renormalization group equations to extrapolate eq. (8.10.9) to predict the values of λ and h_t, at low energies. Since $h_t(v/\sqrt{2}) = m_t$, this will lead to a prediction of the t-quark mass. Similarly, the value of λ leads to a prediction of the Higgs boson mass.

Two questions immediately come to mind:

1. What numerical value to choose for λ and h_t, at Λ?

2. Can one trust the renormalization group equations when λ and h_t, become large?

The answer to the first question is given by earlier studies of the renormalization group evolution of coupling constants in the standard model [35]. These works have established that, regardless of what the numerical values of λ and h at $\mu = \Lambda$ are, they go to an almost fixed point at low energies. The second question can also be answered using the same studies, which show that a rise to infinity at $\mu = \Lambda$ is quite steep, so that an error made by using perturbative renormalization group equations is not large.

To see the further implications of these hypotheses, let us write the one-loop renormalization group equations for coupling parameters of the standard model ($t = \ln \mu$):

$$\frac{dg_i}{dt} = \beta_i(g_i), \qquad i = \mathrm{SU}(3), \mathrm{SU}(2), \text{ or } \mathrm{U}(1),$$
$$\frac{dh}{dt} = \beta_h(g_i, h), \tag{8.10.10}$$
$$\frac{d\lambda}{dt} = \beta_\lambda(g_i, h, \lambda).$$

Ignoring the $\mathrm{SU}(2)_L$ and $\mathrm{U}(1)$ couplings, we can write the evolution equation for h_t,

$$\frac{dh_t}{dt} = \frac{h_t}{16\pi^2}(\tfrac{9}{2}h_t^2 - 8g_3^2). \tag{8.10.11}$$

We see from eq. (8.10.11) that, if at $\mu = M_W$, $h_t^2 < \frac{16}{9}g_3^2$, then h_t is asymptotically free and h_t, never goes to infinity. This means that, for the idea of the $t\bar{t}$ condensates to be useful, the t-quark must be heavy. In fact, detailed analysis of eq. (8.10.10) shows that we must have $m_t > 95$ GeV.

Using eq. (8.10.10), one can predict m_t and m_H, as a function of Λ. As Λ increases, m_t goes down; but if we keep $\Lambda \leq M_{P1}$, we get $m_t \gtrsim 250$ GeV in the minimal model. Such high values are of course not acceptable in view of the recent Z-boson data from LEP. However, this does not mean that this idea is wrong. It probably means that the minimal scheme is incomplete. Secondly, the large values of Λ also require a fine tuning, which probably means that the true underlying theory may be supersymmetric.

Many ideas have been proposed to improve the minimal scheme [36–38], and in some of these schemes lower m_t values indeed can arise without pushing Λ above M_{p1}. This, therefore, appears to be a promising way to understand the origin of electroweak symmetry breaking.

Exercises

8.1. Construct a technicolor model where the low-energy electroweak group is $SU(2)_L \times SU(2)_R \times U(1)_{B-L}$.

   1. Choose a set of technifermions and a technicolor group, as well as an extended technicolor group.
   2. Discuss the spectrum of pseudo-Goldstone bosons in this theory.
   3. Discuss the fermion masses and the problem of flavor changing neutral currents.

8.2. Consider the composite models where quarks and leptons are identified with quasi-Nambu–Goldstone fermions, resulting from the spontaneous symmetry breaking of a global symmetry in a supersymmetric preon model. If a supersymmetric version of the color flavor factorized is considered:

   (a) Discuss the global symmetries of the model.
   (b) To get a realistic quark–lepton spectrum, what must be the unbroken symmetry of the model?
   (c) Discuss anomaly matching for this case.
   (d) Is it possible to obtain the generation number in this case? [Not much knowledge of supersymmetry is required to answer this problem.]

References

[1] L. Susskind, *Phys. Rev.* **D20**, 2619 (1979);
    S. Weinberg, *Phys. Rev.* **D19**, 1277 (1979).

[2] E. Fahri and L. Susskind, *Phys. Rev.* **D20**, 3404 (1979);
S. Dimopoulos, *Nucl. Phys.* **B169**, 69 (1980).

[3] S. Simopoulos and L. Susskind, *Nucl. Phys.* **B155**, 237 (1979);
E. Eichten and K. Lane, *Phys. Lett.* **90B**, 125 (1980).

[4] S. Dimopoulos and J. Ellis, *Nucl. Phys.* **BI82**, 505 (1981).

[5] J. C. Pati, in *Superstrings, Compositeness and Cosmology* (edited by S.
Gates and R. N. Mohapatra), World Scientific, Singapore, 1987, p. 462;

[6] J. Calmet, S. Narison, M. Perrottet, and E. DeRafael, *Rev. Mod. Phys.*
49, 21 (1977);
T. Kinoshita and W. B. Lindquist, *Phys. Rev. Lett.* **41**, 1573 (1981);
T. Kinoshita, B. Nizic, and Y. Okamoto, *Phys. Rev. Lett.* **53**, 717 (1984).

[7] R. Barbieri, L. Maiani, and R. Petronzio, *Phys. Lett.* **96B**, 63 (1980);
S. J. Brodsky and S. D. Drell, *Phys. Rev.* **D22**, 2236 (1980).

[8] See, for instance, models of O. W. Greenberg and J. Sucher, *Phys. Lett.*
99B, 339 (1981);
J. C. Pati and A. Salam, *Phys. Rev.* **D10**, 275 (1974);
R. Barbieri, R. N. Mohapatra, and A. Masiero, *Phys. Lett.* **105B**, 369
(1981).

[9] O. W. Greenberg, R. N. Mohapatra, and M. Yasue, *Phys. Rev. Lett.* **51**,
1737 (1983).

[10] E. J. Eichten, K. D. Lane, and M. E. Peshkin, *Phys. Rev. Lett.* **50**, 811
(1983).

[11] M. K. Gaillard and B. W. Lee, *Phys. Rev.* **D10**, 897 (1974);
G. Beall, M. Bender, and A. Soni, *Phys. Rev. Lett.* **48**, 848 (1982).
These bounds have been discussed in composite model framework by I.
Bars, *Nucl. Phys.* **B198**, 269 (1982);
and for a recent discussion of the conditions under which these bounds
may be evaded, see
O. W. Greenberg, R. N. Mohapatra, and S. Nussinov, *Phys. Lett.* **148B**,
465 (1984).

[12] B. Weinstein, in *TSIMESS Workshop Proceedings, 1983* (edited by T.
Goldman et al.), American Institute of Physics, New York, 1983.

[13] G. 't Hooft, in *Recent Developments in Gauge Theories*, Plenum, New
York, 1980, p. 135.

[14] S. L. Alder, *Phys. Rev.* **177**, 2426 (1969);
R. Jackiw and J. S. Bell, *Nuovo Cimeno*, **60A**, 47 (1969);
S. L. Alder and W. Bardeen, *Phys. Rev.* **182**, 1517 (1969).

[15] Y. Frishman, A. Schwimmer, T. Banks, and S. Yankielowicz, *Nucl. Phys.*
B177, 157 (1981).

[16] J. C. Pati and A. Salam, *Phys. Rev.* **D10**, 275 (1974).

[17] J. C. Pati, O. W. Greenberg, and J. Sucher (Ref. [8]).

[18] T. Appelequist and J. Carrazone, *Phys. Rev.* **D11**, 2856 (1975).

[19] J. Preskill and S. Weinberg, Texas preprint, 1981.

[20] D. Weingarten, *Phys. Rev. Lett.* **51**, 1830 (1983);
E. Witten, *Phys. Rev. Lett.* **51**, 2351 (1983);
S. Nussinov, *Phys. Rev. Lett.* **51**, 2081 (1983).

[21] They are analogous to the massless Goldstone–Majoron boson suggested
by
Y. Chikashige, R. N. Mohapatra, and R. D. Peccei, *Phys. Lett.* **98B**, 265
(1981).

[22] For an apparent exception to this argument see an $E(6)$ hypercolor model
by
Y. Tosa, J. Gibson, and R. E. Marshak, Private communication, 1984.

[23] S. Weinberg and E. Witten, *Phys. Lett.* **96B**, 59 (1980);
See also E. C.G. Sudarshan, *Phys. Rev.* **D** (1981).

[24] D. A. Dicus, E. Kolb, V. Teplitz, and R. Wagoner, *Phys. Rev.* **17**, 1529
(1978);
M. Fukugita, S. Watamura, and M. Yoshimura, *Phys. Rev. Lett.* **48**, 1522
(1982).

[25] L. Abbott and E. Farhi, *Phys. Lett.* **101B**, 69 (1981);
H. Fritzsch and G. Mandelbaum, *Phys. Lett.* **102B**, 319 (1981);
R. Barbieri, R. N. Mohapatra, and A. Masiero, *Phys. Lett.* **105B**, 369
(1981);
For a review see R. N. Mohapatra, *Proceedings of the Telemark Neutrino
Mass Mini-Conference*, 1982, American Institute of Physics, New York,
1982.

[26] J. D. Bjorken, *Phys. Rev.* **D19**, 335 (1979);
P. Q. Hung and J. J. Sakurai, *Nucl. Phys.* **B143**, 81 (1978).

[27] R. Barbieri and R. N. Mohapatra, *Phys. Lett.* **120B**, 195 (1982).

[28] D. Schildknecht, in *Proceedings of the Europhysics Study Conference on
Electro weak Effects at High Energies* (edited by H. Newman), Plenum,
New York,1983.

[29] U. Baur, H. Fritzsch, and H. Faissner, *Phys. Lett.* **135B**, 313 (1984).

[30] A. Masiero, R. N. Mohapatra, and R. D. Peccei, *Nucl. Phys.* **B192**, 66
(1981).

[31] A. Masiero and R. N. Mohapatra, *Phys. Lett.* **103B**, 343 (1981).

[32] Y. Nambu, *1988 International Workshop on New Trends in Strongly
Coupled Gauge Theories* (edited by M. Bando et al.), World Scientific,
Singapore, 1989, p. 3.

[33] V. Miransky, M. Tanabashi, and K. Yarnawaki, *Phys. Lett.* **B221** 177
(1989).

[34] W. Bardeen, C. Hill, and M. Lindner, *Phys. Rev.* **D41**, 1647 (1990).

[35] B. Pendleton and G. G. Ross, *Phys. Lett.* **98B** 291 (1981);
C. Hill, *Phys. Rev.* **D24**, 691 (1981);
C. Hill, C. Leung, and S. Rao, *Nucl. Phys.* **B262**, 517 (1985).

[36] M. Luty, *Phys. Rev.* **D41** 2893 (1990);
      M. Suzuki, *Phys. Rev.* **D41** 3457 (090).

[37] T. Clark, S. Love, and W. Bardeen, *Phys. Lett.* **B237**, 235 (1990).

[38] K. S. Babu and R. N. Mohapatra, *Phys. Rev. Lett.* **66** 556 (1991).

9
Global Supersymmetry

§9.1 Supersymmetry

Ultimate unification of all particles and all interactions is the eternal dream of theoretical physicists. The unified gauge theories have taken us a step closer to realizing the second goal. However, since known "elementary" particles consist of both fermions (q, l) and bosons (photons γ, W, Z color octet of gluons), their ultimate unification would require them either to be composites of some basic set of fermions that can be unified within a Lie group framework or that there must exist a new symmetry that transforms bosons to fermions. In this chapter, we begin discussion of this latter kind of symmetry [1], known as supersymmetry. Supersymmetry was invented in the early 1970s by Golfand, Likhtman, Wess, and Zumino [2] and in a nonlinear realization by Volkov and Akulov [3]. In the context of two-dimensional theories, it was discueed by Sakita and Gervais [3a].

To see why symmetry between bosons and fermions may be of interest to the study of elementary particle physics, we point out that renormalizable quantum field theories with scalar particles (such as the Higgs particles of unified gauge theories) have a very disturbing feature in that the scalar masses have quadratic divergences in one- and higher-loop orders. Unlike the logarithmic divergences associated with fermion masses, which can be eliminated by taking advantage of chiral symmetries, there is no apparent symmetry that can control the divergences associated with scalar field masses. If we assume that loop integrals are cut-off at a scale $\Lambda \gg M_W$, where new physics appears, a natural value for the scalar mass could be

Λ, and it will be hard to understand in a natural manner why the Higgs mechanism leads to a mass scale $\approx M_W/g$. In fact, the problem gets very severe if there is no new physics until the Planck scale, since in that case $\Lambda = M_{P1}$, and extreme fine tuning is needed to understand the electroweak scale. In the technicolor scenario, the scale of technicolor interaction provides a natural cutoff Λ; but in the absence of such possibilities, one must look for ways to eliminate the quadratic divergence. On the other hand, if we have a theory that couples fermions and bosons, the scalar masses have two sources for their quadratic divergences, one from a scalar loop that comes with a positive sign and another from a fermion loop with a negative sign. It is then suggestive that, if there were a symmetry that related the couplings and masses of fermions and bosons, all divergences from scalar field masses could be eliminated. Supersymmetry provides such an opportunity.

One of the first requirements of supersymmetry is an equal number of bosonic and fermionic degrees of freedom in one multiplet. We demonstrate this with a simple example. Consider two pairs of creation annihilation operators: $(a, a^\dagger)$ and $(b, b^\dagger)$ with a being bosonic and b being fermionic. They satisfy the following commutation and anticommutation relations, respectively:

$$[a, a^\dagger] = \{b, b^\dagger\} = 1. \tag{9.1.1}$$

The Hamiltonian for this system can be written in general as

$$H = \omega_a a^\dagger a + \omega_b b^\dagger b. \tag{9.1.2}$$

If we define a fermionic operator

$$Q = b^\dagger a, \tag{9.1.3}$$

then

$$[Q, a^\dagger] = +b^\dagger,$$
$$\{Q, b^\dagger\} = a^\dagger. \tag{9.1.4}$$

Thus, if $a^\dagger|0\rangle$ and $b^\dagger|0\rangle$ represent bosonic and fermionic states, respectively, Q will take bosons to fermions and vice versa. Moreover,

$$[Q, H] = (\omega_a - \omega_b)Q. \tag{9.1.5}$$

So, for $\omega_a = \omega_b = \omega$ (i.e., equal energy for the bosonic and fermionic states), H is supersymmetric. Furthermore, in this case

$$\{Q, Q^\dagger\} = \frac{2}{\omega}H. \tag{9.1.6}$$

Thus, the algebra of Q, $Q^\dagger$, and H closes under anticommutation. If there is more than one a and b, then there must be an equal number of them, otherwise eqs. (9.1.5) and (9.1.6) cannot be satisfied together.

Another point that distinguishes this symmetry from other known symmetries is that the anticommutator of Q, $Q^\dagger$ involves the Hamiltonian; for any other bosonic symmetry, the charge commutation never involves the Hamiltonian. This, as we will see later, has important implications for physics. Now we proceed to the consideration of field theories with supersymmetry.

§9.2   A Supersymmetric Field Theory

Consider the free Lagrangian for the massless complex bosonic field ϕ

$$\phi = \frac{1}{\sqrt{2}}(A + iB)$$

and a Majorana field $\psi(\psi = C\bar\psi^T)$ and $C = -i\gamma_2\gamma_0$

$$\mathcal{L} = (\partial^\mu \phi^*)(\partial_\mu \phi) + i\tfrac{1}{2}\bar\psi\gamma^\mu\partial_\mu\psi. \tag{9.2.1}$$

The action for this Lagrangian (not the Lagrangian itself) is invariant under the transformations

$$\delta A = \bar\varepsilon\psi,$$
$$\delta B = i\bar\varepsilon\gamma_5\psi,$$
$$\delta\psi = i\gamma^\mu(\partial_\mu A + i\gamma_5\partial_\mu B)\varepsilon, \tag{9.2.2}$$

where ε is an anticommuting, Majorana spinor, which is an element of the Grassmann algebra. One can show that under these transformations,

$$\delta\mathcal{L} = \partial^\mu S_\mu, \tag{9.2.3}$$

where S_μ is the supersymmetric spinor-vector current:

$$S_\mu = \gamma_\mu(\gamma^\nu\partial_\nu A + i\gamma_5\gamma^\nu\partial_\nu B)\psi. \tag{9.2.4}$$

Let us demonstrate this for the A part of the Lagrangian, i.e., keeping only the A terms:

$$\delta(\tfrac{1}{2}\partial^\mu A\partial_\mu A) = \partial^\mu A\bar\varepsilon\partial_\mu\psi$$
$$\frac{i}{2}\delta(\bar\psi\gamma^\mu\partial_\mu\psi) = -\frac{1}{2}\bar\varepsilon\gamma^\alpha\partial_\alpha A\gamma^\beta\partial_\beta\psi + \frac{1}{2}\bar\psi\partial^2 A\varepsilon$$
$$= \tfrac{1}{2}[\partial^\mu\bar\varepsilon S_\mu^A] - \bar\varepsilon\partial^\mu\psi\partial_\mu A. \tag{9.2.5}$$

Note that the last term in the last line of eq. (9.2.5) precisely cancels the variation of the $(\partial A)^2$ term in the first line of the equation. In deriving the final step, we have used the identity, $\bar\varepsilon\gamma_{\mu_1}\gamma_{\mu_2}\cdots\gamma_{\mu_n}\psi = \bar\psi\gamma_{\mu_n}\gamma_{\mu_{n-1}}\cdots$ $\varepsilon(-1)^n$. To obtain the supersymmetry algebra we consider the commutator of successive supersymmetry operations, i.e.,

$$[\delta_{\varepsilon_1}, \delta_{\varepsilon_2}]A = ?.$$

Let us evaluate this commutator explicitly using eq. (9.2.2):

$$\delta_{\varepsilon_1}\delta_{\varepsilon_2}A - \delta_{\varepsilon_2}\delta_{\varepsilon_1}A = \delta_{\varepsilon_1}\bar{\varepsilon}_2\psi - \delta_{\varepsilon_2}\bar{\varepsilon}_1\psi$$
$$= i\bar{\varepsilon}_2\gamma^\mu(\partial_\mu A + i\gamma_5\partial_\mu B)\varepsilon_1 - i\bar{\varepsilon}_1\gamma^\mu(\partial_\mu A + i\gamma_5\partial_\mu B)\varepsilon_2$$
$$= 2i\bar{\varepsilon}_2\gamma^\mu\varepsilon_1\partial_\mu A, \qquad (9.2.6)$$

since

$$\bar{\varepsilon}_2\gamma_\mu\gamma_5\varepsilon_1 = \bar{\varepsilon}_1\gamma_\mu\gamma_5\varepsilon_2$$

and

$$\bar{\varepsilon}_2\gamma_\mu\varepsilon_1 = -\bar{\varepsilon}_1\gamma_\mu\varepsilon_2.$$

We can therefore write

$$[\bar{\varepsilon}_1 S, \bar{S}\varepsilon_2]A = +2i\bar{\varepsilon}_1\gamma^\mu\varepsilon_2\partial_\mu A, \qquad (9.2.7a)$$

where $S = i\int d^3x S_0$ and S_0 is the time component of supersymmtry spinor-vector current. Similarly, we can show that

$$[\bar{\varepsilon}_1 S, \bar{S}\varepsilon_2]B = 2i\bar{\varepsilon}_1\gamma^\mu\varepsilon_2\partial_\mu B. \qquad (9.2.7b)$$

Finally, let us calculate $[\delta_{\varepsilon_1}, \delta_{\varepsilon_2}]\psi$. We obtain

$$(\delta_{\varepsilon_1}\delta_{\varepsilon_2} - \delta_{\varepsilon_2}\delta_{\varepsilon_1})\psi = i\delta_{\varepsilon_1}\gamma^\mu(\partial_\mu A + i\gamma_5\partial_\mu B)\varepsilon_2 - i\delta_{\varepsilon_2}\gamma^\mu(\partial_\mu A + i\gamma_5\partial_\mu B)\varepsilon_1$$
$$= i\gamma^\mu\varepsilon_2\bar{\varepsilon}_1\partial_\mu\psi - i\gamma^\mu\gamma_5\varepsilon_2\bar{\varepsilon}_1\gamma_5\partial_\mu\psi$$
$$- i\gamma^\mu\varepsilon_1\bar{\varepsilon}_2\partial_\mu\psi + i\gamma^\mu\gamma_5\varepsilon_1\bar{\varepsilon}_2\gamma_5\partial_\mu\psi. \qquad (9.2.8)$$

To evaluate the last term we use the identity for products of four-component spinors

$$\bar{\psi}_1 M\psi_3\bar{\psi}_4 N\psi_2 = -\frac{1}{4}\sum_j \sigma(j)\bar{\psi}_1 O_j\psi_2\bar{\psi}_4 NO_j M\psi_3,$$

where

$$\sigma(j) = \begin{cases} +1 \\ +1 \\ +1 \end{cases} \quad \text{for} \quad O_j = \begin{cases} 1, \gamma_5, \gamma_\mu, \\ i\gamma_5\gamma_\mu, \\ \sigma_{\mu\nu} \end{cases} \qquad (9.2.9)$$

where $\sigma_{\mu\nu} = \frac{i}{2}[\gamma_\mu, \gamma_\nu]$. To make the evaluation of eq. (9.2.8) simpler, we first note that $[\delta_{\varepsilon_1}, \delta_{\varepsilon_2}]$ is antisymmetric in the interchange of ε_1 and ε_2 and only antisymmetric products involving ε_1 and ε_2 are $\bar{\varepsilon}_1\gamma_\mu\varepsilon_2$ and $\bar{\varepsilon}_1\sigma_{\mu\nu}\varepsilon_2$. We then get

$$[\delta_{\varepsilon_1}\delta_{\varepsilon_2} - \delta_{\varepsilon_2}\delta_{\varepsilon_1}]\psi$$
$$= -\frac{1}{4}\sum_j \sigma(j)\bar{\varepsilon}_1 O_j\varepsilon_2[\gamma_\mu O_j\partial_\mu\psi - \gamma_\mu\gamma_5 O_j\gamma_5\partial_\mu\psi] - (\varepsilon_1 \leftrightarrow \varepsilon_2), \quad (9.2.10)$$

where $O_j = \gamma_\alpha$ or $\sigma_{\alpha\beta}$, but $\gamma_5\sigma_{\alpha\beta}\gamma_5 = \sigma_{\alpha\beta}$ so that the two terms within the bracket involving $\sigma_{\alpha\beta}$ cancel. Thus we get

$$[\delta_{\varepsilon_1}, \delta_{\varepsilon_2}]\psi = +i\bar{\varepsilon}_1\gamma^\sigma\varepsilon_2\gamma^\mu\gamma_\sigma\partial_\mu\psi$$

$$= 2i\bar{\varepsilon}_1\gamma^\mu\varepsilon_2\partial_\mu\psi + i\bar{\varepsilon}_1\gamma^\sigma\varepsilon_2\gamma_\sigma\gamma\cdot\partial\psi. \tag{9.2.11}$$

If we use field equations, the last term is zero and we get

$$[\delta_{\varepsilon_1},\delta_{\varepsilon_2}]\psi = +2i\bar{\varepsilon}_1\gamma^\mu\varepsilon_2\partial_\mu\psi. \tag{9.2.12}$$

Noting that $\delta_{\varepsilon_1} \equiv i\bar{\varepsilon}_1 S$ and $\delta a_\mu \equiv -ia^\mu P_\mu \equiv a^\mu\partial_\mu$, we obtain

$$\{S,\bar{S}\} = 2\gamma^\mu P_\mu. \tag{9.2.13}$$

Thus, the algebra of the supersymmetry generators closes with the algebra of translation. We can, therefore, write down the super-Poincaré algebra as one that includes (9.2.13) and the following equations:

$$[M_{\mu\nu}, M_{\alpha\beta}] = i[g_{\nu\alpha}M_{\mu\beta} - g_{\mu\alpha}M_{\nu\beta} - g_{\nu\beta}M_{\mu\alpha} + g_{\mu\beta}M_{\nu\alpha}],$$
$$[M_{\mu\nu}, P_\alpha] = i[g_{\nu\alpha}P_\mu - g_{\mu\alpha}P_\nu],$$
$$[P_\alpha, P_\beta] = 0,$$
$$[P_\alpha, \bar{\varepsilon}S] = 0,$$
$$[M_{\mu\nu}, \bar{\varepsilon}S] = \bar{\varepsilon}\sigma_{\mu\nu}S. \tag{9.2.14}$$

An important point to note is that it was essential for the closure of the algebra to use the field equation for ψ. This raises the question as to whether supersymmetry leaves the Lagrangian invariant for arbitrary values of the fields. This can actually be achieved by adding two more fields, F and G. They are called auxiliary fields and are required to transform under supersymmetry transformations as

$$\delta F = \bar{\varepsilon}\gamma^\mu\partial_\mu\psi$$

and

$$\delta G = i\bar{\varepsilon}\gamma_5\gamma^\mu\partial_\mu\psi, \tag{9.2.15}$$
$$\delta\psi = i\gamma^\mu(\partial_\mu A + i\gamma_5\partial_\mu B)\varepsilon + i(F + i\gamma_5 G)\varepsilon. \tag{9.2.16}$$

It is then easily verified that the extra term in eq. (9.2.8) cancels and also

$$[\delta_{\varepsilon_1},\delta_{\varepsilon_2}](F,G) = +2i\bar{\varepsilon}_1\gamma^\mu\varepsilon_2(\partial_\mu F, \partial_\mu G). \tag{9.2.17}$$

Thus, supersymmetry can indeed be a symmetry for arbitrary field configurations. The deeper reason for the necessity of the auxiliary fields is that supersymmetry requires the fermionic and bosonic degrees of freedom to be equal. Since ψ has four fermionic degrees of freedom, A and B are not enough. However, by imposing field equations, two fermionic degrees of freedom are removed. If we do not impose field equations, two more degrees of freedom F and G must be added.

Another point to note is that the parameter ε has mass dimension $-\frac{1}{2}$ so that F and G have mass dimension two. Therefore, their kinetic energy terms will have mass dimension 6 and cannot lead to consistent equal-time commutation relations; as a result, they cannot be dynamical fields but must be expressible in terms of other fields of the theory. In fact, in the presence of F and G, the supersymmetric Lagrangian of eq. (9.2.1) gets modified as follows:

$$\mathcal{L} = \tfrac{1}{2}[\partial^\mu A \partial_\mu A + \partial^\mu B \partial_\mu B] + \frac{i}{2}\bar{\psi}\gamma^\mu\partial_\mu\psi - F^2 - G^2, \qquad (9.2.18)$$

which leads to new field equations $F = G = 0$ in addition to the usual ones for A, B, and ψ.

§9.3 Two-Component Notation

We saw in the previous section that the parameter $\bar{\varepsilon}$, which is needed to describe supersymmetry transformations, is a four-component Majorana spinor, i.e., it has four independent Grassmann variables that transform as an irreducible representation of the SO(3, 1) group. But SO(3, 1) is locally isomorphic to the SL(2, C) group consisting of arbitrary complex 2×2 matrices, Z with determinant one, i.e.,

$$Z = \left(\begin{array}{cc} Z_1{}^1 & Z_1{}^2 \\ Z_2{}^1 & Z_2{}^2 \end{array} \right), \qquad (9.3.1)$$

with

$$Z_1{}^1 Z_2{}^2 - Z_1{}^2 Z_2{}^1 = 1$$

or

$$\varepsilon^{ab} Z_a{}^c Z_b{}^d = \varepsilon^{cd}, \qquad (9.3.2)$$

where $\varepsilon^{ab} = -\varepsilon_{ab}$ (with $\varepsilon_{ab}\varepsilon^{bc} = \delta_a^c$) is the antisymmetric Levi–Civita symbol. We take $\varepsilon^{12} = +1$. Equation (9.3.2) implies that this is invariant under the SL(2, C) group.

The fundamental representation of the SL(2, C) group act on the two-component complex column vector ϕ_a, i.e.,

$$\phi_a \to \phi_a' = Z_a{}^b \phi_b. \qquad (9.3.3)$$

If we call ϕ_a a covariant vector, the contravariant vector ϕ^a can be defined using ε^{ab} as follows:

$$\phi^a = \varepsilon^{ab}\phi_b. \qquad (9.3.4)$$

The complex conjugate of ϕ_a, i.e., ϕ_a^* transforms as follows:

$$\phi_a^* \to \phi_a^{*\prime} = Z_a{}^{b*}\phi_b^*. \qquad (9.3.5)$$

The matrices Z and Z^* are not equivalent. Thus, ϕ_a^* belongs to a different two-dimensional representation of SL(2, C) and we will denote its indices by $\bar{\phi}_{\dot{a}}$ to distinguish it from ϕ_a.

Now, we are ready to write the Majorana spinor in terms of two-component vectors: to see this let us write an arbitrary complex four-component Dirac spinor ψ as $\psi = \binom{\phi_1}{\phi_2}$ and impose the Majorana condition

$$\psi = C\bar\psi^T. \tag{9.3.6}$$

In the choice of γ matrices

$$\gamma^k = \begin{pmatrix} 0 & \sigma_k \\ -\sigma_k & 0 \end{pmatrix}, \qquad \gamma^0 = \begin{pmatrix} 0 & 1 \\ 1 & 0 \end{pmatrix},$$

and

$$\gamma_5 = \begin{pmatrix} 1 & 0 \\ 0 & -1 \end{pmatrix} = -i\gamma^0\gamma^1\gamma^2\gamma^3 \tag{9.3.7}$$

we can write

$$\psi = \begin{pmatrix} \phi \\ i\sigma_2\phi^* \end{pmatrix} \tag{9.3.8}$$

Noting that $i\sigma_2 \equiv \varepsilon^{ab}$ we can write a Majorana spinor ψ as

$$\psi = \begin{pmatrix} \phi_a \\ \bar\phi^{\dot a} \end{pmatrix}, \tag{9.3.9}$$

where the overbar implies its transformation as a complex two-dimensional representation.

To see the connection with Lorentz transformations consider the four-vector σ_μ, i.e.,

$$\sigma_\mu = (\sigma_k, 1) \tag{9.3.10}$$

and the Lorentz scalar $\sigma^\mu P_\mu$

$$\sigma^\mu P_\mu = \begin{pmatrix} P_0 + P_3 & P_1 = iP_2 \\ P_1 + iP_2 & P_0 - P_3 \end{pmatrix}, \tag{9.3.11}$$

Note that Det $\sigma^\mu P_\mu = (P_0^2 - P_1^2 - P_2^2 - P_3^2) = -P^2$ and is invariant under the transformation

$$\sigma^\mu P_\mu \to Z\sigma^\mu P_\mu Z^\dagger.$$

If we write

$$\sigma^\mu P'_\mu = Z\sigma^\mu P_\mu Z^\dagger, \tag{9.3.12}$$

then $P'^2 = P^2$, and therefore Z, is a Lorentz transformation. We also note that the matrices σ_μ, P_μ transform as $(\frac{1}{2}, \frac{1}{2})$ under the SL$(2, C)$ transformations, i.e., the SL$(2, C)$ indices of σ_μ and P_μ are $\sigma_{\mu,a\dot a}$ and $P_{\mu,a\dot a}$.

The ϕ and $\bar\chi$ are Weyl spinors and have the following SL$(2, C)$ invariants:

$$\phi\chi \equiv \phi^a\chi_a = \varepsilon^{ab}\phi_b\chi_a$$

and

$$\bar\phi\bar\chi = \bar\phi_{\dot a}\bar\chi^{\dot a} = \varepsilon^{\dot a\dot b}\bar\phi_{\dot a}\bar\chi_{\dot b}. \tag{9.3.13}$$

The familiar Dirac bilinear covariants can be expressed in terms of the $SL(2,C)$ invariant products. For instance, for an arbitrary Dirac spinor ψ

$$\bar{\psi}\psi = (\bar{\phi}_{\dot{a}}\ \chi^a)\begin{pmatrix} 0 & 1 \\ 1 & 0 \end{pmatrix}\begin{pmatrix} \phi_a \\ \bar{\chi}^{\dot{a}} \end{pmatrix}$$

$$= (\chi\phi + \bar{\phi}\bar{\chi}). \tag{9.3.14}$$

Similarly,

$$\bar{\psi}\gamma_k\psi = (\bar{\phi}_{\dot{a}}\chi^a)\begin{pmatrix} \sigma_k & 0 \\ 0 & -\sigma_k \end{pmatrix}\begin{pmatrix} \phi_a \\ \bar{\chi}^{\dot{a}} \end{pmatrix}$$

$$= \bar{\phi}_{\dot{a}}\sigma_k^{a\dot{a}}\phi_a - \chi^a\sigma_{k,a\dot{a}}\bar{\chi}^{\dot{a}},$$

$$\bar{\psi}\gamma_0\psi = \bar{\phi}_{\dot{a}}(1)^{\dot{a}a}\phi_a + \chi^a(1)_{a\dot{a}}\bar{\chi}^{\dot{a}}. \tag{9.3.15}$$

We can group them to write

$$\bar{\psi}\gamma_\mu\psi = +\chi\bar{\sigma}_\mu\bar{\chi} - \bar{\phi}\sigma_\mu\phi,$$

where $\sigma_\mu = (1,\sigma_k)$; $\bar{\sigma}_\mu = (1,-\sigma_k)$.

§9.4   Superfields

Soon after the discovery of supersymmetry, Salam and Strathdee [4] proposed the concept of the superfield as the generator of supersymmetric multiplets. This is a very profound concept and has far-reaching implications. To discuss superfields we introduce the concept of superspace, which is an extension of ordinary space–time by the inclusion of additional fermionic coordinates. We would like to maintain symmetry between ordinary space and fermionic space, so we introduce four extra fermionic coordinates to match the four space–time dimensions. We can describe the fermionic coordinates as elements of a Majorana spinor or as a pair of two-component Weyl spinors. Points in superspace are then identified by the coordinates

$$z^M = (x^\mu, \theta^a, \bar{\theta}_{\dot{a}}). \tag{9.4.1}$$

where θ's are anticommuting spinors with the following properties:

$$\{\theta^a, \theta^b\} = \{\bar{\theta}_{\dot{a}}, \bar{\theta}_{\dot{b}}\} = \{\theta^a, \bar{\theta}_{\dot{a}}\} = 0,$$

$$[x^\mu, \theta^a] = [x^\mu, \bar{\theta}_{\dot{a}}] = 0. \tag{9.4.2}$$

The indices a and $\dot{a}$ go over 1 and 2 and $\theta^{a*} = \bar{\theta}^{\dot{a}}$. The scalar product of θ's is given as follows:

$$\theta\theta = \theta^a\theta_a = \varepsilon^{ab}\theta_b\theta_a = \theta^1\theta_1 + \theta^2\theta_2 = -\theta_1\theta^1 - \theta_2\theta^2. \tag{9.4.3}$$

The summation convention for $\bar{\theta}$ will be $\bar{\theta}\bar{\theta} = \bar{\theta}_{\dot{a}}\bar{\theta}^{\dot{a}}$; the product of more than two θ's vanishes since $(\theta_1)^2 = (\theta_2)^2 = 0$. It also follows that

$$\theta^a\theta^b = -\tfrac{1}{2}\varepsilon^{ab}\theta\theta,$$

$$\bar{\theta}_{\dot{a}}\bar{\theta}_{\dot{b}} = -\tfrac{1}{2}\varepsilon_{\dot{a}\dot{b}}\overline{\theta\theta}. \tag{9.4.4}$$

Salam and Strathdee [4] proposed that a function $\Phi(x,\theta,\bar{\theta})$ of the superspace coordinates, called superfield, which has a finite number of terms in its expansion in terms of θ and $\bar{\theta}$ owing to their anticommuting property, be considered as the generator of the various components of the supermultiplets. Φ could belong to arbitrary representation of the Lorentz group but, for the moment, we will consider only scalar superfields. We can now expand Φ in a power series in θ and $\bar{\theta}$ to get the various components of the supermultiplet

$$\Phi(x,\theta,\bar{\theta}) = \phi(x) + \theta\psi(x) + \bar{\theta}_{\bar{\chi}}(x) + \theta^2 M(x) + \bar{\theta}^2 N(x) + \theta\sigma^\mu\bar{\theta}V_\mu(x)$$
$$+ \theta^2\bar{\theta}\lambda(x) + \bar{\theta}^2\theta\alpha(x) + \theta^2\bar{\theta}^2 D(x), \tag{9.4.5}$$

ϕ, ψ, $\bar{\chi}$, M, N, V_μ, $\bar{\lambda}$, α, and D are the component fields. There are sixteen real bosonic and sixteen fermionic degrees of freedom.

To exhibit the supersymmetry transformations of various fields we define an element of the supergroup as (we will use the Pauli metric)

$$\sigma(x_\mu,\theta,\bar{\theta}) = \exp[i(-x^\mu P_\mu) + \theta^a Q_a + \bar{\theta}_{\dot{a}}\bar{Q}^{\dot{a}}]. \tag{9.4.6}$$

Here we have taken the supersymmetry generators in two-component notation. Supersymmetry transformation will be defined as a translation in superspace, of the form $\sigma(0,\varepsilon,\bar{\varepsilon})$. To study the effect of supersymmetry transformations we consider

$$\sigma(0,\varepsilon,\bar{\varepsilon})\sigma(x_\mu,\theta,\bar{\theta}) = \exp[(\varepsilon Q + \bar{\varepsilon}\bar{Q}]\exp[i(-x^\mu P_\mu) + \theta Q + \bar{\theta}\bar{Q}]. \tag{9.4.7}$$

We then use the Baker–Campbell–Hausdorf formula

$$e^A e^B = \exp(A + B + \tfrac{1}{2}[A,B] + \cdots). \tag{9.4.8}$$

To obtain any result from eq. (9.4.7) we need the commutation in terms of Q and $\bar{Q}$. To obtain this from (9.2.13) we write

$$S = \begin{pmatrix} Q_a \\ \bar{Q}^{\dot{a}} \end{pmatrix}$$

and we obtain

$$\left\{ \begin{pmatrix} Q_a \\ \bar{Q}^{\dot{a}} \end{pmatrix}, (\bar{Q}_{\dot{b}}Q^b) \right\} = 2\begin{pmatrix} +\sigma_{\mu,ab} & 0 \\ 0 & -\bar{\sigma}_\mu^{\dot{a}b} \end{pmatrix} P^\mu \tag{9.4.9}$$

leading to

$$\{Q_a,\bar{Q}_{\dot{b}}\} = 2\sigma_{\mu,a\dot{b}}P^\mu$$

and

$$\{Q_a,Q_b\} = \{\bar{Q}_{\dot{a}},\bar{Q}_{\dot{b}}\} = 0. \tag{9.4.10}$$

To familiarize the reader with manipulations involving σ_μ and $\bar{\sigma}_\mu$ we see that eq. (9.3.9) also implies

$$\{\bar{Q}^{\dot{a}},Q^b\} = -2\bar{\sigma}_\mu^{\dot{a}b}P^\mu. \tag{9.4.11}$$

Let us derive eq. (9.4.11) using eq. (9.4.9), which also leads to

$$\{\bar{Q}_{\dot{a}}, Q_b\} = 2\sigma_{\mu,b\dot{a}}P^\mu. \tag{9.4.11a}$$

To raise indices we multiply both sides by $\varepsilon^{\dot{c}\dot{b}}\varepsilon^{ba}$ and the right-hand side of eq. (9.4.11a) then gives

$$\varepsilon^{ba}\sigma_{\mu,a\dot{c}}\varepsilon^{\dot{c}\dot{b}} = (\sigma_2\sigma_\mu\sigma_2)^{\dot{b}b}$$
$$= -(1, -\sigma_k^T) = \bar{\sigma}_\mu^{\dot{b}b}, \tag{9.4.12}$$

which leads to eq. (9.4.11). From eqs. (9.4.7) and (9.4.8) it follows that

$$\sigma(0, \varepsilon, \bar{\varepsilon})\sigma(x^\mu, \theta, \bar{\theta}) = \exp\{[-i(x^\mu + \xi^\mu)P_\mu + (\theta + \varepsilon)Q + (\bar{\theta} + \bar{\varepsilon})\bar{Q}]\},$$

where

$$\xi_\mu = -i\varepsilon\sigma_\mu\bar{\theta} + i\bar{\varepsilon}\bar{\sigma}_\mu\theta. \tag{9.4.13}$$

Thus, under supersymmetry transformation

$$\Phi(x, \theta, \bar{\theta}) \to \Phi(x^\mu - i(\varepsilon\sigma^\mu\bar{\theta} - \bar{\varepsilon}\bar{\sigma}^\mu\theta), \theta + \varepsilon, \bar{\theta} + \bar{\varepsilon}). \tag{9.4.14}$$

To find the supersymmetry transformation of the component fields we can Taylor expand the right-hand side and compare coefficients of θ, $\bar{\theta}$, etc.

$$\Phi(x^\mu + \xi^\mu, \theta + \varepsilon, \bar{\theta} + \bar{\varepsilon}) = \Phi(x, \theta, \bar{\theta}) + \xi^\mu\partial_\mu\Phi + \varepsilon\frac{\partial\Phi}{\partial\theta} + \bar{\varepsilon}\frac{\partial\Phi}{\partial\bar{\theta}}. \tag{9.4.15}$$

We get

$$\delta\phi(x) = \varepsilon\psi + \bar{\varepsilon}\bar{\chi},$$
$$\delta\psi = i\bar{\varepsilon}\bar{\sigma}^\mu\partial_\mu\phi + \varepsilon M,$$
$$\delta\bar{\chi} = +i\varepsilon\sigma^\mu\partial_\mu\phi + \bar{\varepsilon}N,$$
$$\delta M = \frac{i}{2}\bar{\varepsilon}\bar{\sigma}^\mu\partial_\mu\psi + \bar{\varepsilon}\bar{\lambda},$$
$$\delta N = \frac{i}{2}\varepsilon\sigma^\mu\partial_\mu\bar{\chi} + \varepsilon\alpha,$$
$$\delta V_\mu = i\varepsilon\delta_\mu\bar{\lambda} + \varepsilon\sigma^\mu\partial_\mu\bar{\chi} + \bar{\varepsilon}\bar{\sigma}_\mu\partial_\mu\psi,$$
$$\delta\bar{\lambda} = \bar{\varepsilon}D + (\partial^\mu V^\nu - \partial^\nu V^\mu)\bar{\varepsilon}\bar{\sigma}_\mu\sigma_\nu,$$
$$\delta\alpha = \varepsilon D + (\partial^\mu V^\nu - \partial^\nu V^\mu)\sigma_\mu\bar{\sigma}_\nu\varepsilon,$$
$$\delta D = \frac{i}{4}\varepsilon\sigma^\mu\partial^\mu\bar{\lambda} - \frac{i}{4}\bar{\varepsilon}\bar{\sigma}^\mu\partial^\mu\alpha. \tag{9.4.16}$$

Now we are ready to write down the generators of supersymmetry transformations Q_a and $\bar{Q}_{\dot{a}}$ as differential operators in the superspace

$$Q_a = \frac{\partial}{\partial\theta^a} - i\sigma^\mu_{a\dot{a}}\bar{\theta}^{\dot{a}}\partial_\mu,$$

$$\bar{Q}_{\dot{a}} = \frac{\partial}{\partial\bar{\theta}^{\dot{a}}} - i\theta^a\sigma^\mu_{a\dot{a}}\partial_\mu. \tag{9.4.17}$$

It can be checked that $\bar{Q}_{\dot{a}} = (Q_a)^*$. Using these equations we can then verify eq. (9.4.11a).

Before proceeding further, we would also like to note that the product of two superfields $\Phi_1(x, \theta, \bar{\theta})$ and $\Phi_2(x, \theta, \bar{\theta})$ is also a superfield with components

$$(\phi_1\phi_2, \phi_1\psi_2 + \phi_2\psi_1, \phi_1\bar{\chi}_2 + \phi_2\bar{\chi}_1, \phi_1 M_2 + \phi_2 M_1 - \tfrac{1}{2}\psi\psi,$$

$$\phi_1 N_2 + \phi_2 N_1 - \tfrac{1}{2}\bar{\chi}\bar{\chi}, \phi_1 V_{2\mu} + \phi_2 V_{1\mu} - i\psi\sigma_\mu\bar{\chi}_\mu,$$

$$\phi_1\bar{\lambda}_2 + \phi_2\bar{\lambda}_1 + \bar{\chi}_1 M_2 + \bar{\chi}_2 M_1). \tag{9.4.18}$$

§9.5   Vector and Chiral Superfields

The scalar multiplet discussed in the previous section is a reducible multiplet and we can take subsets of the component fields in eq. (9.3.5) to make irreducible multiplets. The first irreducible multiple we can construct is by demanding reality of Φ, i.e.,

$$\Phi = \Phi^*. \tag{9.5.1}$$

Because $\theta^* = \bar{\theta}$, we find the various component fields in eq. (9.4.5) related to each other, i.e.,

$$\phi = \phi^*, \qquad \psi = \bar{\chi}, \qquad M = N^*, \qquad V_\mu = V_\mu^*, \qquad \bar{\lambda} + \alpha, \qquad D = D^*. \tag{9.5.2}$$

This multiplet is also closed under multiplication. We will call this the vector multiplet and denote it by the symbol $V(x, \theta, \bar{\theta})$.

We will now discuss another irreducible supermultiplet called the chiral superfield. The important thing here is to note that there exists an operator

$$D_a = \frac{\partial}{\partial\theta^a} + i\sigma^\mu_{a\dot{a}}\bar{\theta}^{\dot{a}}\partial_\mu,$$

$$\bar{D}_{\dot{a}} = +\frac{\partial}{\partial\bar{\theta}^{\dot{a}}} + i\theta^a\sigma^\mu_{a\dot{a}}\partial_\mu, \tag{9.5.3}$$

which commutes with Q_a and $\bar{Q}_{\dot{a}}$ and with all the Lorentz generators. Therefore we can impose the requirement that the superfields Φ_+ and Φ_- (respectively, the chiral and antichiral superfields) satisfy the following supersymmetric invariant constraints:

$$D_a\Phi_- = 0 \quad \text{and} \quad \bar{D}_{\dot{a}}\Phi_+ = 0. \tag{9.5.3a}$$

Examining eq. (9.5.3) we can work out its implications. The solution to eq. (9.5.3a) is that

$$\Phi_-(\bar{\theta}, x^\mu - i\theta\sigma^\mu\bar{\theta}) \equiv \Phi_-(\bar{\theta}, y),$$

where

$$y^\mu = x^\mu - i\theta\sigma^\mu\bar{\theta}.$$

Since the Φ_- depends only on $\bar{\theta}$, it has only the following components:

$$\Phi_- = A(y) + \sqrt{2}\theta\psi(y) + \theta^2 F(y)$$
$$\equiv \exp(-i\theta\sigma^\mu\bar{\theta}\partial_\mu)(A_-(x) + \sqrt{2}\theta\psi_-(x) + \theta^2 F_-(x)). \quad (9.5.4)$$

Under the supersymmetry transformation $(\theta \to \theta + \varepsilon, \bar{\theta} \to \bar{\theta} + \bar{\varepsilon})$ the components of the chiral field transform as follows:

$$\delta A = \sqrt{2}\varepsilon\psi,$$
$$\delta\psi = \sqrt{2}\varepsilon F + \sqrt{2}i(\sigma^\mu\partial_\mu)_{a\dot{a}}\bar{\varepsilon}^{\dot{a}}A,$$
$$\delta F = -i\sqrt{2}\partial_\mu\psi^a(\sigma^\mu)_{a\dot{a}}\bar{\varepsilon}^{\dot{a}}. \quad (9.5.4a)$$

The chiral superfield Φ_+ is similarly given by Φ_+

$$\Phi_+(\theta, x^\mu + i\theta\sigma^\mu\bar{\theta}) = \exp(+i\theta\sigma^\mu\bar{\theta}\partial_\mu)(A_+(z) + \sqrt{2}\theta\psi_+ + \theta^2 F_+). \quad (9.5.5)$$

It is clear that $\Phi_+^\dagger = \Phi_-$. It is easy to check that the product of two chiral (two antichiral) fields is also a chiral (antichiral) field. However, products of a chiral and an antichiral field give a general vector field. Note further that we could isolate multiplets by imposing the D or $\bar{D}$ operators more often, i.e.,

$$DD\Phi = 0 \qquad \text{or} \qquad \bar{D}\bar{D}\Phi = 0. \quad (9.5.6)$$

This multiplet is called the complex linear multiplet L and has the following component expansion for $\bar{D}\bar{D}L = 0$, i.e.,

$$L = C(y) + \theta\beta(y) + \bar{\theta}\bar{\sigma}(y) + \theta^2 g(y) + \theta\sigma^\mu\bar{\theta}V_\mu(y) + \theta^2\bar{\theta}\bar{\rho}(y), \quad (9.5.7)$$

where C, V_μ, and g are complex bosonic fields and β, σ, and ρ are Weyl spinors and $y_\mu = x_\mu + i\theta\sigma_\mu\bar{\theta}$. By demanding $L = L^*$ we can obtain a real linear multiplet for which we have $C = C^*$, $\beta = \sigma$, $g = 0$, $V_\mu = V_\mu^*$, with $\partial^\mu V_\mu = 0$, $\bar{\rho} = (\sigma^\mu)\partial_\mu\beta$. In fact, we show below that Φ does contain precisely such a multiplet which satisfies both constraints in eq. (9.5.6).

We now wish to note some properties of the D and $\bar{D}$ operators

$$D_a D_b D_c - \bar{D}_{\dot{a}}D_{\dot{b}}D_{\dot{c}} = 0$$

and

$$D^a\bar{D}^2 D_a = \bar{D}_{\dot{b}}D^2\bar{D}^{\dot{b}}. \quad (9.5.8)$$

Equation (9.5.8) is easily proved

$$D^a\bar{D}^2 D_a = \bar{D}^a\bar{D}_{\dot{b}}\bar{D}^{\dot{b}}D_a$$
$$= \bar{D}^a\bar{D}_{\dot{b}}[2i(\sigma\partial)_a^{\dot{b}} - D_a\bar{D}^{\dot{b}}]$$
$$= [2i(\sigma\partial)_{\dot{b}}^{\dot{a}} - \bar{D}_{\dot{b}}\bar{D}^a][2i(\sigma\partial)_a^{\dot{b}} - D_a\bar{D}^{\dot{b}}]$$
$$= -8\Box + \bar{D}_{\dot{b}}D^2\bar{D}^{\dot{b}} - 2i(\sigma\partial)_a^{\dot{b}}\{D^a, \bar{D}_{\dot{b}}\}$$
$$= -8\Box + \bar{D}_{\dot{b}}D^2\bar{D}^{\dot{b}} - 2i(\sigma\partial)_a^{\dot{b}}2i(\sigma \cdot \partial_{\dot{b}}^a)$$

$$= \bar{D}_{\dot{b}} D^2 \bar{D}^{\dot{b}}. \tag{9.5.9}$$

Some other identities involving D and $\bar{D}$ are

$$D^2 \bar{D}^2 + \bar{D}^2 D^2 - 2D^a \bar{D}^2 D_a = 16\square, \tag{9.5.10a}$$

$$D^2 \bar{D}^2 D^2 = 16\square D^2, \tag{9.5.10b}$$

$$\bar{D}^2 D^2 \bar{D}^2 = 16\square \bar{D}^2, \tag{9.5.10c}$$

Using eqs. (9.5.8) and (9.5.10a–c) we can define the projection operators

$$\pi_{0+} = \frac{\bar{D}^2 D^2}{16\square}, \qquad \pi_{0-} = \frac{D^2 \bar{D}^2}{16\square}, \qquad \text{and} \qquad \pi_{1/2} = -\frac{2D^a \bar{D}^2 D_a}{16\square}, \tag{9.5.11}$$

$$\pi_{0+} + \pi_{0-} + \pi_{1/2} = 1. \tag{9.5.12}$$

In fact, operating on a scalar superfield π_{0+} and π_{0-} project out the chiral and antichiral parts and $\pi_{1/2}$ projects out a piece called the linear multiplet. It is now clear that the linear multiplet $L \equiv \pi_{1/2}\Phi$ satisfies the constraints in eq. (9.5.6). Because it satisfies both the constraints in eq. (9.5.6), it can be written as two real linear multiplets with the following independent components each $L \equiv (C, \beta, B_\mu)$ with $\partial^\mu B_\mu = 0$ and C a real field and β a Majorana spinor as pointed out.

We are now ready to give the supersymmetric generalization of the gauge transformation. Before doing that, let us give the various components of the field. For that purpose, we first realize that V must transform in some way as $\Phi_+ \cdot \Phi_-$.

Let us study the effect of the following transformation on the real vector multiplet

$$V \to V + \Lambda_+ + \Lambda_+^\dagger. \tag{9.5.13}$$

If we write

$$V = c + i\theta_\chi - i\bar{\theta}_{\bar{\chi}} + \frac{i}{2}\theta^2(M + iN) - \frac{i}{2}\bar{\theta}^2(M - iN)$$

$$- \theta\sigma^\mu\bar{\theta}V_\mu + i\theta\theta\bar{\theta}\left[\bar{\lambda} + \frac{i}{2}\bar{\sigma}^\mu\partial_\mu\chi\right] - i\bar{\theta}\bar{\theta}\theta\left[\lambda + \frac{i}{2}\sigma^\mu\partial_\mu\bar{\chi}\right]$$

$$+ \theta^2\bar{\theta}^2\left[D + \frac{1}{2}\square C\right] \tag{9.5.14}$$

and

$$\Lambda_+ + \Lambda_+^\dagger = A + A^* + \sqrt{2}(\theta\psi + \bar{\theta}\bar{\psi}) + \theta\theta F + \bar{\theta}\bar{\theta}F^*$$

$$- i\theta\sigma^\mu\bar{\theta}\partial_\mu(A - A^*) + \frac{1}{\sqrt{2}}\theta\theta\bar{\theta}\bar{\sigma}^\mu\partial_\mu\psi$$

$$+ \frac{i}{\sqrt{2}}\bar{\theta}^2\theta\sigma^\mu\partial_\mu\bar{\psi} + \frac{1}{4}\theta^2\bar{\theta}^2\square(A + A^*),$$

we find the components transforming as follows:

$$C \to C + A + A^*,$$
$$\chi \to \chi - i\sqrt{2}\psi,$$
$$M + iN \to M + iN - 2iF,$$
$$V_\mu \to V_\mu - i\partial_\mu(A - A^*),$$
$$\lambda \to \lambda,$$
$$D \to D, \tag{9.5.15}$$

It is thus clear that the transformation given in eq. (9.5.14) is the supersymmetric generalization of the ordinary gauge transformations. We can now choose

$$\text{Re } A = -C,$$
$$\psi = -\frac{i}{\sqrt{2}}\chi,$$

and

$$F = \frac{1}{2i}(M + iN), \tag{9.5.16}$$

so as to write the vector multiplet in the form

$$V = (0, 0, 0, 0, V_\mu, \lambda, D). \tag{9.5.17}$$

This gauge is known as the Wess–Zumino gauge, which contains, along with the gauge field V_μ, a Majorana spinor partner. Again, we see that off-shell, the number of bosonic and fermionic components match (since V_μ, due to gauge invariance, has off-shell only three degrees of freedom). On-shell D and one more component of V_μ is removed so that we have two real bosonic degrees of freedom. Since the coefficient of V_μ is quadratic in the Grassmann variable, for V_μ and λ to be dynamical fields with canonical dimension 1 and 3/2, the superfield must have canonical dimension 0.

Having given the gauge transformations we can write down the supersymmetric generalization of the gauge covariant (invariant in the abelian case) tensor $F_{\mu\nu}$ as follows: let us consider the abelian group first. The quantity

$$W_a = -\tfrac{1}{4}\bar{D}\bar{D}D_aV \tag{9.5.18}$$

can be shown to be invariant under the transformation (9.5.13). *Proof:*

$$\bar{D}\bar{D}D_a(\Lambda_+ + \Lambda_+^\dagger) = \bar{D}\bar{D}D_a\Lambda_+$$
$$= \bar{D}_{\dot{b}}[i(\sigma P)_a^{\dot{b}}\Lambda_+ - D_a\bar{D}^{\dot{b}}\Lambda_+]$$
$$= i(\sigma P)_a^{\dot{b}}\bar{D}_{\dot{b}}\Lambda_+ = 0. \tag{9.5.19}$$

It is also clear from eq. (9.5.18) that W_a is a chiral field. Similarly, we can define a gauge invariant antichiral field as follows:

$$\bar{W}_{\dot{a}} = -\tfrac{1}{4}DD\bar{D}_{\dot{a}}V. \qquad (9.5.20)$$

We will now show that the components of W contain the gauge covariant field tensor $F_{\mu\nu}$. Remembering that

$$\frac{\partial}{\partial\theta^{\dot{c}}}\frac{\partial}{\partial\bar{\theta}_{\dot{c}}}\bar{\theta}^2 = 4 \qquad (9.5.21)$$

we get

$$W_a = -i\lambda_a + \theta_b[\delta_a^b D - i(\sigma^\mu\bar{\sigma}^\nu)_a^b F_{\mu\nu}] + \theta^2(\sigma\partial)_{\dot{a}}\bar{\lambda}^{\dot{a}} + \text{ other terms.} \quad (9.5.22)$$

Since we have already shown that W_a is a chiral field, the other terms are dictated by this to arise from the exponential $e^{i\theta\sigma_\mu\bar{\theta}\partial_\mu}$ operating on W with the three terms shown above.

To complete this section we give the gauge-invariant coupling of the matter fields to the gauge fields. For this purpose, we note that under a gauge transformation, a matter field Φ transforms as follows:

$$\Phi \to e^{-g\Lambda}\Phi,$$
$$\Phi^\dagger \to \Phi^\dagger e^{-g\Lambda^\dagger}. \qquad (9.5.23)$$

It then follows that the gauge invariant coupling of Φ and V is

$$\mathcal{L}_\Phi = \Phi^\dagger e^{gV}\Phi. \qquad (9.5.24)$$

We will see, in the following chapter, that this gives rise to the gauge couplings of the matter fields after we expand the exponential and note that in the Wess–Zumino gauge, $V^n = 0$ for $n \geq 3$, i.e.,

$$\mathcal{L}_\Phi = \Phi^\dagger\Phi + g\Phi^\dagger V\Phi + \tfrac{1}{2}g^2\Phi^\dagger V^2\Phi. \qquad (9.5.25)$$

Exercises

9.1. Check by explicit calculation that the product of a chiral and antichiral field leads to a vector field.

9.2. Obtain the product rule for the real linear field. Show that $\pi_{1/2}$ actually projects out a linear field.

9.3. Obtain the non-abelian analog of (9.5.25), by writing down the non-abelian analog of the gauge transformations.

9.4. Make an explicit decomposition of the $\bar{W}_{\dot{a}}$, for an SU(2) gauge group and check the different coefficients.

References

[1] For excellent reviews see
A. Salam and J. Strathdee, *Fortschr Phys*, **26**, 57 (1978);
P. van Niuwenhuizen, *Phys. Rep.* **68**, 189 (1981);
P. Fayet and S. Ferrara, *Phys. Rep.* **32**, 249 (1977);
S. J. Gates, M. T. Grisaru, M. Rocek, and W. Siegel, *Superspace*, Benjamin Cummings, New York 1983;
J. Wess and J. Bagger, *Introduction to Supersymmetry*, Princeton University Press, Princeton, NJ, 1983;
Some more recent reviews are
H. P. Nilles, *Phys. Rep.* **110**, 1 (1984);
H. Haber and G. Kane, *Phys. Rep.* **117**, 76 (1984);
A. Chamseddine, P. Nath, and R. Arnowitt, *Applied $N = 1$ Supergravity*, World Scientific, Singapore, 1984;
B. Ovrut, Lecture Notes by S. Kalara and M. Yamawaki, 1982.

[2] Y. A. Golfand and E. A. Likhtman, *JETP Letters*, **13**, 452 (1971);
J. Wess and B. Zumino, *Nucl. Phys.* **B70**, 39 (1974); *Phys. Lett.* **49B**, 52 (1974).

[3] D. Volkov and V. P. Akulov, *JETP Lett.* **16**, 438 (1972).

[3a] J. Gervais and B. Sakita, *Nucl. Phys.* **B34**, 632 (1971).

[4] A. Salam and J. Strathdee, *Nucl. Phys.* **B76**, 477 (1974); *Phys. Lett.* **51B**, 353 (1974).

10

Field Theories with Global Supersymmetry

§10.1 Supersymmetry Action

To apply supersymmetry to describe particle interaction, we have to construct field theories that are invariant under supersymmetry transformations. We will then obtain certain constraints among the parameters of the bosonic and fermionic sectors of the theory and compare them with observations. The kind of field theories we are interested in will involve matter fields, which will be given by the chiral superfields and gauge fields. which in turn will be given by the real gauge superfield V. We will always work in the Wess–Zumino gauge for V. These matter and gauge superfields may (and, in general, will) belong to some irreducible representations of compact internal symmetry groups (local or global). Before going on to the discussion of the most general case, we first consider the simple case of a matter field Φ and illustrate how we can write a general interacting field theory for this.

A field theory consists of two parts: the kinetic energy and the potential energy, and both must be invariant under the supersymmetry transformations described in the previous chapter. As noted in Chapter 9, supersymmetry transformation corresponds to translations in a superspace with coordinates $(x_\mu, \theta, \bar{\theta})$. The volume element in superspace is $d^8z = d^4x\, d^2\theta\, d^2\bar{\theta}$; this is translation invariant. Therefore, the supervolume integrals of products of superfields will lead to a supersymmetric action as follows:

$$\int d^8 z f(\Phi, \Phi^\dagger) \overset{\text{SUSY}}{\longrightarrow} \int d^8 z f(\Phi(x_\mu + a_\mu, \theta + \varepsilon, \bar\theta + \bar\varepsilon), \Phi^\dagger(x + a_\mu, \theta + \varepsilon, \bar\theta + \bar\varepsilon)).$$

$$(10.1.1)$$

If we now redefine the coordinates $z \to z' = z + z_0$ where $z_0 \equiv (a_\mu, \varepsilon, \bar\varepsilon)$, then $d^8 z = d^8 z'$ and the action is invariant. Furthermore, if we have the additional property for a particular product of fields that either θ or $\bar\theta$ multiplies only terms that are space derivatives, we can define a six-dimensional volume integral $d^6 z \equiv d^4 x d^2 \theta$ or $d^6 \bar z \equiv d^4 x d^3 \bar\theta$ that can also lead to a supersymmetric action. As an example of why this is so, consider an arbitrary product of chiral ($\phi_{1+}\phi_{2+}\phi_{3+}...$) or antichiral fields ($\phi_{1+}^* \phi_{2+}^* ...$). We remind the reader of the form of a chiral field Φ (we drop the subscript $\pm$; instead field $\Phi^\dagger$ will be used to denote the antichiral field, whereas a field without † will denote chiral fields):

$$\Phi(x, \theta, \bar\theta) = \exp(i\theta\sigma^\mu\bar\theta\partial_\mu)[A(x) + \sqrt{2}\theta\psi + \theta^2 F]$$

$$= A(x) + \sqrt{2}\theta\psi + \theta^2 F + i\theta\sigma^\mu\bar\theta\partial_\mu A$$

$$- \frac{i}{\sqrt{2}}\theta^2 \partial_\mu \psi \sigma^\mu \bar\theta + \frac{1}{4}\theta^2 \bar\theta\partial^2 A. \qquad (10.1.2)$$

Any product of all chiral fields also has this expression, where we see that all terms involving $\bar\theta$ have space derivatives in them so that they will vanish after integration over $d^4 x$. For this case, the superspace becomes effectively six dimensional. So we can write a supersymmetric Lagrangian as follows:

$$S^2 = \int d^6 z W(\Phi) + \int d^6 \bar z [W(\Phi)]^\dagger. \qquad (10.1.3)$$

The type of action in eq. (10.1.1) is called D-type action, whereas the one in (10.1.3) is called F-type action. The reasons for the names will become obvious soon. $W(\Phi)$ is called superpotential.

Since we have given volume integrals in the space of Grassman coordinates, we must give the rules for integration and precise definition of measure. The rules of integration are as follows:

$$\int \theta_a d\theta_a = \delta_{ab}, \qquad \int d\theta_a = 0, \qquad (10.1.4)$$

and similarly for $\bar\theta$.

$$d^2 \theta = -\frac{1}{4} d\theta^a d\theta_a$$

$$= \frac{1}{4}\varepsilon_{ab} d\theta^a d\theta^b = +\frac{1}{2} d\theta^1 d\theta^2, \qquad (10.1.5)$$

$$\int d^2\theta \theta^2 = -\frac{1}{2}\int d\theta^1 d\theta^2 \varepsilon_{ab}\theta^a\theta^b$$

$$= \frac{1}{2} d\theta^1 d\theta^2 (-2\theta^1\theta^2)$$

$$= +\int d\theta^1 \theta^1 \int d\theta^2 \theta^2 = 1. \qquad (10.1.6)$$

Similarly, we defines

$$d^2\bar{\theta} = -\tfrac{1}{4}d\bar{\theta}_{\dot{a}}d\bar{\theta}^{\dot{a}}$$

$$= -\tfrac{1}{4}\varepsilon_{\dot{a}\dot{b}}d\bar{\theta}^{\dot{b}}d\bar{\theta}^{\dot{a}} = -\tfrac{1}{2}d\bar{\theta}^{\dot{i}}d\bar{\theta}^2,$$

$$\int d^2\bar{\theta}\bar{\theta}^2 = -\tfrac{1}{2}\int d\bar{\theta}^{\dot{i}}d\bar{\theta}^2\varepsilon_{\dot{a}\dot{b}}\bar{\theta}^{\dot{b}}\bar{\theta}^{\dot{a}}$$

$$= \int d\bar{\theta}^{\dot{i}}d\bar{\theta}^2\bar{\theta}^{\dot{i}} = 1. \tag{10.1.7}$$

Let us also note some other properties of θ integration. Suppose we have a function f of θ, $\bar{\theta}$, and x. Then the following identity holds:

$$\int d^8z f(x,\theta,\bar{\theta}) = -\tfrac{1}{4}\int d^6z \bar{D}f$$

$$= -\tfrac{1}{4}\int d^6\bar{z}D^2 f. \tag{10.1.8}$$

This follows because

$$\bar{D} = \frac{\partial}{\partial\bar{\theta}} + i\theta(\sigma \cdot \partial)$$

and since the second term in $\bar{D}$ is a total space divergence, its volume integral is zero by the Gauss theorem. So, inside a volume integral, D and $\bar{D}$ behave as if they only have the first term. Then we note that integrating and differentiating twice with respect to θ or $\bar{\theta}$ amounts to the same thing, i.e., picking up the coefficient of θ^2 or $\bar{\theta}^2$. Hence the proof.

Another property of θ space that follows from the integration rules is that

$$\theta^2 = \delta^2(\theta) \qquad \text{and} \qquad \bar{\theta}^2 = \delta^2(\bar{\theta}). \tag{10.1.9}$$

Also the mass dimension of $\int d\theta$ is $+\tfrac{1}{2}$.

Now we can give the action that describes the interacting field theory of a chiral superfield

$$S = \int d^8z \Phi^\dagger\Phi + \int d^6z W(\Phi) + \int d^6\bar{z}W(\Phi)^\dagger. \tag{10.1.10}$$

The first point we note is that, in units where $hbar = c = 1$, S must be dimensionless. Since the mass dimension of $(d\theta) = +\tfrac{1}{2}$ and that of Φ is $+1$ the first term of eq. (10.1.10) is clearly dimensionless. As far as the second and third terms go, the cubic terms in Φ will have dimensionless coupling and any lower power of Φ will have the powers of mass in the coupling, and any higher power will be suppressed by inverse powers of mass.

Now let us verify that the first term indeed yields the correct form for the kinetic energy term. We have to evaluate

$$\int d^2\theta d^2\bar{\theta}\Phi^\dagger\Phi = ?.$$

Since, by the integration rules given earlier, $\int d^4\theta$ projects out only the coefficient of $\theta^2\bar{\theta}^2$, it is like the D term in the expansion of a vector superfield. Another way to see that this term is supersymmetric is to note that under supersymmetric variation

$$\delta D = \varepsilon\sigma^\mu \cdot \partial_\mu\bar{\lambda} + \bar{\varepsilon}\bar{\sigma}^\mu \cdot \partial_\mu\lambda, \qquad (10.1.11)$$

this being a four divergence which vanishes on integration. From eq. (10.1.2) we can pick out the D term from the product $\Phi^\dagger\Phi$, it is

$$\mathcal{L}_{K.E.} = -\tfrac{1}{2}A^*\Box A + \tfrac{1}{2}\partial^\mu A^*\partial_\mu A - \frac{i}{2}\partial_\mu\psi\sigma^\mu\bar{\psi} + \frac{i}{2}\psi\sigma^\mu\partial_\mu\bar{\psi} + F^*F. \quad (10.1.12)$$

Thus we get precisely the familiar kinetic energy term. Let us now look at the second term in eq. (10.1.10). As discussed earlier in the expansion of the chiral field $W(\Phi)$, we simply have to pick up the coefficient of θ^2, i.e., the F term. To illustrate how it works we choose $W(\Phi) = \lambda\Phi^3 + m\Phi^2$. The F term of this is easily evaluated to be

$$\int d^2\theta W(\Phi) = m(FA - \psi\psi) + \lambda(FA^2 - \psi\psi A). \qquad (10.1.13)$$

Thus, in terms of component fields, the action can be written as

$$\mathcal{L} = +\partial_\mu A^*\partial_\mu A - i\partial_\mu\psi\sigma_\mu\bar{\psi} + F^*F + (FA - \psi\psi) + \lambda(FA^* - \psi\psi A)$$
$$+ (F^*A^* - \bar{\psi}\bar{\psi}) + \lambda(F^*A^{*2} - \bar{\psi}\bar{\psi}A^*). \qquad (10.1.14)$$

In this Lagrangian F is an auxiliary field that has no kinetic energy term associated with it. Therefore, we can eliminate it by writing down the field equation for F obtained by varying the Lagrangian with respect to F:

$$-F = mA^* + \lambda A^{*2}. \qquad (10.1.15)$$

On substituting it into eq. (10.1.14) we get

$$\mathcal{L} = +(\partial^\mu A^*)(\partial_\mu A) - i\partial_\mu\psi\sigma^\mu\bar{\psi} - m(\psi\psi + \bar{\psi}\bar{\psi})$$
$$- \lambda\psi\psi A - \lambda\bar{\psi}\bar{\psi}A^* - |mA + \lambda A^2|^2. \qquad (10.1.16)$$

From this Lagrangian we can easily see the constraints imposed on the parameters of the theory by supersymmetry. For example, it implies

$$m_\psi = m_A$$

and the coupling constant relation

$$g_{A^4} = g_{\psi\psi A}^2. \qquad (10.1.17)$$

Thus, we already note that, in order for supersymmetry to be useful for the description of particle interactions, it must be broken since we do not observe any fermion boson pair degenerate in mass.

For future use we also note that the scalar potential in the Lagrangian is obtained as follows (for theories without gauge fields):

$$V = |F|^2 = \left|\frac{\partial W}{\partial A}\right|^2, \qquad (10.1.18)$$

where $W = W(\theta = 0)$. For any arbitrary theory, not involving gauge fields, the field is generalized to

$$V = \sum_i \left|\frac{\partial W}{\partial A_i}\right|^2. \qquad (10.1.19)$$

§10.2   Supersymmetric Gauge Invariant Lagrangian

In this section we study the gauge-invariant supersymmetric Lagrangian [1]. For simplicity we will consider abelian gauge invariance and gauge coupling of a chiral scalar field with $U(1)$ charge $+1$. The gauge and supersymmetrically invariant action consists of the following two pieces:

$$S = \frac{1}{4}\int d^6 z\, W^a W_a + \int d^8 z\, \Phi^\dagger e^{2gV}\Phi + \text{h.c.} \qquad (10.2.1)$$

We now show that the first term consists of the kinetic energy term for the gauge field, and the second term is the gauge-invariant kinetic energy term for the matter field Φ. To see this, recall that

$$W^a = -i\lambda^a + \theta^b\left[\delta_b^a D - \frac{i}{2}(\sigma_\mu\bar{\sigma}_\nu)_b^a F^{\mu\nu}\right] + \theta^2(\sigma\cdot\partial)_{\dot{a}}^a\bar{\lambda}^{\dot{a}}. \qquad (10.2.2)$$

To obtain the first term we simply pick out the coefficient of θ^2 in the product of $W^a W_a$ and we find, on adding the Hermitian conjugate piece, that

$$S_1 = \int d^4 x\left[-\frac{1}{4}F^{\mu\nu}F_{\mu\nu} - i\lambda\sigma^\mu\partial_\mu\bar{\lambda} + \frac{1}{2}D^2\right]. \qquad (10.2.3a)$$

Now let us look at the second term in eq. (10.2.1) and project out the coefficient of $\theta^2\bar{\theta}^2$ (the D component)

$$S_2 = \int d^8 z\left[\Phi^\dagger\Phi + 2g\Phi^\dagger V\Phi + 2g^2\Phi^\dagger V^2\Phi\right]. \qquad (10.2.3b)$$

The D component of these terms has already been calculated:

$$g\Phi^\dagger V\Phi|_D = igV_\mu(A^*\partial_\mu A - A\partial_\mu A^*) + ig\sqrt{2}(A^*\psi\lambda - A\bar{\psi}\bar{\lambda}) + gDA^*A \qquad (10.2.4)$$

$$g^2\Phi^\dagger V^2\Phi|_D = \frac{g^2}{4}V_\mu^2 A^*A. \qquad (10.2.5)$$

Combining all these we find

$$S_2 = \int d^4x \Big[(D^\mu A)^* (D_\mu A) - i D_\mu \psi \sigma^\mu \bar{\psi} + F^* F$$

$$+ \frac{ig}{\sqrt{2}} (A^* \psi \lambda - A \bar{\psi} \bar{\lambda}) + \frac{g}{2} D A^* A + \frac{g^2}{4} V_\mu^2 A^* A \Big]. \quad (10.2.6)$$

We now point out several important consequences of the Lagrangian in eq. (10.2.6) that give the supersymmetric coupling of matter to gauge fields. Note that the gauge coupling here is chosen to be $g/2$ (instead of the usual definition, g).

(a) There is a gauge fermion λ that is the fermionic partner of the gauge boson. This transforms in the same way under the gauge group as the gauge fields and is a feature common to all supersymmetric gauge theories. This particle will be called gaugino.

(b) In addition to the couplings expected from gauge invariance, there is an additional interaction between the fermionic matter field ψ, its scalar partner A, and the gaugino λ. The strength of this interaction is also given by the gauge coupling g and this is also a general feature of supersymmetric gauge theories.

(c) Gauge invariance implies the masslessness of the gaugino field as well as the gauge fields. Thus, for these theories to be realistic, supersymmetry will have to be broken.

Using the field equation for the D term we can isolate the scalar potential (i.e., that part of the Lagrangian not involving any derivatives or any fermions) as

$$V = \tfrac{1}{2} D^2. \quad (10.2.7)$$

Combining this with eq. (10.1.18) we find that the scalar potential in a theory with matter and gauge fields coupled to each other can be written as

$$V = |F|^2 + \tfrac{1}{2} D^2. \quad (10.2.8)$$

This is the expression we have to analyze in order to study the symmetry breaking in these theories.

These considerations can be generalized to the non-abelian groups [2]. The W^a is then defined as follows:

$$W_a = \tfrac{1}{4} \bar{D} \bar{D} e^{-gV} D_a e^{gV}. \quad (10.2.9)$$

Other definitions remain the same. This leads to the Yang–Mills action for the gauge field with the gaugino belonging to the adjoint representation of the gauge group.

§10.3 Feynman Rules for Supersymmetric Theories [3]

In this section we will derive the Feynman rules for superfields and describe some of their applications. We start by writing down the general form for the supersymmetric action for a chiral field Φ coupled to a gauge field V:

$$S = \tfrac{1}{2} \int d^4x d^4\theta \Phi^\dagger e^{gV} \Phi + \frac{1}{64g^2} \int d^4x d^2\theta W^a W_a$$

$$+ \int d^4x d^2\theta W(\Phi) + \text{ Gauge fixing terms}$$

$$+ \int d^4x d^2\theta J\Phi + \tfrac{1}{2} \int d^4x d^4\theta KV + \text{h.c.} \qquad (10.3.1)$$

We will now use the following identities to convert S into a form in which we can easily invert the kinetic term for matter fields to obtain the propagator:

$$\frac{\delta}{\delta J(z_1)} J(z_2) = -\tfrac{1}{4} \bar{D}_1^2 \delta^8(z_{12}), \qquad (10.3.2a)$$

$$\int d^4x d^2\theta (-\tfrac{1}{4}\bar{D}^2 f) = \int d^8z f, \qquad (10.3.2b)$$

where f is a function of $(x, \theta, \bar{\theta})$. In particular, if f is a chiral field (i.e., $\bar{D}f = 0$), then

$$\bar{D}^2 D^2 f = \bar{D}_{\dot{a}} \bar{D}^{\dot{a}} D^b D_b f$$

$$= [\bar{D}_{\dot{a}} D_b 2i(\bar{\sigma} \cdot \partial)^{\dot{a}b} - \bar{D}_{\dot{a}} D^b (2i\sigma \cdot \partial)^{\dot{a}}_b] f$$

$$= 8(\sigma \cdot \partial)_{b\dot{a}} (\bar{\sigma} \cdot \partial)^{\dot{a}b} f$$

$$= 16\square f, \qquad (10.3.3)$$

where we have used $\{D_a, \bar{D}_{\dot{b}}\} = -2i(\sigma \cdot \partial)_{a\dot{b}}$ and $\{\bar{D}^{\dot{a}}, D^b\} = 2i(\sigma \cdot \partial)^{\dot{a}b}$. This leads to the identity

$$\int d^6z f = \int d^6z \cdot \frac{\bar{D}^2 D^2 f}{16\square}$$

$$= -\int d^8z \frac{D^2}{4\square} f. \qquad (10.3.4)$$

To derive the Feynman rules let us first consider the chiral fields and ignore the gauge fields, which can be treated in a similar manner. Also let us assume that the superpotential has the following simple form:

$$-W(\Phi) = \tfrac{1}{2}m\Phi^2 + \frac{1}{3!}\lambda\Phi^3. \qquad (10.3.5)$$

Let us now try to write $\tfrac{1}{2}m\Phi^2$ in a useful form

$$\tfrac{1}{2}m \int d^4x d^2\theta \Phi^2 = \tfrac{1}{2}m \int d^4x d^2\theta \Phi \frac{\bar{D}^2 D^2 \Phi}{16\square}. \qquad (10.3.6a)$$

But

$$\Phi \bar{D}^2 \frac{D^2}{16\Box} \Phi = +\frac{\bar{D}^2}{4} \left(\Phi \frac{D^2}{4\Box} \Phi \right), \tag{10.3.6b}$$

since Φ is a chiral field and hence obeys the condition $\bar{D}\Phi = 0$. This implies

$$\tfrac{1}{2}m \int d^4x d^2\theta \Phi^2 = +\tfrac{1}{2}m \int d^8z \Phi \left(-\frac{D^2}{4\Box} \Phi \right). \tag{10.3.6c}$$

The bilinear part of the chiral field action can now be written as

$$\begin{aligned} S_0 &= \int d^8z \left[\bar{\Phi}\Phi - \tfrac{1}{2}m\Phi \left(-\frac{D^2}{4\Box}\Phi \right) - \tfrac{1}{2}m\bar{\Phi} \left(-\frac{D^2}{4\Box}\bar{\Phi} \right) \right. \\ &\quad \left. + J \left(-\frac{D^2}{4\Box}\Phi \right) + \bar{J} \left(-\frac{\bar{D}^2}{4\Box}\bar{\Phi} \right) \right. \\ &= \int d^8z \left(\tfrac{1}{2}\psi^T A\psi + \psi^T B \right) \left. \right], \end{aligned} \tag{10.3.7}$$

where

$$\psi^T = (\Phi \; \bar{\Phi}),$$

$$A = \begin{pmatrix} +\frac{mD^2}{4\Box} & 1 \\ 1 & \frac{m\bar{D}^2}{4\Box} \end{pmatrix}, \tag{10.3.8a}$$

$$B = \begin{pmatrix} -\frac{1}{4}\frac{D^2}{\Box}J \\ -\frac{1}{4}\frac{\bar{D}^2}{\Box}\bar{J} \end{pmatrix}, \tag{10.3.8b}$$

To calculate the propagator we follow the procedure employed in the functional approach to conventional nonsupersymmetric field theories, i.e., we write the generating functional

$$Z(J) = \int d\Phi d\bar{\Phi} \exp\left[i \int d^8z S(\Phi, \bar{\Phi}) + i \int d^6z J\Phi + i \int d^6\bar{z}\bar{J}\bar{\Phi} \right]. \tag{10.3.9}$$

Using eq. (10.3.7) we can write

$$Z(J) = \exp\left[i \int d^6z \frac{\lambda}{3!} \frac{\delta^3}{\delta J(z)^3} \right] Z_0(J), \tag{10.3.10}$$

where

$$Z_0(J) = \int d\Phi d\bar{\Phi} \exp\left\{ i \int d^8z \left[\tfrac{1}{2}\psi^T A\psi + \psi^T B \right] \right\}. \tag{10.3.11}$$

Redefining $\psi' = A^{1/2}\psi$ we can integrate the exponential in $Z_0(J)$ to obtain

$$Z_0, (J) = \exp\left(-\frac{i}{2} \int d^8z B^T A^{-1} B \right). \tag{10.3.11a}$$

To obtain the various propagators such as $\langle\Phi\Phi\rangle$, $\langle\bar{\Phi}\Phi\rangle$, and $\langle\bar{\Phi}\bar{\Phi}\rangle$ we rewrite $Z_0(J)$ as follows:

$$Z_0(J) = \exp\left\{-\frac{i}{2}\int d^8z d^8z'\left[\tfrac{1}{2}J(z)\Delta_{\phi\phi}(z,z')J(z')\right.\right.$$

$$\left.\left.+ J(z)\Delta_{\phi\bar{\phi}}(z,z')\bar{J}(z') + \tfrac{1}{2}\bar{J}(z)\Delta_{\bar{\phi}\bar{\phi}}(z,z')\bar{J}(z')\right]\right\}, \quad (10.3.12)$$

where the Δ's represent the correspond propagators. To give their explicit form we have to evaluate $B^T A^{-1} B$

$$B^T A^{-1} B = \left(-\frac{1}{4}\frac{D^2}{\Box}J, -\frac{1}{4}\frac{\bar{D}^2}{\Box}\bar{J}\right)A^{-1}\left(\begin{array}{c}-\frac{1}{4}\frac{D^2}{\Box}J\\-\frac{1}{4}\frac{\bar{D}^2}{\Box}\bar{J}\end{array}\right), \quad (10.3.13)$$

where

$$A^{-1}A = 1.$$

Remembering the identity $D^2\bar{D}^2D^2 = 16\Box D^2$ and $\bar{D}^2D^2\bar{D}^2 = 16\Box\bar{D}^2$ we can easily find A^{-1} to be

$$A^{-1} = \left(\begin{array}{cc}-\frac{1}{4}\frac{m\bar{D}^2}{\Box-m^2} & 1+\frac{m^2\bar{D}^2D^2}{16\Box(\Box-m^2)}\\ 1+\frac{m^2D^2\bar{D}^2}{16\Box(\Box-m^2)} & -\frac{1}{4}\frac{mD^2}{\Box-m^2}\end{array}\right). \quad (10.3.14)$$

This leads to

$$\Delta_{\bar{\phi}\phi}(z,z') = \frac{1}{(\Box - m^2)}, \quad (10.3.15)$$

$$\Delta_{\phi\phi}(z,z') = \frac{1}{4}\frac{mD^2}{\Box(\Box - m^2)}. \quad (10.3.16)$$

In momentum space we can write the propagates as

$$\Delta_{\phi\bar{\phi}}(p,\theta) = \frac{-i}{p^2 - m^2}\delta^4(\theta_1 - \theta_2), \quad (10.3.17)$$

$$\Delta_{\phi\phi}(p,\theta) = \frac{-\frac{i}{4}mD^2(p,\theta)}{p^2(p^2 - m^2)}. \quad (10.3.18)$$

To write the complete set of Feynman rules we have to look at the chiral vertex arising from the $\lambda\phi^3$ term. For this purpose let us write down the generating functional

$$Z(J) = \exp\left[i\int d^6z''\frac{\lambda}{3!}\left(\frac{1}{i}\frac{\delta}{i\delta J(z'')}\right)^3\right] + \text{h.c.}$$

$$\times \exp\left\{-\frac{i}{2}\int d^8z d^8z'[J(z)\Delta_{\phi\phi}(z,z')J(z') + \cdots]\right\}. \quad (10.3.19)$$

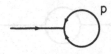

Figure 10.1.

Let us look at the effect of the lowest-order term in λ coming from the first exponent and we find

$$Z_1 = +\frac{i\lambda}{3!} \int d^6 z'' \left[\int d^8 z d^8 z' \delta^6(z - z'') \Delta_{\phi\phi}(z, z') J(z') \right]^3 + \cdots . \quad (10.3.20)$$

Using the identity $\delta^6(z - z'') = -\frac{1}{4}\bar{D}^2_{z''}\delta^8(z - z'')$ and doing partial integration we get

$$Z_1 = \frac{i\lambda}{3!} \int d^6 z'' \left[\int d^8 z' - \frac{1}{4}\bar{D}^2_{z''}\Delta_{\phi\bar\phi}(z'', z') J(z') \right]^3 + \cdots .$$

Now one of the $\frac{1}{4}\bar{D}^2_{z''}$ from the integrand can be removed by converting $\int d^6 z''$ to $\int d^8 z''$. Then, at the vertex, we are left with two $-\frac{1}{4}\bar{D}^2_{z''}$ factors for three legs. Furthermore, since the S matrix is obtained by the following operation

$$S = \exp\left[-i \int \Phi_{in} \Delta_{\phi\bar\phi}^{-1} \frac{\delta}{\delta J} d^6 z \delta^2(\bar\theta) \right] Z(J) \quad (10.3.21)$$

for each external leg, we must remove $-\frac{1}{4}\bar{D}^2\Delta_{\phi\phi}($ or $-\frac{1}{4}D^2\Delta_{\bar\phi\bar\phi})$. We can now state the Feynman rules for supersymmetric field theories in momentum space

$$\Delta_{\phi\bar\phi}(p, \theta_1 - \theta_2) = \frac{-i\delta^4(\theta_1 - \theta_2)}{p^2 - m^2}, \quad (10.3.17)$$

$$\Delta_{\phi\phi}(p, \theta_1 - \theta_2) = \frac{-(i/4)mD^2(p, \theta_1 - \theta_2)\delta^4(\theta_1 - \theta_2)}{p^2(p^2 - m^2)}. \quad (10.3.18)$$

As far as the vertices are concerned each chiral (antichiral) vertex will have a $-\frac{1}{4}\bar{D}^2(-\frac{1}{4}D^2)$ factor for each chiral superfield, but omitting one for converting $d^6 z$ to $d^8 z$ and omitting one for external legs. Each vertex has a $d^4\theta$ integration. Using these rules for the simple theory described, let us evaluate a couple of Feynman diagrams.

The first diagram

$$A_1 = -i\lambda m \frac{1}{16} \int D^2 \bar{D}^2 \delta(\theta_1 - \theta_2)|_{\theta_1=\theta_2} \phi d^4\theta_1 I(p), \quad (10.3.22)$$

where $I(p)$ is the divergent momentum integral. We now note that

$$\frac{1}{16} D^2 \bar{D}^2 \delta^4(\theta_1 - \theta_2)|_{\theta_1=\theta_2} = 1; \quad (10.3.23)$$

this gives $\int \phi d^4\theta = 0$.

Figure 10.2.

The One-Loop Correction to the Propagator

Using the same Feynman rules we can obtain

$$\frac{1}{(2\pi)\phi} \int d^4\theta_1 d^4\theta_2 \tfrac{1}{16} D_1^2 \delta^4(\theta_{12}) \bar{D}_2^2 \delta^4(\theta_{21}) \int \frac{d^4k}{k^2(-k+p)^2}$$

$$= \int d^4\theta_1 d^4\theta_2 \tfrac{1}{16}\delta^4(\theta_{12}) D_1^2 \bar{D}_2^2 \delta^4(\theta_{21}) I(p). \quad (10.3.24)$$

We note that the $\sigma \cdot \partial$ term in D vanishes since it is inside space integration, which helps us to write each D or $\bar{D}$ inside an integral as $\partial/\partial\theta$ or $\partial/\partial\bar{\theta}$, respectively. We then use the identity that

$$\tfrac{1}{16} D_1^2 \bar{D}_2^2 \delta^4(\theta_{21}) = 1.$$

Integrating over $d^4\theta_2$, we find the effective action is $\int d^4\theta \Phi^\dagger \Phi$. This procedure can be repeated for arbitrary loops to show that all loop effects are of the form $\int d^4\theta f(\phi^\dagger, \phi)$. This, therefore, implies that the superpotential is completely unaffected by loop corrections.

§10.4 Allowed Soft-Breaking Terms

We will now discuss the allowed soft-supersymmetry breaking terms that do not disturb the renormalizality of the theory [4]. To study these we first give the rules of power counting in supersymmetric field theories. These rules are different from the conventional field theories and are as follows:

(a) each vertex has a factor D^4, which is $\sim p^2$;

(b) propagators: $\Delta_{\phi\bar{\phi}} - \sim 1/p^2$ and $\Delta_{\phi\phi} \sim 1/p^3$;

(c) the external line has $1/D^2 - 1/p$;

(d) the loop integral $\sim d^4p/p^2$ due to the fact that there is a d^θ integration or due to the fact that four D's are needed to cancel the final $\delta^4(\theta)$ in the loop, thus leaving $\int d^4\theta$ in the

For a graph like that in Fig. 10.2 we can easily count that

$$d = \underset{\text{external leg}}{-2} \quad \underset{\text{two vertices}}{+4} \quad \underset{\text{two propagators}}{-4} \quad \underset{\text{loop integration}}{+2} \quad = 0.$$

It is therefore clear that, if we have additional D factors at the vertices, it will worsen the divergence structure of the theory. This has two implications

(a) $\lambda \Phi^n$, $n > 3$ would imply more than one D^4 at each vertex. Thus, the maximum allowed n is three for the theory to be renormalizable.

(b) To study allowed soft-supersymmetry breaking terms [5], we first note that by introducing constant spurion superfields we can write them in a manifestly supersymmetric form. For instance, if we introduce a spurion $U = \theta^2 \bar{\theta}^2$, we can write a term such as $A^\dagger A = f d^4 \theta U \Phi^\dagger \Phi$. It is then possible to do power counting to see which kind of soft-breaking terms introduce new divergences into the theory.

Allowed soft-breaking terms should not involve D's in their vertices when expressed in manifestly supersymmetric form. This has important implications. For instance, if we want to add a fermion mass term to explicitly break supersymmetry, we can write it as

$$\int d^4 \theta U D^a \Phi | D_a \Phi_2, \qquad (10.4.1)$$

where $U = \theta^2 \bar{\theta}^2$. Note that $D\Phi$ is not a chiral field (i.e., $\bar{D}D\phi \neq 0$), we cannot have a $d^2\theta$ integral and have manifest supersymmetry for the Lagrangian. So we must make it $d^4\theta$. This will add a term $\mu\psi_1\psi_2$, to the Lagrangian without its corresponding superpartner term $F_1^\dagger A_2 + F_2^\dagger A_1$. But this will not be allowed since, at each vertex, it will introduce D^6 (two powers of D from each ϕ and two explicit D's) making the theory nonrenormalizable. However, soft-breaking terms such as $\int d^4 \theta U' \Phi^\dagger \Phi$ that give $\mu^2 A^+ A$ (choosing $U_1 = \mu^2 \theta^2 \bar{\theta}^2$) are allowed. The gaugino mass $\lambda\lambda$ is also allowed since it is of the form $\int U d^4 \theta W^a W_a$.

Exercises

10.1. Using the techniques given in this chapter, derive the propagator for the gauge vector field and the super-Feynman Rules for a matter coupled SU(2)-gauge invariant Lagrangian.

10.2. Write down the Feynman Rules for the super-QED, i.e., super-symmetrized version of quantum electrodynamics. Show that the anomalous magnetic moment of the electron vanishes in the limit of exact supersymmetry.

10.3. In Problem 10.4, what supersymmetry breaking terms can be allowed without destroying renormalizability? In the presence of these terms, obtain the contribution to $g - 2$ of the electron. Using the present measurement of $(g - 2)$, discuss the bounds imposed on the masses of supersymmetry breaking terms.

10.4. In the supersymmetric QED, is there a photon wave function renormalization?

10.5. For a supersymmetric theory of a single chiral field, obtain the supersymmetric generalization of the energy momentum tensor. Expanding it into its components, discuss the physical meaning of the superpartners of the trace of the energy momentum tensor.

10.6. What is the supersymmetric generalization of Noether's theorem? Use it in a toy model with a global U(1) symmetry to obtain the conserved current corresponding to this current. What is the physical meaning of its superpartners?

10.7. In Problem 10.4, arrange the superpotential in such a way as to have the vacuum state break U(1) symmetry spontaneously. Obtain the Goldstone boson superfield. How many massless particles are there in the theory? Are there any hidden symmetries which are spontaneously broken?

References

[1] J. Wess and B. Zumino, *Nucl. Phys.* **B78**, 1 (1974).

[2] S. Ferrara and B. Zumino, *Nucl. Phys.* **B79**, 413 (1974);
    A. Salam and J. Strathdee, *Phys. Lett.* **51B**, 353 (1974).

[3] M. T. Grisara, M. Rocek, and W. Siegel, *Nucl. Phys.* **B159**, 429 (1979).

[4] J. Wess and B. Zumino (Ref. [1]);
    A. Slavnov, *Nucl. Phys.* **B97**, 155 (1975);
    B. DeWit, *Phys. Rev.* **D12**, 1628 (1975);
    S. Ferrara and O. Piguet, *Nucl. Phys.* **B93**, 261 (1975);
    R. Delbourgo, M. Ramon Medrano, *Nucl. Phys.* **B110**, 473 (1976);
    R. Delbourgo, *J. Phys.* **G1**, 800 (1975).

[5] L. Girardello and M.T. Grisaru, *Nucl. Phys.* **B194**, 65 (1982).

11

Broken Supersymmetry and Application to Particle Physics

§11.1 Spontaneous Breaking of Supersymmetry

We pointed out in the previous chapter that in the exact supersymmetric limit fermions and bosons are degenerate in mass, a situation for which there appears to be no evidence in nature. Therefore, in order to apply supersymmetry to particle physics, we must consider models where supersymmetry is broken. There are two ways to break symmetries of Lagrangian field theories (see Chapter 2): first, where extra terms are added to the Lagrangian that are not invariant under the symmetry; and second, the Lagrangian is kept invariant whereas the vacuum is allowed to be noninvariant under the symmetry. The first method introduces an arbitrariness into the theory thereby reducing its predictive power. The condition that the divergence structure should not be altered very much reduces this arbitrariness somewhat; yet it is not a very satisfactory approach. On the other hand, the second method, the Nambu–Goldstone realization of the symmetry provides a unique, appealing, and more predictive way to study the consequences of symmetry noninvariances. We will, therefore, study this approach in this chapter.

In contrast with ordinary bosonic symmetries, supersymmetry breaking requires more careful consideration for the following reason. We have

$$\{Q_a, \bar{Q}_{\dot{b}}\} = 2(\sigma^\mu P_\mu)_{a\dot{b}}. \qquad (11.1.1)$$

Taking vacuum expectation values of both sides we find [1]

$$\langle 0|Q_a\bar{Q}_{\dot{b}} + \bar{Q}_{\dot{b}}Q_a|0\rangle = 2\langle 0|H|0\rangle\delta_{a\dot{b}}$$

or

$$|Q_a|0\rangle|^2 = \langle 0|H|0\rangle. \tag{11.1.2}$$

This equation implies the following.

As long as the vacuum state has zero energy, supersymmetry is unbroken. Thus, to break supersymmetry, we find from eq. (10.2.8) that we must have either $\langle F \rangle \neq 0$ or $\langle D \rangle \neq 0$ or both. The first possibility is called the F-type or O'Raifeartaigh [2] mechanism for supersymmetry breaking, whereas the second mechanism is called the D-type or Fayet-Illiopoulos [3] mechanism. In the subsequent sections we will give examples of both these mechanisms. Right now we show that, in an exactly analogous manner to the case of bosonic symmetries, the spontaneous breaking of supersymmetry leads to the existence of massless fermionic states, which will be called Goldstino.

We note that, for a chiral field, supersymmetric transformation gives

$$\delta\psi_a = \varepsilon_a F - \sigma^\mu_{a\dot a}\bar\varepsilon^{\dot a}\partial_\mu A. \tag{11.1.3}$$

Taking vacuum expectation values of both sides we find

$$\langle\delta\psi_a\rangle = \varepsilon_a\langle F\rangle \neq 0, \tag{11.1.4}$$

which is the condition for supersymmetry breaking. But

$$\delta\psi_a = \varepsilon^b\{Q_b, \psi_a\}, \tag{11.1.5}$$

Equation (11.1.4) implies that

$$\langle 0|\{Q_b, \psi_a\}|0\rangle \neq 0. \tag{11.1.6}$$

Using the supersymmetry current and its conservation condition we can rewrite eq. (11.1.6) as

$$\int d^4x \partial^\mu I_\mu(x) = \int d^4x \frac{\partial}{\partial x_\mu}\langle 0|T\{S_{\mu b}(x), \psi(0)\}|0\rangle \neq 0. \tag{11.1.7}$$

We can convert the above integral into a surface integral and take the surface at infinity. If all fermionic states of the theory are massive, then the T-product falls off as $e^{-m|x|}/x^3$ for large x and leads to vanishing of the surface integral. On the other hand, if there is at least one massless spin-1/2 fermionic state $|X\rangle$ in the theory, two possibilities occur.

(a) We may have

$$\langle 0|S_{\mu,b}|X(P)\rangle = fP_\mu\delta_{ab}. \tag{11.1.8}$$

Then we have

$$\partial^\mu I_\mu(x) = \partial^2\left(\frac{\bar\sigma^\mu \cdot X_\mu}{x^4}\right). \tag{11.1.9}$$

The integral in this case vanishes.

(b) On the other hand, if we have

$$\langle 0|S_{\mu,b}X_a(P)\rangle = (\sigma_\mu)_{b\dot a}fx, \tag{11.1.10}$$

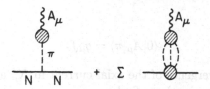

Figure 11.1.

we get

$$\partial^\mu I_\mu(x) = \partial_\mu^{Tr} \frac{(\sigma_\mu \cdot \bar{Q}_\nu)x_\nu}{x^4}. \tag{11.1.11}$$

Clearly, for this case, the integral in eq. (11.1.7) goes like $\int d^4x/x^4$ giving rise to a nonzero value of the integral. In this crude manner we see that spontaneous breaking of supersymmetry leads to massless fermionic states—the Goldstinos.

Similar arguments can be given for the D-terms by noting that under supersymmetry transformation

$$\delta\lambda_a = \varepsilon_a D + \varepsilon^b(\sigma_\mu\bar{\sigma}_\nu)_{ba}F^{\mu\nu}. \tag{11.1.12}$$

Thus, since $\langle 0|F_{\mu\nu}|0\rangle = 0$, $\langle D\rangle \neq 0$ implies $\langle \delta\lambda\rangle \neq 0$. We will see that, in the F-type breaking case, the field ψ is the Goldstino field where, in the D-type case, the corresponding field is λ.

§11.2 Supersymmetric Analog of the Goldberger–Treiman Relation

It is well known in hadronic weak interactions that spontaneous breaking of axial SU(2) symmetry leads to a relation between pion [the Goldstone boson for SU(2)$_A$ symmetry] coupling to nucleons, the pion decay constant, and the value of the current matrix element between nucleons at zero momentum transfer. This is the celebrated Goldberg–Treiman (GT) relation. For the case of spontaneously broken supersymmetry an analogous relation exists. Before deriving this we remind the reader about the derivation of the Goldberg–Treiman relation in hadron physics. Consider the matrix element

$$K_\mu \equiv \langle N(P_1)|A_\mu|N(P_2)\rangle = \bar{u}[\gamma_\mu\gamma_5 G_A(q^2) + \gamma_5 q_\mu F_A(q^2)]u. \tag{11.2.1}$$

Using dispersion relations for the left-hand side we see that K_μ receives contributions from the pion (the massless pole and the continuum are shown in Fig. 11.1). The first term implies that

$$F_A(q^2) = \frac{g_{NN\pi}f_\pi}{q^2} + \text{continuum}, \tag{11.2.2}$$

where

$$\langle 0|A_\mu|\pi\rangle = q_\mu f_\pi. \tag{11.2.3}$$

Now, taking the divergence of the axial current in K_μ and setting it equal to zero (for massless pions), we find

$$2G_A(0)M_N + g_{NN\pi}f\pi = 0, \tag{11.2.4}$$

which is the Goldberg–Treiman relation.

Coming now to supersymmetry we take the matrix element of the supersymmetry current $S_{\mu,a}$ (see Chapter 9) between a fermionic and bosonic state

$$\langle B(P_1)|S_{\mu,a}|F(P_2)\rangle = f_1(q^2)(\bar\sigma_\mu\psi)_a + (P_1 + P_2)_\mu\psi_a f_2(q^2) + q_\mu\psi_a f_3(q^2). \tag{11.2.5}$$

If we assume the supersymmetry to be spontaneously broken, with χ being the associated Goldstino, then eq. (11.1.10) tells us that the matrix element has a fermion pole at $q = 0$ and the pole actually occurs in the function $f_1(q^2)$ and we have

$$f_1(q^2) = \frac{g_{FB_x}f_x}{q} + \text{continuum}. \tag{11.2.6}$$

Now using current conservation, as before, we conclude that

$$g_{FB_x}f_x + q \cdot (p_1 + p_2)f_2 + q^2 f_3 = 0. \tag{11.2.7}$$

(i) If F and B are single-particle states being members of the same supermultiplet, then $q(p_1 + p_2) = M_F^2 - M_B^2$ and taking $q^2 \to 0$ as the limit, we find

$$M_F^2 - M_B^2 \simeq \frac{g_{FB_x}f_x}{f_2(0)}. \tag{11.2.8}$$

This implies that spontaneous breaking of supersymmetry actually lifts the boson–fermion mass degeneracy within a multiplet and as symmetry breaking disappears, i.e., $f_x \to 0$, mass degeneracy is restored.

(ii) If F and B are not single-particle states but multiparticle states, then $g_{FB_x} \equiv A_{FB_x}$, i.e., a scattering amplitude involving Goldstino in the initial state (or final state). Then eq. (11.2.7) is the constraint on that scattering amplitude analogous to the Adler zero for the pion scattering amplitude. To see the usefulness of this formula, let us assume that the neutrino is a Goldstino. Equation (11.2.7) becomes, then, a constraint on all weak decay amplitudes involving the neutrino. It implies that the β-decay amplitudes must vanish as neutrino momentum goes to zero; but we know that it does not behave like that, which means that we cannot interpret neutrino as a Goldstino [4].

§11.3   *D*-Type Breaking of Supersymmetry

Breaking supersymmetry by D terms was first suggested by Fayet–Illiopoulos [3]. The basic idea is that, if the theory has an abelian U(1) gauge invariance, the Lagrangian can include (apart from the terms described earlier) a term linear in the gauge superfield V. As an example consider

$$S = \tfrac{1}{4} \int d^2\theta W^a W_a + \tfrac{1}{4} \int d^2\bar{\theta}\bar{W}_{\dot{a}}\bar{W}^{\dot{a}} + \int [\Phi_+^\dagger e^{gV}\Phi_+ - 2kV]d^4\theta$$

$$+ \int [\Phi_-^\dagger e^{-gV}\Phi_-]d^4\theta + m\int \Phi_+\Phi_- d^2\theta + m\int \Phi_+^\dagger\Phi_-^\dagger d^2\bar{\theta}, \quad (11.3.1)$$

where we have coupled two chiral matter fields $\Phi_\pm$ with equal and opposite U(1) quantum numbers to the U(1) gauge field V. The scalar potential in this model arises from the following term:

$$V = -(\tfrac{1}{2}D^2 + |F_+|^2 + |F_-|^2) - kD - \frac{g}{2}D(A_+^* A_+ - A_-^* A_-)$$

$$- (F_+^* A_+^* - F_-^* A_-^* + F_+ A_+ + F_- A_-). \quad (11.3.2)$$

The field equations for D and $F_\pm$ are given by

$$D + \frac{g}{2}(A_+^* A_+ - A_-^* A_-) + k = 0,$$

$$F_+ + mA_+^* = 0,$$

$$F_- + mA_-^* = 0. \quad (11.3.3)$$

Using this we can rewrite the potential V as follows:

$$V = \tfrac{1}{2}\left\{ k + \frac{g}{2}(A_+^* A_+ - A_-^* A_-)^2 \right\} + m^2(A_+^* A_+ - A_-^* A_-)1$$

$$\equiv \tfrac{1}{2}k^2 \left(m^2 + \frac{kg}{2} \right) A_+^* A_+^* + \left(m^2 - k\frac{kg}{2} \right) A_+^* A_-^*$$

$$+ \tfrac{1}{8}g^2(A_+^* A_+ - A_-^* A_-)^2. \quad (11.3.4)$$

Now, we note that, if $m^2 \pm kg/2 \geq 0$, the minimum of this potential corresponds to $\langle 0|H|0 \rangle > 0$, which means supersymmetry is spontaneously broken. Since $\langle D \rangle \neq 0$ for $\langle F_\pm \rangle = 0$, this is a D-type breaking with $\langle D \rangle = -k$. Since, in this situation, the gaugino field $\lambda \to \lambda + \delta\lambda$ with $\delta\lambda = \epsilon k$, it must remain massless and correspond to the Goldstino. To see the manifestation of supersymmetry breaking in the particle spectrum, note that the fermionic fields of the superfield $\Phi_\pm$ have equal mass $M_{\psi_\pm} = m$ whereas their superpartners $M_{A_\pm}$ have the following masses:

$$m_{A_\pm} = \sqrt{m^2 \pm \frac{kg}{2}}. \quad (11.3.5)$$

In this case $f_x = k$ and $|g_{FBx}| = g/2$, i.e., the gauge coupling. It is worth noting that, since the minimum corresponds to $A_\pm = 0$, the gauge invariance is unbroken even though supersymmetry is broken.

§11.4   O'Raifeartaigh Mechanism or F-Type Breaking of Supersymmetry

We first study this case without any reference to gauge symmetries and subsequently we will include the gauge fields. The most trivial example of this type is to consider a singlet chiral superfield X with the superpotential

$$W = m^2 X. \tag{11.4.1}$$

The potential for this case is given by

$$V = m^4 > 0 \text{ for } m \neq 0 \tag{11.4.2}$$

and $F = m^2$; thus supersymmetry is broken. The Goldstino in this case is ψ_x—the fermion field in X.

To discuss a somewhat nontrivial example we consider the model of the previous section but augment it with the inclusion of two singlets X and Y. Let us consider the following superpotential:

$$W = \lambda_1 X (\Phi_+ \Phi_- - m^2) + \lambda_2 Y \Phi_+ \Phi_-. \tag{11.4.3}$$

This superpotential has a U(1) symmetry under which Φ_+ and Φ_- have equal and opposite charges and X and Y are neutral. The various F terms in this case are the following:

$$F_X^* = \lambda_1 (A_+ A_- - m^2), \tag{11.4.4a}$$

$$F_Y^* = \lambda_2 A_+ A_-, \tag{11.4.4b}$$

$$F_+^* = (\lambda_1 A_X + \lambda_2 A_Y) A_-, \tag{11.4.4c}$$

$$F_-^* = (\lambda_1 A_X + \lambda_2 A_Y) A_+. \tag{11.4.4d}$$

The value of the potential energy of the ground state is determined by $V = \Sigma_i |F_i|^2$, where $i = X, Y, +, -$. The fields will choose that value for which V is minimum. It is clear that, at the minimum, $F_+ = F_- = 0$, which has two solutions

$$\lambda_1 A_x + \lambda_2 A_Y = 0 \tag{11.4.5a}$$

or

$$A_+ = A_- = 0. \tag{11.4.5b}$$

For the second choice $F_Y = 0$ but $F_x \neq 0$. If both A_+, $A_- \neq 0$, then (11.4.5a) must hold and the fields A_+ must be such so as to minimize V, i.e.,

$$\frac{\partial V}{\partial A_+} = \frac{\partial V}{\partial A_-} = 0. \tag{11.4.6}$$

This implies

$$\lambda_1 (A_+ A_- - m^2) + \lambda^2 A_+ A_- = 0, \tag{11.4.7}$$

since W is symmetric in A_+ and A_-, at the ground state $\langle A_+ \rangle = \langle A_- \rangle = k$, where

$$k^2 = \frac{\lambda_1^2 m^2}{(\lambda_1^2 + \lambda_2^2)} \qquad (11.4.8)$$

and $\langle V \rangle_{\text{vac}} \neq 0$ and both supersymmetry and the U(1) symmetry of W are broken.

To study the mass spectrum of the model, and isolate the Goldstino and Goldstone bosons, we have to write down the Yukawa coupling and V, and the resulting fermion and boson masses at $\langle A_\pm \rangle_{\text{vac}} = k$ and $\langle A_X \rangle_{\text{vac}} = \langle A_Y \rangle_{\text{vac}} = 0$:

$$\mathcal{L}_Y = \lambda_1(\psi_X \phi_+ A_- + \psi_- \psi_X A_+ + \psi_+ \psi_- A_X) \\ + \lambda_2(\psi_Y \psi_+ A_- + \psi_- \psi_Y A_+ + \psi_+ \psi_- A_Y). \qquad (11.4.9)$$

At the minimum, we find that $[1/(\lambda_1^2+\lambda_2^2)](\lambda_1\psi_X+\lambda_2\psi_Y)$ and $(1/\sqrt{2})(\psi_+ + \psi_-)$ form a Dirac spinor with mass $(\lambda_1^2 + \lambda_2^2)k$. The remaining fermions $\chi \equiv (\lambda_2\psi_X - \lambda_1\psi_Y)/\lambda_1^2 + \lambda_2^2$ and $(\psi_+ - \psi_-)/\sqrt{2} \equiv 0$ remain massless. Since under supersymmetry $\delta\psi_X \to \langle F_X \rangle \varepsilon$ and $\delta\psi_Y \to \varepsilon \langle F_Y \rangle$ where ε is the constant spinor, the massless fermion χ is the Goldstino and θ is an additional massless fermion, which we will soon see is the fermionic partner of the Goldstone boson corresponding to spontaneous breakdown of the U(1) symmetry.

From the bosonic mass spectrum we infer that $G = (\text{Im } A_+ - \text{Im } A_-)/\sqrt{2}$ and

$$G \equiv \frac{\text{Im } A_+ - \text{Im } A_-}{\sqrt{2}}$$

and

$$R = \frac{\text{Re } A_+ - \text{Re } A_-}{\sqrt{2}} \qquad (11.4.10)$$

has zero mass. Clearly, G is the Nambu–Goldstone boson corresponding to U(1) symmetry breakdown.

The field R is also a massless particle that exists to complete the supersymmetric multiplet corresponding to G and θ, i.e.,

$$(R + iG, \theta, F') \qquad (11.4.11)$$

form a massless supermultiplet. The particle R, henceforth, will be called a quasi-Nambu–Goldstone (QNG) boson, and θ the QNG fermion.

To study this model further we make U(1) a local symmetry, so that there is a Higgs phenomenon along with supersymmetry breaking. Let us denote the components of the gauge multiplet by (V_μ, λ, D). As is clear, from Chapter 2 on the Higgs mechanism, the massless boson G becomes the longitudinal mode of the gauge boson $V\mu$. To see the impact of supersymmetry let us see what happens to the gaugino. As discussed in eq. (10.2.4) the gaugino couples to the matter chiral multiplet as follows:

$$\mathcal{L}_\lambda = \frac{ig}{\sqrt{2}}(\lambda\psi_+ A_+^* - \lambda\psi_- A_-^*) + \text{h.c.} \tag{11.4.12}$$

On substituting the vacuum expectation values for $A_\pm$, this leads to a fermion bilinear as follows:

$$\mathcal{L}_\lambda = igk\lambda\theta + \text{h.c.} \tag{11.4.13}$$

Thus, the gaugino acquires a Dirac mass by "eating" up the superpartner of the Goldstone boson field and its mass is the same as the mass of the gauge boson after the Higgs mechanism. This will be a general feature of supersymmetric gauge theories with spontaneous symmetry breaking. The gaugino picking up mass is independent of the question of supersymmetry breaking. We also note that the quasi-Nambu–Goldstone boson field R now acquires mass gk from the D term of the Lagrangian. Thus, the three components of the massive gauge field V_μ, and massive R field make up the four real massive superpartners of the massive four-component Dirac spinor $(\lambda, \bar{\theta})$.

§11.5   A Mass Formula for Supersymmetric Theories and the Need for Soft Breaking

As we have argued at the beginning of this chapter, if supersymmetry is to have a role in understanding the world of quarks and leptons, it must be a broken symmetry because no fermions and bosons are observed with degenerate mass. It is for this reason that we started the study of spontaneous breakdown of supersymmetry. In this section we show that the spontaneous breaking of supersymmetry by F or D terms does not lead to an acceptable particle spectrum. To show this, we derive a mass formula relating the bosons and fermions in a supersymmetric theory with or without spontaneous symmetry breaking. We will do this in two steps: first, we consider a theory without gauge fields; and second, we will generalize the formula by adding the gauge fields.

Let us consider a set of chiral fields $\Phi_a, a = 1, \ldots, N$, and consider a superpotential $W(\Phi_a)$ that is an arbitrary polynomial (usually of degree ≤ 3), which is an analytic function of Φ_a. The Lagrangian can then be written as [where (A, ψ) denote the scalar and fermion fields]

$$\mathcal{L} = \mathcal{L}_{\text{K.E.}} + \mathcal{L}_Y - V, \tag{11.5.1}$$

where

$$\mathcal{L}_Y = \sum_{a,b} \frac{\partial W}{\partial A_a \partial A_b} \psi_a \psi_b \tag{11.5.2}$$

and

$$V = \sum_a \left(\frac{\partial W}{\partial A_a} \right) \left(\frac{\partial W^*}{\partial A_a^*} \right). \tag{11.5.3}$$

Taking the second derivative of V with respect to the bosonic fields, we get the general boson mass matrix as a function of fields

$$M_0^2 = \begin{matrix} & A_b^* & A_b \\ A_a & \\ A_a^* & \end{matrix} \begin{pmatrix} \frac{1}{2} W_{ac}'' W''^{cb^*} & W_{abc}'' W'^{c*} \\ W'''^{abc^*} W_c' & \frac{1}{2} W_{ac}'' W''^{cb^*}. \end{pmatrix}. \tag{11.5.4}$$

From eq. (11.5.2) we get the fermion mass matrix

$$(M_{1/2})_{ab} = W_{ab}''. \tag{11.5.5}$$

It follows from the above two equations that (choosing all parameters in the Lagrangian to be real)

$$T_R(M_0^2 - M_{1/2}^2) = 0. \tag{11.5.6}$$

Let us now couple the chiral fields to a general non-abelian gauge field denoted by (V_μ, λ, D). The gauge kinetic term contributes to M_0^2, as well as to $M_{1/2}$, via the gaugino matter field coupling. For simplicity of discussion we see the gauge coupling to one (i.e., $g/2 = 1$). We can then write

$$\mathcal{L}_Y = \sum_{a,b} W_{ab}'' \psi_a \psi_b + i\sqrt{2} \lambda_i \psi_a T_i A_a^* + \text{h.c.} \tag{11.5.7}$$

and

$$V = \sum_a W_a' W'^{,a^*} + \tfrac{1}{2}(A^* T_i A)^2. \tag{11.5.8}$$

The diagonal elements of M_0^2 then get modified to

$$(M_0^2)_{ab^*} = \tfrac{1}{2} W_{ac}'' W''^{cb^*} + \tfrac{1}{2}(T_i)_a^b A^* T_i A + \tfrac{1}{2}(T_i A)^b (A^* T_i)_a, \tag{11.5.9}$$

$$M_{1/2} = \begin{matrix} & \psi & \lambda \\ \psi & \\ \lambda & \end{matrix} \begin{pmatrix} W_{ab}'' & \sqrt{2}(A^* T_i) \\ \sqrt{2} T_i A & 0 \end{pmatrix}. \tag{11.5.10}$$

The mass matrix for the gauge bosons is

$$M_{ij}^2 = (A^* T_i, T_j A). \tag{11.5.11}$$

It is clear from this that, using trace condition $\text{Tr} T_i = 0$ for non-abelian groups, we obtain

$$\text{Tr} (M_0^2 - M_{1/2}^2 + 3M_1^2) = 0. \tag{11.5.12}$$

This mass formula was derived by Ferrara, Girardello, and Palumbo [5] in 1979 and has important implications for model building. The generality of the mass formula implies that, regardless of whether the supersymmetry is broken by F terms or whether the internal symmetry is broken, all the boson masses cannot be heavier than the fermion masses as would be required for useful model building. If there is a linear D term present in the theory, the right-hand side of eq. (11.5.12) is proportional (becomes proportional) to $D_i \mathrm{Tr}\,(T_i)$, which may improve the situation somewhat except that, in order to cancel gauge anomalies, it is often desirable to have $\mathrm{Tr}\,T_i = 0$ so that, again in realistic gauge models, eq. (11.5.12) is the actual constraint.

Therefore, spontaneous breaking of supersymmetry would appear to be too restrictive a requirement for model building. We may argue that radiative corrections [6] may induce changes in the above mass formula but studies of semirealistic models in this regard do not appear very promising. Therefore, we will now proceed to a discussion of soft breaking of supersymmetry.

As we saw in the previous chapter, requirements of renormalizability allow us to introduce only three classes of soft-breaking terms.

(i) Scalar mass terms, $\mu^2 A^* A$-type.

(ii) Trilinear scalar interactions of type $W(\Phi)|_{\theta=0}$.

(iii) Mass terms for the gaugino, i.e., $\lambda\lambda$.

We will see in the next chapter that these three kinds of terms are enough to lead to realistic models for particle physics; in fact, without additional constraints they lead to a large proliferation of parameters.

Exercises

11.1. Reconsider Problem 10.4 to generate supersymmetry breaking in super-QED, not explicitly but spontaneously by appropriately modifying the theory. Consider both F- and D-type SUSY breaking mechanisms.

References

[1] J. Illiopoulos and B. Zumino, *Nucl. Phys.* **B76**, 310 (1974).

[2] L. O'Raifeartaigh, *Nuel. Phys.* **B96**, 331 (1975).

[3] P. Fayet and J. Illiopoulos, *Phys. Lett.* **51B**, 461 (1974).

[4] B. deWit and D. Freedman, *Phys. Rev. Lett.* **35**, 827 (1975);
W. Bardeen, unpublished.

[5] S. Ferrara, Ll Girardello, and F. Palumbo, *Phys. Rev.* **D20**, 403 (1979).

[6] For radiative corrections to supersymmetry breaking see
B. Zumino, *Nucl. Phys.* **B89**, 535 (1975);
S. Weinberg, *Phys. Lett.* **62B**, 111 (1976);

C. Nappi and B. A. Ovrut, *Phys. Lett.* **113B**, 175 (1982);

M Dine and W. Fischler, *Nucl. Phys.* **B204**, 346 (1982);

M Huq, *Phys. Rev.* **D14**, 3548 (1976);

E. Witten, Trieste lectures, 1981; *Nucl. Phys.* **B195**, 481 (1982).

12
Minimal Supersymmetric Standard Model

§12.1 Introduction, Field Content and the Lagrangian

In the previous three chapters we have laid the foundation for applying the ideas of supersymmetry to building models of particle physics. At present there exists a successful (at low energies) model of electroweak and strong interactions—the standard $SU(2)_L \times U(1)_Y \times SU(3)_c$ model. The discovery of W and Z bosons at the CERN $Sp\bar{p}S$ machine as well as the that of the top quark at the Fermilab accelerator has proved the correctness of the essential elements of this theory. As discussed in Chapter 3, the model however, leaves many questions unresolved. It is therefore widely believed that there is a considerable amount of new physics beyond the standard model that remains to be discovered. In Chapters 6, 7, and 8 we have discussed some interesting classes of models that provide examples of possible new physics that address one or other of these unresolved issues. Two of the key issues that are central to the discussion in this chapter are: (i) why is not the weak scale pushed to the Planck scale due to the quadratically divergent radiative corrections to the Higgs mass that determines the scale of weak interactions and (ii) what turns the Higgs mass-square parameter in the Lagrangian to be negative so that the Higgs potential leads to the breakdown of the electroweak symmetry? We will see in this chapter that supersymmetry can successfully address both these issues and is therefore a prime candidate for new physics at the TeV scale. The first has to with the mild divergence structure of the supersymmetric theories already discussed in Chapter 10, where it is pointed out that as long as supersymmetry is broken by soft terms, then the superpotential of the theory remains free of quadratic divergences, thereby providing stability to whatever mass term is chosen. We will therefore not provide any explicit demonstration of this in this chapter but instead write down the softly broken Lagrangian that has this property. Secondly, we will see that in many supersymmetric scenarios, the scale at which the soft breaking terms are determined is much higher than the weak scale. One must therefore use renormalization group equations to extrapolate the soft breaking terms to the weak scale. It turns out that due to the large mass of the top quark, this process of extrapolation can change the sign of the Higgs mass and lead to symmetry breaking.

To begin the discussion, we will present [1, 2] a supersymmetric extension of the standard model. All fermions and bosons of the standard model must be accompanied by their supersymmetric partners, which are bosons and fermions, respectively. Moreover, since supersymmetry commutes with the electroweak symmetry, the transformation properties of the known particles and their superpartners must be the same, and the part of the Lagrangian that breaks supersymmetry must be invariant under the local electroweak symmetry. These are the general guidelines that dictate the construction of supersymmetric models. In this chapter, we will use them to construct the minimal supersymmetric extension of the standard model (to be called MSSM below).

First, a word about the notation: in the rest of this chapter, we will call the supersymmetric partners of quarks, leptons, etc., squarks, sleptons, ... and denote them with a tilde over the symbol representing the corresponding particle. In Table 12.1 we give the particle spectrum along with their electroweak quantum numbers for one generation. We will also choose all particles to be left-handed (or chiral) particles so that a right-handed field (i.e., u_R) will be denoted as a left-handed antiparticle field (i.e., u_L^c). This is a convention in all discussions of supersymmetric models and we will adhere to this throughout the rest of the book.

The field content of the minimal supersymmetric standard model (to be called MSSM from now on) is the same as the standard model in the matter sector with the superpartners of all fields added. In the Higgs sector, however, two doublets are needed (instead of one in the standard model). The Higgs superfields are denoted as $H_u(2, 1, 1)$ and $H_d(2, -1, 1)$, where the numbers in the parenthesis represent the quantum numbers under the $SU(2)_L \times U(1)_Y \times SU(3)_c$ gauge group. The fields along with their superpartners are shown explicitly in Table 12.1.

The need for the extra Higgs field is because the fermionic partners of the Higgs fields induce gauge anomalies into the theory and the presence of the second Higgs field with opposite $U(1)_Y$ quantum number helps to cancel this anomaly. The second reason is that the holomorphy of the superpotential requires that it cannot contain any complex conjugate superfield. Therefore, owing to the different $U(1)_Y$ quantum numbers of the superfields u^c $(Y = -\frac{4}{3})$ and d^c $(Y = +\frac{2}{3})$, one needs two Higgs fields to give invariant Yukawa couplings of type $QH_u u^c$ and $QH_d d^c$ so that once the $H_{u,d}$ acquire v.e.v.s, the up and down quarks will acquire nonzero mass as required by observations.

We are now ready to write down the action for MSSM. We will use the notation of superfields to denote not only the supersymmetric part of the action but also the explicit soft-breaking terms. We will introduce the constant superfield $\eta = \theta^2$ to write the latter terms. We split the action into two parts:

$$S = S_0 + S_1, \tag{12.1.1}$$

$$S_0 = S_g + \int W d^2\theta + \text{h.c.}, \tag{12.1.2}$$

where S_g represents the gauge part of the action which can be written down following the rules in Chapter 10 and we will give this in component notation soon. (Again, we carry out our discussion for one generation for simplicity.)

$$W = h_u Q^T \tau_2 H_u u^c + h_d Q^T \tau_2 H_d d^c + h_e L^T \tau_2 H_d e^c + \mu(H_u^T \tau_2 H_d). \tag{12.1.3}$$

It is important to emphasize at this stage that we have not included in the superpotential all possible terms allowed by the electroweak symmetry. For instance, there are the so-called R-parity-violating terms that are allowed by both gauge invariance and supersymmetry such as

$$W = MLH_u + \lambda_{ijk} L_i L_j e^c_k + \lambda'_{ijk} Q_i L_j d^c_k + \lambda''_{ijk} u^c_i d^c j d^c_k,$$

which lead to lepton and baryon number violation. They are considered an undesirable feature of the MSSM and are generally dropped in the belief that the embeddings of MSSM in higher unified theories will prevent them from appearing.

The soft-breaking part of the action, S_1, is chosen to include all allowed terms that break supersymmetry softly

$$S_1 = \int d^2\theta \eta [A_0 W' + M_2 \mathbf{W}^a \mathbf{W}_a + M_1 W_B W_B + M_3 W_G W_G] + \text{h.c.}$$
$$+ d^4\theta \eta^\dagger \eta [m_Q^2 Q^* Q + m_U^2 U^* U + m_D^2 D^* D + m_L^2 L^* L + m_E^2 E^* E],$$
$$\tag{12.1.4}$$

where $\eta = \theta^2$ and the terms in W' have the same structure as W but different coupling coefficients. As we will see below, all but the first term within the bracket are needed to give arbitrary masses to the superpartners. Using the component field notation of Table 12.1 we can write down the Lagrangian for this model

$$\mathcal{L} = \mathcal{L}_{\text{gauge}} + \mathcal{L}_{\text{matter}} + \mathcal{L}_Y - V + \mathcal{L}_1. \tag{12.1.5}$$

Table 12.1.

Superfield	Component fields	$SU(2)_L \times U(1) \times SU(3)$ quantum number	Name
	Matter fields		
Q	$\begin{pmatrix} u_L \\ d_L \end{pmatrix} \equiv Q_L$	$(2, \frac{1}{3}, 3)$	Quark
	$\begin{pmatrix} \tilde{u}_L \\ \tilde{d}_L \end{pmatrix} \equiv \tilde{Q}$		Squark
u^c	u_L^c	$(1, -\frac{4}{3}, 3^*)$	Quark
	$\tilde{u}_L^c$		(denotes right-handed up-quark)
d^c	d_L^c	$(1, +\frac{2}{3}, 3^*)$	Squark
	$\tilde{d}_L^c$		(denotes right-handed down-quark)
L	$\begin{pmatrix} \nu_L \\ e_L^- \end{pmatrix}$	$(2, -1, 1)$	Lepton
	$\begin{pmatrix} \tilde{\nu}_L \\ \tilde{e}_L \end{pmatrix}$		Slepton
e^c	e_L^c	$(1, +2, 1)$	Antilepton
	$\tilde{e}^c$		Antislepton
Gauge fields			
V	$\begin{pmatrix} W^\pm \\ W^3 \end{pmatrix}$	$(3, 0, 1)$	Gauge bosons
	$\begin{pmatrix} \tilde{W}^\pm \\ \tilde{W}^3 \end{pmatrix}$		Wino
B	B	$(1, 0, 1)$	Gauge bosons
	$\tilde{B}$		Bino
G	G	$(1, 0, 8)$	Gluon
	$\tilde{G}$		Gluino
	Higgs fields		
H_u	$\begin{pmatrix} \phi_u^+ \\ \phi_u^0 \end{pmatrix}$	$(2, +1, 1)$	Higgs field
	$\begin{pmatrix} \tilde{\phi}_u^+ \\ \tilde{\phi}_u^0 \end{pmatrix}$		Higgsino
H_d	$\begin{pmatrix} \phi_d^0 \\ \phi_d^- \end{pmatrix}$	$(2, -1, 1)$	Higgs field
	$\begin{pmatrix} \tilde{\phi}_d^0 \\ \tilde{\phi}_d^- \end{pmatrix}$		Higgsino

We will have the familiar four-component notation

$$\mathcal{L}_{\text{gauge}} = -\tfrac{1}{4}\mathbf{W}^{\mu\nu} \cdot \mathbf{W}_{\mu\nu} - \tfrac{1}{4}B_{\mu\nu}B^{\mu\nu} - i\tfrac{1}{2}\bar{\tilde{W}}\gamma \cdot \nabla\tilde{W}$$
$$- i\tfrac{1}{2}\bar{\tilde{B}}\gamma \cdot \partial\tilde{B}i\tfrac{1}{2}\bar{\tilde{G}}\gamma \cdot \partial\tilde{G}, \tag{12.1.6a}$$
$$\mathcal{L}_{\text{matter}} = \Sigma - \bar{\psi}\gamma_\mu(D_\mu)\psi - \Sigma(D^\mu A_\varphi)^+(D_\mu A_\varphi) + i\frac{g}{\sqrt{2}}\Sigma\bar{\psi}_L\tau \cdot \tilde{\mathbf{W}}\tilde{\psi}$$

$$+ i\frac{g'}{\sqrt{2}}\Sigma\bar{\psi}_L\tilde{B}Y\tilde{\psi} + i\frac{g_3}{\sqrt{2}}\sum_{\text{quarks}} \bar{\psi}\lambda_a\tilde{G}^a\tilde{\psi} + \text{h.c.} \tag{12.1.6b}$$

Summation goes over $\psi = Q, u^c, d^c, L, e^c, \tilde{H}_u, \tilde{H}_d$, and

$$\tilde{\psi} = \tilde{Q}, \tilde{u}^c, \tilde{d}^c, \tilde{L}, \tilde{e}^c, H_u, H_d,$$
$$D_\mu = \partial_\mu - \frac{ig}{2}\tau \cdot \mathbf{W}_\mu - \frac{ig'}{2}YB_\mu, \tag{12.1.6c}$$

Y being the appropriate U(1)$_Y$ quantum number. The $\tau \cdot \mathbf{W}$ term will be absent when a particle is SU(2)$_L$ singlet. Similarly, for a field that carries SU(3)-color quantum number such as quarks, the covariant derivative will contain a term $-i\tfrac{1}{2}\lambda_a G_\mu$. ∇ denotes the covariant derivative for the appropriate gauge fields

$$\mathcal{L}_Y = h_u(Q_L^T C^{-1}\tau_2 H_u u_L^c + Q_L^T C^{-1}\tau_2 \tilde{H}_u \tilde{u}^c + \tilde{u}^c \tilde{H}_u^T C^{-1}\tau_2\tilde{Q}) + (u \to d)$$
$$+ h_e(L^T C^{-1}\tau_2 H_d e^c + L^T C^{-1}\tau_2 \tilde{H}_d\tilde{e}_c + \tilde{H}_d^T C^{-1}\tau_2\tilde{L}e_L^c)$$
$$+ \mu\tilde{H}_u^T C^{-1}\tau_2\tilde{H}_d + \text{h.c.} \tag{12.1.7}$$

Here C is the Dirac charge conjugation matrix and τ_2 is the second Pauli matrix. We identify the first terms in each of the first three bracketed expressions in eq. (12.1.7) as the Yukawa couplings present in a two Higgs extension of the standard model (see Chapter 3). The remaining term involving fermions is the soft-supersymmetry-breaking Majorana mass term, i.e.,

$$\mathcal{L}_{\text{soft}} = M_2\tilde{\mathbf{W}}^T C^{-1}\tilde{\mathbf{W}} + M_1\tilde{B}^T C^{-1}\tilde{B} + M_3\tilde{G}^T C^{-1}\tilde{G} + \text{h.c.} \tag{12.1.8}$$

Let us now turn to the potential

$$V = |F|^2 + \tfrac{1}{2}D^2 + V_{\text{soft}}, \tag{12.1.9}$$
$$|F|^2 = |h_u\tilde{Q}\tilde{u}^c + \mu H_d|^2 + |h_d\tilde{Q}\tilde{d}^c + h_e\tilde{L}\tilde{e}^c + \mu H_u|^2$$
$$+ |h_u H_u\tilde{u}^c + h_d H_d\tilde{d}^c|^2 + h_u^2|\tilde{Q}^T\tau_2 H_u|^2 + h_d^2|\tilde{Q}^T\tau_2 H_d|^2$$
$$+ h_e^2(H_d^+ H_d\tilde{e}^{c^*}\tilde{e}^c + |\tilde{L}^T\tau_2 H_d|^2) + \lambda^2|H_u^T\tau_2 H_d - \mu^2|^2, \tag{12.1.10a}$$
$$V_{\text{soft}} = A_u\tilde{Q}^T\tau_2 H_u\tilde{u}^c + A_d\tilde{Q}^T\tau_2 H_d\tilde{d}^c + A_\ell\tilde{L}^T\tau_2 H_d\tilde{e}^c$$
$$+ \mu B H_u^T\tau_2 H_d + \text{h.c.} + m_Q^2\tilde{Q}^+\tilde{Q} + m_L^2\tilde{L}^+\tilde{L} + m_u^2\tilde{u}^{c^*}\tilde{u}^c$$
$$+ m_D^2\tilde{d}^{c^*}\tilde{d}^c + m_E^2\tilde{e}^{c^*}\tilde{e}^c, \tag{12.1.10b}$$

$$\tfrac{1}{2}D^2 = \tfrac{1}{8}g^2\sum_a \left|\sum_\psi \tilde{\psi}^\dagger\tau_a\tilde{\psi}\right|^2 + \tfrac{1}{8}g'^2\left|\sum_\psi \tilde{\psi}^\dagger Y\tilde{\psi}\right|^2. \tag{12.1.10c}$$

Let us now study the spontaneous breaking of the gauge symmetry. A look at eqs. (12.1.9) and (12.1.10) makes it clear that for m_Q^2, m_L^2, m_U^2, m_D^2, m_E^2 positive, at the minimum of the potential V, the squarks and sleptons have zero v.e.v., which corresponds to

$$\langle \tilde{Q} \rangle = \langle \tilde{L} \rangle = \langle \tilde{e}^c \rangle = \langle \tilde{u}^c \rangle = \langle \tilde{d}^c \rangle = 0. \tag{12.1.11}$$

In order to find the vacuum expectation values (v.e.v.) of the Higgs fields, we have to do some more analysis. The part of the potential involving H_u and H_d is given by:

$$V_{\text{tree}}(H_u, H_d) = m_1^2 |H_u|^2 + m_2^2 |H_d|^2 - m_3^2 (H_u H_d + \text{h.c.})$$
$$+ \frac{g^2 + g'^2}{8} (|H_u|^2 - |H_d|^2) + g^2 |H_u^\dagger H_d|^2, \tag{12.1.12}$$

where $m_i^2 = m_{H_u}^2 + \mu^2$, $m_2^2 = m_{H_d}^2 + \mu^2$, and $m_3^2 = -B\mu$. Note the first difference from the standard model, i.e., the quadratic terms in the Higgs field have a coupling strength given by known gauge couplings rather than an unknown self-scalar coupling λ. As we will see shortly, this gives an upper limit on the Higgs mass in the MSSM.

To find the minimum of the potential, we give vacuum expectation values to the neutral components of the $H_{u,d}$ and call them $v_{u,d}$, respectively. The extremization conditions can then be written as:

$$\frac{1}{2} \frac{\partial V}{\partial H_u} = m_1^2 v_u - m_3^2 v_d + \frac{g^2 + g'^2}{4} (v_u^2 - v_d^2) v_u - 0,$$

$$\frac{1}{2} \frac{\partial V}{\partial H_d} = m_2^2 v_d - m_3^2 v_u - \frac{g^2 + g'^2}{4} (v_u^2 - v_d^2) v_d = 0. \tag{12.1.13}$$

Let us choose the parameterization $v_u = v \sin \beta$ and $v_d = v \cos \beta$ ($v = 174.5$ GeV). Using the fact that $M_Z^2 = \frac{g^2 + g'^2}{2} v^2$, we can write the solutions of the above equations as:

$$\frac{M_Z^2}{2} = \frac{m_{H_u}^2 - m_{H_d}^2 \tan^2 \beta}{\tan^2 \beta - 1} - \mu^2$$

$$\sin 2\beta = \frac{m_3^2}{m_1^2 + m_2^2}. \tag{12.1.14}$$

Note that if $m_1^2 = m_2^2$, then the first of the above two conditions cannot be satisfied and therefore there will be no symmetry breaking. In fact the conditions for symmetry breaking to occur are:

$$m^2 + m_2^2 > 2m_3^2; m_1^2 m_2^2 < m_3^4. \tag{12.1.15}$$

After symmetry breaking, the physical Higgs fields that survive are:
(i) a charged Higgs bosons, $H^+ = \sin \beta H_u^+ + \cos \beta H_d^+$;
(ii) two CP-even neutral Higgs bosons: $h = -\sin \alpha ReH_u^0 + \cos \alpha ReH_d^0$ and $H = \cos \alpha ReH_u^0 + \sin \alpha ReH_d^0$. The first of these is lighter and as the supersymmetry becomes higher, it becomes the standard model Higgs field.

(iii) one CP-odd neutral Higgs field A.

To express their masses in terms of the symmetry, breaking parameters, let us write down the mass matrix of the different fields using the notation where the fields are parametrized as $H_{u,d}^0 = v_{u,d} + \frac{\sigma_{u,d} + i\chi_{u,d}}{\sqrt{2}}$:

For σ fields, we have

$$M_\sigma^2 = \frac{1}{2} \begin{pmatrix} m_3^2 \cot \beta + M_Z^2 \sin^2 \beta & -m_3^2 - M_Z^2 \cos \beta \sin \beta \\ -m_3^2 - M_Z^2 \cos \beta \sin \beta & m_3^2 \tan \beta + M_Z^2 \cos^2 \beta \end{pmatrix} \quad (12.1.16)$$

For the χ fields, we have

$$M_\chi^2 = \begin{pmatrix} m_3^2 \cot \beta & m_3^2 \\ m_3^2 & m_3^2 \tan \beta \end{pmatrix}. \quad (12.1.17)$$

For the charged fields we have

$$M_+^2 = (m_3^2 + M_W^2 \cos \beta \sin \beta) \begin{pmatrix} \cot \beta & 1 \\ 1 & \tan \beta \end{pmatrix}. \quad (12.1.18)$$

Diagonalizing these matrices, we see that $M_A^2 = \frac{m^2}{\cos \beta \sin \beta}$; $M_{H+}^2 = M_W^2 + \frac{m_3^2}{\cos \beta \sin \beta}$ and the two CP-even neutral states have masses:

$$M_{H,h}^2 = \frac{1}{2} \left[M_A^2 + M_Z^2 \pm \sqrt{(M_A^2 + M_Z^2)^2 - 4M_A^2 M_Z^2 \cos^2 2\beta} \right]. \quad (12.1.19)$$

For large values of M_A compared to M_Z, we get for the lighter of the two CP-even neutral Higgs field masses $M_h^2 \simeq M_Z^2 \cos^2 2\beta$ implying that the lightest Higgs field mass in the MSSM is less than the mass of the Z boson at the tree level. This is a very interesting prediction of the MSSM, which has been a hallmark in the pursuit of supersymmetry.

Because the light Higgs mass is such an important prediction of the MSSM, radiative corrections to this have been calculated [3]:

$$M_h^2 \simeq M_Z^2 \cos^2 2\beta + \frac{3g_2^2 m_t^4}{16\pi^2 M_W^2} \log \frac{m_{\tilde{t}_1}^2 m_{\tilde{t}_2}^2}{m_t^4} \quad (12.1.20)$$

where $\tilde{t}_{1,2}$ are the two stop mass eigenstates, which are linear combinations of $\tilde{t}$ and $\tilde{t}^c$. This can lead to a mass of the light Higgs boson to near 135 GeV or so for plausible values of the parameters of the MSSM. For further details of this discussion see Ref. 4.

Fermion Spectrum

Let us now turn to the masses of those fermions that arise as supersymmetric partners of known particles. By our choice of soft-breaking terms, the winos ($\tilde{W}$), bino ($\tilde{B}$) already have a Majorana mass. The Higgsino masses, on the other hand, cannot be put in as part of the soft-breaking Lagrangian because they spoil renormalizability of the theory. To study their mass spectra, subsequent to spontaneous breakdown, we look at the last terms in eq. (12.1.6b) and find the following fermion bilinears in the Lagrangian:

$$\mathcal{L}_{\text{mass}} = \frac{iv}{2}\left\{ \sin\beta\tilde{H}^0{}_u(g\tilde{W}^3 + g'\tilde{B}) - \cos\beta\tilde{H}^0{}_d(g\tilde{W}_3 + g'\tilde{B}) \right.$$
$$\left. + \sqrt{2}gv(\sin\beta\tilde{H}^-_u\tilde{W}^- + \cos\beta\tilde{H}^-_d\tilde{W}^+) \right\} + \text{h.c.},\qquad (12.1.21)$$

where

$$\tilde{W}^{\pm} = \frac{1}{\sqrt{2}}(\tilde{W}^1 \mp i\tilde{W}^2).$$

As mentioned in eq. (12.1.9) supersymmetry breaking brings in Majorana mass terms for the gauginos, which can be rewritten as follows:

$$\mathcal{L}_{\text{soft}} = M_2\tilde{W}^{+T}C^{-1}\tilde{W}^- + M_2\tilde{W}^{3T}C^{-1}\tilde{W}^3 + M_1\tilde{B}^TC^{-1}\tilde{B} + \text{h.c.} \quad (12.1.22)$$

Combining these, one can write the mass matrix for the neutral Higgsino–gaugino system in the basis $(\tilde{B}, \tilde{W}_3, \tilde{H}_d, \tilde{H}_u)$ as follows:

$$\begin{pmatrix} M_1 & 0 & -M_Z\cos\beta\sin\theta_W & M_Z\sin\beta\sin\theta_W \\ 0 & M_2 & M_Z\cos\beta\cos\theta_W & -M_Z\sin\beta\cos\theta_W \\ -M_Z\cos\beta\sin\theta_W & M_Z\cos\beta\cos\theta_W & 0 & -\mu \\ M_Z\sin\beta\sin\theta_W & -M_Z\sin\beta\cos\theta_W & -\mu & 0 \end{pmatrix}.$$
$$(12.1.23)$$

For charginos, the mass matrix is given in a basis where the rows are $(\tilde{W}^+, \tilde{H}_u^+)$ and the column are $(\tilde{H}_d^-, \tilde{W}^-)$:

$$M^+ = \begin{pmatrix} M_2 & \sqrt{2}M_W\sin\beta \\ \sqrt{2}M_W\cos\beta & \mu \end{pmatrix}. \qquad (12.1.24)$$

Finally, we can explicitly write down the $SU(3)_c$ sector. This involves the interactions of the quarks and squarks with gluons, G, and gluinos, $\tilde{G}$. The general form of their interaction is similar to the electroweak gauge interactions with appropriate modifications, i.e.,

$$\mathcal{L}^{\text{color}}_{\text{gauge+matter}} = -\tfrac{1}{4}G^i_{\mu\nu_j}G^j_{\mu\nu_i} - \bar{\tilde{G}}^i_j\gamma\cdot(\nabla_G\tilde{G})^j_i - \bar{Q}^i\gamma_\mu D^j_{\mu,i}Q_j - (D_\mu\tilde{Q})^{+i}(D_\mu\tilde{Q})_i$$
$$+ \frac{ig_S}{\sqrt{2}}\bar{\tilde{Q}}^i(\tilde{G})^j_iQ_j + \text{h.c.} + \tfrac{1}{4}(\tilde{Q}^{+i}\tilde{Q}_j)(\tilde{Q}^{\perp j}\tilde{Q}_i)$$
$$+ \text{similar terms replacing } Q \text{ by } U, D. \qquad (12.1.25)$$

Without supersymmetry breaking the gluino fields have no mass. As in the previous case, the supersymmetry breaking introduces Majorana masses for the gluinos, $\tilde{G}$, which we choose as $m_{\tilde{G}}$. By $SU(3)_C$ invariance all eight gluinos have the same Majorana mass

$$\mathcal{L}_{\tilde{G}} = m_{\tilde{G}}(\tilde{G}^{T_i}_jC^{-1}\tilde{G}^j_i + \text{h.c.}). \qquad (12.1.26)$$

§12.2   Constraints on the Masses of Superparticles

In this section we will discuss the phenomenological constraints on the masses of the superpartners of the "known" particles. (We emphasize that, by "known," we do not necessarily mean that a particle has been observed but simply that it is familiar to us from our study of nonsupersymmetric models.) We will use the Lagrangian discussed in the previous section as representing the broad features of the interactions between particles and superparticles, although we will let the parameters be arbitrary. An important point to remember is that the interaction between particles and their superpartners obeys a selection rule, even in the presence of the most general supersymmetry-breaking terms. Let us call it R parity [5]. It is given in terms of recognizable quantum numbers as $(-1)^{(3(B-L)+2S)}$. We can assign R-parity $+1$ to "normal" particles and -1 to their superpartners. Then all interaction vertices must respect R parity invariance, i.e., they must contain an even number of super- (tilded) particles. One way to break R-parity is to give a vacuum expectation value to the scalar neutrino [6] $\langle \tilde{\nu} \rangle \neq 0$. But this also gives rise to lepton number violation implying that $\langle \tilde{\nu} \rangle \leq 10^{-6} m_W$. Therefore, in general, its effects will be very small.

To get an idea about phenomenologically allowed masses for various s particles, we look at their effects on known processes, such as e^+e^- annihilation and low-energy weak processes. We consider them one by one and we will also assume that the lightest s particles are $\tilde{\nu}$, $\tilde{G}$, and $\tilde{\gamma}$. Before we proceed further it is also important to point out that the interactions of $\tilde{\gamma}$, $\tilde{\nu}$, and $\tilde{G}$ are all extremely weak. For instance, $\tilde{\gamma}$-interaction with matter proceeds through the diagrams in Fig. 12.1 and is given by [1]

$$\sigma_{\tilde{\gamma}- \text{ matt}} \approx 2 \times 10^{-37} (E_{\tilde{\gamma}} \text{ in G3V }) \left(\frac{m_W}{M_{\tilde{q}}} \right)^4 F \text{ cm}^2, \qquad (12.2.1)$$

where F is a structure factor. Thus, $\tilde{\gamma}$ interaction is similar to ν interaction in cross sectional though different in final-state characteristics. Thus, $\tilde{\gamma}$ interacts only very weakly with matter.

(a) Scalar Leptons

There are two kinds of charged scalar leptons: $\tilde{l}_L$, and $\tilde{l}_R$ ($l = e, \mu, \tau$). In general, supersymmetry breaking will mix these states. These states will be produced in e^+e^- annihilation: $e^+e^- \rightarrow \tilde{l}^+\tilde{l}^-$. Since these are spin-0 bosons, their cross section will be $\sim (\beta^3/4)\sigma(e^+e^- \rightarrow l^+l^-)$, where $\beta = v/c$, the P-ware threshold factor. Because of the β^3 factor, the rise in the ratio $R = \sigma(e^+e^- \rightarrow \tilde{l}^+\tilde{l}^-)/\sigma_{\text{tot}}$ will only be observable far above the threshold. However, in general,

$$\tilde{l}^\pm \rightarrow l^\pm \tilde{\gamma}. \qquad (12.2.2)$$

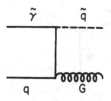

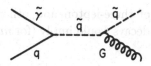

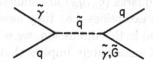

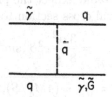

Figure 12.1.

Therefore, above the threshold for the production of $\tilde{l}^{\pm}$, we would expect the product of $l^{\pm}$ with a lot of missing energy similar to the product of $\tau^{\pm}$. Because $\tau^{\pm}$ is a calculable background, this can be calculated and subtracted. So far, no such events have been observed in e^+e^- colliders leading to bounds of 100 GeV on $\tilde{e}$ and $\tilde{\mu}$.

(b) Scalar Neutrino

In contrast to the superpartner of the charged lepton, there is only one superpartner of ν_L to be denoted by $\tilde{\nu}_L$. This, being a neutral particle, will be produced in e^+e^- collision only via the Z_0 intermediate state. Its cross section will therefore be extremely small ($\sim G_F^2$) and e^+e^- scattering does not lead to any bound on its mass. One way to obtain a lower bound on its mass is to study τ decays. If $m_{\tilde{\nu}_e} + m_{\tilde{\nu}_\tau} < (m_\tau - m_e)$, then there will be a supersymmetric contribution to the τ decay arising from the diagram shown in Fig. 12.2. For expected values of $\tilde{W}^{\pm}$ mass, this amplitude will

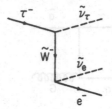

Figure 12.2.

be comparable to that of three-lepton and other decay modes. Present observations on τ-lepton decays would imply (for $m_{W^+} \simeq m_{\tilde{W}^+}$) that $m_{\tilde{\nu}_e} + m_{\tilde{\nu}_\tau} > (m_\tau - m_e)$.

(c) Scalar Quarks

The properties of scalar quarks $(\tilde{q}_L, \tilde{q}_R)$ are similar to those of scalar leptons, with two important exceptions, i.e., they have fractional charges and strong interactions. Again, in this discussion, we will assume photino mass to be less than $m_{\tilde{q}}$. This immediately implies that $\tilde{q} \to q\tilde{\gamma}$. If $m_{\tilde{G}} < m_{\tilde{q}}$, there exists the additional decay mode $\tilde{q} \to q\tilde{g}$. In this case the production of $\tilde{q}$ will be indicated by a quark jet with missing energy.

As in the case of sleptons, we would expect the production of squarks in e^+e^- annihilation, with differential cross section to be given by

$$\left(\frac{d\sigma}{d\Omega}\right)_{e^+e^- \to \tilde{q}\bar{\tilde{q}}} = \frac{3}{8}\frac{\alpha^2}{s}e_q^2\beta^3 \sin^2\theta, \tag{12.2.3}$$

where e_q is the quark charge, $\beta = \sqrt{1 - (4M_{\tilde{q}}^2/S)}$, and θ is the center-of-mass production angle.

Present experimental limits from e^+e^- collisions give 3.1 GeV $< M_{\tilde{q}} <$ 17.8 GeV and 7.4 GeV $< M_{\tilde{q}} <$ 16.0 GeV, assuming $\tilde{q} \to q\tilde{\gamma}$ to be the dominant decay mode.

Because the squarks carry strong-interaction quantum numbers, they can be produced in $p\bar{p}$ collision (through the Drell–Yan mechanism) through collisions of quarks, antiquarks, gluons, or quarks and gluons through the diagrams shown in Fig. 12.3. The quark produced in the final state will decay to $q + \tilde{\gamma}$ which would appear as a monojet in $p\bar{p}$ collision.

(d) Gluino Physics

Gluino $(\tilde{G})$ is the spin-1/2 superpartner of the gluon and is similar to the photino except that there are eight of them corresponding to the eight generators of $SU(3)_c$, and they have only strong interactions and no weak and electromagnetic interactions. We will again assume that it is more massive than $\tilde{\gamma}$. A dominant decay mode of the gluino is $\tilde{G} \to q\bar{q}\tilde{\gamma}$ arising from the Feynman diagram in Fig. 12.4. This is similar to a weak decay since $\Gamma \sim (\alpha\alpha_s M_{\tilde{G}}^5/M_{\tilde{q}}^4)$.

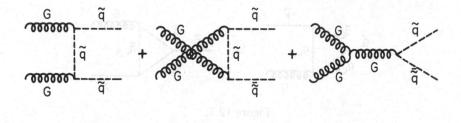

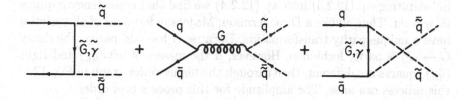

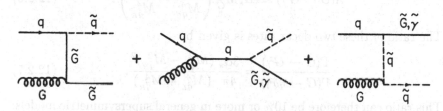

Figure 12.3.

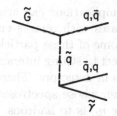

Figure 12.4.

An interesting decay mode of the gluino is $\tilde{G} \to G\tilde{\gamma}$. This violates both parity and charge conjugation. To see why it violates parity, let us note that both $\tilde{G}$ and $\tilde{\gamma}$ are Majorana particles that satisfy the condition

$$\psi = C\bar{\psi}^T,$$

where

$$\psi = \tilde{G}, \tilde{\gamma}. \tag{12.2.4}$$

Let

$$\psi \xrightarrow{P} \eta\gamma_4\psi. \tag{12.2.5}$$

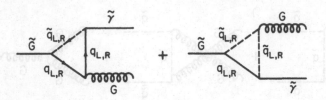

Figure 12.5.

Substituting eq. (12.2.5) into eq. (12.2.4) we find that consistency requires [8] $\eta = \pm i$. Thus, unlike a Dirac fermion, Majorana fermions of all varieties have a unique parity transformation. Because G has odd parity, the decay $\tilde{G} \to G\tilde{\gamma}$ is parity forbidden. However, if the masses of left ($\tilde{q}_L$) and right ($\tilde{q}_R$) squarks are different, then through the higher-order graph in Fig. 12.5, this process can arise. The amplitude for this process is of order

$$ A(\tilde{G} \to G\tilde{\gamma}) \approx e\alpha_s M_{\tilde{G}}^2 \left(\frac{1}{M_{\tilde{q}_L}^4} - \frac{1}{M_{\tilde{q}_R}^4} \right). \tag{12.2.6} $$

The ratio of these two decar rates is given by

$$ \frac{\Gamma(\tilde{G} \to G\tilde{\gamma})}{\Gamma(\tilde{G} \to q\bar{q}\tilde{\gamma})} = \frac{3\alpha_s}{4\pi} \frac{(M_{\tilde{q}_L}^2 - M_{\tilde{q}_R}^2)^2}{(M_{\tilde{q}_L}^4 + M_{\tilde{q}_R}^4)}. \tag{12.2.7} $$

This ratio can therefore be 10% or more in general supersymmetric models. This would be important in experimental searches for the gluino in high-energy hadron experiments.

Interesting spectroscopy implications can arise if the gluino is light because it can form bound states with quarks and antiquarks: $q\bar{q}\tilde{G}$. These particles will be fermions. Some of these particles (both charged and neutral) will be stable with respect to strong interactions and will be long lived and can leave tracks or gaps in detectors. There can also exist two gluino bound states, the gluinonium, whose spectroscopy has been studied. The dominant decay mode of the $\eta_{\tilde{G}}$ is to hadrons. Because of the color octet nature of $\tilde{G}$, production of $\eta_{\tilde{G}}$ is enhanced by a factor of $\frac{27}{2}$ with respect to η_c and would thus be copiously produced in $p\bar{p}$ collision.

(e) Wino

As we saw in the previous section supersymmetric models of weak interactions lead to new fermions $\tilde{W}^{\pm}$, which are Dirac particles consisting of the weak gaugino and the Higgsino as its components. The Winos will couple to $q\bar{q}$, $l\bar{l}$, etc. If we assume that they are heavier than $\tilde{q}$, their decay models will be of type

$$ \tilde{W}^{\pm} \to q\bar{q}\tilde{G} $$

$$ \to q\bar{q}\tilde{\gamma} $$

$$ \to q\bar{q}\tilde{Z}, $$

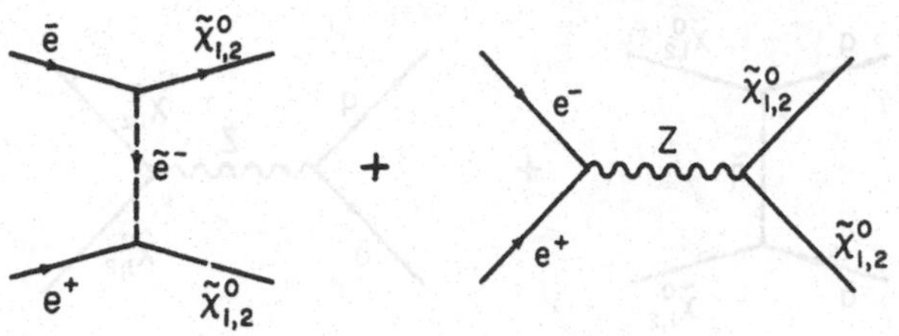

Figure 12.6.

if

$$m_{\tilde{Z}} \to m_{\tilde{W}+}.$$

Their lifetimes are of order $\sim (M^4\tilde{g}/g^4 M_{\tilde{W}}^5)$ and we therefore expected them to be too short to be directly observable. Their production in lepton and hadron collisions has been extensively studied. In e^+e^- collision no evidence for $\tilde{W}^\pm$ production exists yet, leading to a lower bound on their mass of about 100 GeV. Production of $\tilde{W}^+\tilde{W}^-$ would lead to an increase in R by one unit and should therefore be observable. The final-state characteristics can then be used to distinguish them from heavier leptons or quarks.

(f) Zino and Neutral Higgsino

The existence of fermionic partners of Z boson and netural Higgs bosons are also characteristic of supersymmetric extensions of electroweak models. The general mass matrix was discussed in Eq. (12.1.21). As we saw before, $M_Z^D = M_Z$ and $m_{\tilde{\gamma}\tilde{z}}$ can be small or zero. This matrix has to be diagonalized to find the mass eigenstates as well as the eigenvalues. This has been analyzed in great detail in Ref. 1. These particles will have typical decay modes of the following type:

$$\tilde{\chi}_{1,2}^0 \to \begin{cases} \bar{u}_{L,R} u_{L,R} \tilde{\gamma} \\ \bar{d}_{L,R} d_{L,R} \tilde{\gamma} \end{cases}.$$

They will be produced in e^+e^- collision through Z- and slepton-exchange graphs as shown in Fig. 12.6. Similarly, in $p\bar{p}$ collision, they can be produced by the constituent quark–antiquark annihilation diagrams shown in Fig. 12.7.

§12.3   Other Effects of Superparticles

The existence of superparticles in the 1–100 GeV mass range can give rise to new effects in processes where they are exchanged as virtual particles. In this section we enumerate a few of them.

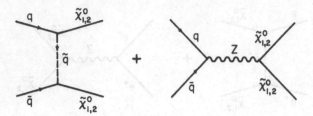

Figure 12.7.

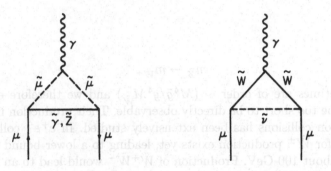

Figure 12.8.

(a) $g - 2$ of the Electron and the Muon

Two of the most precisely measured parameters of elementary particle physics are the $g - 2$ of the electron and the muon. Therefore, they provide useful constraints on the parameters of any theoretical extension of the QED. The quantity $a_\mu \equiv (g - 2)_\mu/2$ is so precisely measured and calculated that any new contribution to a_μ (called Δa_μ) must be bounded as follows [9]:

$$\Delta a_\mu = (0.26 \pm .16) \times 10^{-8}. \qquad (12.3.1)$$

In supersymmetric models new contributions to a_μ from the diagrams shown in Fig. 12.8 are [10]

$$\Delta a_\mu \simeq \frac{\alpha}{12\pi} \frac{\cos\theta \sin\theta m_\mu (M_{\tilde{\mu}_L}^2 - M_{\tilde{\mu}_R}^2)}{M_{\tilde{\chi}}^3}, \qquad (12.3.2)$$

where θ is a mixing angle between $\tilde{\mu}_L$ and $\tilde{\mu}_R$, and $M_{\tilde{x}}$ is a typical superparticle mass. These constraints are typically of the form $\Delta a_\mu \sim (\alpha/12\pi)(m_\mu/m_{\tilde{x}})^2\theta$, which leads to bounds on masses $M_{\tilde{\mu}} \geq 60$ GeV.

(b) Flavor-Changing Neutral Currents

The existence of superparticles of arbitrary masses upsets the delicate cancellations implied by the Glashow–Illiopoulos–Mariani mechanism for understanding the observed suppression of flavor-changing neutral currents (FCNC) as well as the $\Delta S = 2$ effective Hamiltonians. The new diagrams

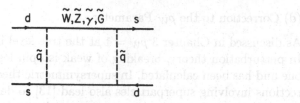

Figure 12.9.

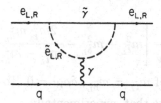

Figure 12.10.

to $K^0 - \bar{K}^0$ mixing are shown in Fig. 12.9. These and other FCNC effects have been estimated by several groups [11]. Typically they contribute to $\Delta S = 2$ matrix elements an amount

$$M_{\text{SUSY}}^{\Delta S=2} \simeq \frac{G_F \alpha}{4\pi} \left(\frac{\Delta M_{\tilde{q}}^2}{M_{\tilde{q}}^2} \right) \leq 10^{-12} \text{ GeV}^{-2}. \tag{12.3.3}$$

This implies

$$\frac{\Delta M_{\tilde{q}}^2}{2\tilde{q}} \leq 10^{-4}. \tag{12.3.4}$$

This constraints the mass splitting between squarks of different flavor.

(c) Parity Violation in Atomic Physics

The existence of mass splitting between left- and right-handed sleptons can lead to parity-violating effects in atomic physics arising from graphs of the type shown in Fig. 12.10. The effective parity-violating Hamiltonian is of the form

$$H_{\text{wk}} = \frac{\alpha^2}{q^2} (q^2 \delta_{\mu\nu} - q_\mu q_\nu) \bar{e} \gamma_\mu \gamma_5 e \bar{q} \gamma_\mu q \left(\frac{1}{M_L^2} - \frac{1}{M_R^2} \right). \tag{12.3.5}$$

This implies that

$$\frac{\alpha^2 \Delta M_{LR}^2}{M_{L,R}^4} \leq G_F/10. \tag{12.3.6}$$

For strong interactions [12] the factor α is replaced by α_s, which leads to a somewhat stronger bound on splitting between left- and right-handed squarks.

(d) Correction to the ρ_W Parameter

As discussed in Chapter 3 $\rho_W = 1$ at the tree level in the standard model. In perturbation theory, breaking of weak isospin leads to deviations from one and has been calculated. In supersymmetric theories the radiative corrections involving superparticles also lead [13] to deviations from $\rho_W = 1$. The corrections δ_{ρ_W} are sensitive to mass splitting between the left squarks $\tilde{t}_L$, and $\tilde{b}_L$ and are given by

$$\delta\rho_W = \frac{3\alpha}{16\pi \sin^2 \theta m_W^2} \left[m_{\tilde{t}_L}^2 + m_{\tilde{b}_L}^2 - \frac{2m_{\tilde{t}_L}^2 m_{\tilde{b}_L}^2}{m_{\tilde{t}_L}^2 - m_{\tilde{b}_L}^2} \ln \frac{m_{\tilde{t}_L}^2}{m_{\tilde{b}_L}^2} \right]. \qquad (12.3.7)$$

As is clear in the limit of $m_{\tilde{t}_L} = m_{\tilde{b}_L}$, $\delta\rho_W = 0$. To obtain the constraints on $m_{\tilde{t}_L}$, and $m^{\tilde{b}_L}$ we note that, experimentally,

$$\rho_W^{\text{expt}} = 1.002 \pm 0.015. \qquad (12.3.8)$$

If $m_{\tilde{b}_L} = 0$, this leads to a constraint on $m_{\tilde{t}_L} < 300$ GeV. This implies that splitting between members of a weak isospin multiplet must be less than 100 GeV and is not a very severe limit on model building.

§12.4   Why Go beyond the MSSM?

Even though the MSSM solves two outstanding peoblems of the standard model, i.e., the stabilization of the Higgs mass and the breaking of the electroweak symmetry, it brings in a lot of undesirable consequences. They are:

(a) Presence of arbitrary baryon- and lepton-number-violating couplings, i.e., the λ, λ', and λ'' couplings described above. In fact, a combination of λ' and λ'' couplings lead to proton decay. Present lower limits on the proton lifetime then imply that $\lambda'\lambda'' \leq 10^{-25}$ for squark masses of order of a TeV. Recall that a very attractive feature of the standard model is the automatic conservation of baryon and lepton numbers. The presence of R-parity breaking terms [6, 7] also makes it impossible to use the LSP as the Cold Dark Matter of the universe, since it is not stable and will therefore decay away in the very early moments of the universe. We will see that as we proceed to discuss the various grand unified theories, keeping the R-parity violating terms under control will provide a major constraint on model building.

(b) The different mixing matrices in the quark and squark sectors leads to arbitrary amount of flavor violation manifesting in such phenomena as $K_L - K_S$ mass difference, etc. Using present experimental information and the fact that the standard model more or less accounts for the observed magnitude of these processes implies that there must be strong constraints on the mass splittings among squarks. Detailed calculations indicate [11] that one must have $\Delta m_{\tilde{q}}^2/m_{\tilde{q}}^2 \leq 10^{-3}$ or so. Again recall that this undoes another nice feature of the standard model.

(c) The presence of new couplings involving the superpartners allows for the existence of extra CP phases. In particular the presence of the phase in the gluino mass leads to a large electric dipole moment of the neutron unless this phase is assumed to be suppressed by two to three orders of magnitude [14]. This is generally referred to in the literature as the SUSY CP problem. In addition, there is of course the famous strong CP problem, which neither the standard model nor the MSSM provide a solution to.

In order to cure these problems as well as to understand the origin of the soft-SUSY-breaking terms, one must seek new physics beyond the MSSM. Below, we pursue two directions for new physics: one that analyses schemes that generate soft-breaking terms and a second one that leads to automatic B and L conservation as well as solves the SUSY CP problem. The second model also provides a solution to the strong CP problem without the need for an axion under certain circumstances.

§12.5 Mechanisms for Supersymmetry Breaking

One of the major focuses of research in supersymmetry is to understand the mechanism for supersymmetry breaking. The usual strategy employed is to assume that SUSY is broken in a hidden sector that does not involve any of the matter or forces of the standard model (or the visible sector) and this SUSY breaking is transmitted to the visible sector via some intermediary, to be called the messenger sector.

There are generally two ways to set up the hidden sector—a less ambitious one where one writes an effective Lagrangian (or superpotential) in terms of a certain set of hidden sector fields that lead to supersymmetry breaking in the ground state and another more ambitious one where the SUSY breaking arises from the dynamics of the hidden sector interactions. For our purpose we will use the simpler schemes of the first kind. As far as the messenger sector goes, there are three possibilities as already referred to earlier: (i) gravity mediated (see chapter 14 and 15), (ii) gauge mediated [15], and (iii) anomalous U(1) mediated [16]. Below we give examples of each class.

(i) Gravity-Mediated SUSY Breaking

The scenario that uses gravity to transmit the supersymmetry breaking is one of the earliest hidden sector scenarios for SUSY breaking and forms much of the basis for the discussion in current supersymmetry phenomenology. In order to discuss these models one needs to know the supergravity couplings to matter. This is given in the classic paper of Cremmer et al.

discussed in Chapter 14. An essential feature of supergravity coupling is the generalized kinetic energy term in gravity-coupled theories called the Kahler potential, K. We will denote this by G and it is a Hermitian operator that is a function of the matter fields in the theory and their complex conjugates. The effect of supergravity coupling in the matter and the gauge sector of the theory is given in terms of G and its derivatives as follows:

$$L(z) = G_{zz^*}|\partial_\mu z|^2 + e^{-G}[G_z G_{z^*} G_{zz^*}^{-1} + 3], \qquad (12.5.1)$$

where z is the bosonic component of a typical chiral field (e.g., we would have $z \equiv \tilde{q}, \tilde{l}$, etc.) and $G = 3\ln(\frac{-K}{3}) - \ln|W(z)|^2$. A superscript implies derivative with respect to that field. The simplest choice for the Kahler potential K is $K = -3e^{-\frac{|z|^2}{3M_{P\ell}^2}}$ that normalizes the kinetic energy term properly. Using this, one can write the effective potential for supergravity-coupled theories to be

$$V(z, z^*) = e^{\frac{|z|^2}{M_{P\ell}^2}}[|W_z + \frac{z^*}{M_{P\ell}^2}W|^2 - \frac{3}{M_{P\ell}^2}|W|^2] + D - \text{terms.} \qquad (12.5.2)$$

The gravitino mass is given in terms of the Kahler potential as

$$m_{3/2} = M_{P\ell}e^{-G/2}. \qquad (12.5.3)$$

A popular scenario suggested by Polonyi is based upon the following hidden sector consisting of a gauge singlet field, denoted by z, and the superpotential W_H given by

$$W_H = \mu^2(z + \beta), \qquad (12.5.4)$$

where μ and β are mass parameters to be fixed by various physical considerations described in Chapter 15 in detail. It is clear that this superpotential leads to an F term that is always nonvanishing and therefore breaks supersymmetry. Requiring the cosmological constant to vanish fixes $\beta = (2 - \sqrt{3})M_{P\ell}$. Given this potential and the choice of the Kahler potential as discussed earlier, supergravity calculus predicts universal soft-breaking parameters m given by $m_0 \sim \mu^2/M_{P\ell}$. Requiring m_0 to be in the TeV range implies that $\mu \sim 10^{11}$ GeV. The complete potential to zeroth order in $M_{P\ell}^{-1}$ in this model is given by

$$V(\phi_a) = [\Sigma_a|\frac{\partial W}{\partial \phi_a}|^2 + V_D] \qquad (12.5.5)$$

$$+[m_0^2\Sigma_a\phi_a^*\phi_a + (AW^{(3)} + BW^{(2)} + \text{h.c.}), \qquad (12.5.6)$$

where $W^{(3,2)}$ denote the dimension-three and -two terms in the superpotential, respectively. The values of the parameters A and B at $M_{P\ell}$ are related to each other in this example as $B = A - 1$. The gaugino masses in these models arise out of a separate term in the Lagrangian depending on a new function of the hidden sector singlet fields, z:

$$\int d^4x d^2\theta f(z) W_\lambda^\alpha W_{\lambda,\alpha}. \tag{12.5.7}$$

If we choose $f(z) = \frac{z}{M_{P\ell}}$, then gaugino masses come out to be of order $m_{3/2} \sim \frac{\mu^2}{M_{P\ell}}$, which is also of order m_0, i.e., the electroweak scale. Furthermore, in order to avoid undesirable color and electric charge breaking by the SUSY models, one must require that $m_0^2 \geq 0$.

It is important to point out that the superHiggs mechanism operates at the Planck scale. Therefore all parameters derived at the tree level of this model need to be extrapolated to the electroweak scale. Therefore, after the soft-breaking Lagrangian is extrapolated to the weak scale, it will look like:

$$\mathcal{L}^{SB} = m_a^2 \phi_a^* \phi_a + m\Sigma_{i,j,k} A_{ijk} \phi_i \phi_j \phi_k + \Sigma_{i,j} B_{ij} \phi_i \phi_j. \tag{12.5.8}$$

These extrapolations depend among other things on the Yukawa couplings of the model. As a result of this the universality of the various SUSY breaking terms is no more apparent at the electroweak scale. Moreover, since the top Yukawa coupling is now known to be of order one, its effect turns the mass-squared of the H_u negative at the electroweak scale even starting from a positive value at the Planck scale. This provides a natural mechanism for the breaking of electroweak symmetry adding to the attractiveness of supersymmetric models. In the lowest-order approximation, one gets,

$$m_{H_u}^2(M_Z) \sim m_{H_u}^2(M_{P\ell}) - \frac{3h_t^2}{8\pi^2} ln\left(\frac{M_{P\ell}}{M_Z}\right)(m_{H_u}^2 + m_{\tilde{q}}^2 + m_{u^c}^2)|_{\mu = M_{P\ell}}$$

$$\tag{12.5.9}$$

Figure 12.15 depicts the actual evolution of the superpartner masses from the Planck scale to the weak scale and in particular how the mass-square of the H_u Higgs field turns negative at the weak scale leading to the breakdown of electroweak symmetry.

Before leaving this section it is worth pointing out that despite the simplicity and the attractiveness of this mechanism for SUSY breaking, there are several serious problems that arise in the phenomenological study of the model that have led to the exploration of other alternatives. For instance, the observed constraints on the flavor-changing neutral currents require that the squarks of the first and the second generation must be nearly degenerate, which is satisfied if one assumes the universality of the spartner masses at the Planck scale. However, this universality depends on the choice of the Kahler potential, which is *ad hoc*.

(ii) Gauge-Mediated SUSY Breaking [15]

This mechanism for the SUSY breaking has recently been quite popular in the literature and involves different hidden as well as messenger sectors.

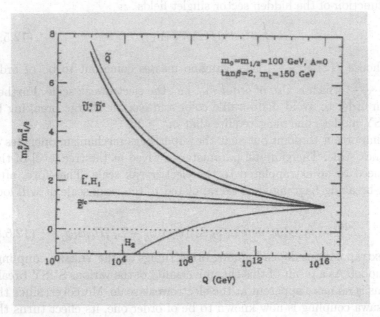

Figure 12.11. This figure shows the running of the superpartner masses from their GUT scale value; in particular, note how the mass-square of H_u turns negative at the weak scale triggering the breakdown of electroweak symmetry.

In particular, it proposes to use the known gauge forces as the messengers of supersymmetry breaking. As an example, consider a unified hidden messenger sector toy model of the following kind, consisting of the fields $\Phi_{1,2}$ and $\bar{\Phi}_{1,2}$, which have the standard model gauge quantum numbers and a singlet field S and with the following superpotential:

$$W = \lambda S(M_0^2 - \bar{\Phi}_1\Phi_1) + M_1(\bar{\Phi}_1\Phi_2 + \Phi_1\bar{\Phi}_2) + M_2\bar{\Phi}_1\Phi_1. \quad (12.5.10)$$

The F terms of this model are given by:

$$F_S = \lambda(M_0^2 - \bar{\Phi}_1\Phi_1), \quad (12.5.11)$$
$$F_{\Phi_2} = M_1\Phi_1; \; F_{\bar{\Phi}_2} = M_1\Phi_1,$$
$$F_{\Phi_1} = M_2\bar{\Phi}_1 + M_1\bar{\Phi}_2 - \lambda S\bar{\Phi}_1.$$

It is easy to see from the above equation that for $M_1 \gg M_0, M_2$, the minimum of the potential corresponds to all Φ's having zero v.e.v. and $F_S = \lambda M_0^2$, thus breaking supersymmetry. The same superpotential responsible for SUSY-breaking also transmits the SUSY-breaking information to the visible sector. Although the spirit of this model is similar to the original papers on the subject, this unified construction is different and has its characteristic predictions.

The SUSY breaking to the visible sector is transmitted via one and two loop diagrams. The gaugino masses arise from the one-loop diagram where a gaugino decomposes into the SUSY partners ϕ_1 and $\tilde{\phi}$ and the loop is completed as ϕ_1 and $\bar{\phi}_1$ mix through the SUSY-breaking term, F_S and the fermionic partners mix via the mass term M_2. The squark and slepton masses arise from the two-loop diagram where the squark–squark gauge boson–gauge boson coupling begins the first loop and one of the gauge boson couples to the two ϕ_1's and another to the two $\bar{\phi}_1$'s which in turn mix via the F terms for S to complete the two-loop diagram. This is only one typical diagram and there are many more that contribute in the same order.

It is then easy to see that their magnitudes are given by

$$m_\lambda \simeq \frac{\alpha}{4\pi} \frac{\langle F_S \rangle}{M_2}, \tag{12.5.12}$$

$$m_{\tilde{q}}^2 \simeq \left(\frac{\alpha}{4\pi}\right)^2 \left(\frac{\langle F_S \rangle}{M_2}\right)^2.$$

The first point to note is that the gaugino and squark masses are roughly of the same order, and requiring the squark masses to be around 100 GeV, we get for $F_S/M_2 \simeq 100$ TeV. Of course, $\langle F_S \rangle$ and M_2 need not be of same order, in which case the numerics will be different. Another important point to note is that by choosing the quantum numbers of the messengers Φ_i appropriately, one can have widely differing spectra for the superpartners.

A distinguishing feature of this approach is that because of the low scale for SUSY breaking, the gravitino mass is always in the milli-eV to kilo-eV range and therefore is always the LSP. Thus these models cannot lead to a supersymmetric CDM.

The attractive property of these models is that they lead naturally to near degeneracy of the squark and sleptons thus alleviating the FCNC problem of the MSSM and have therefore been the focus of intense scrutiny.

This class of models however suffer from the fact that the messenger sector is too *ad hoc*.

(iii) Anomalous U(1)-Mediated Supersymmetry Breaking

This class of models owe their origin to the string models, which after compactification can often leave anomalous U(1) gauge groups. Since the original string model is anomaly free, the anomaly cancellation must take place via the Green–Schwarz mechanism as follows. Consider a U(1) gauge theory with a single chiral fermion that carries a U(1) quantum number. This theory has an anomaly. Therefore, under a gauge transformation, the low-energy Lagrangian is not invariant and changes as

$$L \to L + \frac{\alpha}{4\pi} F\tilde{F} \tag{12.5.13}$$

where $F_{\mu\nu} = \partial_\mu A_\nu - \partial_\nu A_\mu$ and $\tilde{F}$ is the dual of $F_{\mu\nu}$. The last term is the anomaly term. To restore gauge invariance, we can add to the Lagrangian the Green–Schwarz term and rewrite the effective Lagrangian as

$$L' = L + \frac{a}{M}F\tilde{F} \tag{12.5.14}$$

where under the gauge transformation $a \rightarrow a - M\alpha/4\pi$. In order to obtain the supersymmetric version of the Green–Schwarz term, we have to add a dilaton term to the axion a to make a complex chiral superfield. Let us denote the dilaton field by ϕ and the complex chiral field containing it as $S = \phi + ia$. The gauge-invariant action containing the S and the gauge superfield V has terms of the following form:

$$A = \int d^4\theta \ln(S + S^\dagger - V) + \int d^2\theta SW^\alpha W_\alpha + \text{matter field parts.} \tag{12.5.15}$$

It is clear that in order to get a gauge field Lagrangian out of this, the dilaton S must have a v.e.v. with the identification that $\rangle S\langle (= g^{-2}$ and it is a fundamental unanswered question in superstring theory as to how this v.e.v. arises. If we assume that this v.e.v. has been generated, then, one can see that the first term in the Lagrangian when expanded around the dilaton v.e.v., leads to a term $\frac{1}{\rangle 2S\langle} \int d^4\theta V$, which is nothing but a linear Fayet–Illiopoulos D term. Combining this with other matter field terms with nonzero U(1) charge, one can then write the D term of the Lagrangian. As an example that can lead to realistic model building, we take two fields with equal and opposite U(1) charges ± 1 in addition to the squark and slepton fields. The D term can then be written as

$$V_D = \frac{g^2}{2}(n_q^2|\tilde{Q}|^2 + n_L^2|\tilde{L}|^2 + |\phi_+|^2 - |\phi_-|^2 + \zeta)^2. \tag{12.5.16}$$

This term when minimized does not break supersymmetry. However, if we add to the superpotential a term of the form $W_\phi = m\phi_+\phi_-$, then there is another term in low-energy effective potential that leads to the combined potential as

$$V = V_D + m^2(|\phi_+|^2 + |\phi_-|^2). \tag{12.5.17}$$

The minimum of this potential corresponds to:

$$\rangle\phi_+\langle = 0; \rangle\phi_-\langle = (\zeta - \frac{m^2}{g^2})^{1/2} \simeq \varepsilon M_{P\ell} : F_{\phi_+} = mM_{P\ell}\varepsilon \tag{12.5.18}$$

where we have assumed that $\zeta = \varepsilon^2 M_{P\ell}^2$. This then leads to nonzero squark masses $m_{\tilde{Q}}^2 \simeq n_Q^2 m^2$. Thus supersymmetry is broken and superpartners pick up mass. In the simplest model it turns out that the gaugino masses may be too low and one must seek ways around this. However, the A and B terms are also likely to be small in this model and that may provide certain advantages. On the whole, this approach has great potential for model building and has not been thoroughly exploited—for instance, it can be used to solve the FCNC problems, SUSY CP problem, to study the fermion mass hierarchies, etc.

(iv) Conformal Anomaly-Mediated Supersymmetry Breaking [17]

During the past year, a very interesting supersymmetry-breaking mechanism has been uncovered. This is based on the observation that in the absence of mass terms, a supergravity-coupled Yang–Mills theory has a conformal invariance. However, the process of renormalization always introduces a mass scale into the theory, which therefore breaks this symmetry. This leads to conformal anomaly, which leads to soft-breaking terms with a very definite pattern. We do not go into detailed derivation of the result but simply present a sketch of how to understand the origin of the result and the formulae for the SUSY-breaking squark mass-square term and the gaugino mass terms in this theory. Note that in supersymmetric theories, the only renormalization is that of the wave function, denoted by $Z(\mu)$, where μ is the renormalization scale. The conformally anomaly is therefore going to manifest as a modification of the wave function renormalization as $Z(\frac{\mu^2}{\Sigma^\dagger \Sigma})$, where Σ is the compensator superfield in the superconformal calculus and superconformal gauge is fixed by choosing $\Sigma = 1 + \theta^2 m_{3/2}$. Expanding in powers of θ^2 and noting the properties of θ's, we get

$$\ln Z(\frac{\mu^2}{\Sigma^\dagger \Sigma}) = \ln Z(\mu) - \frac{\gamma}{2} m_{3/2} \theta^2 + \text{h.c.} + \frac{d\gamma/dt}{4} m_{3/2}^2 \theta^2 \bar{\theta}^2. \quad (12.5.19)$$

Similarly, conformal anomaly also changes the dependence of the gauge coupling on mass μ to the form $g^2(\frac{\mu^2}{\Sigma^\dagger \Sigma})$, from which one gets a formula for the gaugino mass after fixing of superconformal anomaly. Denoting $m_{3/2}$ as the gravitino mass, one gets for the soft-breaking parameters

$$m_\lambda = -\frac{\beta(g^2)}{2g^2} m_{3/2}, \quad (12.5.20)$$

$$A_{ijk} = -y_{ijk} \frac{(\gamma_i + \gamma_j + \gamma_k)}{2} m_{3/2},$$

$$m_{\tilde{f}}^2 = -\frac{1}{4} \left(\frac{\partial \gamma}{\partial g} \beta(g) + \frac{\partial \gamma}{\partial y} \beta y \right) m_{3/2}^2,$$

where β is the usual beta function that determines the running of gauge couplings and γ is the anomalous dimension of the particular scalar field under question; y's denote the Yukawa coupling in the superpotential. For instance, if there are no Yukawa interactions, we can set $y = 0$ and get for the sfermion mass-square in a theory the expression

$$m_{\tilde{f}}^2 = -\frac{1}{2} c_0 b_0 g^4 m_{3/2}^2. \quad (12.5.21)$$

A very important consequence of this equation is that exactly like the gauge-mediated models, the sfermion masses are horizontally degenerate, thereby helping to solve the flavor-changing neutral-current problem. A down side to this formula is, however, the fact that for MSSM, b_0 is positive (i.e., non-asymptotically free) for both the SU(2) as well as U(1) groups. Since c_0 is always positive, this implies that any superpartner field that does not have color will have a tachyonic mass that is unacceptable. More needs to be done to cure this problem in a satisfactory way.

It is interesting to note that the minimal attempts to realize the anomalous $U(1)$ models ran into difficulty with small gaugino masses. One could therefore perhaps invoke a combination of conformal anomaly mediation with $U(1)$ anomaly mediation to construct viable models. Another generic feature of these models is that since the gravitino mass generates the susy breaking mass terms via gauge loop corrections, for superpartner masses in the 100 GeV range, one would expect the gravitino mass to be in the 10 TeV range or higher. This makes its lifetime ($\tau_{3/2} \sim M_{P\ell}^2/m_{3/2}^3$) of the order of a few seconds making it relatively safe from constraints of big bang nucleosynthesis.

§12.6 Renormalization of Soft Supersymmetry-Breaking Parameters

In all the mechanisms for supersymmetry breaking discussed so far, generally the true scale where supersymmetry breaks is much higher than the scale of superpartners needed to stabilize the weak scale, i.e., around a TeV or less. Because of this, the vaules of superpartner masses are generally defined at that scale and, in order to find what their values are at the weak scale, we must use the renormalization group equations. Below we give list of the one-loop renormalization group equations for the soft-breaking parameters: $m^2_{\tilde{Q}, \tilde{L}...}$; gaugino masses M_i, μ parameter, $B\mu$ parameter; the trilinear scalar couplings $A_{u,d}$ terms, etc. [17a]. Define $t = \ln \mu$ (μ is the mass scale and is different from the μ parameter used below.

$$\frac{d\mu}{dt} = \frac{1}{16\pi^2} \mu \left[\text{Tr}\,(3h_u^\dagger h_u + 3h_d^\dagger h_d + h_\ell^\dagger h_\ell) - 3g_2^2 - \frac{3}{5}g_1^2 \right], \qquad (12.6.1)$$

$$\frac{dB\mu}{dt} = \frac{1}{16\pi^2} B\mu \left[\text{Tr}\,(3h_u^\dagger h_u + 3h_d^\dagger h_d + h_\ell^\dagger h_\ell) - 3g_2^2 - \frac{3}{5}g_1^2 \right]$$
$$+ \mu \left[\text{Tr}\,(6A_u h_u^\dagger + 6A_d h_d^\dagger + 2A_\ell h_\ell^\dagger) + 6g_2^2 M_2 + \frac{6}{3}g_1^2 M_1 \right], (12.6.2)$$

$$\frac{dM_a}{dt} = \frac{2g_a^2}{16\pi^2} B_a M_a, \qquad (12.6.3)$$

where $B_a = (33/5, 1, -3)$ for U$(1)_Y$, SU$(2)_L$ and SU$(3)_c$, respectively.

$$\frac{dA_{u,d,\ell}}{dt} = \frac{1}{16\pi^2} \beta_{u,d,\ell} \qquad (12.6.4)$$

where the beta functions are

$$\beta_u = A_u \left[3\text{Tr}\,(h_u h_u^\dagger) + 5h_u^\dagger h_u + h_d^\dagger h_d - \frac{16}{3}g_3^2 - 3g_2^2 - \frac{13}{15}g_1^2 \right]$$

$$+ h_u \left[6\mathrm{Tr}\,(A_u h_u^\dagger) + 4h_u^\dagger A_u + 2h_d^\dagger A_d + \frac{32}{3}g_3^2 M_3 + 6g_2^2 M_2 + \frac{26}{15}g_1^2 M_1 \right],$$

$$\tag{12.6.5}$$

$$\beta_d = A_d \left[3\mathrm{Tr}\,(h_d h_d^\dagger + h_e h_e^\dagger) + 5h_d^\dagger h_d + h_u^\dagger h_u - \frac{16}{3}g_3^2 - 3g_2^2 - \frac{7}{15}g_1^2 \right]$$

$$+ h_d \left[6\mathrm{Tr}\,(A_d h_d^\dagger + A_e h_e^\dagger) + 4h_d^\dagger A_d + 2h_u^\dagger A_u + \frac{32}{3}g_3^2 M_3 \right.$$

$$\left. + 6g_2^2 M_2 + \frac{14}{15}g_1^2 M_1 \right], \tag{12.6.6}$$

$$\beta_\ell = A_\ell \left[3\mathrm{Tr}\,(h_d h_d^\dagger + h_e h_e^\dagger) + 5h_\ell^\dagger h_\ell - 3g_2^2 - \frac{9}{5}g_1^2 \right]$$

$$+ h_\ell \left[6\mathrm{Tr}\,(A_d h_d^\dagger + A_e h_e^\dagger) + 4h_\ell^\dagger A_\ell + 6g_2^2 M_2 + \frac{18}{5}g_1^2 M_1 \right], \tag{12.6.7}$$

$$\frac{dm^2}{dt} = \frac{1}{16\pi^2}\beta_{m^2}, \tag{12.6.8}$$

where

$$\beta_{m_{H_u}^2} = 6\mathrm{Tr}\,[(m_{H_u}^2 + m_Q^2)h_u^\dagger h_u + h_u^\dagger m_u^2 h_u + A_u^\dagger A_u]$$

$$- 6g_2^2|M_2|^2 - \frac{6}{5}g_1^2|M_1|^2 + \frac{3}{5}g_1^2 \mathcal{S}, \tag{12.6.9}$$

where $\mathcal{S} = m_{H_u}^2 - m_{H_d}^2 + \mathrm{Tr}\,[m_Q^2 - m_L^2 - 2m_u^2 + m_d^2 + m_e^2]$ and we have dropped the ˜ over the matter fields even though they denote the superpartners.

$$\beta_{m_{H_d}^2} = \mathrm{Tr}\,[6(m_{H_d}^2 + m_Q^2)h_d^\dagger h_d + 6h_d^\dagger m_d^2 h_d + 6A_d^\dagger A_d$$

$$+ 2(m_{H_d}^2 + m_L^2)h_\ell^\dagger h_\ell + 2h_\ell^\dagger m_e^2 h_\ell + 2A_\ell^\dagger A_\ell]$$

$$- 6g_2^2|M_2|^2 - \frac{6}{5}g_1^2|M_1|^2 - \frac{3}{5}g_1^2 \mathcal{S}, \tag{12.6.10}$$

$$\beta_{m_Q^2} = (2m_{H_u}^2 + m_Q^2)h_u^\dagger h_u + 2h_d^\dagger m_d^2 h_d + 2h_u^\dagger m_u^2 h_u + 2A_d^\dagger A_d$$

$$+ 2A_u^\dagger A_u + (2m_{H_d}^2 + m_Q^2)h_d^\dagger h_d + (h^\dagger d h_d + h_u^\dagger h_u)m_Q^2$$

$$- \frac{32}{3}g_3^2 M_3^2 - 6g_2^2|M_2|^2 - \frac{2}{15}g_1^2|M_1|^2 + \frac{1}{5}g_1^2 \mathcal{S}, \tag{12.6.11}$$

$$\beta_{m_L^2} = (2m_{H_d}^2 + m_L^2)h_\ell^\dagger h_\ell + 2h_\ell^\dagger m_e^2 h_\ell + 2h_\ell^\dagger h_\ell m_L^2$$

$$+ 2A_\ell^\dagger A_\ell + -6g_2^2|M_2|^2 - \frac{6}{5}g_1^2|M_1|^2 - \frac{3}{5}g_1^2 \mathcal{S}, \tag{12.6.12}$$

$$\beta_{m_u^2} = (4m_{H_u}^2 + 2m_u^2)h_u h_u^\dagger + 4h_u m_Q^2 h_u^\dagger + 2h_u h_u^\dagger m_u^2$$

$$+ 4A_u A_u^\dagger - \frac{32}{3}g_3^2|M_3|^2 - \frac{32}{15}g_1^2|M_1|^2 - \frac{4}{5}g_1^2 \mathcal{S}, \tag{12.6.13}$$

$$\beta_{m_d^2} = (4m_{H_d}^2 + 2m_d^2)h_d h_d^\dagger + 4h_d m_Q^2 h_d^\dagger + 2h_d h_d^\dagger m_d^2$$

$$+ 4A_d A_d^\dagger - \frac{32}{3}g_3^2|M_3|^2 - \frac{8}{15}g_1^2|M_1|^2 + \frac{2}{5}g_1^2 \mathcal{S}, \tag{12.6.14}$$

$$\beta_{m_e^2} = (4m_{H_d}^2 + 2m_e^2)h_\ell h_\ell^\dagger + 4h_\ell m_L^2 h_\ell^\dagger + 2h_\ell h_\ell^\dagger m_e^2$$
$$+ 4A_\ell A_\ell^\dagger - \frac{24}{5}g_1^2|M_1|^2 + \frac{6}{5}g_1^2 \mathcal{S}. \tag{12.6.15}$$

Using these equations and using approximate values for the gauge couplings one can deduce that the squark masses increase as we approach the low scale, whereas the slepton masses remain practically unchanged. The extrapolation of the gaugino masses is purely determined by the beta functions of the gauge groups. If they start from a common mass $m_{1/2}$, then at the weak scale one gets

$$M_3 \simeq 2.7m_{1/2}; M_2 \simeq 0.8m_{1/2}; M_1 \simeq 0.4m_{1/2}. \tag{12.6.16}$$

Furthermore, we see that the $m_{H_u}^2$ goes down to smaller values due to the large top Yukawa coupling, whereas for the $m_{H_d}^2$ there is no such large effect. This is at the heart of the understanding of the electroweak symmetry breaking in the MSSM and is of course one of its most attractive features. We come back to this discussion in Chapter 15. For completeness, we also give the renormalization evolution equation for the Yukawa matrices $h_{u,d}$ for the MSSM:

$$\frac{dh_u}{dt} = \frac{1}{16\pi^2}h_u\left[3\text{Tr}\left(h_u h_u^\dagger\right) + 3h_u^\dagger h_u + h_d^\dagger h_d - \frac{16}{3}g_3^2 - 3g_2^2 - \frac{13}{15}g_1^2\right],$$
$$\frac{dh_d}{dt} = \frac{1}{16\pi^2}h_u\left[3\text{Tr}\left(h_u h_u^\dagger + h_\ell h_\ell^\dagger\right) + 3h_d^\dagger h_d + h_u^\dagger h_u - \frac{16}{3}g_3^2 - 3g_2^2 - \frac{7}{15}g_1^2\right]$$

$$\tag{12.6.17}$$

These equations are useful in studying the extrapolation of the quark masses from the high scales such as the SUSY-breaking scale down to the weak scale and vice versa.

§12.7 Supersymmetric Left–Right Model

One of the attractive features of the supersymmetric models is its ability to provide a candidate for the cold dark matter of the universe. This however relies on the theory obeying R-parity conservation [with $R \equiv (-1)^{3(B-L)+2S}$]. It is easy to check that particles of the standard model are even under R, whereas their superpartners are odd. The lightest superpartner is then absolutely stable and can become the dark matter of the universe. In the MSSM, R-parity symmetry is not automatic and is achieved by imposing global baryon- and lepton-number conservation on the theory as additional requirements. First of all, this takes us one step back from the nonsupersymmetric standard model where the conservation B and L arise automatically from the gauge symmetry and the field content of the model. Second, there is a prevalent lore supported by some calculations that in the presence of nonperturbative gravitational effects such as black holes or worm holes, any externally imposed global symmetry must be violated by Planck-suppressed operators. In this case, the R-parity violating effects again become strong enough to cause rapid decay of the lightest R-odd neutralino so that there is no dark matter particle in the minimal supersymmetric standard model. It is therefore desirable to seek supersymmetric theories where, like the standard model, R-parity conservation (hence baryon and lepton number conservation) becomes automatic, i.e., guaranteed by the field content and gauge symmetry. It was realized in mid-1980's [18] that such is the case in the supersymmetric version of the left–right model that implements the seesaw mechanism for neutrino masses. We briefly discuss this model in the section.

The gauge group for this model is $SU(2)_L \times SU(2)_R \times U(1)_{B-L} \times SU(3)_c$. The chiral superfields denoting left-handed and right-handed quark superfields are denoted by $Q \equiv (u, d)$ and $Q^c \equiv (d^c, -u^c)$, respectively, and similarly the lepton superfields are given by $L \equiv (\nu, e)$ and $L^c \equiv (e^c, -\nu^c)$. The Q and L transform as left-handed doublets with the obvious values for the $B - L$ and the Q^c and L^c transform as the right-handed doublets with opposite $B - L$ values. The symmetry breaking is achieved by the following set of Higgs superfields: $\phi_a(2, 2, 0, 1)$ $(a = 1, 2)$; $\Delta(3, 1, +2, 1)$; $\bar{\Delta}(3, 1, -2, 1)$; $\Delta^c(1, 3, -2, 1)$ and $\bar{\Delta}^c(1, 3, +2, 1)$. There are alternative Higgs multiplets that can be employed to break the right-handed $SU(2)$; however, this way of breaking the $SU(2)_R \times U(1)_{B-L}$ symmetry automatically leads to the seesaw mechanism for small neutrino masses discussed in Chapter 6.

The superpotential for this theory has only a very limited number of terms and is given by (we have suppressed the generation index)

$$W = \mathbf{Y}_q^{(i)} Q^T \tau_2 \Phi_i \tau_2 Q^c + \mathbf{Y}_l^{(i)} L^T \tau_2 \Phi_i \tau_2 L^c$$
$$+ i(\mathbf{f} L^T \tau_2 \Delta L + \mathbf{f}_c L^{cT} \tau_2 \Delta^c L^c)$$
$$+ \mu_\Delta \text{Tr}(\Delta \bar{\Delta}) + \mu_{\Delta^c} \text{Tr}(\Delta^c \bar{\Delta}^c) + \mu_{ij} \text{Tr}(\tau_2 \Phi_i^T \tau_2 \Phi_j)$$
$$+ W_{NR}, \tag{12.7.1}$$

where W_{NR} denotes nonrenormalizable terms arising from higher-scale physics such as grand unified theories or Planck-scale effects. At this stage all couplings $\mathbf{Y}_{q,l}^{(i)}$, μ_{ij}, μ_Δ, μ_{Δ^c}, $\mathbf{f}$, $\mathbf{f}_c$ are complex with μ_{ij}, $\mathbf{f}$, and $\mathbf{f}_c$ being symmetric matrices.

The part of the supersymmetric action that arises from this is given by

$$S_W = \int d^4x \int d^2\theta \, W + \int d^4x \int d^2\bar{\theta} \, W^\dagger. \tag{12.7.2}$$

It is clear from the above equation that this theory has no baryon- or lepton-number-violating terms. Since all other terms in the theory automatically conserve B and L, R-parity symmetry $(-1)^{3(B-L)+2S}$ is automatically conserved in the SUSYLR model. As a result, it allows for a dark matter particle provided the vacuum state of the theory respects R

parity. The desired vacuum state of the theory that breaks parity and pre-serves R-parity corresponds to $\langle \Delta^c \rangle \equiv v_R \neq 0$; $\langle \bar{\Delta}^c \rangle \neq 0$ and $\langle < \tilde{\nu}^c \rangle = 0$. This reduces the gauge symmetry to that of the standard model, which is then broken via the v.e.v.'s of the ϕ fields. These two together via the seesaw mechanism lead to a formula for neutrino masses of the form $m_\nu \simeq \frac{m_f^2}{f v_R}$. Thus we see that the suppression of the $V + A$ currents at low energies and the smallness of the neutrino masses are intimately connected.

It turns out that left–right symmetry imposes rather strong constraints on the ground state of this model. It was pointed out in 1993 [19] that if we take the minimal version of this model, the ground state leaves the gauge symmetry unbroken. To break gauge symmetry one must include singlets in the theory. However, in this case, the ground state breaks electric charge unless R parity is spontaneously broken. Furthermore, R parity can be spontaneously broken only if $M_{W_R} \leq$ few TeV's. Thus the conclusion is that the renormalizable version of the SUSYLR model with only singlets, $B - L = \pm 2$ triplets, and bidoublets can have a consistent electric-charge-conserving vacuum only if the W_R mass is in the TeV range and R parity is spontaneously broken. This conclusion can however be avoided either by making some very minimal extensions of the model such as adding super-fields $\delta(3, 1, 0, 1) + \bar{\delta}(1, 3, 0, 1)$[19] or by adding nonrenormalizable terms to the theory [20]. Such extra fields often emerge if the model is embedded into a grand unified theory or is a consequence of an underlying composite model.

In order to get a R-parity conserving vacuum (as would be needed if we want the LSP to play the role of the cold dark matter) without introducing the extra fields mentioned earlier, one must add the nonrenormalizable terms. In this case, the doubly charged Higgs bosons and Higgsinos become very light unless the W_R scale is above 10^{10} GeV or so [21]. This implies that the neutrino masses must be in the eV range, as would be required if they have to play the role of the hot dark matter. Thus an interesting connection between the cold and hot dark matter emerges in this model in a natural manner.

This model solves two other problems of the MSSM: (i) one is the SUSY CP problem and (ii) the other is the strong CP problem when the W_R scale is low. To see how this happens, let us define the transformation of the fields under left–right symmetry as follows and observe the resulting constraints on the parameters of the model:

$$Q \leftrightarrow Q^{c*},$$
$$L \leftrightarrow L^{c*},$$
$$\Phi_i \leftrightarrow \Phi_i^\dagger,$$
$$\Delta \leftrightarrow \Delta^{ct},$$
$$\bar{\Delta} \leftrightarrow \bar{\Delta}^{ct},$$
$$\theta \leftrightarrow \bar{\theta},$$

$$\tilde{W}_{SU(2)_L} \leftrightarrow \tilde{W}^*_{SU(2)_R},$$

$$\tilde{W}_{B-L,SU(3)_C} \leftrightarrow \tilde{W}^*_{B-L,SU(3)_C}. \tag{12.7.3}$$

Note that this corresponds to the usual definition $Q_L \leftrightarrow Q_R$, etc. To study its implications on the parameters of the theory, let us write down the most general soft supersymmetry terms allowed by the symmetry of the model (which make the theory realistic):

$$\mathcal{L}_{\text{soft}} = \int d^4\theta \sum_i m_i^2 \phi_i^\dagger \phi_i + \int d^2\theta\,\theta^2 \sum_i A_i W_i + \int d^2\bar{\theta}\,\bar{\theta}^2 \sum_i A_i^* W_i^\dagger$$

$$+ \int d^2\theta\,\theta^2 \sum_p m_{\lambda_p} \tilde{W}_p \tilde{W}_p + \int d^2\bar{\theta}\,\bar{\theta}^2 \sum_p m_{\lambda_p}^* \tilde{W}_p^* \tilde{W}_p^*. \tag{12.7.4}$$

In eq. 12.6.4, $\tilde{W}_p$ denotes the gauge-covariant chiral superfield that contains the $F_{\mu\nu}$-type terms with the subscript going over the gauge groups of the theory including $SU(3)_c$. W_i denote the various terms in the superpotential, with all superfields replaced by their scalar components and with coupling matrices which are not identical to those in W. Equation 12.6.4 gives the most general set of soft breaking terms for this model.

With the above definition of L-R symmetry, it is easy to check that [22]

$$\mathbf{Y}_{q,l}^{(i)} = \mathbf{Y}_{q,l}^{(i)\dagger},$$

$$\mu_{ij} - \mu_{ij}^*,$$

$$\mu_\Delta - \mu_{\Delta^c}^*,$$

$$\mathbf{f} = \mathbf{f}_c^*,$$

$$m_{\lambda_{SU(2)_L}} = m_{\lambda_{SU(2)_R}}^*,$$

$$m_{\lambda_{B-L,SU(3)_C}} = m_{\lambda_{B-L,SU(3)_C}}^*,$$

$$A_i = A_i^\dagger. \tag{12.7.5}$$

Note that the phase of the gluino mass term is zero due to the constraint of parity symmetry. As a result the one-loop contribution to the electric dipole moment of neutron from this source vanishes in the lowest-order [21]. The higher-order loop contributions that emerge after left–right symmetry breaking can be shown to be small, thus solving the SUSYCP problem. Furthermore, since the constraints of left–right symmetry imply that the quark Yukawa matrices are Hermitian, if the vaccum expectation values of the $\langle\phi\rangle$ fields are real, then the Θ parameter of QCD vanishes naturally at the tree level. This can provide a solution to the strong CP problem [23]. It however turns out that to keep the one-loop finite contributions to the Θ less than 10^{-9}, the W_R scale must be in the TeV range. Such models generally predict the electric dipole moment of neutron of order 10^{-26} [24] ecm, which can be probed in the next round of neutron dipole moment searches.

An important subclass of the SUSYLR models is the one that has only one bidoublet Higgs field in addition to the fields that break left–right symmetry such as the triplets (Δ's) or the doublets $(2, 1, +1) + (1, 2, -1)$. These models have the property that above the W_R scale the up and the down Yukawas unify to Yukawa matrix. We call these models up–down unification models [25]. The interesting point about these models is that since up–down unification at the tree level implies that the quark mixing angles must vanish at the tree level, all observed mixings must emerge out of the one-loop corrections. This restricts the allowed ranges of the SUSY-breaking parameters such as the A parameters or the squark mixings as well as the squark and gluino masses. This has the advantage of being testable. The model also provide a new way to understand the CP-violating phenomena purely out of the supersymmetry-breaking sector. This model also has the potential to solve the strong CP problem without the need for an axion. Many phenomenological implications of this model have been studied in order to provide collider tests of these models.

To conclude this chapter, we have condensed one of the active areas of present day research in particle theory within a few sections to give a flavor for some of the main ideas in the hope that the reader can follow the literature and his/her own ideas on the subject to expand this still widening frontier of research.

Exercises

12.1. Write all possible terms in the superpotential allowed by gauge invariance and describe the kind of new baryon- and lepton-number-violating processes induced by these new terms. From the existing experimental upper limits on rare baryon- and lepton-number-violating processes, obtain the constraints on the strengths of these new interactions.

12.2. Show that, if the gauge group is extended to $SU(2)_L \times U(1)_{I_{3R}} \times U(1)_{B-L}$, baryon and lepton number conservation in the Lagrangian becomes automatic. Can you show how R-parity violating interactions may be induced in these theories? Write down a reasonable model to discuss this effect.

12.3. Consider the minimal supersymmetric standard model without the singlet chiral field S. How can you implement the breakdown of electroweak symmetry in this model? Discuss the neutral fermion masses and Higgs boson masses in this model.

12.4. Discuss the decay of the lightest supersymmetric partners (LSP) in the R-parity violating theories. Consider both the photino and S neutrino as the LSP's.

12.5. Which R-parity violating interaction gives rise to neutron–antineutron oscillation? Obtain the $N - \bar{N}$ mixing time for a reasonable choice of parameters.

References

[1] There exist several excellent recent reviews of the subject:
H. Haber and G. Kane, *Phys. Rep.* **117**, 76 (1984).

[2] H. P. Nilles, *Phys. Rep.* **110**, 1 (1984);
R. Arnowitt, A. Chamseddine, and P. Nath, $N = 1$ *Supergravity*, World Scientific, Singapore, 1984;
S. Martin, *Perspectives in Supersymmetry*, ed. G. Kane (World Scientific, 1998).

[3] H. E. Haber and R. Hempfling, *Phys. Rev. Lett.* **66**, 1815 (1991);
J. Ellis, G. Ridolfi and F. Zwirner, *Phys. Lett.* **B257**, 83 (1991);
R. Barbieri, M. Frigeni and F. Caravaglios, *Phys. Lett.* **B258**, 167 (1991).

[4] H. Haber, *Perspectives on Higgs Physics*, ed. G. Kane (World Scientific, 1993).

[5] P. Fayet, *Nucl. Phys.* **B90**, 104 (1975);

[6] C. S. Aulakh and R. N. Mohapatra, *Phys. Lett.* **121B**, 147 (1983);

[7] L. Hall and M. Suzuki, *Nucl. Phys.* **B231**, 419 (1984).

[8] B. Kayser, Private communication, 1983.

[9] H. N. Brown et al., Muon (g-2) Collaboration, hep-ex/0102017.

[10] P. Fayet, in *Unification of the Fundamental Particle Interactions*, edited by S. Ferrara et al. (Plenum, New York, 1980), p. 587;
J. Ellis, J. Hagelin, and D. V. Nanopoulos, *Phys. Lett.* **116B**, 283 (1982);
R. Barbieri and L. Maiani, *Phys. Lett.* **117B**, 203 (1982);
For a recent discussion, see P. Nath, U. Chattopadhyay, A. Corsetti, hep-ph/0202275.

[11] J. Ellis an D. V. Nanopoulos, *Phys. Lett.* **110B**, 44 (1982);
R. Barbieri and R. Gatto, *Phys. Lett.* **110B**, 211 (1982);
T. Inami and C. S. Lim, *Nucl. Phys.* **B207**, 533 (1982);
B. A. Cambell, *Phys. Rev.* **D28**, 209 (1983);
J. Donoghue, H. P. Nilles, and D. Wyler, *Phys. Lett.* **128B**, 55 (1983);
M. Suzuki, *Phys. Lett.* **115B**, 40 (1982);
E. Franco and M. Mangano, *Phys. Lett.* **135B**, 40 (1982).
F. Gabbiani, E. Gabrielli, A. Masiero and L. Silvestrini, *Nucl. Phys.* **B 477**, 321 (1996).

[12] M. Suzuki, *Phys. Lett.* **115B**, 40 (1982);
A Duncan, *Nucl. Phys.* **B214**, 21 (1983).

[13] T. K. Kuo and N. Nakagawa, *Nuovo Cim. Lett.* **36**, 560 (1983);
R. Barbieri and L. Maiani, *Nucl. Phys.* **B224**, 32 (1983);
C. S. Lim, T. Inami, and N. Sakai, *Phys. Rev.* **D29**,1488 (1984).

[14] J. Ellis, S. Farrara and D. Nanopoulos, *Phys. Lett.* **B 114**, 231 (1982);
J. Polchinski and M. Wise, *Phys. Lett.* **B 125**, 393 (1983);
T. Ibrahim amnd P. Nath, *Phys. Lett.* **B 418**, 98 (1998) and references therein.

[15] M Dine, A Nelson: *Phys. Rev.* **D48**, 1277 (1993);
M. Dine, A.E. Nelson, Y. Nir, and Y. Shirman, *Phys. Rev.* **D53** (1996) 2658;
M. Dine and A.E. Nelson, *Phys. Rev.* **D48**, 1277 (1993);
M. Dine, A.E. Nelson, and Y. Shirman, *Phys. Rev.* **D51**, 1362 (1995).

[16] P Binetruy, E Dudas, *Phys. Lett.* **B389**, 503 (1996);
G Dvali, A Pomarol, *Phys. Rev. Lett.* **77**, 3728 (1996);
R N Mohapatra, A Riotto, *Phys. Rev.* **D55**, 4262 (1997).

[17] L. Randall and R. Sundrum, hep-th/9810155;
G. F. Giudice, M. A. Luty, H. Murayama and R. Rattazi, hep-ph/9810442.

[17a] For a review, see S. Martin and M. Vaughn, *Phys. Rev.* **D50**, 2282 (1994).

[18] R N Mohapatra: *Phys. Rev.* **34**, 3457 (1986); A. Font, L. Ibanez and F. Quevedo, *Phys. Lett.* **B228**, 79 (1989); S. P. Martin, *Phys. Rev.* **D46**, 2769 (1992).

[19] R Kuchimanchi, R N Mohapatra, *Phys. Rev.* **48**, 4352 (1993);
R. Kuchimanchi and R. N. Mohapatra, *Phys. Rev. Lett.* **75**, 3989 (1995);
C. Aulakh, K. Benakli, and G. Senjanović, *Phys. Rev. Lett.* **79**, 2188 (1997).

[20] R. N. Mohapatra and A. Rasin, *Phys. Rev. Lett.* **76**, 3490 (1996);
C. S. Aulakh, A. Melfo and G. Senjanović, hep-ph/9707256;
C. S. Aulakh, A. Melfo, A. Rasin and G. Senjanović, hep-ph/9712551.
Z. Chacko and R. N. Mohapatra, hep-ph/9712359.

[21] R. N. Mohapatra and A. Rasin, *Phys. Rev.* **D54**, 5835 (1996).

[22] R. N. Mohapatra and A. Rasin, *Phys. Rev. Lett.* **76**, 3490 (1996);
R. Kuchimanchi, *Phys. Rev. Lett.* **76**, 3486 (1996);
R. N. Mohapatra, A. Rasin, and G. Senjanović, *Phys. Rev. Lett.* **79**, 4744 (1997).

[23] M. Pospelov, *Phys. Lett.* **B391**, 324 (1997).

[24] K. S. Babu, B. Dutta, and R. N. Mohapatra, hep-ph/9812421, hep-ph/9904366 and hep-ph/9905464.

[25] M Francis, M Frank, C S Kalman, *Phys. Rev.* **D43**, 2369 (1991);
K Huitu, J Malaampi, M Raidal, *Nucl. Phys.* **B 420**, 449 (1994).

13
Supersymmetric Grand Unification

§13.1   The Supersymmetric SU(5)

One of the original motivations for the application of supersymmetry to particle physics was to solve the gauge hierarchy problem that arises in the grand unification program. As has been emphasized in Chapter 5, the tree level parameters must be fine tuned to an accuracy of 10^{-26} or so, to generate the mass ratio $M_x/m_W \simeq 10^{12}$ in the SU(5) model. In other models, owing to the presence to intermediate mass scales, the problem of fine tuning is not as severe, but a lesser degree of fine tuning is always required. Because a nonsupersymmetric theory with scalar bosons is plagued with quadratic divergences, such tree-level fine tunings are upset in higher orders. This need not happen in supersymmetric theories due to the nonrenormalization theorem of Grisaru, Rocek, and Siegel described in Chapter 10. According to this theorem, the parameters of the superpotentials not only become free of any infinite renormalization but they also do not receive finite renormalization in higher orders. Supersymmetry can, therefore, be used to solve one aspect of the gauge hierarchy problem, i.e., once we fine tune parameters at the tree level, the radiative corrections do not disturb the hierarchy. This point was utilized by Dimopoulos and Georgi [1] and Sakai [2] to construct supersymmetric SU(5) models with partial solutions to the gauge hierarchy problem. We discuss this model below.

Chiral superfields ψ (belonging to the {$\bar{5}$}-dimensional representation) and T (belonging to the {10}-dimensional representation) are chosen to denote the matter fields. As in the supersymmetric generalization of the $SU(2)_L \times U(1)$ model, we will need two sets of Higgs superfields denoted by H_u and H_d belonging to the {5}- and {5̄}-dimensional representations of $SU(5)$ to generate fermion masses as well as to cancel gauge anomalies. We also choose a {24}-dimensional Higgs superfield Φ to break the $SU(5)$ group down to $SU(3)_c \times SU(2)_L \times U(1)_Y$. Except for an additional Higgs superfield (H_u or H_d) all other fields are supersymmetric generalizations of fields present in the $SU(5)$ model of Chapter 5.

The first stage in the construction of the model is to give the superpotential $W(H_u, H_d, \phi, \psi, T)$. The superpotential has to fulfill two functions:

(i) to give realistic breaking for the $SU(5)$ symmetry down to the $SU(3)_c \times SU(2)_L \times U(1)_Y$ group and subsequently to the $SU(3)_c \times U(1)_{em}$ group; and

(ii) to give masses to the quarks and leptons.

We choose W to be the following:

$$W = h_u \epsilon^{ijklm} T_{ij} T_{kl} H_{u,m} + h_d T_{ij} \psi^i H_d^j + z \operatorname{Tr} \Phi + x \operatorname{Tr} \Phi^2$$
$$+ y \operatorname{Tr} \Phi^3 + \lambda_1 (H_{ui} H_d^j \phi_j^i + m' H_{u,i} H_d^i). \tag{13.1.1}$$

Since our aim is to make the fine tuning of parameters natural, we would like supersymmetry to stay unbroken down to the weak-interaction scale of $\sim m_W / g$. This requires that $F_\phi = 0 + O(m_W^2 / g^2)$. From eq. (13.1.1) we get

$$\langle F_{\phi_j}^i \rangle = z \delta_j^i + 2x \langle \phi_j^i \rangle + 3y \langle \phi_k^i \phi_j^k \rangle. \tag{13.1.2}$$

Since $\operatorname{Tr} \langle F_\phi \rangle = 0$, this implies that

$$z = -\tfrac{3}{5} y \operatorname{Tr} \langle \Phi^2 \rangle. \tag{13.1.3}$$

Now, $F_{\phi j}^i$ has the following three solutions:

(a) $\langle \Phi_j^i \rangle = 0.$ (13.1.4a)

(b) $\langle \Phi_j^i \rangle = a \begin{pmatrix} 1 & & & & \\ & 1 & & & \\ & & 1 & & \\ & & & -3/2 & \\ & & & & -3/2 \end{pmatrix},$

with

$$a = \frac{4x}{3y}, \tag{13.1.4b}$$

(c) $\langle \Phi_j^i \rangle = b \begin{pmatrix} 1 & & & & \\ & 1 & & & \\ & & 1 & & \\ & & & 1 & \\ & & & & -4 \end{pmatrix},$

with

$$b = \frac{10}{33}\frac{x}{y}. \tag{13.1.4c}$$

This implies that the superpotential in eq. (13.1.1) leads to a three-fold vacuum degeneracy, one corresponding to unbroken SU(5), SU(3)$_c \times$ SU(2)$_L \times$ U(1), and SU(4) $\times$ U(1). This creates cosmological problems since, as the universe cools below temperature $T \sim M_X/g$, it finds three degenerate minima including the one it is in, i.e., the SU(5) symmetric one. Thus we fail to understand why, below the SU(5) scale, SU(3)$_c \times$ SU(2) $\times$ U(1) is the symmetry of electroweak physics in this model. Again, radiative corrections cannot change this picture because of the nonrenormalization theorem. This is the first price we pay for the introduction of supersymmetry. The first minimum is easily removed by the introduction of a singlet field X and rewriting the part of the superpotential with ϕ's only as

$$W(\phi) = \lambda_2 X(\mathrm{Tr}\Phi^2 - \mu^2). \tag{13.1.5}$$

However, there is no way to split the SU(3)$_c \times$ SU(2)$_L \times$ U(1)$_Y$ and SU(4) $\times$ U(1) vacuum degeneracy. As we will see in Chapter 15 introduction of supergravitational interactions does help to remove this degeneracy. We will therefore assume, for now, that somehow the vacuum symmetry is uniquely chosen to be SU(3)$_c \times$ SU(2)$_L \times$ U(1)$_Y$, below the grand unification scale.

Let us now look at the breaking of SU(2)$_L \times$ U(1)$_Y$ down to U(1)$_{em}$ symmetry. In the most general SU(5) invariant superpotential given in eq. (13.1.1) we now focus on the terms involving $H_{u,d}$ and Φ. We see that, subsequent to SU(5) breaking, the last term leads to an effective low-energy Higgs potential of the form

$$|F^2| = \lambda_1^2 \left(m' - \frac{3a}{2}\right)^2 (\phi_u^+\phi_u + \phi_d^+\phi_d) + (\text{triplet terms}), \tag{13.1.6}$$

where $\phi_{u,d}$ denote the SU(2)$_L$, doublet piece of the Higgs fields $H_{u,d}$. Note that if the fine tune m' is such that

$$m' - \frac{3a}{2} = \mathrm{O}(m_W/g), \tag{13.1.7}$$

then weak-interaction-breaking scales will emerge. [The sign of the mass term in eq. (13.1.6) is, however, positive; but soft-supersymmetry-breaking terms with negative (mass)^2 of the same order can fix the sign.] Since m' and a are determined in terms of parameters present in the superpotential, they are not affected by radiative corrections at all. This partially resolves the gauge hierarchy problem.

Before we leave this section we wish to point out that as in the supersymmetry extension of the SU(3)$_c \times$ SU(2)$_L \times$ U(1) model, we will add appropriate soft-breaking terms that do not disturb renormalizability to make the model phenomenologically acceptable. The supersymmetry breaking terms must, of course, respect SU(5) symmetry.

§13.2 Proton Decay in the Supersymmetry SU(5) Model

In the supersymmetric grand unified models there are several sources for proton decay: (i) the conventional lepto–quark (X, Y) vector boson exchange; and (ii) the new supersymmetric contributions [3]. From our discussion in Chapter 5 we learn that the conventional X, Y gauge boson exchange contributions depend on the magnitude of the unification mass. Since supersymmetric models involve new fields, there will be new contributions to the β function that will affect the evolution of the various gauge coupling constants, and this will change the value of the unification mass M_U. This has been studied by various authors [4]. To study these effects we note that for the supersymmetric $SU(N)$ gauge group the β function is given by the following formula:

$$\beta_i(g_i) = b_{0i} \frac{g_i^3}{16\pi^2}, \tag{13.2.1}$$

where

$$b_{0i} = -3N + T_F + T_H, \tag{13.2.2}$$

where T_F and T_H represent the contributions of the fermionic and Higgs fields. The gauge contribution changes from $-11N/3$ to $-3N$ due to the contribution of the gauginos and, similarly, the changes in fermionic and Higgs contributions reflect the new contributions from supersymmetric partners. We must, however, be careful to use this formula only in the domain where supersymmetry is exact. In the case under discussion in this chapter, this formula will be applicable between $m_W < \mu < M_X$. We note that this reduces the slope in the evolution of the α_i^{-1} in its approach to the grand unified value α_i^{-1} and, as a result, the unification is "delayed," i.e., Fig 13.1 displays the evolution of the gauge couplings in the supersymmetric model. Form this we deduce that the unification scale $M_U \equiv M_X$ is given by

$$M_{x,\text{SUSY}} \simeq 6 \times 10^{16} \Lambda_{\overline{\text{MS}}} \approx 10^{16} \text{ GeV} \tag{13.2.3}$$

and we get

$$\sin^2 \theta_W(m_W) = 0.236 \pm 0.003. \tag{13.2.4}$$

Because of the larger M_X, the proton lifetime is predicted to be $\tau_p \simeq 10^{35}$ yr. Thus, gauge-boson-induced baryon nonconservation in the supersymmetric SU(5) model is not measurable. This is not, however, the whole story of baryon nonconservation in supersymmetric models and there exist other potentially large sources of baryon number violation in these theories.

To understand these new class of phenomena we note that, in supersymmetric theories, there exist two classes of possible operators for baryon-number-violating processes:

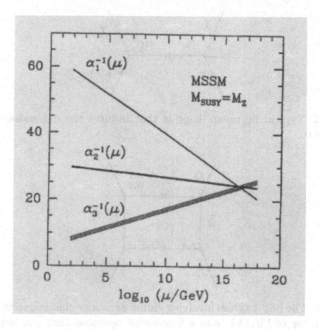

Figure 13.1. This figure shows the unification of gauge couplings with supersymmetric model spectrum. α_i^{-1} is plotted against the mass scale and the values at the weak scale are the measures values from LEP and SLC as well as other experiments.

(a) D type

$$\frac{1}{M_U^2} \int d^4\theta \Phi^+ \Phi \Phi^+ \Phi, \tag{13.2.5}$$

where Φ denotes the matter superfields, ψ, T, etc.

(b) F type

$$\frac{1}{M_U} \int d^2\theta \Phi^4. \tag{13.2.6}$$

The first class is of the type that arises from gauge boson exchange and has already been discussed and shown to be small. The second class of terms has lower dimension and *a priori* could be very large, leading to rapid proton disintegration. They must, therefore, be carefully examined in order to uncover any additional suppression. In the component language the F type terms give rise to the baryon-number-violating operator of dimension 5 typically of the following type:

$$O_{\Delta B \neq 0} = \frac{1}{M_U} Q_i^T C^{-1} \tau_2 Q_j \tilde{Q}_k^T \tau_2 \tilde{L} \varepsilon^{ijk}. \tag{13.2.7}$$

This operator could arise from the Feynman diagram shown in Fig. 13.2. Note that, as it stands, the above operator involves squark and slepton

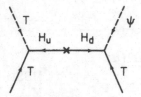

Figure 13.2. Typical Feynman diagram that induces the dimension-5 operator with $\Delta B \neq 0$.

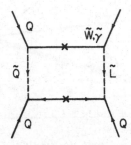

Figure 13.3. The box diagram involving gluino exchange that converts the dimension-5 operator of Fig.13.1. into a four-Fermi operator that can induce proton decay.

fields and, therefore, cannot induce proton decay in the lowest-order. The scalar quark and the scalar lepton fields could, however, turn into quarks and lepton fields on emitting weak gauginos as in Fig. 13.3.

The dominant contribution to proton decay [5] will arise from the graph where a gluino is exchanged to convert the squarks to quarks and has the following magnitude:

$$m_{\tilde{W}} > m_{\tilde{q}},$$

$$M_{qqql} \simeq \frac{g_2^2}{32\pi^2} \cdot \frac{Q^T C^{-1} \tau_2 h_u Q \cdot Q^T C^{-1} \tau_2 h_d L}{m_{\tilde{W}} m_H} \tag{13.2.8}$$

and

$$m_{\tilde{G}} \ll m_{\tilde{q}},$$

$$M_{qqql} \simeq \frac{8 g_3^2}{48\pi^2} \cdot \frac{m_{\tilde{G}}}{m_{\tilde{q}}^2 M_H}. \tag{13.2.9}$$

The above operators involve purely left-handed quark–lepton fields. We can construct similar operators involving right-handed fields. To calculate the proton decay amplitude we have to find the dominant contributions from (13.2.8) and (13.2.9). We note that $h_u = G_F^{1/2} M_u$ and $h_d = G_F^{1/2} M_d$. It is easy to find that the leading operator is of the form

$$\sim G_F m_u m_s \frac{2\alpha_s m_{\tilde{G}} \theta_c u_L^T C^{-1} d_L s_L^T C^{-1} \nu_{\mu L}}{3\pi m_{\tilde{q}}^2 M_H}. \tag{13.2.10}$$

This is of order

$$\sim 10^{-26} \frac{(m_{\tilde{G}} \text{ in GeV})}{m_{\tilde{q}}^2} \text{ GeV}^{-2}. \tag{13.2.11}$$

For $m_{\tilde{q}} \simeq 10^2$ GeV and $M_G \leq 1$ GeV, this amplitude $\simeq 10^{-30}$ GeV^2, which is compatible with presently known bounds on the proton lifetime. An important outcome of supersymmetry, however, is that the dominant decay mode is different from that of the SU(5) model and is

$$p \to K^+ \bar{\nu}_\mu,$$
$$n \to K^0 \bar{\nu}_\mu.$$

The discovery of the above decay modes will therefore be an indication of supersymmetric nature, or at least the important role played by the Higgs fields in proton decay.

Another important point to note is that the selection rule $\Delta(B - L) = 0$ holds for proton decay amplitudes as in the nonsupersymmetric model.

13.2.1 Problems and Prospects for SUSY SU(5)

While the simple SUSY SU(5) model exemplifies the power and utility of the idea of SUSY GUTs, it also brings to the surface some of the problems one must solve if the idea eventually has to be useful. Let us enumerate them one by one and also discuss the various ideas proposed to overcome them.

(i) R-parity breaking:

There are renormalizable terms in the superpotential that break baryon and lepton number:

$$W' = \lambda_{abc} T_a \bar{F}_b \bar{F}_c. \tag{13.2.12}$$

When written in terms of the component fields, this leads to R-parity breaking terms of the MSSM such as $L_a L_b e_c^c$, QLd^c as well as $u^c d^c d^c$, etc. The new point that results from grand unification is that there is only one coupling parameter that describes all three types of terms and also the coupling λ satisfies the antisymmetry in the two generation indices b, c. The total number of parameters that break R parity are nine instead of 45 in the MSSM. There are also nonrenormalizable terms of the form $T\bar{F}\bar{F}(\Phi/M_{P\ell})^n$, which are significant for $n = 1, 2, 3, 4$ and can add different complexion to the R-parity violation. Thus, the SUSY SU(5) model does not lead to an LSP that is naturally stable to lead to a CDM candidate. As we will see in the next section, the SO(10) model provides a natural solution to this problem if only certain Higgs superfields are chosen.

(ii) Doublet–triplet splitting problem:

We saw earlier that to generate the light doublets of the MSSM, one needs a fine tuning between the two parameters $3/2\lambda b$ and M in the superpotential. However, once SUSY breaking is implemented via the hidden sector mechanism, one gets a SUSY-breaking Lagrangian of the form

$$L_{SB} = A\lambda\bar{H}\Phi H + BM\bar{H}H + \text{h.c.,} \tag{13.2.13}$$

where the symbols in this equation are only the scalar components of the superfields. In general supergravity scenarios, $A \neq B$. As a result, when the Higgsinos are fine tuned to have mass in the weak-scale range, the same fine tuning does not leave the scalar doublets at the weak scale.

There are two possible ways out of this problem: we discuss them below.

(iiA) Sliding singlet

The first way out of this is to introduce a singlet field S and choose the superpotential of the form

$$W_{DT} = 2\bar{H}\Phi H + S\bar{H}H. \tag{13.2.14}$$

The supersymmetric minimum of this theory is given by

$$F_H = H_u(-3b + \langle S \rangle) = 0. \tag{13.2.15}$$

The F_ζ equation is automatically satisfied when color is unbroken as is required to make the theory physically acceptable. We see that one then automatically gets $\langle S \rangle = 3b$, which is precisely the condition that keeps the doublets light. Thus the doublets remain naturally of the weak scale without any need for fine tuning. This is called the sliding singlet mechanism. In this case the supersymmetry breaking at the tree level maintains the masslessness of the MSSM doublets for both the fermion as well as the bosonic components. There is however a problem that arises once one-loop corrections are included because they lead to corrections for the $\langle S \rangle$ v.e.v. of order $\frac{1}{16\pi^2}m_{3/2}M_U$, which then produces a mismatch in the cancellation of the bosonic Higgs masses. One is back to square one!

(iiB) Missing partner mechanism:

A second mechanism that works better than the previous one is the so called missing partner mechanism where one chooses to break the GUT symmetry by a multiplet that has coupling to the H and $\bar{H}$ and other multiplets in such a way that once SU(5) symmetry is broken, only the color triplets in them have multiplets in the field it couples to pair up with but not weak doublets. As a result, the doublet is naturally light. An example is provided by adding the **50**, **50** dimensional representations (denoted by $\Theta^{\alpha\beta}_{\gamma\delta\sigma}$ and $\bar{\Theta}^{\gamma\delta\sigma}_{\alpha\beta}$, respectively) and replacing **24** by the **75** (denoted Σ) dimensional multiplet. Note that **75** dimensional multiplet has a standard model singlet in it so that it breaks the SU(5) down to the standard model gauge group. At the same time **50** has a color triplet only and no doublet. The **50.75.5** coupling enables the color triplet in **50** and $\bar{5}$ to pair up leaving the weak doublet in $\bar{H}$ light. The superpotential in this case can be given by

$$W_G = \lambda_1 \Theta\Sigma H + \lambda_2 \bar{\Theta}\Sigma\bar{H} + M\Theta\bar{\Theta} + f(\Sigma). \qquad (13.2.16)$$

This mechanism can be applied in the case of other groups too.

(iii) Neutrino masses

Finally, in the SU(5) model there seems to no natural mechanism for generating neutrino masses although using the R-parity violating interactions for such a purpose has often been suggested. One would then have to accept that the required smallness of their couplings has to be put in by hand.

§13.3   Supersymmetric SO(10)

In this section, we like to discuss supersymmetric SO(10) models, which have a number of additional desirable features over SU(5) model. For instance, all the matter fermions fit into one spinor representation of SO(10); second, the SO(10) spinor being 16-dimensional contains the right-handed neutrino leading to nonzero neutrino masses. The gauge group of SO(10) is left–right symmetric which has the consequence that it can solve the SUSY CP problem and R-parity problem, etc., of the MSSM unlike the SU(5) model.

13.3.1   Symmetry Breaking and Fermion Masses

Let us now proceed to discuss the breaking of SO(10) down to the standard model. Much of this discussion is similar to the ones in Chapter 7. SO(10) contains the maximal subgroups SU(5) × U(1) and SU(4)$_c$ × SU(2)$_L$ × SU(2)$_R$ × Z_2, where the Z_2 group corresponds to charge conjugation. The SU(4)$_c$ group contains the subgroup SU(3)$_c$ × U(1)$_{B-L}$. There are therefore many ways to break SO(10) down to the standard model. Below we list a few of the interesting breaking chains along with the SO(10) multiplets whose v.e.v.'s lead to that pattern.

(A) SO(10) $\rightarrow$ SU(5) $\rightarrow$ G_{STD}

The Higgs multiplet responsible for the breaking at the first stage is a **16**-dimensional multiplet (to be denoted ψ_H) that has a field with the quantum number of ν^c, which is an SU(5) singlet but with nonzero $B - L$ quantum number. The second stage can be achieved by

$$16_H \rightarrow 1_{-5} + 10_{-1} + \bar{5}_{+3}. \tag{13.3.1}$$

The breaking of the SU(5) group down to the standard model is implemented by the **45**-dimensional multiplet which contains the **24** dimension. representation of SU(5), which as we saw in the previous section contains a singlet of the standard model group. In the matrix notation, we can write breaking by **45** as $\langle A \rangle = i\tau_2 \times \mathrm{Diag}\,(a, a, a, b, b,)$ where $a \neq 0$, whereas we could have $b = 0$ or nonzero.

A second symmetry breaking chain of physical interest is

(B) $SO(10) \rightarrow G_{224D} \rightarrow G_{STD}$

where we have denoted $G_{224D} \equiv SU(2)_L \times SU(2)_R \times SU(4)_c \times Z_2$. We will use this obvious shorthand for the different subgroups. This breaking is achieved by the Higgs multiplet

$$54 = (1, 1, 1) + (3, 3, 1) + (1, 1, 20') + (2, 2, 6).$$

The second stage of the breaking of G_{224D} down to G_{STD} is achieved in one of two ways and the physics in both cases are very different as we will see later: (i) **16**+**1̄6** or (ii) **126**+ **1̄26**. For clarity, let us give the G_{224D} decomposition of the **16** and **126**.

$$16 = (2, 1, 4) + (1, 2, \bar{4}), \tag{13.3.2}$$
$$126 = (3, 1, 10) + (1, 3, \bar{10}) + (2, 2, 15) + (1, 1, 6).$$

In matrix notation, we have

$$\langle 54 \rangle = \mathrm{Diag}\,(2a, 2a, 2a, 2a, 2a, 2a, -3a, -3a, -3a, -3a) \tag{13.3.3}$$

and for the **126** case it is the $\nu^c \nu^c$ component that has nonzero v.e.v.

It is important to point out that since the supersymmetry has to be maintained down to the electroweak scale, we must consider the Higgs bosons that reduce the rank of the group in pairs (such as **16**+ **1̄6**). Then the D terms will cancel among themselves. However, such a requirement does not apply if a particular Higgs boson v.e.v. does not reduce the rank.

(C) $SO(10) \rightarrow G_{2231} \rightarrow G_{STD}$

This breaking is achieved by a combination of **54**- and **45**-dimensional Higgs representations. Note the absence of the Z_2 symmetry after the first stage of breaking. This is because the $(1,1,15)$ (under G_{224}) submultiplet that breaks the SO(10) symmetry is odd under the D parity. The second stage breaking is as in the case (B).

(D) $SO(10) \rightarrow G_{224} \rightarrow G_{STD}$

Note the absence of the D parity in the second stage. This is achieved by the Higgs multiplet **210**, which decomposes under G_{224} as follows:

$$\mathbf{210} = (1,1,15) + (1,1,1) + (2,2,10) \qquad (13.3.4)$$
$$+(2,2,\overline{10}) + (1,3,15) + (3,1,15) + (2,2,6).$$

The component that acquires v.e.v. is $\langle \Sigma_{78910} \rangle \neq 0$.

It is important to point out that since the supersymmetry has to be maintained down to the electroweak scale, we must consider the Higgs bosons that reduce the rank of the group in pairs (such as **16**+ **16**). Then the D terms will cancel among themselves. However, such a requirement does not apply if a particular Higgs boson v.e.v. does not reduce the rank.

Let us now proceed to the discussion of fermion masses. As in all gauge models, they will arise out of the Yukawa couplings after spontaneous symmetry breaking. To obtain the Yukawa couplings, we first note that $\mathbf{16} \times \mathbf{16} = \mathbf{10} + \mathbf{120} + \mathbf{126}$. Therefore the gauge-invariant couplings are of the form $\mathbf{16.16.10} \equiv \Psi^T C^{-1} \Gamma_a \Psi H_a$; $\mathbf{16.16.120} \equiv \Psi \Gamma_a \Gamma_b \Gamma_c \Psi \Lambda_{abc}$, and $\mathbf{16.16.1\overline{26}} \equiv \Psi \Gamma_a \Gamma_b \Gamma_c \Gamma_d \Gamma_e \Psi \Delta_{abcde}$. We have suppressed the generation indices. Treating the Yukawa couplings as matrices in the generation space, one gets the following symmetry properties for them: $h_{10} = h_{10}^T$, $h_{120} = -h_{120}^T$, and $h_{126} = h_{126}^T$, where the subscripts denote the Yukawa couplings of the spinors with the respective Higgs fields.

To obtain fermion masses after electroweak symmetry breaking, one has to give v.e.v.s to the following components of the fields in different cases: $\langle H_{9,10} \rangle \neq 0$, $\Lambda_{789,7810} \neq 0$ or $\Lambda_{129} = \Lambda_{349} = \Lambda_{569} \neq 0$ (or with 9 replaced by 10), and similarly $\Delta_{12789} = \Delta_{34789} = \Delta_{56789} \neq 0$, etc. Several important constraints on fermion masses implied in the SO(10) model are:

(i) If there is only one **10** Higgs responsible for the masses, then only $\langle H_{10} \rangle \neq 0$ and one has the relation $M_u = M_d = M_e = M_{\nu D}$; where the M_F denote the mass matrix for the F-type fermion.

(ii) If there are two **10**'s, then one has $M_d = M_e$ and $M_u = M_{\nu D}$.

(iii) If the fermion masses are generated by a **126**, then we have the mass relation following from $SU(4)$ symmetry i.e. $3M_d = -M_e$ and $3M_u = -M_{\nu D}$.

It is then clear that, if we have only **10**'s generating fermion masses, we have the bad mass relations for the first two generations in the down-electron sector. On the other hand, it provides the good $b - \tau$ relation. One way to cure it would be to bring in contributions from the **126**, which split the quark masses from the lepton masses—since in the G_{224} language, it contains $(2,2,15)$ component that gives the mass relation $m_e = -3m_d$. This combined with the **10** contribution can perhaps provide phenomenologically viable fermion masses. One possibility is to have the M_d and M_e of the following forms to avoid the bad mass relations among the first generations while keeping $b - \tau$ unification (known as Georgi–Jarlskog mechanism):

$$M_d = \begin{pmatrix} 0 & d & 0 \\ d & f & 0 \\ 0 & 0 & g \end{pmatrix}, \quad M_e = \begin{pmatrix} 0 & d & 0 \\ d & -3f & 0 \\ 0 & 0 & g \end{pmatrix}, \quad (13.3.5)$$

$$M_u = \begin{pmatrix} 0 & a & 0 \\ a & 0 & b \\ 0 & b & c \end{pmatrix}. \quad (13.3.6)$$

These mass matrices lead to $m_b = m_\tau$ at the GUT scale and $\frac{m_e}{m_\mu} \simeq \frac{1}{9} \frac{m_d}{m_s}$, which are in much better agreement with observations.

13.3.2 Neutrino Masses, R-Parity Breaking, 126 vrs. 16

One of the attractive aspects of the SO(10) models is the left–right symmetry inherent in the model. A consequence of this is the complete quark–lepton symmetry in the spectrum. This implies the existence of the right-handed neutrino, which as we saw in §6.4, is crucial to our understanding of the small neutrino masses. This comes about via the seesaw mechanism mentioned earlier (§6.4). The generic seesaw mechanism for one generation can be seen in the context of the standard model with the inclusion of an extra right-handed neutrino, which is a singlet of the standard model group. As is easy to see, if there is a right-handed neutrino denoted as ν^c, then we have additional terms in the MSSM superpotential of the form $h_- \nu L H_u \nu^c + M \nu^c \nu^c$. After electroweak symmetry breaking, there emerges a 2×2 mass matrix for the (ν, ν^c) system of the following form:

$$M_\nu = \begin{pmatrix} 0 & h_u v_u \\ h_u^T v_u & M \end{pmatrix}. \quad (13.3.7)$$

This is the seesaw matrix discussed in Chapter 6 and leads to small neutrino masses as follows: using the fact that $m_f \simeq h_u v_u$, we get $m_\nu \simeq m_f^2/M \ll m_f$.

To see the implications of embedding the seesaw matrix in the SO(10) model, let us first note that if the only source for the quark and charged lepton masses is the 10-dimensional representation of SO(10), then we have a relation between the Dirac mass of the neutrino and the up-quark masses: $M_u = M_{\nu^D}$. Let us now note that the $\nu^c \nu^c$ mass term M arises from the vacuum expectation value of the $\nu^c \nu^c$ component of $\overline{126}$ and therefore corresponds to a fundamental gauge symmetry-breaking scale in the theory, which can be determined from the unification hypothesis. Thus apart from the coupling matrix of the $\overline{126}$ denoted by f, everything can be determined. This gives predictive power to the SO(10) model in the neutrino sector. For instance, if we take typical values for the f coupling to be one and ignore the mixing among generations, then, we get

$$m_{\nu_e} \simeq m_u^2/10 f v_{B-L}, \qquad (13.3.8)$$
$$m_{\nu_\mu} \simeq m_c^2/10 f v_{B-L},$$
$$m_{\nu_\tau} \simeq m_t^2/10 f v_{B-L}.$$

If we take $v_{B-L} \simeq 10^{12}$ GeV, then we get $m_{\nu_e} \simeq 10^{-8}$ eV $m_{\nu_\mu} \simeq 10^{-4}$ eV, and $m_{\nu_\tau} \simeq$ eV. These values for the neutrino masses are of great interest in connection with the solutions to the solar neutrino problem as well as to the hot dark matter of the universe. Things in the SO(10) model are therefore so attractive that one can go further in this discussion and about the prediction for v_{B-L} in the SO(10) model. The situation here however is more complex and we summarize the situation below.

If the particle spectrum all the way until the GUT scale is that of the MSSM, then both the M_U and the v_{B-L} are same and $\simeq 2 \times 10^{16}$ GeV. On the other hand, if above the v_{B-L} scale, the symmetry is G_{2213} and the spectrum has two bidoublets of the SUSYLR theory, $B-L = \pm 2$ triplets of both the left- and the right-handed groups and a color octet, then one can easily see that in the one-loop approximation the $v_{B-L} \simeq 10^{13}$ GeV or so. On the other hand, with a slightly more complex system described in section 2, we could get v_{B-L} almost down to a few TeV's. Thus unfortunately, the magnitude of the scale v_{B-L} is quite model dependent.

Finally, it is also worth pointing out that the equality of M_u and M_{ν^D} is not true in more realistic models. The reason is that if the charged fermion masses arise purely from the **10**-dimensional representations, then one has the undesirable relation $m_d/m_s = m_e/m_\mu$, which was recognized in the SU(5) model to be in contradiction with observations. Therefore in order to predict neutrino masses in the SO(10) model, one needs additional assumptions than simply the hypothesis of grand unification.

(i) Neutrino masses in the case $B-L$ breaking by $\mathbf{16}_H$

As has been noted, it is possible to break $B-L$ symmetry in the SO(10) model by using the $\mathbf{16} + \overline{\mathbf{16}}$ pair. This line of model building has been inspired by string models, which in the old fashioned fermionic compactification do not seem to lead to **126**-type representations at any level [6]. There have been several realistic models constructed along these lines [7]. In this case, one must use higher-dimensional operators to obtain the ν^c mass. For instance, the operator $\mathbf{16}_m\mathbf{16}_m\overline{\mathbf{16}}_H\overline{\mathbf{16}}_H/M_{Pl}$ after $B-L$ breaking would give rise to a ν^c mass $\sim v_{B-L}^2/M_{Pl}$. For $v_{B-L} \simeq M_U$, this will lead to $M_{\nu^c} \simeq 10^{13}$ GeV. This then leads to the neutrino spectrum of the above type.

Another way to get small neutrino masses in SO(10) models with **16**'s rather than **126**'s without invoking higher-dimensional operators is to use the 3×3 seesaw [8] rather than the 2×2 one discussed above. To implement the 3×3 seesaw, one needs an extra singlet fermion and write the following superpotential:

$$W_{33} = h\Psi H\Psi + f\Psi\bar\Psi_H S + \mu S^2. \tag{13.3.9}$$

After symmetry breaking, one gets the following mass matrix in the basis (ν, ν^c, S):

$$M_\nu = \begin{pmatrix} 0 & hv_u & 0 \\ hv_u & 0 & f\bar v_R \\ 0 & f\bar v_R & \mu \end{pmatrix}, \tag{13.3.10}$$

where $\bar v_R$ is the v.e.v. of the $\mathbf{\overline{16}}_H$. On diagonalizing for the case $v_u \simeq \mu \ll v_R$, one finds the lightest neutrino mass to be $m_\nu \simeq \frac{\mu h^2 v_u^2}{f v_R}$ and two other heavy eigenstates with masses of order $f v_R$.

(ii) R-parity conservation: automatic vs. enforced

One distinct advantage of $\mathbf{126}$ over $\mathbf{16}$ is in the property that the former leads to a theory that conserves R parity automatically even after $B - L$ symmetry is broken. This is very easy to see as was emphasized in section 1. Recall that $R = (-1)^{3(B-L)+2S}$. Since the $\mathbf{126}$ breaks $B - L$ symmetry via the $\nu^c\nu^c$ component, it obeys the selection rule $B - L = 2$. Putting this in the formula for R, we see clearly that R parity remains exact even after symmetry breaking. On the other hand, when $\mathbf{16}$ is employed, $B - L$ is broken by the ν^c component, which has $B - L = 1$. As a result R parity is broken after symmetry breaking. To see some explicit examples, note that with $\mathbf{16}_H$ one can write renormalizable operators in the superpotential of the form $\Psi\Psi_H H$, which after $\langle \nu^c \rangle \neq 0$ leads to R-parity breaking terms of the form LH_u discussed in the section 1. When one goes to the nonrenormalizable operators, many other examples arise: e.g., $\Psi\Psi\Psi\Psi_H/M_{Pl}$ after symmetry breaking lead to QLd^c, LLe^c as well as $u^c d^c d^c$ type terms.

13.3.3   Doublet–Triplet Splitting (D-T-S):

As we noted in section 3, splitting the weak doublets from the color triplets appearing in the same multiplet of the GUT group is a very generic problem of all grand unification theories. Since in the SO(10) models, the fermion masses are sensitive to the GUT multiplets which lead to the low-energy doublets, the problem of D-T-S acquires an added complexity. What we mean is the following: as noted earlier, if there are only $\mathbf{10}$ Higgses giving fermion masses, then we have the bad relation $m_e/m_\mu = m_d/m_s$ that contradicts observations. One way to cure this is to have either an $\mathbf{126}$ which leaves a doublet from it in the low-energy MSSM in conjunction with the doublet from the **10**'s or to have only **10**'s and have nonrenormalizable operators give an effective operator that transforms like **126**. This means that the process of doublet–triplet splitting must be done in a way that accomplishes this goal.

One of the simplest ways to implement D-T-S is to employ the missing v.e.v. mechanism[9], where one takes two **10**'s (denoted by $H_{1,2}$) and couple them to the **45** as AH_1H_2. If one then gives v.e.v. to A as $\langle A \rangle = i\tau_2 \times$ Diag$(a, a, a, 0, 0)$, then it is easy to verify that the doublets (four of them) remain light. This model without further ado does not lead to MSSM. So one must somehow make two of the four doublets heavy. This was discussed in great detail in Ref. 9. A second problem also tackled in Ref. 9 is the question that once the SO(10) model is made realistic by the addition of say $16 + \overline{16}$, then new couplings of the form $16.\overline{16}.45$ exist in the theory that give nonzero entries at the missing v.e.v. position thus destroying the whole suggestion. There are however solutions to this problem by increasing the number of **45**'s.

Another more practical problem with this method is the following. As mentioned before, the low-energy doublets in this method are coming from **10**'s only and is problematic for fermion mass relations. This problem was tackled in two papers [10]. In the first paper, it was shown how one can mix in a doublet from the **126** so that the bad fermion mass relation can be corrected. To show the bare essentials of this technique, let consider a model with a single H, single pair $\Delta + \bar{\Delta}$, and a A and $S \equiv \mathbf{54}$ and write the following superpotential:

$$W = M\Delta\bar{\Delta} + \Delta A\bar{\Delta} + HA^2\Delta/M + SH^2 + M'H^2. \quad (13.3.11)$$

After symmetry breaking this leads to a matrix of the following form among the three pairs of weak doublets in the theory, i.e., $H_{u,10}, H_{u,\Delta}, H_{u,\bar{\Delta}}$ and the corresponding H_d's. In the basis where the column is given by $(\mathbf{10}, \mathbf{126}, \overline{\mathbf{126}})$ and similarly for the row, we have the doublet matrix:

$$M_D = \begin{pmatrix} 0 & \langle A \rangle^2/M & 0 \\ \langle A \rangle^2/M & 0 & M \\ 0 & M & 0 \end{pmatrix}, \quad (13.3.12)$$

where the direct $H_{u,10}H_{d,10}$ mass term is fine tuned to zero. This kind of a 3×3 mass matrix allows the low-energy doublets to have components from both the **10** and **126** and thus avoid the mass relations. It is easy to check that the triplet mass matrix in this case makes all of them heavy.

To summarize, we have given only two examples of the simplest grand unification models. There are many more possibilities; hopefully, the reader will have sufficient resources now to scan the literature for other ones. The two models we have chosen have canonical normalization of the generators as we embed the gauge groups into the GUT group, i.e., $\text{Tr}\Theta_{2L}^2 = 2$, etc. This normalization can change for other groups. This can have profound effect on grand unification. A simple example of this is $SU(5) \times SU(5)$ model discussed in [11].

Exercises

13.1. Construct a supersymmetric $SO(10)$ model and show in this model that the low-energy constraints of $\sin^2 \theta_W$ and α_{strong} do not allow the right-handed scale to be in the TeV range unless there is a breakdown of D-parity symmetry as defined in Chapter 7.

13.2. Derive eq. (13.2.3) using the renormalization group evolution of gauge couplings in the supersymmetric $SU(5)$ model.

13.3. Construct a supersymmetric $SO(10)$ model with the inverse gauge-hierarchy mechanism and evaluate the GUT scale induced in this method. Discuss possible problems with this model. (See Ref. 10.)

13.4. Consider the supersymmetric $SU(5) \times U(1)$ GUT model. Discuss the superpotential, symmetric breaking and the gauge hierarchy problem in this model. Discuss the $\sin^2 \theta_W$ constraints on the GUT scale in this model.

References

[1] S. Dimopoulos and H. Georgi, *Nucl. Phys.* **B193**, 150 (1981).

[2] N. Sakai, Z. *Phys.* **C11**, 153 (1981).

[3] S. Weinberg, *Phys. Rev.* **D25**, 287 (1982);
    N. Sakai and T. Yanagida, *Nucl. Phys.* **B197**, 533 (1982).

[4] M. B. Einhorn and D. R. T. Jones, *Nucl. Phys.* **B196**, 475 (1982);
    W. Marciano and G. Senjanovic, *Phys. Rev.* **D25**, 3092 (1982).

[5] J. Ellis, D. V. Nanopoulos, and S. Rudaz, *Nucl. Phys.* **202**, 43 (1982);
    J. Hisano, H. Murayama and T. Yanagida, *Nucl. Phys.* **B402**, 46 (1993);
    R. Arnowitt and P. Nath, *Phys. Rev.* **D 49**, 1479 (1994).

[6] S. Choudhuri, S. Chung and J. Lykken, hep-ph/9511456; G. Cleaver,hep-th/9708023;
    S. Choudhuri, S. Chung, G. Hockney and J.Lykken, *Nucl. Phys.* **456**, 89 (1995);
    G. Aldazabal, A. Font, L. Ibanez and A. M. Uranga, *Nucl. Phys.* **B452**, 3 (1995).

[7] K. S.Babu and S. Barr, *Phys. Rev.* **D50**, 3529 (1994);
    K. S. Babu and R. N. Mohapatra, *Phys. Rev. Lett.* **74**, 2418 (1995);
    L. Hall and S. Raby, *Phys. Rev.* **D51**, 6524 (1995); D. G. Lee and R. N. Mohapatra, *Phys. Rev.* **D51**, 1353 (1995);
    S. Barr and S. Raby, hep-ph/9705366;

C. Albright and S. Barr, hep-ph/9712488;
Z. Berezhiani, *Phys. Lett.* **B355**, 178 (1995).

[8] R. N. Mohapatra, *Phys. Rev. Lett.* **56**, 561 (1986);
R. N. Mohapatra and J. W. F. Valle, *Phys. Rev.* **D34**, 1642 (1986).

[9] S. Dimopoulos and F. Wilczek, I.T.P preprint (unpublished);
K. S. Babu and S. Barr, *Phys. Rev.* **D48**, 5354 (1993).

[10] D. G. Lee and R. N. Mohapatra, *Phys. Rev.* **D51**, 1353 (1995).
K. S. Babu and R. N. Mohapatra, *Phys. Rev. Lett.* **74**, 2418 (1995).

[11] R. N. Mohapatra, *Phys. Rev.* **D 54**, 5728 (1998).

14
Local Supersymmetry ($N = 1$)

§14.1 Connection Between Local Supersymmetry and Gravity

In this chapter we will study the implications of the hypothesis that the parameters of supersymmetry transformation ε become a function of space-time, i.e., $\varepsilon = \varepsilon(x)$. We know, from Chapter 1, that invariance under local symmetry requires new fields in the theory that have spin 1, and have the same number of components as the number of independent parameters in the group. In analogy, local supersymmetry will require the introduction of the spin-3/2 field, which is the Majorana type. This will bring us into a completely new domain of particle physics where new spin-3/2 elementary fields interact with ordinary matter fields. Furthermore, there will also be analogs of the Higgs mechanism once supersymmetry is spontaneously broken (the so-called super-Higgs effect). There is, however, a much more profound aspect to local supersymmetry. Once the spin-3/2 fields are introduced, to make the theory supersymmetric in the high-spin sector, it will turn out that we will require a massless spin-2 field, which can be identified with the graviton field $g_{\mu\nu}$ thus "unifying" gravitation with the other three forces of nature. This discovery was made independently by Freedman, Ferrara, and Van Niuenhuizen [1], and by Deser and Zumino [1], and opened up a whole new possibility, not only of unification of gravity with particle physics [2] but also of new consequences for particle physics with supersymmetry. We will call the spin-3/2 particle gravitino and denote it by a Majorana field ψ_μ.

To see the need for gravity in local supersymmetry, we first give some heuristic intuitive arguments, which we follow up in the subsequent sections with more rigorous formulation. Let us look at the supersymmetry transformations for matter fields (ϕ, ψ, F):

$$\delta\phi = \varepsilon\psi,$$

$$\delta\psi = \varepsilon F - i\partial^\mu\phi\sigma_\mu\bar\varepsilon,$$

$$\delta F = \frac{i}{2}\partial^\mu\psi\sigma_\mu\bar\varepsilon. \tag{14.1.1}$$

If ε is a function of x, then we can look at the trivial free supersymmetric Lagrangian in (9.2.1) (with the addition of the F^*F term) and see what modifications it will require. To see this, note that

$$\delta(\partial_\mu\phi) = \varepsilon\partial_\mu\psi + \partial_\mu\varepsilon \cdot \psi,$$

$$\delta(\partial_\mu\psi) = \varepsilon\partial_\mu F + \partial_\mu\varepsilon F - i\partial_\mu\partial^\nu\phi\sigma_\nu\bar\varepsilon - i\partial_\nu\phi\sigma^\nu\partial_\mu\bar\varepsilon. \tag{14.1.2}$$

Since global supersymmetry $[\varepsilon(x) = \text{const}]$ is a special case of this, the change in the action can be written as

$$\delta S = \int d^4x\partial_\mu\varepsilon^a K^\mu_a + \text{h.c.}, \tag{14.1.3}$$

where

$$K^a_\mu \equiv -\partial_\mu\phi^* \cdot \psi^a - \frac{i}{2}\psi^b(\sigma_\mu\bar\sigma_\nu)^a_b\partial^\nu\phi^* \tag{14.1.4}$$

(a, b are spinor indices). To keep the action invariant we must add the corresponding gauge field (ψ_μ) coupling, i.e.,

$$S' = \kappa \int K^a_\mu\psi^\mu_a d^4x, \tag{14.1.5}$$

such that, under the local supersymmetry transformation,

$$\psi^\mu_a \to \psi^\mu_a + \kappa^{-1}\partial^\mu\varepsilon_a. \tag{14.1.6}$$

ψ^μ_a is called the gravitino. What is the parameter κ? Looking at (14.1.5) and (14.1.6) we can conclude that κ has dimension of M^{-1}. So, in contrast with local bosonic symmetries, local supersymmetry requires a dimensional coupling. Since the gravitational constant $G_N = M_P^{-2}$ provides such a dimensional parameter, we could define $\kappa = \sqrt{8\pi G_N}$ and use this as the universal gauge coupling associated with local supersymmetry. Furthermore, we will see that the spin-2 boson, which carries the gravitational force, provides the superpartner of the gravitino.

A second heuristic argument that implies a connection between local supersymmetry and gravity is the following: if we look at the commutator of two successive local supersymmetry transformations

$$[\delta(\varepsilon_1(x)), \delta(\varepsilon_2(x))]B \sim \bar\varepsilon_1(x)\gamma^\mu\varepsilon_2(x)\partial_\mu B. \tag{14.1.7}$$

(From this point on we will switch to four-component notation.) The right-hand side implies that the translation depends on the space–time point but this precisely represents the coordinate transformations that lead to general relativity. A more powerful argument is to consider the variation of $(S + S')$ when we find (from the variation of K_μ^a under supersymmetry transformation)

$$\delta(S + S') = \int d^4x\, \bar{\psi}_\mu \gamma_\nu \varepsilon T^{\mu\nu}, \tag{14.1.8}$$

where $T^{\mu\nu}$ is precisely the energy momentum tensor. The only way to cancel this is to add to the Lagrangian a new term $T^{\mu\nu} g_{\mu\nu}$ provided under supersymmetry transformation, $\delta g_{\mu\nu} = \bar{\psi}_\mu \gamma_\nu \varepsilon$; this is the usual way to couple gravity to matter fields. Before proceeding to this discussion we will give a brief introduction to free-field theory of the spin-3/2 Rarita–Schwinger field and review the elements of general relativity.

§14.2   Rarita–Schwinger Formulation of the Massless Spin-3/2 Field

The massless Rarita–Schwinger field is described by a Majorana spinor with a Lorentz index, i.e., ψ_μ. The Lagrangian for this system is given by

$$\mathcal{L} = -\tfrac{1}{2}\varepsilon^{\mu\nu\rho\sigma}\bar{\psi}_\mu \gamma_5 \gamma_\nu \partial_\rho \psi_\sigma. \tag{14.2.1}$$

We see that it is invariant under the gauge transformation $\psi_\sigma \to \psi_\sigma + \kappa^{-1}\partial_\sigma\varepsilon$. The field equations are given by

$$-\varepsilon^{\mu\nu\rho\sigma}\gamma_5 \gamma_\nu \partial_\rho \psi_\sigma = 0. \tag{14.2.2}$$

We can evaluate the canonical momenta $\pi_\nu = \delta L/\delta(\partial_0 \psi_\mu)$ and we find $\psi_0 = 0$. Using the Dirac quantization procedure, the gauge condition corresponding to this can be written as $\psi_0 = 0$; otherwise, $\pi_0 = 0$ cannot be maintained at all times. Then calculating π_k, we obtain $\pi_k^\dagger = \varepsilon^{ijk}\gamma_5\gamma_4\gamma_i\psi_j$ which implies that

$$\partial_\kappa \pi_k^\dagger - \varepsilon^{ijk}\gamma_5\gamma_4\gamma_i\partial_k\psi_j = 0. \tag{14.2.3}$$

To maintain this condition at all times we can choose a gauge condition

$$\partial_i\psi^i = 0. \tag{14.2.4}$$

After these conditions are implemented, we are left with only two independent Majorana spinors. We, furthermore, note that there is a set of second class constraints

$$\pi_k^\dagger - \varepsilon^{ijk}\gamma_5\gamma_4\gamma_i\psi_j = 0. \tag{14.2.5}$$

We can now count the number of degrees of freedom by subtracting all the momentum and coordinate constraints from the total number of coordinates and momenta, i.e., 16 (coordinates) +16 (momentum) $-4(\pi^0 = 0) - 4(\psi^0 = 0) - 4$ [eq. (14.2.3)] -4 [eq. (14.2.4)] -12 [eq. (14.2.5)] $= 4$ which corresponds to the two helicity states.

§14.3 Elementary General Relativity

The basic ideas of general relativity [3] can be described either in terms of a metric $g_{\mu\nu}$, which is a symmetric second rank tensor with $\mu, \nu = 0, 1, 2, 3$ or in terms of a vierbein e_μ^m. We start by introducing the properties of $g_{\mu\nu}$. The basic ingredient of the theory of relativity is the requirement that, under generalized coordinate transformations

$$x^\mu \to x^{\mu'}, \tag{14.3.1}$$

the physics remains invariant. The line element $ds^2 = g_{\mu\nu}dx^\mu dx^\nu$ is invariant under the coordinate transformations

$$dx'^\mu = \frac{\partial x'^\mu}{\partial x^\nu}dx^\nu, \tag{14.3.2}$$

which implies that

$$g'_{\mu\nu} = \frac{\partial x^\rho}{\partial x'^\mu}\frac{\partial x^\sigma}{\partial x'^\nu}g_{\rho\sigma}. \tag{14.3.3}$$

We can define a contravariant $g^{\lambda\rho}$ that has the property

$$g_{\mu\lambda}g^{\lambda\rho} = \delta_\mu^\rho \tag{14.3.4a}$$

and under coordinate transformation

$$g'^{\mu\nu} = \frac{\partial x'^\mu}{\partial x^\rho}\frac{\partial x'^\nu}{\partial x^\sigma}g^{\rho\sigma}. \tag{14.3.4b}$$

We can then define the Christoffel symbol (or the affine connection) $\Gamma^\mu_{\nu\rho}$,

$$\Gamma^\mu_{\nu\rho} = \tfrac{1}{2}g^{\mu\lambda}\left[\frac{\partial g^{\lambda\nu}}{\partial x^\rho} + \frac{\partial g_{\lambda\rho}}{\partial x^\nu} - \frac{\partial g_{\rho\nu}}{\partial \chi^\lambda}\right]. \tag{14.3.5}$$

Using this, we can define the covariant derivative of tensors

$$D_\sigma V^\mu \equiv V^\mu_{;\sigma} = \frac{\partial V^\mu}{\partial x^\sigma} + \Gamma^\nu_{\rho\sigma}V^\rho. \tag{14.3.6}$$

The curvature (Ricci) tensor is then given as follows:

$$R^\lambda_{\mu\nu\sigma} = \frac{\partial \Gamma^\lambda_{\mu\nu}}{\partial x^\sigma} - \frac{\partial \Gamma^\lambda_{\mu\sigma}}{\partial x^\nu} + \Gamma^\rho_{\mu\nu}\Gamma^\lambda_{\sigma\rho} - \Gamma^\rho_{\mu\sigma}\Gamma^\lambda_{\nu\rho}. \tag{14.3.7}$$

The gravitational Lagrangian can then be written as

$$\mathcal{L}_G = -\frac{1}{2\kappa^2}\sqrt{g}R, \tag{14.3.8}$$

where $R = g^{\mu\nu} R^{\lambda}_{\mu\nu\lambda}$. To this, we can add the matter piece.

An alternative formulation, more suitable for the discussion of local supersymmetry, is the vierbein (or tetrad) formalism, where we define the vierbein e^m_{μ} where m transforms like a flat space coordinate while μ transforms like a curved space coordinate. This alternative formulation is possible because of the principle of equivalence, which states that at each space–time point we can choose a locally inertial coordinate system to describe physics in the presence of gravitational fields. Of course, this coordinate system will differ from point to point. We can then write the metric tensor $g_{\mu\nu}$ as follows:

$$g_{\mu\nu} = e^m_{\mu}(x) e^n_{\nu}(x) \eta_{mn}, \tag{14.3.9}$$

where η is metric in flat space–time. Under coordinate transformations

$$e^m_{\mu} \to e^{m\prime}_{\nu} = \frac{\partial x^{\nu}}{\partial x^{\prime\mu}} e^m_{\nu}. \tag{14.3.10}$$

We can use the vierbein to express any contravariant vector A^{μ} as a vector in the locally inertial coordinate system

$$A^m = e^m_{\mu} A^{\mu}. \tag{14.3.11}$$

We can raise and lower indices on e^m_{μ} by using $g^{\mu\nu}$ and η_{mn} to obtain e^{μ}_m, where e^{μ}_m satisfies the following property:

$$e^{\mu}_m e^n_{\mu} = \delta^n_m \quad \text{and} \quad e^{\mu}_m e^m_{\nu} = \delta^{\mu}_{\nu}. \tag{14.3.12}$$

The principle of equivalence requires that special relativity should apply in locally inertial frames; therefore, the index m will transform as a flat space vector index under Lorentz transformation, i.e.,

$$e^m_{\mu} \to \Lambda^m_n e^n_{\mu}. \tag{14.3.13}$$

Let us proceed to the construction of the Lagrangian in terms of e^m_{μ}. We must require it to be invariant under both coordinate as well as *local* Lorentz transformations (the latter because, at each space–time, point, the inertial coordinate system is different). There is an excellent discussion of this point in the book by Weinberg [3]. We refer the reader to this book for details.

To write down the Lagrangian in this formalism we will have to express the curvature tensor in terms of the vierbeins. For this purpose we will need to express $\Gamma^{\mu}_{\nu\alpha}$ in terms of e^m_{μ}. This is done by using the spin connection defined as follows.

The vierbein formalism also requires the definition of a covariant derivative, which must be such that it must not only transform appropriately under coordinate transformation, but also transform appropriately under local Lorentz transformation. As is obvious, this will require the introduction of a new quantity ω^{mn}_{ν}, called the spin connection, which will transform under local Lorentz transformations as

$$M_{mn}\omega_\nu^{mn} \to D(\Lambda)M_{mn}\omega_\mu^{mn}D^{-1}(\Lambda) - \partial_\mu D(\Lambda)D^{-1}(\Lambda). \qquad (14.3.14)$$

The ω_ν^{mn} are the gauge fields corresponding to local Lorentz transformation. We now note that since the covariant derivative of the metric tensor is zero, that of the vierbein must also be zero, i.e.,

$$D_\rho e_\sigma^m = 0 \qquad (14.3.15)$$

or

$$\partial_\rho e_\sigma^m + \omega_\rho^{mn}e_{n\sigma} - \Gamma_{\rho\sigma}^\alpha e_\alpha^m = 0. \qquad (14.3.16)$$

Using this equation we can reexpress $\Gamma_{\rho\sigma}^\alpha$ in terms of ω and e and we can define the following curvature tensor:

$$R_{\mu\nu}^{mn} = \partial_\mu\omega_\nu^{mn} - \partial_\nu\omega_\mu^{mn} + \omega_\mu^{mp}\omega_{\nu\rho}^n - \omega_\nu^{mp}\omega_{\mu p}^n. \qquad (14.3.17)$$

It is interesting to note that this is in the form of $F_{\mu\nu}$ for usual gauge fields. The usual curvature tensor is then defined as

$$R_{\tau\mu\nu}^\sigma e_{m\sigma}e_n^t = R_{\mu\nu,mn}. \qquad (14.3.18)$$

The action for pure gravity is then given by

$$S^{(2)} = -\frac{1}{2\kappa^2}\int eR(e,\omega)d^4x. \qquad (14.3.19)$$

Before closing this section we would like to cast the coordinate transformations, and the transformation of $g_{\mu\nu}$ under them, in a form that is reminiscent of the gauge theories. For this purpose define

$$x_\mu' = x_\mu + \zeta_\mu(x). \qquad (14.3.20)$$

Note a very powerful implication of this equation, which is coordinate transformations can be thought of as a set of local translations. It would then be possible to think of gravitation as the gauge theory of translation and Lorentz transformations.

To obtain the transformation property of $g^{\mu\nu}$ we use eq. (14.3.4b) and expand for infinitesimal ζ_μ

$$g^{\mu\nu\prime} = \frac{\partial(x^\mu + \zeta^\mu)}{\partial x^\rho}\frac{\partial(x^\nu + \zeta^\nu)}{\partial x^\sigma}g^{\rho\sigma}(x + \zeta)$$
$$= g^{\mu\nu} + g^{\mu\sigma}\partial_\sigma\zeta^\nu + g^{\rho\nu}\partial_\rho\zeta^\mu + \zeta^\rho\partial_\rho g^{\mu\nu} + O(\zeta^2). \qquad (14.3.21)$$

For the vierbein e_a^μ this implies

$$e_a^{\mu\prime} = e_a^\mu + e_a^\lambda\partial^\mu\zeta_\lambda + \zeta^\rho\partial_\rho e_a^\mu + O(\zeta^2). \qquad (14.3.22)$$

§14.4 $N = 1$ Supergravity Lagrangian

At this stage we are ready to write down the supergravity Lagrangian

$$\mathcal{L} = \mathcal{L}^{(2)} + \mathcal{L}^{(3/2)}, \tag{14.4.1}$$

where

$$\mathcal{L}^{(2)} = -\frac{1}{2\kappa^2} e R(e, \omega)$$

and

$$\mathcal{L}^{(3/2)} = -\tfrac{1}{2}\varepsilon^{\mu\nu\rho\sigma}\bar{\psi}_\mu \gamma_5 \gamma_\nu D_\rho \psi_\sigma, \tag{14.4.2}$$

where

$$D_\rho = \partial_\rho + \tfrac{1}{2}\sigma_{mn}\omega_\rho^{mn}. \tag{14.4.3}$$

Note that the absence of $\sqrt{g} \equiv e$ in eq. (14.4.2) $\mathcal{L}$ is invariant under the following set of supersymmetry transformations:

$$\delta e_\mu^m = \frac{\kappa}{2}\bar{\varepsilon}\gamma^m \psi_\mu,$$

$$\delta \psi_\mu = \frac{1}{\kappa}(D_\mu \varepsilon),$$

$$\delta \omega_{\mu,ab} = \tfrac{1}{4}\bar{\varepsilon}\gamma_5 \gamma_\mu \tilde{\psi}_{ab} + \tfrac{1}{8}\bar{\varepsilon}\gamma_5(\gamma^\lambda \tilde{\psi}_{\lambda b} e_{a\mu} - \gamma^\lambda \tilde{\psi}_{\lambda a} e_{b\mu}), \tag{14.4.4}$$

where

$$\psi_{ab} = e_a^\mu e_b^\nu (D_\mu \psi_\nu - D_\nu \psi_\mu).$$

As in the case of global supersymmetry, if we check the commutator of two supersymmetry transformations, we find that they do not close. This is merely a reflection of the fact that the number of bosonic and fermionic degrees of freedom off-shell do not match, and auxiliary fields must be introduced to compensate for this mismatch. To see how many auxiliary fields are needed, we can count the fermionic degrees of freedom: since, by local supersymmetry transformation, we can remove four fermionic degrees of freedom, we have, off-shell twelve fermionic degrees of freedom. On the other hand, from the *a priori* sixteen bosonic degrees of freedom in e_μ^a, by translation and Lorentz gauge transformation we can remove $4 + 6 = 10$ degrees of freedom. This leaves us with six bosonic degrees of freedom. So we need six auxiliary fields, which we call (S, P, A_μ). For completeness, we give the supersymmetry transformation in the presence of auxiliary fields:

$$\delta e_\mu^m = \frac{\kappa}{2}\bar{\varepsilon}\gamma^m \psi_\mu,$$

$$\delta \psi_\mu = \frac{1}{\kappa}(D_\mu + \frac{i\kappa}{2}A_\mu \gamma_5)\varepsilon - \tfrac{1}{2}\gamma_\mu \eta\varepsilon,$$

$$\delta S = \tfrac{1}{4}\bar{\varepsilon}\gamma^\mu R_\mu^{\text{cov}},$$

$$\delta P = -\frac{i}{4}\bar{\varepsilon}\gamma_5 \gamma^\mu R_\mu^{\text{cov}},$$

$$\delta A_m = \frac{3i}{4}\bar{\varepsilon}\gamma_5(R_m^{\text{cov}} - \tfrac{1}{8}\gamma_m\gamma \cdot R^{\text{cov}}),$$

where

$$\eta = -\tfrac{1}{8}(S - i\gamma_5 P - iA\gamma_5),$$

$$R_\mu^{\text{cov}} = \varepsilon^{\mu\nu\rho\sigma}\gamma_5\gamma_\nu\left(D_\rho\psi\sigma - \frac{i}{2}A_\sigma\gamma_5\psi_5\psi_\rho + \tfrac{1}{2}\gamma_\sigma\eta\psi_\rho\right) \qquad (14.4.5)$$

and the supersymmetric Lagrangian is

$$\mathcal{L} = \mathcal{L}^{(2)}(e,\omega) + \mathcal{L}^{(3/2)}(e,\psi,\omega) - \frac{e}{3}(S^2 + P^2 - A_m^2). \qquad (14.4.6)$$

So far we have considered only pure supergravity. Now we can try to couple supergravity to matter field systems. One way is to look at the variation of various fields under local supersymmetry and construct covariant derivatives; for instance, writing ϕ in eq. (14.1.1) as $\phi = A + iB$ we observe that local supersymmetry would require $D_\mu A = (\partial_\mu A - i\kappa\bar{\psi}\psi_\mu)$, etc. In the next section we will develop a systematic set of rules to couple matter fields and supergravity. This is known as tensor calculus.

§14.5 Group Theory of Gravity and Supergravity Theories

We will adopt a gauge-theoretic approach to supergravity where not only the gravitino field but also the gravitational field will arise as gauge potentials (or connections) corresponding to an underlying local symmetry. This will make it easier to obtain the coupling of the matter fields to supergravity simply by following the rules for the usual Yang–Mills theory. Important for this purpose is the discovery of the underlying symmetry group.

To get an idea about the underlying group, we recall the algebra of supersymmetric generators Q in Chapter 9 and we find that they involve the momentum. Therefore, the gravitational field must be a result of gauging translational symmetry. As mentioned in this chapter, making the translation parameter local is equivalent to a coordinate transformation. This, therefore, appears to be an immensely satisfying framework for the study of gravity. Below we note some of the well-known facts about the algebraic structure of gravity theories [2].

(a) Poincaré Algebra

This is the algebra of angular momentum operators $M_{\mu\nu}$ and momentum P_μ and is given by ($g_{np} = + + + -$)

$$[M_{mn}, M_{pq}] = g_{np}M_{mq} - g_{mp}M_{nq} - g_{nq}M_{mp} + g_{mq} + M_{np}, (14.5.1)$$

$$[M_{mn}, P_q] = g_{nq}P_m - g_{mq}P_n, \tag{14.5.2}$$
$$[P_m, P_n] = 0. \tag{14.5.3}$$

A coordinate space representation for M_{mn} and P_n is

$$M_{mn} = x_m \frac{\partial}{\partial x_n} - x_n \frac{\partial}{\partial x_m},$$

$$P_m = \frac{\partial}{\partial x_m}. \tag{14.5.4}$$

(b) de Sitter Algebra

We can define $O(4,1)$ algebra by considering a five-dimensional space and writing generalized angular momentum operators M_{AB} satisfying the algebra ($g_{AB} = +++--$)

$$[M_{AB}, M_{CD}] = g_{BC}M_{AD} - g_{AC}M_{BD} - g_{BD}M_{AC} + g_{AD}M_{BC}. \tag{14.5.5}$$

We can identify $M_{\mu 5} = P'_\mu$; we then see that

$$[P'_\mu, P'_\nu] = M_{\mu\nu}. \tag{14.5.6}$$

To obtain the Poincaré algebra we perform the Inonu–Wigner contraction by defining $P = \varepsilon P'$ and letting $\varepsilon \to 0$. The right-hand side of (14.5.6) then vanishes.

(c) Conformal Algebra

The Poincaré algebra can be extended by including the generators of scale (S) and conformal transformations (K_m) that have the following coordinate space representations:

$$S = x_m \frac{\partial}{\partial x_m}$$

and

$$K_m = 2x_m x_n \partial_n - x^2 \partial_m. \tag{14.5.7}$$

The full conformal algebra is obtained by adding the following commutation relation to eqs. (14.5.1), (14.5.2), and (14.5.3):

$$[K_m, K_n] = 0,$$
$$[K_m, P_n] = -2(q_{mn}D + M_{mn}),$$
$$[K_m, D] = -K_m,$$
$$[P_m, D] = P_m. \tag{14.5.8}$$

This algebra has 15 elements and can also be obtained from the algebra of the $O(4,2)$ group by performing the Inonu–Wigner contraction. $SU(2,2)$ is also locally isomorphic to the $O(4,2)$ group.

(d) Super-Poincaré Algebra

In order to study the algebraic structure of supersymmetry, we must extend the Poincaré algebra by including the spinorial (or odd) elements Q^a in the algebra, and which supplement the algebra in (14.5.1),(14.5.2), and (14.5.3) by addition of the following. In four-component notation

$$\{Q^a, \bar{Q}^b\} = (\gamma^m)^{ab} P_m, \tag{14.5.9}$$

$$[P_m, Q^a] = 0, \tag{14.5.10}$$

$$[M_{mn}, Q^a] = \tfrac{1}{2}(\sigma_{mn} Q^a). \tag{14.5.11}$$

In terms of superfield coordinates (x, θ) we can write the generators as follows:

$$P_m = \partial_m,$$

$$M_{mn} = x_m \partial_n - x_n \partial_m + \bar{\theta}\sigma_{mn}\frac{\partial}{\partial\bar{\theta}},$$

$$Q^a = \frac{1}{2}\left(\frac{\partial}{\partial\bar{\theta}} - \gamma^m \partial_m\theta\right)^a. \tag{14.5.12}$$

In fact, the de Sitter algebra can also be extended in a similar fashion to include supersymmetry. The graded Lie algebra [i.e., an algebra consisting of fermionic (odd) and bosonic (even) elements] associated with this is called OSP(4/1), which leaves the following line element invariant:

$$ds^2 = x_M x_M + \theta_a(C^{-1})_{ab}\theta_b, \tag{14.5.13}$$

where $M = 1, \ldots, 5$ with $g_{MN} = (+++--)$. The group on bosonic coordinates is O(3, 2) and on fermionic coordinates is SP(4) which are isomorphic to each other. By group contraction we can obtain the super-Poincaré algebra from this.

(e) Superconformal Algebra

The super-Poinearé algebra can be further extended to include the entire conformal group in ordinary bosonic coordinates. To close the algebra we will need one new spinorial element, S^a, and another bosonic element, denoted by A, called the superchiral symmetry. The graded Lie algebra for the superconformal group can be written as

$$[M_{mn}, M_{pq}] = [g_{np}M_{mq} - g_{mp}M_{nq} - g_{nq}M_{mp} + g_{mq}M_{np}],$$

$$[M_{mn}, P_q] = [g_{nq}P_m - g_{mq}P_n],$$

$$[P_m, P_n] = 0,$$

$$[K_m, K_n] = 0, \qquad [K_m, P_n] = -2(g_{mn}D + M_{mn}),$$

$$[K_m, D] = -K_m, \qquad [P_m, D] = P_m,$$

$$\{Q^a, \bar{Q}^b\} = 2(\gamma \cdot P)^{ab},$$

$$[P_m, Q^a] = 0, \qquad [M_{mn}, Q^a] = -(\sigma_{mn}Q)^a,$$

$$\{S^a, \bar{S}^b\} = -2(\gamma \cdot K)^{ab},$$

$$\{Q^a, \bar{S}^b\} = 2[D\sigma^{ab} - (\sigma^{mn})^{ab} M_{mn} - i(\gamma_5)^{ab} A],$$

$$[M_{mn}, S^a] = -(\sigma_{mn})^{ab} S^b,$$

$$[Q^a, A] = -\frac{3i}{4}(\gamma_5)^a_b Q^b,$$

$$[S^a, A] = \frac{3i}{4}(\gamma_5)^a_b S^b,$$

$$[Q^a, D] = \tfrac{1}{2} Q^a, \qquad [S^a, D] = -\tfrac{1}{2} S^a,$$

$$[S^a, P_m] = (\gamma_m)^a_b Q^b,$$

$$[Q^a, K_m] = -(\gamma_m)^a_b S^b. \qquad (14.5.14)$$

These elements also satisfy the Jacobi identities. We can now write down the super-coordinate representation for K, S, and A:

$$K_m = 2x_m x \cdot \partial - x^2 \partial_m - \bar{\theta}\gamma \cdot x\gamma_m \frac{\partial}{\partial \bar{\theta}} - \tfrac{1}{2}(\bar{\theta}\theta)^2 \partial_m - \bar{\theta}\theta(\bar{\theta}\gamma_m \partial/\partial \bar{\theta}),$$

$$S = \frac{1}{2}\left(\bar{\theta}\theta \frac{\partial}{\partial \bar{\theta}} + \bar{\theta}\gamma_5\theta\gamma_5 \frac{\partial}{\partial \bar{\theta}}\right) + \tfrac{1}{4}\bar{\theta}\gamma_5\gamma_m\theta\gamma_5\gamma_m \frac{\partial}{\partial \bar{\theta}}$$

$$+ \left(-\tfrac{1}{2}\gamma \cdot x\frac{\partial}{\partial \bar{\theta}} + \tfrac{1}{2}\gamma \cdot x\gamma \cdot \partial\theta + \tfrac{1}{2}\bar{\theta}\theta \; \partial\!\!\!/\theta\right),$$

$$A = -\tfrac{1}{4}i\bar{\theta}\gamma_5 \frac{\partial}{\partial \bar{\theta}}. \qquad (14.5.15)$$

Another representation for these operators can be given in terms of 4×4 matrices of the following type:

$$P_m = -\tfrac{1}{2}\gamma_m(1 - \gamma_5),$$

$$K_m = -\tfrac{1}{2}\gamma_m(1 + \gamma_5),$$

$$D = -\tfrac{1}{2}\gamma_5, \qquad M_{mn} = (\sigma_{mn}),$$

$$(Q^a)^\kappa_{4+i} = \{-\tfrac{1}{2}(1 + \gamma_5)c^{-1}\}^{\kappa a},$$

$$(S^a)^k_{4+i} = \{-\tfrac{1}{2}(1 + \gamma_5)c^{-1}\}^{ka},$$

$$(Q^a)^{4+i}_k = -\tfrac{1}{2}(1 - \gamma_5)^a_k,$$

$$(S^a)^{4+i}_k = -\tfrac{1}{2}(1 - \gamma_5)^a_k. \qquad (14.5.16)$$

The algebra of superconformal symmetry can be obtained from the graded Lie algebra $SU(2, 2/1)$, which leads to the following metric invariant:

$$ds^2 = |z_1|^2 + |z_2|^2 - |z_3|^2 - |z_\phi|^2 + \theta_1^* \theta^1 + \theta_2^* \theta^2 - \theta_3^* \theta^3 - \theta_4^* \theta^4. \quad (14.5.17)$$

§14.6   Local Conformal Symmetry and Gravity

Before we study supergravity with the new algebraic approach developed, we would like to discuss how gravitational theory can emerge from the gaug-

ing of conformal symmetry. For this purpose we briefly present the general
notation for constructing gauge covariant fields. The general procedure is
to start with the Lie algebra of generators X_A of a group

$$[X_A, X_B] = f_{AB}^C X_c, \qquad (14.6.1)$$

where f_{AB}^c are structure constants of the group. We can then introduce a
gauge field connection h_μ^A as follows:

$$h_\mu \equiv h_\mu^A X_A. \qquad (14.6.2)$$

Let us denote the parameter associated with X_A by ε^A. The gauge
transformations on the fields h_μ^A are given as follows:

$$\delta h_\mu^A = \partial_\mu \varepsilon^A + h_\mu^B \varepsilon^C f_{CB}^A \equiv (D_\mu \varepsilon)^A. \qquad (14.6.3)$$

We can then define a covariant curvature

$$R_{\mu\nu}^A = \partial_\nu h_\mu^A - \partial_\mu h_\nu^A + h_\nu^B h_\nu^C f_{CB}^A. \qquad (14.6.4)$$

Under a gauge transformation

$$\delta_{\text{gauge}} R_{\mu\nu}^A = R_{\mu\nu}^B \varepsilon^C f_{CB}^A. \qquad (14.6.5)$$

We can then write the general gauge invariant action as follows:

$$I = \int d^4 x Q_{AB}^{\mu\nu\rho\sigma} R_{\mu\nu}^A R_{\rho\sigma}^B. \qquad (14.6.6)$$

Let us now apply this formalism to conformal gravity. In this case

$$h_\mu = P_m e_\mu^m + M_{mn} \omega_\mu^{mn} + K_m f_\mu^m + D b_\mu. \qquad (14.6.7)$$

The various $R_{\mu\nu}$ are

$$R_{\mu\nu}(P) = \partial_\nu e_\mu^m - \partial_\mu e_\nu^m + \omega_\mu^{mn} e_\nu^n - \omega_\nu^{mn} e_\mu^m + b_\mu e_\nu^m + b_\nu e_\mu^m, \qquad (14.6.8)$$

$$R_{\mu\nu}(M) = \partial_\nu \omega_\mu^{mn} - \partial_\mu \omega_\nu^{mn} - \omega_\nu^{mp} \omega_{\mu,p}^n - \omega_\mu^{mp} \omega_{\nu,p}^n$$
$$- 4(e_\mu^m f_\nu^n - e_\nu^m f_\mu^n), \qquad (14.6.9)$$

$$R_{\mu\nu}(K) = \partial_\nu f_\mu^m - \partial_\mu f_\nu^m - b_\mu f_\nu^m + b_\nu f_\mu^m + \omega_\mu^{mn} f_\nu^n - \omega_\nu^{mn} f_\mu^n, \qquad (14.6.10)$$

$$R_{\mu\nu}(D) = \partial_\nu b_\mu - \partial_\mu b_\nu + 2e_\mu^m f_\nu^m - 2e_\nu^m f_\mu^m. \qquad (14.6.11)$$

The gauge-invariant Lagrangian for the gravitational field can now be
written down, using eq. (14.6.6), as

$$S = \int d^4 x \varepsilon_{mnrs} \varepsilon^{\mu\nu\rho\sigma} R_{\mu\nu}^{mn}(M) R_{\rho\sigma}^{rs}(M). \qquad (14.6.12)$$

We also impose the constraint that

$$R_{\mu\nu}(P) = 0, \qquad (14.6.13)$$

which expresses ω_μ^{mn} as a function of (e, b). The reason for imposing this
constraint has to do with the fact that P_m transformations must be eventually identified with coordinate transformation. To see this point more

explicitly let us consider the vierbein e_μ^m. Under coordinate transformations

$$\delta_{GC}(\xi^\nu)e_\mu^m = \partial_\mu \xi^\lambda e_\lambda^m + \xi^\lambda \partial_\lambda e_\mu^m. \qquad (14.6.14)$$

Using eq. (14.6.8) we can rewrite

$$\delta_{GC}(\xi^\nu)e_\mu^m = \delta_P(\xi e^n)e_\mu^m + \delta_\mu(\xi\omega^{mn})e_\mu^m + \delta_D(\xi b)e_\mu^m + \xi^\nu R_{\mu\nu}^m(P),$$

where

$$\delta_P(\xi^n)e_\mu^m = \partial_\mu \xi^m + \xi^n \omega_\mu^{mn} + \xi^m b_\mu. \qquad (14.6.15)$$

If $R^{\mu\nu}(P) = 0$, the general coordinate transformation becomes related to a set of gauge transformations via eq. (14.6.15).

At this point we also wish to point out how we can define the covariant derivative. In the case of internal symmetries $D_\mu = \partial_\mu - iX_A h_\mu^A$; now since momentum is treated as an internal symmetry, we have to give a rule. This follows from eq. (14.6.15) by writing a redefined translation generator $\tilde{P}$ such that

$$\delta_{\tilde{P}}(\xi) = \delta_{GC}(\xi^\nu) - \sum_{A\prime} \delta_{A\prime}(\xi^m h_m^A), \qquad (14.6.16)$$

where $A\prime$ goes over all gauge transformations excluding translation. The rule is

$$\delta_{\tilde{P}}(\xi^m)\phi = \xi^m D_m^C \phi. \qquad (14.6.17)$$

We also wish to point out that for fields that carry spin or conformal charge, only the intrinsic parts contribute to D_m^C and the orbital parts do not play any rule.

Coming back to the constraints we can then vary the action with respect to f_μ^m to get an expression for it, i.e.,

$$e_\nu^m f_{\mu m} = -\tfrac{1}{4}[e_m^\lambda e_{n\nu} R_{\mu\lambda}^{mn} - \tfrac{1}{6} g_{\mu\nu} R], \qquad (14.6.18)$$

where f_μ^m has been set to zero in R written in the right-hand side.

This eliminates (from the theory the degrees of freedom) ω_μ^{mn} and f_μ^m and we are left with e_μ^m and b_μ. Furthermore, these constraints will change the transformation laws for the dependent fields so that the constraints do not change.

Let us now look at the matter coupling to see how the familiar gravity theory emerges from this version. Consider a scalar field ϕ. It has conformal weight $\lambda = 1$. So we can write a convariant derivative for it, eq. (14.6.17),

$$D_\mu^C \phi = \partial_\mu \phi - \phi b_\mu. \qquad (14.6.19)$$

We note that the conformal charge of ϕ can be assumed to be zero since $K_m = x^2 \partial$ and is the dimension of inverse mass. In order to calculate $\Box^c \phi$ we start with the expression for d'Alembertian in general relativity

$$\frac{1}{e} \partial_\nu (g^{\mu\nu} e D_\mu^C \phi). \qquad (14.6.20)$$

The only transformations we have to compensate for are the conformal transformations and the scale transformations. Since

$$\delta b_\mu = -2\xi_k^m e_\mu, \qquad \delta(\phi b_\mu) = \phi \delta b_\mu = -2\phi f_\mu^m e_m^\mu + \tfrac{2}{12}\phi R, \qquad (14.6.21)$$

where, in the last step, we have used the constraint equation (14.6.18). Putting all these together we find

$$\Box^C \phi = \frac{1}{e}\partial_\nu(g^{\mu\nu} e D_\mu^c \phi) + b_\mu D_\mu^c \phi + \tfrac{2}{12}\phi R. \qquad (14.6.22)$$

Thus, the Lagrangian for conformal gravity coupled to matter fields can be written as

$$S = \int e d^4 x \tfrac{1}{2}\phi \Box^C \phi. \qquad (14.6.23)$$

Now we can use conformal transformation to gauge $b_\mu = 0$ and local-scale transformation to set $\phi = \kappa^{-1}$ leading to the usual Hilbert action for gravity. To summarize, we start with a Lagrangian invariant under full local conformal symmetry and fix conformal and scale gauge to obtain the usual action for gravity. We will adopt the same procedure for supergravity. An important technical point to remember is that, $\Box^C$, the conformal d'Alembertian contains R, which for constant ϕ, leads to gravity. We may call ϕ the auxiliary field.

§14.7 Conformal Supergravity and Matter Couplings

Our primary goal in this chapter is to write a Lagrangian for the interaction of quarks and leptons that not only respects local electroweak symmetries but also invariances under local supersymmetry. This provides a truly unified theory of all forces in nature. In the previous section the strategy was outlined for obtaining the Hilbert action for gravity. In this chapter we would like to follow the same procedure for supergravity, i.e., we will start with conformal supergravity so that a full conformal invariant Lagrangian can be written down for matter fields using the standard methods familiar from non-abelian gauge theories. Then we would reduce the gauge freedom by the use of auxiliary multiplets, thereby introducing the Planck constant into physics and obtaining the theory of supergravity coupled to matter. This approach was pioneered by Kaku, Townsend, and Van Nieuwenhuizen [4].

The Lie algebra valued connection and parameter for conformal supergravity are given by

$$h_\mu = P_m e_\mu^m + M_{mn}\omega_\mu^{mn} + D b_\mu + K_m f_\mu^m + \bar{Q}\psi_\mu + \bar{S}\phi_\mu + AA_\mu, \qquad (14.7.1)$$

$$\varepsilon = P_m \xi^m + \theta^{mn} M_{mn} + \bar{\varepsilon}Q + \bar{S}\xi + \xi_k^m K_m + \lambda_D D + \lambda_A A. \qquad (14.7.2)$$

The covariant derivative D_μ and the curvature $R^A_{\mu\nu}$ can be written following the rules outlined in the previous section once the quantum numbers of a particular field are known under conformal transformation. However, before we do that, we have to give the constraint equations involving $R_{\mu\nu}$:

$$R_{\mu\nu}(P) = 0, \qquad (14.7.3a)$$

$$R_{\mu\nu}(Q)^{\Gamma^\nu} = 0, \qquad (14.7.3b)$$

$$R^{mn}_{\mu\lambda}(M)e^\lambda_m e_{n\nu} - \tfrac{1}{2}R_{\lambda\nu}(Q)\gamma_\nu\psi^2 - \frac{i}{4}e\varepsilon_{\mu\nu\rho\sigma}R^{\rho\sigma}(A) = 0. \quad (14.7.3c)$$

The first and third constraints are, of course, the same as in the non-supersymmetric case, whereas the second constraint is specific to the supersymmetric theories. These are necessary in order to convert the P_m transformations to generalized coordinate transformations as discussed earlier. We also define the covariant derivative in a manner similar to the one given in eq. (14.6.17), except that A' goes over to new gauge degrees of freedom of the theory.

Moreover, we note that eq. (14.7.3) can be used to reexpress ω^{mn}_μ, ϕ_μ, and f^m_μ gauge fields in terms of other fields of the theory and the Q- and P-transformation rules get changed. We do not go into these details here and refer the reader to the review article by Van Nieuwenhuizen [2] and a recent review article by Kugo and Uehara [5]. (Important earlier references are given in Ref. 6.)

(a) Transformation Rules for Matter Fields

To study matter couplings to supergravity we need transformation rules of matter fields under the conformal group. Let us consider the chiral multiplet $\Sigma \equiv (A, \chi_L, \mathcal{F})$. It transforms under the various superconformal transformations as follows:

Q supersymmetry

$$\delta_Q A = \tfrac{1}{2}\bar{\varepsilon}\chi_L,$$

$$\delta_Q \chi_L = -\frac{i}{2}(1 + \gamma_5)(\mathcal{F} + \gamma^m D^C_m A)\varepsilon,$$

$$\delta_Q \mathcal{F} = \frac{i}{2}\bar{\varepsilon}\lambda^m D^C_m \chi_L. \qquad (14.7.4)$$

Note that the only difference from global supersymmetry is that ∂_m is replaced by the conformal covariant derivative D^C_m.

S supersymmetry

Let the corresponding parameter be ξ

$$\delta_S A = 0,$$

$$\delta_S \chi_L = -i\lambda \mathcal{A}(1 + \lambda_5)\xi,$$
$$\delta_S \mathcal{F} = i(\lambda - 1)\bar{\xi}\chi_L, \tag{14.7.5}$$

where λ is the scale dimension of the $\mathcal{A}$-component of the superfield Σ.

D (scale) transformations

$$\delta_D \mathcal{A} = \lambda\lambda_D \mathcal{A},$$
$$\delta_D \chi_L = (\lambda + \tfrac{1}{2})\lambda_D \chi_L,$$
$$\delta_D \mathcal{F} = (\lambda + 1)\lambda_D \mathcal{F}. \tag{14.7.6}$$

K_m (conformal) transformations

$$\delta_K \mathcal{A} = \delta_K \chi_L = \delta_K \mathcal{F} = 0. \tag{14.7.7}$$

Chiral transformations

$$\delta_A \mathcal{A} = \frac{i\lambda}{2}\lambda_A \mathcal{A},$$

$$\delta_A \chi_L = i\left(\frac{2\lambda - 3}{4}\right)\lambda_A \chi_L,$$

$$\delta_A \mathcal{F} = i\left(\frac{\lambda - 3}{2}\right)\lambda_A \mathcal{F}. \tag{14.7.8}$$

Similar rules can be written for the vector multiplet with component $(C, Z, H, K, V_\mu, \Lambda, D)$. Now we are ready to write the rules for constructing the matter-coupled Lagrangian for the case of supergravity [5, 6]. In complete analogy with the case of global supersymmetry, there are two kinds of Lagrangians for which the action is invariant under supersymmetry transformations: (i) F type and (ii) $\hat{D}$ type. But we see from eq. (14.7.4) that, due to local symmetries, $\delta_Q F$ is no more a full divergence but has additional terms that must be compensated. Furthermore, to maintain scale (D) and chiral (A) invariance we must have scale dimension $\lambda = 3$; and to maintain S invariance something must be added. Taking all this into account, it has been noted that the conformal invariant F-type Lagrangian is given by [6]

$$S_F = \int d^4x e[F + \tfrac{1}{2}\bar{\psi}_\mu \gamma^\mu \chi_L + \tfrac{1}{2}\bar{\psi}_\mu \sigma^{\mu\nu}(A - iB\gamma_5)\psi_\nu], \tag{14.7.9}$$

where we have rewritten $\mathcal{A} = A + iB$, A, B being real fields. Similarly, the locally conformal invariant D-type Lagrangian is

$$S_D = \int d^4x e\left[D + \Box^c C - \frac{i}{2}\bar{\psi}\cdot\gamma\gamma_5(\not{D}^c Z + \Lambda) = \tfrac{1}{2}\bar{\psi}_\mu \sigma^{\mu\nu}(H + i\gamma_5 K)\phi_\nu\right], \tag{14.7.10}$$

where $\Box^c$ is the conformal d'Alembertain, which is a supersymmetric generalization of eq. (14.6.22) and is given by

$$\Box^c = \partial_m D^{mc} - \sum_{A\prime} \delta_{A\prime}(h_m^{A\prime}) D_m^c \qquad (14.7.11)$$

and

$$D_m^c \mathcal{A} = \partial_m \mathcal{A} - \frac{i}{2}\bar{\psi}_m \chi_L - \lambda b_m \mathcal{A} - \frac{i\lambda}{2} A_m \mathcal{A}. \qquad (14.7.12)$$

We saw that in eq. (14.6.22) the Hilbert action arose because of the conformal gauge variation of the dilatation gauge field b_μ. In a similar manner, the variation of the term $\bar{\psi}_m \chi_L$ yields the gravitino counterpart of the Hilbert action, i.e.,

$$\Box^c = -\tfrac{1}{6}|\mathcal{A}|^2 \left(R + \frac{1}{e}\bar{\psi}_\mu \gamma_5 \gamma_\rho \partial_\sigma \psi_\lambda \varepsilon^{\mu\rho\sigma\lambda} \right) + D_\mu \mathcal{A}^* D^\mu \mathcal{A} + \cdots. \qquad (14.7.13)$$

We recognize the coefficient of $\mathcal{A}^* \mathcal{A}$ as the Lagrangian for pure Poincaré supergravity.

If we start with chiral fields of arbitrary scale dimension, we must multiply them until we get their total scale dimension (to be three) to construct S_F. To construct S_D from chiral fields we must multiply them with antichiral fields. We also often have to multiply several vector multiplets. We therefore need the conformal generalization of the multiplication rules for global supersymmetry. For chiral fields there is no change. For vector fields

$$V_1 \times V_2 = V_3,$$

where

$$V_3 = \big[C_1, C_2, C_1 Z_2 + C_2 Z_1, C_1 H_2 + C_2 H_1 - \tfrac{1}{2} Z_1^T C^{-1} Z_2,$$
$$C_1 K_2 + C_2 K_1 + \frac{i}{2} Z_1^T C^{-1} \gamma_5 Z_2, C_1 V_m^2 + C_2 V_m^1 + \frac{i}{2} Z_1^T C^{-1} \gamma_5 \gamma_m Z_2,$$
$$\{C_1 \Lambda_2 + \tfrac{1}{2}(H_1 - i\gamma_5 K_1 + i\gamma_5 \not{V}_1 - D^c Z_2) + (1 \leftrightarrow 2)\},$$
$$C_1 D_2 + C_2 D_1 + H_1 H_2 + K_1 K_2 - V_M^1 V^{m2} - D_m^C C_1 D^{cm} C_2$$
$$-Z_1^T C^{-1} \Lambda_2 - Z_2^T C^{-1} \Lambda_1 - \tfrac{1}{2} Z_1^t C^{-1} D^c Z_2 - \tfrac{1}{2} Z_2^T C^{-1} \not{D}^c Z_1 \big].$$
$$(14.7.14)$$

Similarly, we give the formula for the vector multiplet obtained from the product of a chiral and antichiral multiple Φ and Φ^* where $\Phi = (A + iB, \chi_L, F + iG)$

$$\Phi_1 \Phi_2^* = \Big[A_1 A_2 + B_1 B_2 - \{(B_1 + i\gamma_5 A_1)\chi_2 + 1 \leftrightarrow 2\},$$
$$- \{(A_1 F_2 + B_1 G_2) + 1 \leftrightarrow 2\}, \{(B_1 F_2 - A_1 G_2) + 1 \leftrightarrow 2\},$$
$$\Big\{ B_1 D_m^c A_2 - A_1 D_m^c B_2 + \frac{i}{2}\bar{\chi}_{1L} \gamma_m \chi_{2L} + 1 \leftrightarrow 2 \Big\},$$

$$\{(G_1 + i\gamma_5 F_1)\chi_{2L} + \slashed{D}^c(B_1 + i\gamma_5 A_1)\chi_2 + 1 \leftrightarrow 2\},$$

$$\{2F_1 F_2 + 2G_1 G_2 - 2D_m^c A_1 D_m^c A_2 - 2\slashed{D}_m^c B_1 D_m^c B_2 - \chi_{1L}^T C^{-1} \slashed{D}^c \chi_{2L}$$

$$- \chi_{2L}^T C^{-1} \slashed{D}^c \chi_{1L}\}\Big]. \tag{14.7.15}$$

(b) The Pure Supergravity Lagrangian

We illustrate the use of the action formula by constructing the Lagrangian for pure supergravity. We consider an auxiliary chiral multiplet Σ with conformal and chiral weight 1 and construct a real vector multiple out of it taking $\Sigma^*\Sigma$. This has weight 2 and therefore its D term has got weight 4 and we can use the S_D formula to obtain (writing $\mathcal{F} = F + iG$)

$$S_D = \frac{1}{2} \int d^4x e \Big[F^2 + G^2 + \Box^c(A^2 + B^2) - \frac{i}{2}\bar{\psi} \cdot \gamma\gamma_5(D^c(A\chi_L) + \chi\mathcal{F})$$

$$+ \frac{1}{2}\bar{\psi}_\mu \sigma^{\mu\nu}(1 + \gamma_5)(A\mathcal{F} + \cdots)\phi_\nu \Big]. \tag{14.7.16}$$

Now, we can fix the conformal gauge by setting $b_m = 0$ and the S gauge by setting $\chi = 0$; $A = \kappa^{-1}$ by using chiral and scale transformations. The resulting action, then, is of the form [using eq. (14.7.14)]

$$S_D = \frac{1}{2} \int d^4x e \Big[-\frac{1}{6\kappa^2}\mathcal{L}_{SG} + \frac{A_\mu^2}{4\kappa^2} + F^2 + G^2 \Big]. \tag{14.7.17}$$

We thus obtain the supergravity Lagrangian in a very intuitive manner using familiar methods of gauge theories. Furthermore, we see that the auxiliary fields F, G, A_μ of supergravity are related to an underlying conformal symmetry.

Another important thing to be noted is that the ψ_μ field appearing in the connection formula has dimension $+1/2$ rather than $+3/2$, which is to be expected for a propagating spinor. However, after the conformal gauges are fixed, a new spin-3/2 field can be defined such that $\psi_\mu^{\text{new}} = (1/\kappa)\psi_\mu$; the ψ_μ^{new} has dimension 3/2 and is the actual dynamical field.

§14.8 Matter Couplings and the Scalar Potential in Supergravity

To discuss the application of $N = 1$ supergravity, we must couple a system of the Yang–Mills matter system to supergravity. The rules for the construction are given in the previous section. We illustrate it for the case of a matter field assigned to a chiral superfield $S(z, \chi_L, h)$ where components are denoted in increasing order of the power of θ. We ignore the gauge fields to start with.

The starting point is to assign zero conformal weight ($\lambda = 0$) to the superfield S and choose an auxiliary multiplet $\Sigma(\mathcal{A}, \chi'_L, \mathcal{F})$ with $\lambda = +1$. In global (or rigid) supersymmetry the superpotential part denotes the self-interactions. Let us take a superpotential $g(S)$. Let us choose another function $f(S, S^\dagger)$ to obtain the kinetic energy part. It is clear that $g(s)$ will become part of the F-type action whereas $f(S, S^\dagger)$ will be used in constructing the D-type action. Note that we have chosen an arbitrary function $f(S, S^\dagger)$; each term in its power series expansion must have at least one S and one $S^\dagger$.

To construct the F-type Lagrangian we consider the $(g(S)\Sigma^3)$ superfield since it has scale dimension 3 and, remember that by using the s transformation, D and A transformations, we can write

$$\Sigma = \left(\kappa^{-1}, 0, \frac{u}{3}\right), \tag{14.8.1}$$

so that

$$\Sigma^3 = (\kappa^{-3}, 0, u), \tag{14.8.2}$$

$$g(s)\Sigma^3 = (g(z)\kappa^{-3}, \chi_L g'(z), ug(z)\kappa^{-1}$$
$$- g''\kappa^{-2}\chi_L^T C^{-1}\chi_L^\dagger + g'(z)h\kappa^{-3}). \tag{14.8.3}$$

So we have

$$S_F = \tfrac{1}{2}\int d^4x e\Big[\kappa^{-1}ug(z) + hg'(z)\kappa^{-3} - g''\chi_L^T C\chi_L\kappa^{-2}$$
$$+ \tfrac{1}{2}\bar{\psi}\cdot\gamma\chi_{Li}g'(z)\kappa^{-2} + \frac{\kappa^{-2}}{2}g(z)\bar{\psi}_\mu\sigma^{\mu\nu}(1 - \gamma_5)\psi_\nu\Big] + \text{h.c.} \tag{14.8.4}$$

To discuss the kinetic energy part we write down the real vector multiplet $\Sigma^*\Sigma$ using the formula in eq. (14.7.6):

$$\Sigma^*\Sigma = (\kappa^{-2}, 0, -\tfrac{2}{3}u, -\tfrac{2}{3}A_m\kappa^{-1}, 0, \tfrac{2}{9}(uu^* - A_m A^m\kappa^{-2})). \tag{14.8.5}$$

Note that the origin of the A_m term is the covariant conformal derivative $D^c_m \cdot 1$. Using formula (14.7.5) we can also expand $f(S, S^\dagger)$ into its component fields

$$f(S, S^\dagger) = [f(z, z^*), -2if'^i\chi_{iL}, -2f'^i h_i + 2f''^{ij}\chi_{Li}^T C^{-1}\chi_{Lj},$$
$$if'^i D_m z_i - if'_i D_m z^{*i} - 2if''^{ij}\chi_L^{-i}\gamma_m\chi_{Lj},$$
$$- 2if'''_i h_j\chi_L^i + 2if'''^{ij}_k\chi_{L}^k\chi_{Li}^T C^{-1}\chi_{Lj} + 2if''^{ij}_i \not{D}z^{*i}\chi_{Lj},$$
$$2f'''^i_j h_i h^{*j} - 2f'''^{ij}_k\chi_{Li}^T C^{-1}\chi_{Lj}h^{*k} - 2f'''^k_{ij}\chi_{Lj}^T C\chi_{Li}^* C\chi_{Li}^* h_k$$
$$+ 2f''''^{ij}_{kl}\chi_{LK}^\dagger C\chi_{iL}^* - 2f''^j_i D_m z^{*i} D_m z_j + \chi_{Li}^T C^{-1}\chi_{Lj}$$
$$- 2f''^j_i\bar{\chi}_{Lj}D\chi_{Li} - 2(f'''^i_{ji} D_\mu z^{*k} + f'''^{ik}_j D_\mu z_k\bar{\chi}_{Li}\gamma^\mu\chi_{Lj})]. \tag{14.8.6}$$

To finally write down the action, we have to multiply $f(S, S^\dagger)\Sigma\Sigma^*$ and use the formula in eq. (14.7.10). The Lagrangian has a long complicated form, but let us write down some of the most important terms

$$
\begin{aligned}
\kappa^{-2}e^{-1}\mathcal{L}_{\text{K.E.}} = &-\tfrac{1}{6}f(z,z^*)e^{-1}\mathcal{L}_{SG} \\
&+ f_j''^i(-\tfrac{1}{2}D_\mu z_i D^\mu z^{j*} - \bar{\chi}_{Li}\, \slashed{D}\chi_L^j + \tfrac{1}{2}h_i h^{*j}) \\
&+ \tfrac{1}{3}u^*(f'^i h_i - f''^{ij}\chi_{Li}^T C^{-1}\chi_{Lj}) \\
&+ \frac{i}{3}A^\mu(\tfrac{1}{2}f_j''^i \bar{\chi}_L^j \gamma_\mu \chi_{Li} + f'^i(D_\mu z_i - \bar{\psi}_{\mu L}\chi_{Li})), \quad (14.8.7)
\end{aligned}
$$

where

$$
\begin{aligned}
\mathcal{L}_{SG} &= (R + R_\psi + \tfrac{1}{3}(u^*u - A_m A_m)), \\
e^{-1}\mathcal{L}_{\text{P.E.}} &= -\tfrac{1}{2}g''^{ig}\chi_{Li}^T C^{-1}\chi_{Lj} + \tfrac{1}{2}g'^i h_i + \tfrac{1}{2}gu \\
&\quad - \tfrac{1}{2}\bar{\psi}_L \cdot \gamma\chi_{Li}g'^i + \tfrac{1}{2}g\bar{\psi}_{\mu L}\sigma^{\mu\nu}\psi_{\nu R}. \quad (14.8.8)
\end{aligned}
$$

An important point to note is that even though we wrote for the action

$$
S = \int d^4\theta(f(s,s^\dagger)\Sigma\Sigma^*) + \int d^2\theta(g(s)\Sigma^3) + \text{h.c.}, \quad (14.8.9)
$$

we can rescale the auxiliary field by $\Sigma g^{-1/3}$ to argue that the action will depend only on the function $f(z,z^*)/g^{2/3}$. We can define a new function for this:

$$
e^{\mathcal{G}(z,z^*)/3} = \frac{f(z,z^*)}{g^{2/3}(z)}. \quad (14.8.10)
$$

We would now like to discuss two important implications of the supergravity Lagrangian written as: (i) the supergravity modifications of the effect scalar potential; and (ii) the spontaneous breaking of supersymmetry and the super-Higgs effects.

Effective Scalar Potential in Supergravity Theories

To address this question we have to eliminate the auxiliary fields and cast the kinetic energy terms in canonical form. Let us therefore isolate the relevant parts of the Lagrangian from eqs. (14.8.7) and (14.8.8). First we focus on the part of the Lagrangian that involves only the auxiliary fields and bosonic fields

$$
\begin{aligned}
e^{-1}\mathcal{L}_{\text{aux}} = &\tfrac{1}{18}f(z,z^*)(uu^* - A_m A^m) + \tfrac{1}{2}f_j''^i h_i h^{*j} \\
&+ \tfrac{1}{2}g'^i h_i + \tfrac{1}{3}u^*f'^i h_i + \tfrac{1}{2}g^*u^* + f'^i D_\mu z_i + \text{h.c.} \quad (14.8.11)
\end{aligned}
$$

Let us define a new expression

$$
T = 3\ln\left(-\left(\frac{f}{3}\right)\right) \quad \text{and} \quad \tilde{u} = u + T''^i h_i. \quad (14.8.12)
$$

We can then rewrite $e^{-1}\mathcal{L}_{\text{aux}}$ as follows:

$$
e^{-1}\mathcal{L}_{\text{aux}} = \tfrac{1}{18}f(\tilde{u}\tilde{u}^* - A_m A^m) + \frac{1}{6}fT_j''^i h_i h^{*j}
$$

$$+ \tfrac{1}{2}g \left[\tilde{u} - h_i \left(T'^i - \frac{g'^i}{g} \right) \right] + f'^i D_\mu Z_i + \text{h.c.,} \quad (14.8.13)$$

this leads to the following field equations:

$$\tilde{u} = \frac{9g^*}{2f},$$

$$\frac{f}{3} h_i T_k''^i = -\tfrac{1}{2} g^* \left(\frac{g_k^*}{g^*} - T_k'^* \right),$$

$$\tfrac{2}{3} f A_\mu = i f'^I D_\mu Z_i. \quad (14.8.14)$$

Substituting back into (14.8.13) we find

$$e^{-1} \mathcal{L}_{\text{aux}} = -\frac{9}{4f} |g|^2 - \frac{3}{f} (T''^{-1})_i^k \left[\frac{g^*}{2} \left(\frac{g_k'^*}{g^*} - T_k' \right) \times \frac{g}{2} \left(\frac{g'^l}{g} - T''^l \right) \right]$$

$$- \frac{1}{4f} [f'^i D_\mu z_i - f_i' D_\mu z^{*i}]^2. \quad (14.8.14a)$$

Equation (14.8.14a) has no auxiliary fields and gives the form of the effective potential in terms of the dynamical fields and their functions.

Let us now look at other terms in the Lagrangian and try to bring them to canonical form by adjusting the arbitrary function $f(z, z^*)$. First, we see that to get the Hilbert action for gravity, we have to do Weyl resealing of the vierbein field $e_{m\mu}$ as follows:

$$e_{m\nu} \to e^\sigma e_{m\mu}$$

and

$$e_m^\mu \to e^{-\sigma} e_m^\mu, \qquad e \to e^{4\sigma} e,$$

$$\chi \to e^{-\sigma/2} \chi,$$

$$\psi_\mu \to e^{\sigma/2} \psi_\mu,$$

where

$$\sigma = \tfrac{1}{2} \ln \left(-\frac{3}{f\kappa^2} \right) = -\frac{T}{6}. \quad (14.8.15)$$

Under this transformation

$$\tfrac{1}{6} e f R \to -\frac{1}{2\kappa^2} e R - \tfrac{3}{4} e (\partial_\mu \ln f)^2 + \text{four-divergence,}$$

$$e \varepsilon^{\mu\nu\rho\sigma} \bar{\psi}_\mu \gamma_5 \gamma_\rho \partial_\sigma \psi_\nu \to -\frac{1}{2\kappa^2} \bar{\psi}_\mu \gamma_5 \gamma_\rho D_\sigma \psi_\nu. \quad (14.8.16)$$

We now call $\kappa^{-1} \psi_\mu \equiv \psi_\mu^g$ the dynamical gravitino field with dimension $3/2$. This gives us the correct form of the gravitational part of the Lagrangian. Next we wish to get the canonical form for the kinetic energy part of the scalar fields. For that purpose, first note that using the $(\partial_\mu \log f)$ term in eq. (14.8.16), we can cast the kinetic energy part of the z field in a suitable form. We can now rewrite the entire bosonic part of the Lagrangian as follows:

$$e^{-1}\mathcal{L}_B = \frac{\kappa^2}{4}e^{-T}[3gg^* + T_l''^{-1k}(g_k^{*\prime} - T_k'g^*)(g_l' - T_g'g)]$$

$$-\frac{1}{2\kappa^2}R + \frac{2}{\kappa^2}T_j''^i D_\mu z_i D_\mu^* z^j. \qquad (14.8.17)$$

It is now clear that, if we choose $T = -(\kappa^2/2)z^*z$, the effective potential takes the form

$$V_{\text{eff}} = \frac{1}{2}\exp\left(+\frac{\kappa^2}{2}z^*z\right)\left[\left|\frac{\partial g}{\partial z} + \frac{\kappa^2}{2}z^*g\right|^2 - \frac{3}{2}\kappa^2|g|^2\right]. \qquad (14.8.18)$$

When we include gauge fields, we must add D terms to this potential. Equation (14.8.18) is the fundamental equation for model building in $N = 1$ supergravity. The particular form of the potential is intuitively understandable, referring to our equation which says that the Lagrangian can only be a function of $(f/g^{2/3})$; therefore, the effective potential must depend not just on g as in the case of global supersymmetry but on both f and g.

EXERCISE

First redefining $\Sigma \to \Sigma g^{-1/3}$ we can recast the kinetic energy term in the form

$$\int d^4\theta \left(\frac{f}{|g|^{2/3}}\Sigma^*\Sigma\right).$$

Show that this form also leads to the same effective potential.

Another point of major importance to cosmology is the fact that the potential has negative terms in it so that subsequent to spontaneous breaking of the value of the potential need not be positive definite unlike in the case of global supersymmetry. This point will be elaborated upon in the next chapter.

§14.9 Super-Higgs Effect

Since supersymmetry is now a gauge symmetry, it is important to ask whether the analog of the Higgs mechanism exists in this case. If it does, subsequent to the spontaneous breakdown of supersymmetry of the massless Goldstone fermion, the Goldstino and the massless gravitino should combine to give a massive spin-3/2 field. We wish to demonstrate in this section that this indeed happens. Our discussion will be very schematic and we will leave out most of the technical details which can be found in the work of Cremmer et al. [6]. Let us look at the fermionic part of the F-type

Lagrangian in eq. (14.8.8)

$$S_F = \int d^4x e[-\tfrac{1}{2}g''^{ij}\chi_{Li}^T C^{-1}\chi_{Lj} + \tfrac{1}{2}\bar{\psi}_L\gamma\chi_{Li}g'^i + \tfrac{1}{2}g\bar{\psi}_{\mu L}\sigma^{\mu\nu}\psi_{\nu R}]. \quad (14.9.1)$$

When supersymmetry is broken, $g'^i \neq 0$. Therefore, there are bilinears involving $\bar{\psi}\cdot\gamma\chi$, the latter field being the Goldstino. Thus, gravitino acquires a mass whose magnitude is proportional to

$$M_{3/2}^2 = \exp\left(-\frac{\kappa^2}{2}z^*z\right)\frac{\kappa^4|g|^2}{4}. \quad (14.9.2)$$

An important point to note about the gravitino mass is that, before its value can be fixed, the cosmological constant must be set equal to zero to that order. This is because a combination of terms of the term

$$\mathcal{L}' = 3em^2\kappa^{-2} - \frac{i}{2}m\varepsilon^{\mu\nu\lambda\sigma}\bar{\psi}_\mu\Sigma_{\nu\lambda}\psi_\sigma \quad (14.9.3)$$

is invariant under local supersymmetry. This means that a cosmological constant is equivalent to a gravitino mass. Thus, the physical mass of the gravitino is obtained when the cosmological constant is set equal to zero.

For simplicity, we have ignored the Yang–Mills couplings in the above discussion. They can easily be included by replacing $S \rightarrow e^{gV}S$ in the function f and including the following generalized kinetic energy term for the gauge fields

$$\mathcal{L}_g = \int d^4x \ \text{Re} \int d^2\theta(f_{\alpha\beta}(s)W_a^\alpha\varepsilon^{ab}W_b^\beta) + \text{h.c.} \quad (14.9.4)$$

If $f_{\alpha\beta}(S) = \delta_{\alpha\beta}$, this gives the canonical kinetic energy term for the gauge multiplet. It is also clear that if we choose $f_{\alpha\beta} = (1 + S)\delta_{\alpha\beta}$, then, on expanding the Lagrangian, we will have a term of the form $h\lambda^T C^{-1}\lambda$, i.e., a mass term for the gaugino, after rescalings, and spontaneous supersymmetry breakings that can give rise to a gaugino mass term.

§14.10   Different Formulations of Supergravity

We saw in Section 14.4 that to match the bosonic and fermionic degrees of freedom for the gravitino–graviton supermultiplet we need six bosonic auxiliary fields. In the formulation just presented, those auxiliary fields are identifiable with the complex F component of an auxiliary chiral superfield, which is used to fix conformal gauges and the gauge field corresponding to chiral symmetry. However, supergravity can be formulated by different choices of auxiliary fields. We list below the various formulations and their authors.

(a) The New Minimal Formulation developed by Sohnius and West [7] uses a real linear multiplet L, i.e., a multiplet satisfying $D^2L = \bar{D}^2L = 0$.

Its components can be written as (C, η, B_m), where B_m is a transverse vector field, η is a Majorana field, and C is a real field. The auxiliary fields in this case are two transverse vector fields B_m and the transverse chiral field A_m. The chiral field becomes transverse (and therefore has only three components) because the real auxiliary field has zero chiral quantum number and cannot be used to fix chiral gauge. The matter coupling to supergravity in this gauge has been studied and straightforward techniques lead to a scalar potential that is different from the one obtained in the old minimal case [8].

(b) The Breitenlohner Formulation [9]. In this formulation a complex linear multiplet is used as the auxiliary field. The components of a complex linear multiplet are $(\mathcal{A}, \eta_1, +i\eta_2, H, B_m, \Lambda)$. After fixing the gauge, the multiplet becomes $(1, \eta_1, H, B_{m_1} + iB_{m_2}, \Lambda)$. We have fourteen auxiliary bosonic fields ($H_1, H_2, B_{m_1}, B_{m_2}$, and A_m) and eight fermionic auxiliary fields (η_1, Λ), leaving six unpaired bosonic fields to match degrees of freedom in the graviton–gravitino multiplet. There also exists a tensor calculus for this multiplet [5].

(c) The de Wit–Van Niuenhuizen [10] Formulation uses a real vector multiplet as the auxiliary multiplet. Here again, after conformal gauge fixing, six unmatched bosonic fields remain as required.

We can give new formulations taking different sets [8, 11] (chiral, vector, etc.), of multiplets as auxiliary fields, but the most interesting for application to particle physics seems to be the old minimal formulation or any formulation that contains a chiral field in its auxiliary field set. In this case only an effective scalar potential, which is nontrivially different from that of global supersymmetry, emerges.

Exercises

14.1. Consider a coordinate transformation given by

$$\delta x^\mu = \xi_\mu,$$

such that

$$\delta g^{\mu\nu} = 0.$$

1. What conditions are satisfied by ξ^μ? ξ_μ is called the killing vector.
2. For the case of the flat metric, what is the algebra generated by the killing vector field defined as $R(\xi) = \xi^\mu \partial_\mu$?

14.2. If the coordinate transformation $\delta x^\mu = \xi^\mu(x)$ is such that under it

$$\delta g_{\mu\nu} = \lambda(x) g_{\mu\nu},$$

this ξ^μ is called a conformal killing vector.

1. What differential equation is satisfied by ξ^μ in this case?

2. Again choosing the flat metric, obtain the algebra generated by the killing vector field in 2 and 4 dimensions.

14.3. Show that in terms of $\zeta_\mu(x)$, defined by eq. (14.3.20), the change of $g_{\mu\nu}$ under coordinate transformations can be written

$$\delta g_{\mu\nu} = -(D_\mu \zeta_\nu + D_\nu \zeta_\mu),$$

where D_μ is the covariant derivative.

References

[1] D. Freedman, S. Ferrara, and P. Van Nieuwenhuizen, *Phys. Rev.* **D13**, 3214 (1976);
S. Deser and B. Zumino, *Phys. Lett.* **62B**, 335 (1976).

[2] For a review see
P. Van Nieuwenhuizen, *Phys. Rep.* **68**, 189 (1981).

[3] C. Misner, K. S. Thorne, and J. Wheeler, *Gravitation*, Freeman, San Francisco, 1970;
S. Weinberg, *Gravitation and Cosmology*, Wiley, New York, 1972.

[4] M. Kaku, P. K. Townsend, and P. Van Nieuwenhuizen, *Phys. Rev.* **D17**, 3179 (1978).

[5] T. Kugo and S. Uehara, *Nucl. Phys.* **B222**, 125 (1983).

[6] K. S. Stelle and P. C. West, *Nucl. Phys.* **B145**, 175 (1978);
P. Van Nieuwenhuizen and S. Ferrara, *Phys. Lett.* **B76**, 404 (1978);
E. Cremmer, S. Ferrara, L. Girardello and A. van Proeyen, *Phys. Lett.* **116B**, 231 (1982); *Nucl. Phys.* **B212**, 413 (1983);
E. Cremmer, B. Julia, J. Scherk, S. Ferrara, L. Girardello, and P. Van Nieuwenhuizen, *Nucl. Phys.* **B147**, 105 (1979);
A. H. Chamseddine, R. Arnowitt, and P. Nath, *Phys. Rev. Lett.* **49**, 970 (1982).

[7] M. Sohnius and P. West, *Phys. Lett.* **105B**, 353 (1981).

[8] C. S. Aulakh, M. Kaku, and R. N. Mohapatra, *Phys. Lett.* **126B**, 183 (1983);
For recent discussion of theories with multiple compensators, see
K. T. Mahanthappa and G. Stabler, VPI preprint, 1985.

[9] P. Breitenlohner, *Phys. Lett.* **67B**, 49 (1977)

[10] B. de Wit and P. Van Nieuwenhuizen, *Nucl. Phys.* **B139**, 216 (1978).

[11] V. O. Rivelles and J. G. Taylor, *Phys. Lett.* **113B**, 467 (1982).

15

Application of Supergravity ($N = 1$) to Particle Physics

§15.1 Effective Lagrangian from Supergravity

In this chapter we would like to build models of elementary particle interactions using the ideas developed in the previous chapter. To summarize the discussion we note that the general action that couples supergravity ($N = 1$) to the Yang–Mills matter field system can be written as follows:

$$S_{\text{eff}} = \int d^4x e (f(S, e^{gV} s^\dagger) \Sigma^* \Sigma)_D + \int d^4x e (g(S)\Sigma^3)_F + \text{h.c.}$$
$$+ \int d^4x e (f_{\alpha\beta}(S) W^{\alpha,a} W_a^\beta)_F + \text{h.c.}, \qquad (15.1.1)$$

where S and V denote the matter and gauge fields, respectively, and D and F denote the generalized D and F terms invariant under local supersymmetry transformations defined in eqs. (14.7.9) and (14.7.10), respectively.

In terms of these function we can write the Lagrangian for the matter Yang–Mills system as follows [denoting $S_i = (z_i, \chi_i)$ and $V = (V_\mu, \lambda)$]:

$$\mathcal{L}(z_i, \chi_i, V_\mu, \lambda, \psi_\mu) = \mathcal{L}_{\text{K.E.}} - V + \mathcal{L}_g + \mathcal{L}', \qquad (15.1.2)$$

where $\mathcal{L}_{\text{K.E.}}$ denotes the kinetic energy for the matter Yang–Mills fields, $\mathcal{L}_g$ for the graviton–gravitino Lagrangian is given by

$$\mathcal{L}_g = -\frac{e}{2\kappa^2} R - \varepsilon^{\mu\nu\rho\sigma} \bar\psi_\mu \gamma_5 \gamma_\nu D_\rho \psi_\sigma - m_g \bar\psi_\mu \sigma^{\mu\nu} \psi_\nu, \qquad (15.1.3)$$

where

$$m_g^2 = \kappa^{-2} e^{-\mathcal{G}}, \qquad (15.1.4)$$

and where

$$\mathcal{G} = 3 \ln\left(-\frac{\kappa^2 f}{3}\right) - \ln \frac{|g|^2}{3} \kappa^6$$

$$= 3 \ln\left(-\frac{f}{|g|^{2/3}}\right) - \ln 9. \qquad (15.1.5)$$

We note that this is the form suggested by resealing arguments given in the previous chapter.

The effective scalar potential V is given by

$$V(z, z^*) = -\kappa^{-4} e^{-\mathcal{G}}(3 + \mathcal{G}'^i \mathcal{G}_i''^{-1j} \mathcal{G}_j'). \qquad (15.1.6)$$

To obtain the canonical form for the kinetic energy we choose

$$\mathcal{G}(z, z^*) = -\frac{\kappa^2}{2} z_i z^{*i} - \ln \frac{|g|^2}{4} \kappa^6, \qquad (15.1.7)$$

in which case we get

$$V = \frac{1}{2} \exp\left(\frac{\kappa^2}{2} \sum_i z^{*i} z_i\right) \left[\sum_i \left|\frac{\partial g}{\partial z_i} + \frac{\kappa^2}{2} z^{*i} g\right| - \frac{3}{2} \kappa^2 |g|^2\right] \qquad (15.1.8)$$

and

$$m_g^2 = \frac{\kappa^4}{4} \exp\left(-\frac{\kappa^2}{2} \sum_i z^{*i} z_i\right) |g|^2. \qquad (15.1.9)$$

[To compare with the expressions in Chapter 12, Equation (12.5.2), identify $\kappa^2/2 = M_{P\ell}^{-2}$.] $\mathcal{L}'$ consists of terms involving powers of κ multiplying $f_{\alpha\beta}$, etc., and for the simplest choice $f_{\alpha\beta} = \delta_{\alpha\beta}$, such terms are small. A particularly interesting term in $\mathcal{L}'$ is of the form $\kappa^{-1} f' \lambda^T c^{-1} \lambda$, which can be nonzero if $f_{\alpha\beta} = (1 + a\kappa z)\delta_{\alpha\beta}$. This will generate a finite tree level mass for the gauginos without affecting anything else. In the subsequent section we would like to apply these formulas to building models.

An important property of eq. (15.1.6) worth emphasizing is that the potential is invariant under the transformation

$$\mathcal{G}(z, z^*) \rightarrow \mathcal{G}(z, z^*) + f(z) + f'(z^*).$$

This property is known as Kahler invariance. As a first application of this formalism let us choose $g(z)$ as follows:

$$g(z) = \mu z^2 + m^3. \qquad (15.1.10)$$

It is then easy to see that $q'(z) = \mu z$ and a minimum occurs at $z = 0$, which preserves supersymmetry. Thus, we should expect the gravitino mass to vanish. But eq. (15.1.9) tells us that

$$m_g^2 = \frac{\kappa^4}{4} m^6.$$

(15.1.11)

Equation (15.1.8) gives for V at the minimum

$$V_{\min} = -3\kappa^2 m^6.$$

(15.1.12)

This gives the relation

$$V_{\min} = -12\kappa^{-2} m_g^2,$$

(15.1.13)

which is an explicit verification of the connection between the cosmological constant and the gravitino mass given in eq. (14.9.3) provided we write $m_g = m/2$. Thus, in this case, the gravitino is actually massless and the appearance of the mass is due to the cosmological constant being nonzero, as was explained in the previous chapter. This point was emphasized by Deser and Zumino [1].

§15.2   The Polonyi Model of Supersymmetry Breaking

As in the case of global supersymmetry models, the physics discussion of supergravity models starts with the superpotential function $g(z)$. A particularly useful form for this was proposed in an unpublished paper by Polonyi [2], which splits g into a sum of two functions

$$g = g_1(z_a) + g_2(z),$$

(15.2.1)

where z_a denotes the usual matter fields (or the scalar member of the matter chiral multiplets) and z denotes a gauge singlet scalar field, which is the scalar member of a chiral multiplet. The $g_2(z)$ is used to break supersymmetry and involves parameters that are of the order $\kappa^{-1} \equiv M_p$. Polonyi chose

$$g_2(z) = \mu^2(z + \beta).$$

(15.2.2)

To simplify the study of its implications we choose $g_1(z_a) = 0$. It is easy to see that, if the gravitational effects are ignored (i.e., $\kappa = 0$), eq. (15.2.2) leads to a flat potential independent of z (see Fig. 15.1). Once gravitational effects are turned on, the presence of the negative terms produces a minimum of the potential at $z \sim O(M_p)$. We would like to find this minimum and we would also require that the value of the potential at this minimum vanish. This is to ensure that the cosmological constant vanishes. The vanishing of the cosmological constant is an additional attractive feature of supergravity not shared by global supersymmetry models and is, in fact, needed to give a consistent definition of the physical gravitino mass.

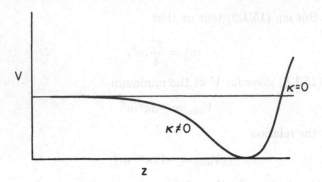

Figure 15.1. Form of the Polonyi potential with and without gravitational interactions.

Using eq. (14.8.17) we can write the effective scalar potential as

$$V(z, z_a) = \frac{1}{2} \exp\left(\frac{\kappa^2}{2}\left[|z|^2 + \sum_a |z_a|^2\right]\right) \left[\left|\mu^2 + \frac{\kappa^2}{2}\mu^2 z(z + \beta)\right|^2\right.$$

$$\left. + \sum_a \frac{\kappa^4}{4}\mu^4|z_a|^2|z + \beta|^2 - \frac{3}{2}\kappa^2\mu^4|z + \beta|^2\right]. \tag{15.2.3}$$

For the choice of the parameter $\mu^2\kappa^2 < 1$, the minimum of V to leading order is given by minimizing the potential ignoring the second term in eq. (15.2.3). We then have the following constraint equations:

$$\left.\frac{\partial V}{\partial z}\right|_{z=z_0} = 0,$$
$$V(z_0) = 0, \tag{15.2.4}$$

choosing real z we obtain

$$1 + \frac{\kappa^2}{2}z(z + \beta) = \pm\sqrt{\frac{3}{2}}\kappa(z + \beta) \tag{15.2.5}$$

and

$$\frac{\kappa^2}{2}(2z + \beta)\left(1 + \frac{\kappa^2}{2}z(z + \beta)\right) = 6\kappa^2(z + \beta). \tag{15.2.6}$$

Choosing the positive sign, this leads to the value of z at the minimum, i.e., z_0 being

$$z_0^{\mp)} = -\sqrt{2}\kappa^{-1}(-\sqrt{3} + 1),$$
$$\beta^{(\mp)} = -\sqrt{2}\kappa^{-1}(-2 + \sqrt{3}). \tag{15.2.7}$$

The corresponding value of the gravitino mass, denoted by m_{g_0} is (choosing the $z_0 = z_{\bar{0}}$)

$$m_{g_0}^2 = e^{2(\sqrt{3}-2)}\mu^4\kappa^2/2. \tag{15.2.8}$$

The numerical value for the lowest-order gravitino mass m_{g_0} is an interesting one if we choose the supersymmetry-breaking scale $\mu \simeq 10^{11}$ GeV, i.e., $M_g \simeq 10^3$ GeV. This is of the same order of magnitude as the weak-interaction symmetry-breaking scale m_W/g. This raises the interesting possibility that the scale of weak interactions could be induced by supergravity. In fact, this suggestion appears one step closer to realization once we evaluate the potential in eq. (15.2.3) and keep terms to zeroth power in (κm_g)

$$V(z_0, z_a) = m_{g_0}^2 \sum_a |z_a|^2. \tag{15.2.8a}$$

Note that z_a are the Lorentz scalar components of the light matter superfields such as quark, leptons, etc., and that among the light fields is included the doublet Higgs field of the Weinberg–Salam models, responsible for the electroweak symmetry breaking whose gravitationally induced mass is also m_{g_0}. If this mass term appeared with a negative sign, the electroweak symmetry breaking would be induced by supergravity and thus would solve a major problem of unified gauge theories.

The next step in trying to explore this possible connection between supergravity and electroweak-symmetry breaking is to include $g_1(z_a)$ and see whether it can trigger symmetry breaking, in spite of the positive $(\text{mass})^2$ term for the Higgs field. Let us write down the effective potential involving the light fields z_a in the presence of nonzero $g_1(z_a)$ at $z = z_0$. To zeroth order in (κm_g) we have [3, 4]

$$V(z_a) = m_g^2 \Sigma z_a^* z_a + \sum_a \left| \frac{\partial g_1}{\partial z_a} \right|^2$$
$$+ m_g \left[\sum_a z_a \frac{\partial g_1}{\partial z_a} + (A-3)g_1 + \text{h.c.} \right] + D\text{-terms}, \tag{15.2.9}$$

where $A = 3 - \sqrt{3}$. It is worth pointing out at this stage that there is no reason to implement supersymmetry breaking by the Polonyi-like superpotential. We can choose other forms. The interesting point is that the effective potential in eq. (15.2.9) retains its form except that the value of A changes provided there are no other intermediate scales in the theory between M_p and μ^2/M_p $(M_p = \kappa^{-1})$. In the next section we will explore the question of relating electroweak symmetry breaking to m_g for arbitrary A.

Another important point regarding the effective potential is that the structure of the supersymmetry-breaking terms is uniquely fixed by supergravity, thus removing a certain degree of arbitrariness from supersymmetry model building. A curious feature of the supersymmetry-breaking terms in eq. (15.2.9) is that, even though the scale of supersymmetry breaking is of order $10^{11} - 10^{12}$ GeV. the mass splitting between the components of

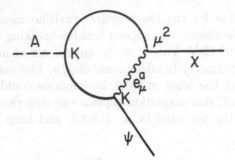

Figure 15.2. The one-loop graph that induces Goldstino coupling to matter.

a supermultiplet is only of order m_g. We may then ask how this can be reconciled with the mass formula in a supermultiplet which says that

$$m_\psi^2 - m_A^2 = f_{\chi\psi A} M_{\text{SUSY}}^2. \tag{15.2.10}$$

The answer is that the Goldstino χ-coupling to matter in this case is extremely small, i.e.,

$$f_{\chi\psi A} = \mu^2 \kappa^2. \tag{15.2.11}$$

The diagram responsible for the Goldstino coupling to matter is shown in Fig. 15.2.

§15.3 Electroweak Symmetry Breaking and Supergravity

In this section we pursue the idea of connecting the scale of electroweak symmetry breaking with the gravitino mass. In the absence of the light-particle superpotential, i.e., $g_1(z_a) = 0$, we saw that (mass)^2 for the Higgs multiplet is positive leading to no electroweak breaking. Let us include a nontrivial superpotential involving two Higgs doublets H_u, H_d and a singlet Y:

$$g_1(H_u, H_d, Y) = \lambda_1 H_u H_d Y + \lambda_2 Y^3. \tag{15.3.1}$$

Recall that H_u and H_d are the two $SU(2)_L$ doublet superfields introduced in globally supersymmetric $SU(2)_L \times U(1)$ model building. The singlet Y is introduced and the form of g_1, in eq. (15.3.1) without any dimensional parameters, is chosen so that the weak symmetry breaking (if it occurs) will automatically be proportional to m_g. Using eq. (15.2.9) we can write the low-energy effective potential as

$$V(z_a) = m_g^2 \sum_{a=H_u, y_d, Y} z_a^* z_a + \sum_a \left| \frac{\partial g_1}{\partial z_a}(z_a) \right|^2 + m_g \frac{A}{3} \sum_{Z_a} \left(z_a \frac{\partial g_1}{\partial z_a} + \text{h.c.} \right) \tag{15.3.2}$$

$$= \left(1 + \frac{A}{3}\right) \sum_a \left| m_g z_a + \frac{\partial g_1}{\partial z_a} \right|^2 + \left(1 - \frac{A}{3}\right) \sum_a \left| m_g z_a - \frac{\partial g_1}{\partial z_a} \right|^2.$$

$$(15.3.3)$$

It then follows that for $|A| < 3$ the absolute minimum of V is at $\langle H_u \rangle = \langle H_d \rangle = \langle Y \rangle = 0$. Thus, a necessary condition for electroweak symmetry breaking is that $|A| \geq 3$ [5]. Clearly the Polonyi potential does not satisfy this condition. Thus, we must go beyond this simple analysis to realize the dream of connecting m_W to m_g. It has been further pointed out that, even if we assume $|A| \geq 3$, the symmetry-breaking solution in realistic examples leads to the breakdown of electric charge. Therefore, alternative approaches must be sought.

An ingenious alternative method has been proposed by Alvarez-Gaume, Polchinski, and Wise [6], who note that the effective potential in eq. (15.2.9) has been defined at the Planck scale, i.e., $m = M_p$. To study their behavior at lower energies the parameters must be extrapolated down to the TeV scale. Since the nature of the extrapolation is determined by radiative corrections, the various parameters in $V(z_a)$ will change from their values at the Planck scale so, at the TeV scale, the potential should be written as

$$V(z_a)|_{\text{TeV}} = \sum_a M_a^2 Z_a^* Z_a + \sum \mu_{abc} Z_a Z_b Z_c + \text{h.c.}$$

$$+ \sum \lambda_{abcd} Z_a^* Z_b^* Z_c Z_d + \text{h.c.} + D \text{ terms}, \quad (15.3.4)$$

with M_a, μ_{abc}, and λ_{abcd} satisfying the following boundary conditions:

$$M_a^2(M_p) = m_g^2,$$

$$\mu_{abc}(M_p) = A,$$

$$\lambda_{abcd}(M_p) = \frac{\partial^3 g_1^*}{\partial Z_a^* \partial Z_b^* \partial Z_c^*} \frac{\partial^3 g_1}{\partial Z_c \partial Z_d \partial Z_e}. \quad (15.3.5)$$

If the process of extrapolation makes the (mass)^2 of the Higgs doublets H_u and H_d negative, then, conspiring with the D terms, it will give rise to electroweak symmetry breaking. To see that this can indeed be realized for certain values of the parameters in the Lagrangian, we look at a realistic $SU(2)_L \times U(1)_Y \times SU(3)_C$ model. As mentioned in Chapter 12 it will have the matter fields as shown in Table 15.1.

The most general gauge-invariant superpotential involving these fields can be written as

$$W = h_{u,ab} Q_a H_u u_b^c + h_{d,ab} Q_a H_d d_b^c + h_{e,ab} L_a H_d e_b^c + \mu H_u H_d. \quad (15.3.6)$$

The last term in eq. (15.3.6) introduces a mass parameter, but we will keep it small and will not adjust it to get electroweak breaking. The significance of this parameter is that it provides a solution to the strong CP-problem which we do not describe here [7]. It is now possible to write the renormalization group equations for the various parameters in order to study their

Table 15.1.

Fields	$\mathrm{SU}(2)_L \times \mathrm{U}(1)_Y \times \mathrm{SU}(3)_C$ quantum number
Q_a ($a = 1, 2, 3$ for generation)	$(\frac{1}{2}, \frac{1}{3}, 3)$
u_a^c ($a = 1, 2, 3$ for generation)	$(0, -\frac{4}{3}, 3^*)$
d_a^c ($a = 1, 2, 3$ for generation)	$(0, +\frac{2}{3}, 3^*)$
L_a ($a = 1, 2, 3$ for generation)	$(\frac{1}{2}, -1, 1)$
e_a^c ($a = 1, 2, 3$ for generation)	$(0, +2, 1)$
H_u ($a = 1, 2, 3$ for generation)	$(\frac{1}{2}, +1, 1)$
H_d ($a = 1, 2, 3$ for generation)	$(\frac{1}{2}, -1, 1)$

extrapolation from the Planck mass down to the TeV scale. The general analysis of the renormalization group equations for softly-broken supersymmetric theories has been carried out in the literature (see Chapter 12.6) and can be used to write the equations for m_a^2 and other parameter equations for m_a^2. Other parameters of the potential are given below:

$$V = \sum_a \left| \frac{\partial W}{\partial Z_a} \right|^2 + \sum_a M_a^2 Z_a^* Z_a + A_c^{ab} h_{e,ab} \tilde{L}_a H_d \tilde{e}_b^c + A_u^{ab} h_{u,ab} \tilde{Q}_a H_i \tilde{u}_b^c$$

$$+ A_d^{ab} h_{d,ab} \tilde{Q}_a H_d \tilde{d}_b^c + B\mu M_g H_u H_d + \text{h.c.}, \tag{15.3.7}$$

with ($t = \ln m$) [6, 8]

$$\frac{\partial M_{H_u}^2}{\partial t} = \frac{1}{8\pi^2}[3h_{u,33}^2(M_{H_u}^2 + M_{U_3^c}^2 + M_{Q_3}^2 + M_g^2|A_u^{33}|^2)$$
$$- (3|\tilde{M}_2|^2 g_2^2 + |\tilde{M}_1|^2 g_1^2)], \tag{15.3.8}$$

$$\frac{\partial M_{U_3^c}^2}{\partial t} = \frac{1}{8\pi^2}[2h_{u,33}^2(M_{H_u}^2 + M_{U_3^c}^2 + M_{Q_3}^2 + M_g^2|A_u^{33}|^2)$$
$$- (\tfrac{16}{3}|\tilde{M}_3|^2 g_3^2 + \tfrac{16}{9}|\tilde{M}_1|^2 g_1^2)], \tag{15.3.9}$$

$$\frac{\partial M_{Q_3}^2}{\partial t} = \frac{1}{8\pi^2}[(h_{u,33})^2(M_{H_u}^2 + M_{U_3^c}^2 + M_{Q_3}^2 + M_g^2|A_u^{33}|^2)$$
$$- (\tfrac{16}{3}|\tilde{M}_3|^2 g_3^2 + 3|\tilde{M}_2|^2 g_2^2 + \tfrac{1}{9}|\tilde{M}_1|^2 g_1^2)], \tag{15.3.10}$$

where $\tilde{M}_a$ are the gaugino masses for the SU(3), SU(2), and U(1) groups, respectively, g_a's being the corresponding gauge couplings. There are other equations that must also be analyzed. However, to illustrate the basic point, we need only the above three equations.

From eq. (15.3.8) we see that, if $\tilde{M}_2$ and $\tilde{M}_1$ are chosen small, the slope $\partial M_H^2/\partial t$ is positive, which means that as we come down from the Planck scale M_H^2 decreases and, therefore, if $h_{u,33}$ is sufficiently large, m_H^2 can be negative at the TeV scale and the SU(2) $\times$ U(1) symmetry will break down.

We must, however, be careful because in the same approximation $m_{U_3^c}^2$ and $m_{Q_3}^2$ also decrease with energy and *could* become negative leading to the lead result that the $SU(3)_C \times U(I)_{em}$ symmetries will break down. This can be prevented since we note that, due to the coefficients 3, 2, and 1 in the slopes of eqs. (15.3.8), (15.3.9), and (15.3.10) (arising, respectively, from three-colors and the doublet and singlet nature of Q and U), the $m_{\frac{2}{H}}$ decreases faster than $m_{Q_3}^2$ and $m_{U_3^c}^2$. So for h_u^{33}, within a certain limited range, at the TeV scale only $m_{\frac{2}{H}}$ becomes negative and $m_{Q_3}^2$ and $m_{U_3^c}^2$ remain positive. Thus, for only $SU(2)_L \times U(1)$ to occur, $h_{u,33}$ and hence m_t must lie within certain limits. It is interesting that m_t lies in the range 100 GeV$\leq m_t \leq$ 190 GeV. The recent discovery of the t-quark in the mass range of 40–60 GeV therefore rules out the simple-minded analysis carried out here. The idea is, however, quite profound and in some variations may provide an understanding of the electroweak scale.

§15.4   Grand Unification and $N = 1$ Supergravity

In this section we would like to study several applications of supergravity to grand unified models. We saw in Chapter 12 that, while global super-symmetry ameliorates somewhat the gauge hierarchy problem, it gives rise to several other problems, such as vacuum degeneracy and the associated cosmological difficulties. It would, therefore, be interesting to see if su-pergravity affects these considerations. More important, since the general grand unification scales are of the order of $10^{15} - 10^{19}$ GeV, we may hope that embedding grand unified models into $N = 1$ supergravity models may explain the grand unification scale in terms of the scale of gravity, i.e., the Planck scale. Combining this with the discussion of the previous section, we may visualize a scenerio where the only independent scales are the scales of supersymmetry breaking μ_s, and the Planck mass, and all other scales are consequences of the unification of gravity with other forces of nature.

It is with the hope of realizing this dream that a number of grand uni-fied models have been embedded into $N = 1$ supergravity theories. While no convincing model of this kind exists, we review in this section some of the attempts in this direction and the difficulties that arise in trying to realize this goal. We will work with the $SU(5)$ model for the purpose of illustration and discuss the following points: (i) supergravity has the po-tential to solve the vacuum degeneracy problem; and (ii) to maintain the tree level hierarchy of mass scales, the superpotential must satisfy certain constraints.

(a) Lifting Vacuum Degeneracy with Supergravity

To study this problem we consider the superpotential of the $SU(5)$ model given in Chapter 13 involving only the {24}-dimensional Higgs superfield

Φ:

$$W = z \operatorname{Tr} \Phi + x \operatorname{Tr} \Phi^3 + y \operatorname{Tr} \Phi^3. \qquad (15.4.1)$$

We recall our discussion in Chapter 13 that, if we keep supersymmetry unbroken at the grand unification scale, i.e., $F_\phi = 0$, we get three degenerate vacuums corresponding to the unbroken symmetries SU(5), SU(4) × U(1), and SU(3) × SU(2) × U(1):

$$\text{SU(5)}: \quad \langle\Phi\rangle = 0,$$

$$\text{SU(4)} \times \text{U(1)}: \quad \langle\Phi\rangle = b \begin{pmatrix} 1 & & & & \\ & 1 & & & \\ & & 1 & & \\ & & & 1 & \\ & & & & -4 \end{pmatrix},$$

$$\text{SU(3)} \times \text{SU(2)} \times \text{U(1)}: \quad \langle\Phi\rangle = a \begin{pmatrix} 1 & & & & \\ & 1 & & & \\ & & 1 & & \\ & & & -3/2 & \\ & & & & -3/2 \end{pmatrix}, \qquad (15.4.2)$$

where

$$b = \frac{10}{33}\frac{x}{y}$$

and

$$a = \frac{4}{3}\frac{x}{y}.$$

The inclusion of supergravity modifies the form of the effective potential at low energies and to the lowest order in powers of κ we have

$$V \simeq \left|\frac{\partial W}{\partial \phi}\right|^2 + \kappa^2 \left(\phi W \frac{\partial W}{\partial \phi} + \text{h.c.}\right) + \kappa^4 \operatorname{Tr} \phi^2 \left|\frac{\partial W}{\partial \phi}\right|^2 - 3\kappa^2 |W(\phi)|^2 + O(\kappa^4). \qquad (15.4.3)$$

Since $\partial W/\partial \phi = F_\phi = 0$ for the minima in eq. (15.4.2) we find at the minimum

$$V \simeq -3\kappa^2 |W(\phi)|^2. \qquad (15.4.4)$$

Using eq. (15.4.2) we find that

$$V_{\text{SU(5)}} = 0, \qquad V_{\text{SU(4)} \times \text{U(1)}} = -\left(\frac{7200}{35937}\right)^2 \frac{x^6}{y^4} 3\kappa^2,$$

and

$$V_{\text{SU(3)} \times \text{SU(2)} \times \text{U(1)}} = -3\left(\frac{112}{9}\right)^2 \frac{x^6}{y^4}\kappa^2.$$

Thus the vacuum degeneracy is lifted by gravitational interaction as was pointed out by Weinberg [8]. A new problem arises. The physical vacuum corresponding to the symmetry $SU(3) \times SU(2) \times U(1)$ has a large cosmological constant. This can, however, be set equal to zero by subtracting from $W(\phi)$ the constant $\sqrt{3}(\frac{112}{9})(x^3/y^2)\kappa$, but it is easy to see that this makes the minima with $SU(4) \times U(1)$ and $SU(5)$ symmetries lower than the desired minimum. It has, however, been argued by Weinberg [8] that, if the universe is in the $SU(3) \times SU(2) \times U(1)$ minimum, the tunneling probability to the lower minima is not significant. The question could be raised as to why the universe chose the highest of the three minima in the first place. It is, however, important that supergravity does remove the utterly high degree of vacuum degeneracy associated with global supersymmetry.

(b) Maintaining the Hierarchy of Mass Scales at the Tree Level

We saw in the previous section that coupling the electroweak model to $N = 1$ supergravity can produce a mass scale m_g of order $10^3 - 10^4$ GeV through supersymmetry breaking. If, instead of an electroweak symmetry whose breaking scale is of order m_g, we consider a grand unified symmetry [such as $SU(5)$, $SO(10)$, etc.] or a partial unified symmetry [such as $SU(2)_L \times SU(2)_R \times SU(4)_C$] which manifest themselves at a scale $\mu \gg m_g$, we can ask whether the gravitino mass can still be considered as inducing the W-boson mass. One technical problem that arises is the following: if we assume $\mu = M_U \approx 10^{15}$ GeV, $\kappa M_U \approx 10^{-4}$ and a priori the gravitino mass could receive tree level corrections of order $M_U(\kappa M_U)^n$, $n = 1, 2, 3, \ldots$. For the tree level mass hierarchy to remain, the theory must be such that $n > 3$. This problem has been analyzed in several papers and it has been observed that, in general, the superpotential has to satisfy certain constraints. To understand these constraints let us classify the fields entering the superpotential according to their masses and vacuum expectation values as (ϕ_α, T_i, D_a), where the subscript α denotes fields that have v.e.v. as well as mass of order M_U, i for fields with v.e.v. of order m_g and mass of order M_U, and a for fields with both v.e.v. and mass of order m_g. The constraints to be satisfied by the superpotential in order to maintain the gauge hierarchy are [3]

$$\frac{\partial^2 g}{\partial T_i \, \partial \phi_\alpha}, \frac{\partial^2 g}{\partial D_a \, \partial \phi_\alpha}, \frac{\partial^2 g}{\partial \phi_\alpha \, \partial \phi_\beta}\bigg|_{\text{minimum}} \sim O(m_g). \tag{15.4.5}$$

The next question is: For a superpotential that satisfies eq. (15.4.5), what is the low-energy form of the effective potential? Using the form in eq. (15.2.1), where z_a goes over z_α, z_i, z_a we can write down the effective potential as

$$V(z_a, z_a^\dagger) = \exp\left(\frac{\kappa^2}{2} \sum_n z_n^* z_n\right) [|\tilde{g}_{1,a}|^2 + m_1^2 z_a^* z_a$$

$$+ (\omega + \omega^\dagger) + m_g^2(\tilde{g}_{1,\alpha} G_\alpha^{(0)} + \tilde{g}_{1,i} G_i^{(0)} + \text{h.c.})], \quad (15.4.6)$$

where

$$\omega = m_2 \tilde{g}_1 + m_3 z^a \tilde{g}_{1,a} \quad (15.4.7)$$

$$\tilde{g}_1 = g_1(z_\alpha, z_i, z_a) - g_1(z_\alpha, z_i, 0), \quad (15.4.8)$$

$$G_m = \frac{\partial g_1}{\partial z^m} + \frac{\kappa^2}{2} z_m g_1, \quad (15.4.9)$$

$$m_1^2 = \tfrac{1}{2} m_g^2 [|\bar{G}_z^{(0)}|^2 - |\bar{g}_2^{(0)}|^2],$$

$$m_2 = \tfrac{1}{2} m_g [Z^{(0)} \bar{G}_z^{(0)} - 3\bar{g}_2^{(0)}],$$

$$m_3 = \tfrac{1}{2} m_g \bar{g}_2^{(0)}, \quad (15.4.10)$$

where

$$\bar{g}_2 = \frac{\kappa}{\mu^2} g_2, \qquad \bar{G} = \frac{\kappa}{\mu^2} G, \quad \text{etc.,}$$

and the superscript zero implies the value of the function at the zeroth order minimum.

As an example of a grand unified model coupled to $N = 1$ supergravity, we consider the SU(5) model discussed in Ref. 3, with the following structure for the superpotential, in terms of the Higgs superfields $\Phi\{24\}$, $H\{5\}$, $H\{\bar{5}\}$, $U\{1\}$ and the matter fields $\bar{F}\{\bar{5}\}$ and $T\{10\}$:

$$g_1(\Phi, H, H', F, T) = \lambda_1 \left(\tfrac{1}{3} \operatorname{Tr} \Phi^3 + \frac{M}{2} \operatorname{Tr} \Phi^2 \right) + \lambda_2 H(\Phi + 3M')H'$$

$$+ \lambda_3 U H' H + T f_1 T H + H' T f_2 \bar{F}. \quad (15.4.11)$$

The role of the various pieces of g_1 is clear. The terms proportional to λ_1 are responsible for SU(5) breaking while keeping supersymmetry unbroken. The λ_2 coupling is the one that keeps the Weinberg–Salam Higgs doublet massless; the f_1 and f_2 terms lead to fermion masses. The supersymmetry breaking is achieved by the usual Polonyi term. We leave it as an exercise to the reader to work out the symmetry-breaking chain as well as other consequences of the model. These ideas have been applied in various papers for the discussion of grand unified models with $N = 1$ supergravity [9].

Exercises

15.1. Instead of eq. (15.1.7), choose

$$G(z, z^*) = +\ln(z + z^*).$$

Obtain the scalar potential and gravitino mass in this case. These are the so-called no-scale models. [See J. Ellis, A. B. Lahanas, D. V. Nanopoulos, and K. A. Tamvakis, *Phys. Lett.* **134B**, 429 (1984).]

Derive the Lagrangian for the bosonic z field implied by the above $G(z, z^*)$. Show that it is invariant under the transformations

$$z \to \frac{\alpha z + i\beta}{i\gamma z + i\delta} \quad \text{with} \quad \alpha\delta + \beta\gamma = 1$$

($\alpha, \beta, \gamma, \delta$ real). Show that these transformations generate the group $SU(1,1)/U(1)$.

15.2. Discuss electroweak as well as supersymmetry breaking for a no-scale Lagrangian with

$$G = \ln(s + s^*) + 3\ln(T + T^* - |c_i|^2 - \ln|g(c_i)|^2),$$

where s and T are singlets under the gauge group which we choose to be E_6, and c_i are matter fields transforming as $\{27\}$-dimensional representations under E_6.

References

[1] S. Deser and B. Zurnino, *Phys. Rev. Lett.* **38**, 1433 (1977).

[2] J. Polonyi, Budapest preprint no. KFKI-1977-93, 1977.

[3] A. H. Chamseddine, R. Arnowitt, and P. Nath, *Phys. Rev. Lett.* **49**, 970 (1982); *Phys. Lett.* **121B**, 33 (1983); *Phys. Lett.* **120B**, 145 (1983).

[4] R. Barbieri, S. Ferrara, and C. A. Savoy, *Phys. Lett.* **119B**, 343 (1982).

[5] H. P. Nilles, M. Srednicki, and D. Wyler, *Phys. Lett.* **124B**, 337 (1983); *Phys. Lett.* **120B**, 346 (1983).

[6] L. Alvarez-Gaume, J. Polchinski, and M. Wise, *Nucl. Phys.* **B221**, 495 (1983).

[7] R. N. Mohapatra, S. Ouvry, and G. Senjanović, *Phys. Lett.* **126B**, 329 (1983).

[8] S. Weinberg, *Phys. Rev. Lett.* **48**, 1303 (1982).

[9] L. Hall, J. Lykken, and S. Weinberg, *Phys. Rev.* **D27**, 2359 (1983); L. Ibanez, *Nucl. Phys.* **B218**, 514 (1983).

16
Beyond $N = 1$ Supergravity

§16.1 Beyond Supergravity

So far, in this book, we have described the philosophical motivations, the mathematical foundations, and working principles for locally supersymmetric unification that described all four forces of nature, weak, electromagnetic, and strong as well as gravitational, within one theoretical framework. A theorist's dream is, however, more ambitious and rightly so, since, even the very elegant $N = 1$ supergravity leaves many questions unanswered: a partial list includes:

(i) The gauge symmetry describing electroweak unification has to be put in by hand. Thus, we really need two fundamental principles to derive the laws of physics: first, the equivalence principle to derive gravitational forces; and second, the local Yang–Mills symmetry to derive the rest of the interactions. It certainly would be more satisfying if both these principles could be combined into one.

(ii) The matter fields are chosen to fit phenomenology rather than being an outcome of the theoretical principles. It would certainly be more desirable if the basic principle that yields the physical laws could also yield the matter multiplets.

(iii) Finally, the age-old problem of divergences that beset the local field theories since their introduction to physics does not get resolved by the $N = 1$ supergravity theories. In fact, this problem is worse for $N = 1$ supergravity theories than either globaly supersymmetric or nonsupersymmetric theories. In that sense, it could be construed as a step backward in the quest for the ultimate theory.

The search for answers to such questions has led to many interesting theoretical advances. In this chapter, we provide a brief introduction to and overview of the extended supersymmetries and higher-dimensional supergravities that have emerged as interesting candidates for this ultimate unification, some of that also seem to have a more fundamental origin in string theories of matter.

§16.2 Extended Supersymmetries ($N = 2$)

We have seen in previous chapters that $N = 1$ global supersymmetry not only improves the divergence structure of the theory but also constrains the parameters in an interesting manner even after soft breakings are introduced. It is, therefore, tempting to go beyond $N = 1$ supersymmetry. The simplest extension is to consider $N = 2$ supersymmetry. Here, instead of considering a superspace with one two-component complex Weyl spinor coordinate θ as the fermionic coordinates, we will consider two sets of θ-coordinates θ_a, $a = 1, 2$. This theory has been studied in detail both for global [1] and local supersymmetry [2, 3] and it turns out that their structures are very different.

Before discussing the representations of $N = 2$ supersymmetry, it is worth pointing out that $N = 2$ supersymmetry algebra can support an additional element, the central charge denoted by Z in addition to the usual supersymmetry generators: Q_a, $a = 1, 2$, i.e., the $N = 2$ SUSY algebra is written as

$$\{Q_{a,\alpha}, Q_{b,\beta}\} = \varepsilon_{ab}\varepsilon_{\alpha\beta}Z, \qquad (16.2.1)$$

$$\{Q_{a,\alpha}, \bar{Q}_{b,\beta}\} = \delta_{ab}(\sigma^\mu)_{\alpha\beta}P_\mu. \qquad (16.2.2)$$

It is obvious that for $N = 1$ supersymmetry, the right-hand side of eq. (16.2.1) is zero. In physical applications, we generally restrict ourselves to the $Z = 0$ sector of the algebra. For this sector, the simplest irreducible multiplets are the hypermultiplet and the Yang–Mills multiplet. Their representation content is given in terms of $N = 1$ superfields as follows:

Hypermultiplet:	(ψ, χ),
Yang–Mills multiplet:	(W, ϕ),

where ψ, χ, and ϕ are chiral multiplets and W is the chiral multiplet constructed out of the $N = 1$ gauge multiplet in the Weiss–Zumino gauge. An important point, however, is that, under an internal symmetry group, ψ and χ transform like $\{N\}$- and $\{\bar{N}\}$-representations, respectively, under the group, whereas W and ϕ both transform like adjoint representations. This property is important for physical applications since it implies that, if quarks and leptons are assigned to the hypermultiplet, there must necessarily be mirror multiplets with $V + A$ interactions [4].

Before discussing the supersymmetric action for this case, we wish to analyze the structure of the hypermultiplet and the Yang–Mills multiplet. For this purpose, we write ψ, χ, W and ϕ in terms of components

$$\psi = (A_\psi, \tilde{\psi}),$$
$$\chi = (A_\chi, \tilde{\chi}),$$
$$W = (\lambda, F_{\mu\nu}),$$
$$\phi = (A_\phi, \tilde{\phi}). \qquad (16.2.3)$$

We can exhibit action of the SUSY generators Q^1 and Q^2 on the components as follows:

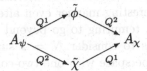

and

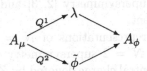

Note that the internal index a of the supersymmetry generator is shared only by the fermions $(\tilde{\psi}, \tilde{\chi})$ and $(\lambda, \tilde{\phi})$. If we define an SU(2) group on these indices, then $(\tilde{\psi}, \tilde{\chi})$ and $(\lambda, \tilde{\phi})$ transform as doublets and all other fields transform as singlets under the group.

The $N = 2$ supersymmetric action can be written in terms of the $N = 1$ supermultiplets: the matter multiplet (ψ, χ) and the gauge multiplet (W, ϕ) as follows:

$$S = \int d^4x \, d^4\theta [\mathrm{Tr}\, \phi^\dagger e^{gV} \phi + \bar{\psi} e^{gV} \psi + \bar{\chi} e^{-gV} \chi]$$

$$+ \frac{1}{64g^2} \int d^4x \, d^2\theta WW + g \int d^4x \, d^2\theta (\psi\phi\chi) + \mathrm{h.c.} \quad (16.2.4)$$

The first point to note about this action is that it is invariant under the internal SU(2) group defined earlier.

It is clear that the structure of the above action is too restrictive to be of much use in model building. However, in the hope that this problem may receive a cure when $N = 2$ breaking terms are included, we proceed to point out one important property of this theory. This has to do with its divergence structure. We look at the β function of this theory:

$$\beta(g) = \frac{g^3}{8\pi^2}\left[\sum_\sigma n_\sigma T(R_\sigma) - C_2(G)\right], \tag{16.2.5}$$

where n_σ is the number of matter multiplets and T is defined as follows:

$$\text{Tr}(\theta_a, \theta_b) = \delta_{ab}T(R_\sigma), \tag{16.2.6}$$

where θ_a are the generators of the gauge group in the representation to which the matter multiplets belong. We find that $\beta(g)$ vanishes if

$$\sum_\sigma n_\sigma T(R_\sigma) = C_2(G). \tag{16.2.7}$$

Since the theory has no other coupling construct, it is finite if $\beta(g) = 0$. Thus, by choosing the number of matter multiplets appropriately, we may hope to construct a fully finite mode of elementary particle interaction. However, no such model has emerged because of the highly restrictive nature of eq. (16.2.6) for known simple groups.

§16.3   Supersymmetries with $N > 2$

The search for the ultimate unification has led to the study of properties of extended supersymmetries with N larger than 2 in four space–time dimensions. The algebra for these supersymmetries reads

$$[Q_\alpha^i, T_K] = i(S^{ij})_K Q_\alpha^j,$$
$$\{Q_\alpha^i, Q_\beta^i\} = i\varepsilon_{\alpha\beta}Z^{ij},$$
$$\{Q_\alpha^i, Q_\beta^{*j}\} = 2\delta^{ij}(\sigma^\mu)_{\alpha\beta}P_\mu, \tag{16.3.1}$$

where i denotes the index whose maximum value denotes the number of supersymmetries in the theory. T_K denotes the generator of the internal symmetry group G of the extended theory, analogous to the SU(2) symmetry of the $N = 2$ theory. For the massless case, the group G has to be U(N) except for $N = 4$ where it could be either U(4) or SU(4) [5].

To study the particle content of N-extended supersymmetric theories, we consider a massless particle with maximum helicity $|\lambda|$ and apply the operator Q_α^i to it successively in which case we find $|\lambda|, |\lambda| - \frac{1}{2}, |\lambda| - 1, \ldots, |\lambda| - N/2$. There is no helicity with $|\lambda| - (N+1)/2$ as, multiplying $(N+1)Q_\alpha^i$'s $(i = 1, \ldots, N)$ gives zero since the Q_α^i's are anticommuting spinorial objects. It is, therefore, clear that if we want to consider matter multiplets with only spin-0 and 1/2, we can have at most $N = 2$. Similarly, for $N = 4$, the smallest irreducible multiplet must contain a spin-1 field along with spin-0 and spin-1/2 fields. Since the highest spin elementary particle of interest is the graviton with spin 2, the above counting argument restricts the N to be ≤ 8.

Table 16.1.

Helicity λ	Number of supersymmetries			
	$N = 1$	$N = 2$	$N = 3$	$N = 4$
1	1	1	1	1
1/2	1	2	$3+1$	4
0		$1+1$	$3+3$	6
$-1/2$	1	2	$1+3$	4
-1	1	1	1	1

In Tables 16.1 and 16.2 respectively, we give the particle content of the various extended supersymmetric theories for highest spin 1 (or the gauge multiplet), highest spin 2 (supergravity multiplet), and zero mass. Note that $N = 3$ and $N = 4$ Yang–Mills multiplets as well as $N = 7$ and $N = 8$ supergravity multiplets have essentially the same number of fields; as a result those theories are identical. Supersymmetric Lagrangians for these theories have been constructed. We do not discuss these here, except to note that as we go to $N = 8$ supergravity neither gravity nor the Yang–Mills multiplet nor the matter multiplet can exist in isolation; they exist together and there mutual interaction is dictated by symmetry. This attractive feature was the basis for the hope that extended supergravity may provide true unification of all interactions as well as constituents. The problem, however, is that SO(8), which is the largest gauge group of these theories, is not large enough to include the realistic strong and electroweak gauge group SU(3)$_c \times$ SU(2)$_L \times$ U(1)$_Y$. Moreover, all $N > 1$ supersymmetry theories are vectorlike, i.e., the interactions are symmetrical with respect to both left- and right-handed chiralities. On the other hand, we know that low-energy weak interactions are left-handed. Thus these models, although very interesting, cannot be realistic models of quark–lepton interactions.

§16.4   Higher-Dimensional Supergravity Theories

Soon after the discovery of extended supersymmetric theories, it was realized that we can obtain these theories from $N = 1$ supersymmetric theories constructed in higher space–time dimensions. As a heuristic example of how this comes about, let us look at $N = 2$ supersymmetry. As discussed in the previous section, it involves two sets of fermionic coordinates θ^1 and θ^2, where θ^i are four-component Majorana spinors. If we define a new spinor ξ such that

$$\xi = \begin{pmatrix} \theta^1 \\ \theta^2 \end{pmatrix},$$

(16.4.1)

Table 16.2.

λ	Number of supersymmetries						
	$N=1$	$N=2$	$N=3$	$N=4$	$N=5$	$N=6$	$N=1$
2	1	1	1	1	1	1	1
3/2	1	2	3	4	5	6	$7+1$
1		1	3	6	10	$15+1$	$21+7$
1/2			1	4	$10+1$	$20+6$	$35+21$
0				$1+1$	$5+5$	$15+15$	$35+35$
$-1/2$			1	4	$1+10$	$6+20$	$21+35$
-1		1	3	6	10	$1+15$	$7+21$
$-3/2$	1	2	3	4	5	6	$1+7$
-2	1	1	1	1	1	1	1

the new eight-component spinor can be thought of as a single spinor in six-dimensional space–time ($d = 6$). A general property of SO(N) spinors is that they decompose into a number of spinors when SO(N) is reduced to one of its subgroups. This is the intuitive way to see how extended supersymmetric theories arise after dimensional reduction of higher-dimensional supersymmetric theories. Of course, we have to be careful since Majorana and Weyl constraints cannot always be imposed in arbitrary dimensions.

Another reason to consider higher-dimensional supersymmetry is that the chirality problem of extended supersymmetric theories may be evaded in the process of dimensional reduction, since in higher dimensions we have to consider only $N = 1$ supersymmetry.

An important ingredient in the discussion of higher-dimensional supersymmetric theories is whether the Majorana spinors can be defined in arbitrary dimensions. We now discuss this question in arbitrary even dimension $d = 2M$. Define the d Γ-matrices Γ^n, $n = 0, 1, \ldots, d-1$ satisfying the Clifford algebra

$$\{\Gamma^n, \Gamma^m\} = 2\eta^{mn}, \tag{16.4.2}$$

where $\eta^{mn} = (+, -, -, -, \cdots)$. We also choose Γ^m such that $\Gamma^0 = \Gamma^{0\dagger}$ and $\Gamma^i = -\Gamma^{i\dagger}, i = 1, \ldots, d-1$. Let us write down the Dirac equation

$$i\Gamma^n(\partial_{n-ieA_n} + m)\psi = 0. \tag{16.4.3}$$

For $m = 0$, we find from eq. (16.4.3)

$$(-i\Gamma^{n^T}\partial_n)\bar\psi^T = 0. \tag{16.4.4}$$

Since Γ^{n^T} also satisfy Clifford algebra, there exists a unitary matrix C such that

$$C\Gamma^{n^T}C^{-1} = \eta\Gamma^n. \tag{16.4.5}$$

Here η can be either $+1$ or -1, $C\bar{\psi}^T$ satisfies the Dirac equation, and C is the familiar charge conjugation matrix. The Majorana condition can therefore be written as

$$C\bar{\psi}^T = \psi. \qquad (16.4.6)$$

However, eq. (16.4.6) can be imposed provided

$$(C\Gamma_0^T)(C^*\Gamma^0) = 1, \qquad (16.4.7)$$

or using eq. (16.4.5),

$$CC^* = \eta. \qquad (16.4.7a)$$

However, we show below that eq. (16.4.7a) need not be satisfied for all values of d. For simplicity, let us choose $\eta = -1$. Let us now use the following identity for the Γ-matrices and their products:

$$2^m \sum_A (\Gamma_A)_{ij}(\Gamma_A^{-1})_{kl} = \sum_A 1 \cdot \delta_{jk}\delta_{il}, \qquad (16.4.8)$$

where

$$\Gamma_A = \Gamma^m, \tfrac{1}{2}[\Gamma^m, \Gamma^n], \ldots .$$

It is easy to convince oneself that $A = 1, \ldots, 2^{2m}$. From eq. (16.4.8), we obtain

$$2^m \sum_A \text{Tr}(C\Gamma_A^T C^{-1}\Gamma_A^{-1}) = \sum_A 1 \cdot \text{Tr}(C^T C^{-1}). \qquad (16.4.9)$$

Let

$$C^T = \lambda C. \qquad (16.4.10)$$

Using $\eta = -1$ in eq. (16.4.5) we obtain

$$\lambda = \cos\frac{\pi m}{2} - \sin\frac{\pi m}{2}. \qquad (16.4.11)$$

It follows from this that, for $m = 2$ (i.e., $d = 4$), $\lambda = -1$; therefore eqs. (16.4.10) and (16.4.7) are consistent with each other. But for $m = 3$ (or $d = 6$), $\lambda = +1$; thus eq. (16.4.7) cannot hold. This means in $d = 6$, mod 8 we cannot define Majorana spinors, whereas in $d = 2, 4$ mod 8, we can.

In $d = 2, 4$ mod 8, we can show that the Γ matrices can be chosen pure imaginary, in which case we can set $C = \Gamma^0$, which on using eq. (16.4.6), leads to $\psi = \psi^*$ as the Majorana condition.

The next question we may ask is: In which dimension can we define a Majorana Weyl spinor?—where both the following conditions are satisfied:

$$\psi = \psi^*, \qquad \Gamma_{\text{FIVE}}\psi = \psi, \qquad (16.4.12)$$

where

$$\Gamma_{\text{FIVE}} = \eta\Gamma^0\Gamma^1, \ldots, \Gamma^{d-1}, \qquad (16.4.13)$$

clearly Γ_{FIVE} must be a real diagonal matrix. This implies $\eta^2 = 1$, which can be true only if $d = 2 \bmod 8$.

We can now determine the number of supersymmetries that will result from dimensional reduction of a D-dimensional theory. In D-dimensions the spinor is $2^{D/2}$ dimensional. However, the Majorana or Majorana–Weyl condition reduces the number of two or four, respectively. We denote this factor by r. Thus the number of supersymmetries N in four dimensions is

$$N = \tfrac{1}{2} r \cdot 2^{D/2}. \tag{16.4.14}$$

Now, coming back to dimensional reduction from $d = 6$ to $d = 4$ to obtain $N = 2$ supersymmetry, we see that we cannot naively define a Majorana spinor. However, a symplectic spinor can be defined by the condition

$$(\psi^i)^* = -\Sigma^{ij} C^{-1} \Gamma^O \psi^j. \tag{16.4.15}$$

We can use them to complete dimensional reduction [6] and using our counting we get $N = \tfrac{1}{2} \cdot \tfrac{1}{2} \cdot 2^3 = 2$.

Having outlined an important subtlety in implementing dimensional reduction, we now pass on to discuss some specific higher-dimensional theories. Two which have received greatest attention are the $d = 11$ and $d = 10$ supergravity theories. We present these theories now.

$d = 11$ Supergravity

We assume the signature of this theory to be $(+ - - - - \cdots)$. The fields of this theory are:

Elfbein $\qquad\qquad E_B^N, \quad N = $ curved space index,

$$B = \text{flat tangent space index}.$$

Gravitino $\qquad\qquad\qquad\qquad\qquad \psi_N$

Antisymmetric Three Index Bosonic Tensor Field A_{MNP}

The Lagrangian invariant under the following supersymmetry transformations

$$\delta E_M^A = -i\kappa \bar{\varepsilon} \Gamma^A \psi_M,$$
$$\delta A_{MNP} = \tfrac{3}{2} \bar{\varepsilon} \Gamma_{[MN;} \psi_{P]},$$
$$\delta \psi_M = \frac{1}{\kappa} \hat{D}_M \varepsilon, \tag{16.4.16}$$

is given by [6]

$$\mathcal{L} = -\frac{E}{4\kappa^2} R(\omega) - \frac{i}{2} \bar{\psi}_M \Gamma^{MNP} D_N \left(\frac{\omega + \hat{\omega}}{2} \right) \psi_P - \frac{E}{48} F_{MNPQ} F^{MNPQ}$$

$$+ \frac{2\kappa}{(144)^2} \varepsilon^{M_1, \ldots, M_{11}} F_{M_1, \ldots, M_4} F_{M_5, \ldots, M_8} A_{M_9, \ldots, M_{11}}$$

$$+ \frac{\kappa E}{192} (\bar{\psi}_M \Gamma^{MNPQRS} \psi_N + 12 \bar{\psi}^P \Gamma^{QR} \psi^S)(F_{PQRS} + \hat{F}_{PQRS}), \tag{16.4.17}$$

where

$$E = \text{Det } E_M^A$$

$$D_M = \partial_M - \tfrac{1}{4}\omega_{M,AB}\Gamma^{AB},$$

and

$$\omega_{MAB} = \tfrac{1}{2}(-\Omega_{MAB} + \Omega_{ABM} - \Omega_{BMA}) + K_{MAB},$$

$$K_{MAB} = \tfrac{i}{4}[-\bar{\psi}_N\Gamma_{MAB}^{NP}\psi_P + 2(\bar{\psi}_M\Gamma_B\psi_A - \bar{\psi}_M\Gamma_A\psi_B + \bar{\psi}_M\Gamma_M\psi_A)],$$

$$\Omega_{MN}^A = 2\partial_{[N}E_{M]}^A,$$

$$\tilde{\omega}_{MAB} = \omega_{MAB} + \tfrac{i}{4}\bar{\psi}_N\Gamma_{MAB}^{NP}\psi_P,$$

$$F_{MNPQ} = 4\partial_{[M}A_{NPQ]},$$

$$\tilde{F}_{MNPQ} = F_{MNPQ} - 3\hat{\psi}_{[M}\Gamma_{NP}\psi_{Q]},$$

$$\tilde{D}_M(\tilde{\omega})\psi_N = D_M(\tilde{\omega})\psi_N + T_M^{PQRS}\tilde{F}_{PQRS}\psi_N, \tag{16.4.18}$$

$$T_{SMNPQ} = \frac{1}{144}[T^{SMNPQ} - 8\Gamma^{[MNP}g^{Q]S}], \tag{16.4.19}$$

Under coordinate transformation by a parameter ξ^M the fields transform as follows:

$$\delta E_M^A = E_N^A\partial_M\xi^N + \xi^N\partial_N E_M^A,$$

$$\delta\psi_M = \psi_N\partial_M\xi^N + \xi^N\partial_N\psi_M,$$

$$\delta A_{MNP} = 3A_{Q[NP}\partial_{M]}\xi^Q, \tag{16.4.20}$$

An interesting result in this model was noted by Freund and Rubin [7], who showed that by choosing a ground state where

$$F_{\mu\nu\rho\sigma} = 3m\varepsilon_{\mu\nu\rho\sigma},$$

$$F_{mnpq} = 0, \tag{16.4.21}$$

where $\mu, \nu = 0, \ldots, 3$ and $m, n, \ldots = 4, \ldots, 11$, we obtain a compactification to the $M_4 \times M_7$ space where M_4 can be identified with space–time. This kind of ground state is however not unique [8], and there exist other kinds of solutions of this model where the four-dimensional theory corresponds to $N = 8$ supergravity. In any case, none of these theories have been shown to lead to realistic models of electroweak interactions.

§16.5 $d = 10$ Super-Yang–Mills Theory

Another higher-dimensional theory of great interest and model building potential is the ten-dimensional super-Yang–Mills theory. This is of particular interest in the superstring models [9, 10], which have recently been proposed as the ultimate unified theory of all forces. The zero slope limit of the superstring theories are also supposed to lead to ten-dimensional super-Yang–Mills theories. Therefore, in this section, we present the particle content and Lagrangian for the $d = 10$ theory [11, 12].

Particles

$$\text{Vielbein: } E_M^A,$$
$$B_{MN} = -B_{NM},$$
$$\text{Spinor } \lambda,$$
$$\text{Real Scalar } \phi,$$
$$\text{Yang–Mills Field } A_M^\alpha,$$
$$\text{Gaugino } \chi^\alpha,$$
$$\alpha \text{ is the index of the gauge group.}$$

The supersymmetric Lagrangian for this model is

$$
\begin{aligned}
E^{-1}\mathcal{L} \\
= -\frac{1}{2\kappa^2}R &- \frac{i}{2}\bar{\psi}_M\Gamma^{MNP}D_N(\omega)\psi_P \\
&+ \tfrac{3}{4}\phi^{-3/2}H_{MNP}H^{MNP} + \frac{i}{2}\bar{\lambda}\Gamma^M D_M(\omega)\lambda \\
&+ \tfrac{9}{16}(\partial_M\phi/\phi)^2 + \tfrac{3}{8}\sqrt{2}\bar{\psi}_M(\partial\!\!\!/\phi/\phi)\Gamma^M\lambda - \frac{\sqrt{2}}{16}\phi^{-3/4}H_{MNP}(i\bar{\psi}_Q\Gamma^{QMNPR}\psi_R \\
&+ \sqrt{2}\bar{\psi}_Q\Gamma^{MNP}\Gamma^Q\lambda + 6i\bar{\psi}^M\Gamma^N\psi^P = \bar{\chi}^\alpha\Gamma^{MMP}\chi^\alpha) - \tfrac{1}{4}\phi^{-3/4}F_{MN}^\alpha F^{\alpha MN} \\
&+ \frac{i}{2}\bar{x}^\alpha\Gamma^M(D_M(\omega)x)^\alpha - \frac{i\kappa}{4}\phi^{-3/8}(\bar{x}\Gamma^M\Gamma^{NP}F_{NP}^\alpha)\left(\psi_M + \frac{i\sqrt{2}}{12}\Gamma_M\lambda\right)
\end{aligned}
$$

$$+ \text{ four Fermi interactions,} \tag{16.5.1}$$

where

$$D_N\psi_P = (\partial_N - \tfrac{1}{2}\omega_{N[RS]}\Gamma^{RS})\psi_P - 2\omega_{NP}^Q\psi_Q, \quad \kappa \text{ is the gravitational constant.}$$
$$\tag{16.5.2}$$

$$D_N\lambda = (\partial_N - \tfrac{1}{2}\omega_{N[RS]}\Gamma^{RS})\lambda$$
$$(D_N\chi)^\alpha = (\partial_N - \tfrac{1}{2}\omega_{N[RS]}\Gamma^{RS})\chi^\alpha - f^{\alpha\beta\delta}A_N^\beta\chi^\delta. \tag{16.5.3}$$

The Yang–Mills field strength

$$F_{MN}^\alpha = \tfrac{1}{2}\partial_{[M}A_{N]}^\alpha + f^{\alpha\beta\delta}A_M^\beta A_N^\gamma, \tag{16.5.4}$$

$f^{\alpha\beta\gamma}$ are structure constants of the gauge group

$$H_{MNP} = \partial_{[M}B_{NP]} - \frac{\kappa}{\sqrt{2}}(\Omega_{MNP}^{YM} - \Omega_{MNP}^L), \tag{16.5.5}$$

where Ω^{YM} and Ω^L are the Yang–Mills and Lorentz Chern–Simons forms given by

$$\Omega^{YM}_{MNP} = \mathrm{Tr}(A_{[M}F_{NP]} - \tfrac{2}{3}A_{[M}A_N A_{P]})$$

and

$$\Omega^L_{MNP} = \mathrm{Tr}(\omega_{[M}R_{NP} - \tfrac{2}{3}\omega_{[M}\omega_N\omega_{P]}). \qquad (16.5.6)$$

It has recently been considered as a serious candidate for the ultimate unification group of quark–lepton interactions including gravity for the following reasons:

(a) This theory is supposed to arise in the zero-slope limit of the superstring theories.

(b) For the case where the Yang–Mills group is chosen to be either SO(32) or $E_8 \times E'_8$ [14], this theory is free [9] of gauge, as well as gravitational, anomalies [13]. This is a unique feature of this model. The anomaly cancellation requires the presence of the Ω-terms in eq. (16.5.5). Their presence, however, breaks the apparent supersymmetry of the Lagrangian but it is expected that once higher-order terms are included, supersymmetry invariance will be restored.

We now comment briefly on the compactification of this model to obtain four-dimensional theories: the physically relevant compactification of this model is of $M_4 \times K_6$ type, where M_4 is the four-dimensional Minkowski space and K_6 is a six-dimensional compact manifold. An interesting compactification scheme that is believed to preserve $N = 1$ supergravity arises when K_6 is a Calabi–Yau manifold with SU(3) holonomy. In this case, the $E_8 \times E'_8$ gauge group breaks down to $E_8 \times E_6$. Since the {248}-dimensional representation of E'_8 when decomposed under E_6, contains {27}-dimensional representations, some of the gauginos of the E'_8 group turn into quarks and leptons. However, in general, we obtain hundreds of massless {27} and {$\overline{27}$} representations. To obtain the number of generations to be less than four, K cannot be simply connected. If K admits a discrete symmetry group that acts freely, then compactifying the ten-dimensional theory to $M_4 \otimes (K/G)$ leads to five {27}-dimensional and one {$\overline{27}$}-dimensional massless superfields (for $G = Z_5 \times Z_5$), which can then be identified with quarks and leptons and Higgs superfields. This model has many features of realistic quark–lepton models because the {27}-dimensional superfield decomposes under the SO(10) subgroup of E_6 [15, 16] as follows:

$$\{27\} = \{16\} + \{10\} + \{1\}. \qquad (16.5.7)$$

The same representation can also serve as a Higgs superfield and can be used to implement gauge symmetry breaking. Two important phenomenological problems of this model are the following:

(a) Since the model has a right-handed neutrino and it has no {126}-dimensional Higgs representation of the SO(10) group (which will be part of the {351}-dimensional representation of E_6), the conventional "seesaw" mechanism used in Chapter 6 to understand tiny neutrino mass does not exist. However, alternative mechanisms [17] have been proposed for this purpose.

(b) *A priori*, there is the possibility of proton decaying rapidly.

Another problem, which may require deeper understanding of the compactification spaces, is how to introduce CP-violation into these models. Thus, even though ten-dimensional $E_8 \times E_8'$ super-Yang–Mills theory has raised expectations for a realistic unification model, a lot of work remains to be done.

References

[1] P. Fayet, *Nucl. Phys.* **B113**, 135 (1976); **B149**, 137 (1979); M. F. Sohnius, K. Stella, and P. West, *Nucl. Phys.* **B173**, 127 (1980).

[2] J. Bagger and E. Witten, *Nucl. Phys.* **B222**, 1 (198 3).

[3] B. deWit, P. G. Lauwers, R. Philippe, S. Q. Su, and A. Van Proyen, preprint NIKHEF-H/83-13 (1983); P. Breitenlohner and M. Sohnius, *Nucl. Phys* **B165**, 483 (1980).

[4] F. del Aguila, B. Brinstein, L. Hall, G. G. Ross, and P. West, HUTP 84/AOOI, 1984; S. Kalara, D. Chang, R. N. Mohapatra, and A. Gangopadhyaya, *Phys. Lett.* **145B**, 323 (1984); J. P. Deredings, S. Ferrara, A. Masiero, and A. Van Proyen, *Phys. Lett.* **140B**, 307 (1984). For physical applications see J. M. Frére, 1. Meznicescu, and Y. P. Yao, *Phys. Rev.* **D29**, 1196 (1984); A. Parkes and P. West, **127B**, 353 (1983).

[5] R. Haag, J. T. Lopuszanski, and M. Sohnius, *Nucl. Phys.* **B88**, 257 (1975).

[6] J. Koller, Cal. Tech. preprint 68-975, 1982; G. Sierra and P. K. Townsend, Ecole Normale preprint, LPTENS 83/26,1983; J. Sherk, Ecole Normale preprint, 1979; M. Duff, Lectures at the GIFT Summer School, San Feliu de Guixols, Spain, 1984.

[7] P. G. 0. Freund and M. Rubin, *Phys. Lett.* **B97**, 233 (1980).

[8] M. Duff and C. N. Pope, in *Supersymmetry and Supergravity* '82 (edited by S. Ferrara et al.), World Scientific, Singapore, 1983.

[9] M. Green and J. H. Schwarz, *Phys. Lett.* **149B**, 117 (1984).

[10] P. Candelas, G. T. Horowitz, A. Strominger, and E. Witten, *Nucl. Phys.* **B258**, 46 (1985); K. Pilch and A. N. Schellekens, Stony Brook preprint, 1985; E. Witten, Princeton preprint, 1985.

[11] E. Bergshoeff, M. de Roo, B. de Wit, and P. Van Niuewenhuizen, *Nucl. Phys.* **B195**, 97 (1982).

[12] G. F. Chapline and N. S. Manton, *Phys. Lett.* **120B**, 105 (1983).

[13] L. Alvarez-Gaume and E. Witten, *Nucl. Phys.* **B234**, 269 (1983).

[14] D. Gross, J. Harvey, E. Martinec, and R. Rohm, *Phys. Rev. Lett.* **54**, 502 (1985).

[15] M. Dine, V. Kaplunovsky, M. Mangano, C. Nappi, and N. Seiberg, *Nucl. Phys.* B (1985) (in press); J. Breit, B. Ovrut, and G. Segre, *Phys. Lett.* **B158**, 33 (1985); A. Sen, *Phys. Rev. Lett.* **55**, 33 (198 5); J. P. Deredings, L. Ibanez, and H. P. Nilles, *Phys. Lett.* **155B**, 65 (1985); and CERN preprint CERN-TH-4228/85,1985; V. S. Kaplunovsky, *Phys. Rev. Lett.* **55**, 1036 (1985); M. Dine and N. Seiberg, *Phys. Rev. Lett.* **55**, 366 (1985); V. S. Kaplunovsky and Chiara Nappi, *Comments on Nuclear and Particle Physics* (1986) (to appear).

[16] F. Geursey, P. Ramond, and P. Sikivie, *Phys. Lett.* **60B**, 117 (1976); Y. Achiman and B. Stech, *Phys. Lett.* **77B**, 389 (1987); Q. Shafi, *Phys. Lett.* **79B**, 301 (1978).

[17] R. N. Mohapatra, *Phys. Rev. Lett.* **56**, 561 (1986); U. Sarkar and S. Nandi, *Phys. Rev. Lett.* **56**, 564 (1986).

17

Superstrings and Quark–Lepton Physics

§17.1   Introduction to Strings

In this chapter, we wish to explore the consequences of the hypothesis that as one probes very small distances ($l \leq 10^{-33}$ cm), the fundamental particles may exhibit a stringlike structure with a size of the order of the Planck size (i.e., 10^{-33} cm), thereby departing from the point particle assumption made throughout the rest of the book. Historically, strings were introduced [1] to describe the world of hadrons, but the appearance of spin-2 particles in the string spectrum, as well as other problems, prompted J. Scherk and J. Schwarz to suggest that they may be relevant for the description of a unified theory of gravity and elementary particles. It is this idea that has been developed into the beautiful superstring theories, which some believe could represent the ultimate theory of everything.

The fundamental objects in superstring theories are one-dimensional strings (rather than zero-dimensional points), and when they evolve, they sweep two-dimensional surfaces. It is amazing that a supersymmetric version of these strings leads to many important ingredients such as the gauge groups, the fermion representations, etc., that form the core of the unified gauge theories, while at the same time fixing the number of space–time dimensions. There are also strong hints that these theories are free of the divergence difficulties that beset the local field theories. It is the purpose of this chapter to summarize the basic ideas and steps involved in extracting, from an abstract string picture, the physics, quarks and leptons.

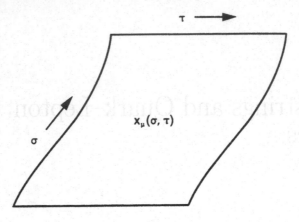

Figure 17.1.

Let us begin our discussion by describing the bosonic string, which is given by the variable $x_\mu(\sigma, \tau)$, where σ parametrizes the position of a point on the string and τ gives its time evolution (Fig. 17.1). $x_\mu(\sigma, \tau)$ describes a surface embedded in the d-dimension space–time, where $\mu = 0, 1, \ldots, d-1$.

To describe the quantum mechanics of this system, we need an action, which we will write down in analogy with the case of point particles. The action of the point particle is given by the distance on the world line. Using this analogy, Nambu [2] and Goto [2] postulated that the action of a string must be given by the area of the surface swept by the string. To calculate the area, we look at a blown-up version of an infinitesimal area on the world surface of the string (Fig. 17.2). The area enclosed by $ABCD$ is given in Minkowski space as

$$\Delta A = \left| \frac{dx_\mu}{d\sigma} \right| \left| \frac{dx_\mu}{d\tau} \right| d\sigma\, d\tau \cdot \sinh\theta$$

$$= \left| \frac{dx_\mu}{d\sigma} \right| \left| \frac{dx_\mu}{d\tau} \right| d\sigma\, d\tau \sqrt{\cosh^2\theta - 1} = d\sigma\, d\tau \{(x' \cdot \dot{x})^2 - x'^2 \dot{x}^2\}^{1/2},$$

$$(17.1.1)$$

where

$$x' \cdot \dot{x} = x'_\mu \dot{x}^\mu; \qquad x'^2 = x'_\mu \cdot x'^\mu; \qquad x'_\mu = \frac{dx_\mu}{d\sigma}; \qquad \dot{x}_\mu = \frac{dx_\mu}{d\tau}.$$

The action for the string is then given by

$$S = T \int d\sigma\, d\tau \{(x' \cdot \dot{x})^2 - x'^2 \dot{x}^2\}^{1/2}, \qquad (17.1.2)$$

where T is the string tension which has mass dimension 2.

From eq. (17.1.2) it is not easy to note a very important symmetry group of the string, which is invariance under the reparametrization (or coordinate transformation) of σ, τ to σ', τ'. To see this invariance, we

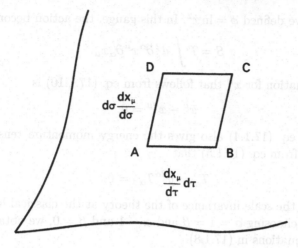

Figure 17.2.

rewrite eq. (17.1.2) as

$$S = T \int d^2\xi \sqrt{-g}g^{\alpha\beta}\partial_\alpha x^\mu \partial_\beta x_\mu \quad \text{where} \quad \xi^0 = \tau \text{ and } \xi^1 = \sigma.$$

$$(17.1.3)$$

To obtain eq. (17.1.2) from eq. (17.1.3), we use the Euler–Lagrange variation principle with respect to $g^{\alpha\beta}$, i.e.,

$$\frac{\delta S}{\delta g^{\alpha\beta}} = 0. \tag{17.1.4}$$

This implies

$$\partial_\alpha x^\mu \partial_\beta x_\mu - \tfrac{1}{2}g_{\alpha\beta}g^{\alpha'\beta'}\partial_{\alpha'}x^\lambda \partial_{\beta'}x_\lambda = 0. \tag{17.1.5}$$

The solution to eq. (17.1.5) is

$$g_{\alpha\beta} = \partial_\alpha x^\mu \partial_\beta x_\mu. \tag{17.1.6}$$

Substituting the expression for $g_{\alpha\beta}$ into eq. (17.1.3) we get eq. (17.1.2). In fact, we can write

$$g_{\alpha\beta} = \begin{pmatrix} \dot{x}^2 & \dot{x} \cdot x' \\ \dot{x} \cdot x' & x'^2 \end{pmatrix}. \tag{17.1.7}$$

We can use the reparametrization invariance of the action in (17.1.3) to fix the following gauge [1]:

$$\dot{x} \cdot x' = 0,$$
$$\dot{x}^2 + x'^2 = 0. \tag{17.1.8}$$

Using this, one gets

$$g_{\alpha\beta} = \dot{x}^2 \eta_{\alpha\beta} = e^\phi \eta_{\alpha\beta}, \tag{17.1.9}$$

where we have defined $\phi = \ln \dot{x}^2$. In this gauge, the action becomes

$$S = T \int d\frac{2}{\xi} \partial^\alpha x^\mu \partial_\alpha x_\mu. \qquad (17.1.10)$$

The field equation for x^μ that follows from eq. (17.1.10) is

$$\ddot{x}^\mu - x''^\mu = 0. \qquad (17.1.11)$$

Note that eq. (17.1.4) also gives the energy momentum tensor $T_{\alpha\beta}$. It then follows from eq. (17.1.5) that

$$T_\alpha^\alpha = g^{\beta\alpha} T_{\alpha\beta} = 0. \qquad (17.1.12)$$

This reflects the scale invariance of the theory at the classical level. Using eq. (17.1.5), choosing $\alpha = 1 = \beta$ and $\alpha = 1$ and $\beta = 0$, we obtain the two constraint equations in (17.1.8).

At this point, we digress to introduce the large conformal symmetry present in a theory with Lorentz and scale invariance in arbitrary dimensions. For this purpose, let the underlying coordinates on which the theory is defined be given by indices $\alpha, \beta, \ldots$ and coordinates ξ^α; the indices $\alpha = 1, \ldots, D$, where D is the underlying "space–time" dimension. In the case of the string Lagrangian defined in eq. (17.1.2), $D = 2$. Note that Lorentz and scale invariance of such a theory imply the existence of a symmetric, conserved, traceless energy momentum tensor $T^{\alpha\beta}$:

$$T^{\alpha\beta} = T^{\beta\alpha}, \qquad (17.1.13a)$$
$$\partial_\alpha T^{\alpha\beta} = 0, \qquad (17.1.13b)$$

and

$$T_\alpha^\alpha = 0. \qquad (17.1.12)$$

Using eqs. (17.1.12) and (17.1.13), one can define a new set of conserved currents

$$J_f^\alpha = T^{\alpha\beta} f_\beta(\xi). \qquad (17.1.14)$$

It then follows that

$$\partial_\alpha J_f^\alpha = T^{\alpha\beta} \partial_\alpha f_\beta(\xi) = \tfrac{1}{2} T^{\alpha\beta} (\partial_\alpha f_\beta + \partial_\beta f_\alpha). \qquad (17.1.15)$$

Note that if the function f_β is such that

$$\partial_\alpha f_\beta + \partial_\beta f_\alpha = g_{\alpha\beta} f(\xi), \qquad (17.1.16)$$

then $\partial_\alpha J_f^\alpha = 0$, leading to a new set of symmetries. These sets of symmetries generated by J_f^α are known as conformal symmetries. Using eq. (17.1.16) we can solve for f_α for the different "dimensions" D. Taking the trace of eq. (17.1.16), we find

$$\partial_\alpha f_\beta + \partial_\beta f_\alpha = \frac{2}{D} g_{\alpha\beta} \partial^\lambda f_\lambda. \qquad (17.1.17)$$

For $D = 4$, the only nontrivial new solution of eq. (17.1.17) is

$$f_\alpha = a_\alpha \xi^2 - \xi_\alpha a_\beta \xi^\beta, \qquad (17.1.18)$$

where a_α is a constant four-vector. On the other hand, for $D = 2$, choosing a Minkowski metric we can easily see that

$$\left[g^{\alpha\beta} = \begin{pmatrix} 1 & \\ & -1 \end{pmatrix} \right],$$
$$\partial_0 f^0 = \partial_1 f^1 \quad \text{and} \quad \partial_0 f^1 = \partial_1 f^0. \qquad (17.1.19)$$

Combining these, we can write

$$(\partial_0 + \partial_1)(f^0 - f^1) = 0,$$
$$(\partial_0 - \partial_1)(f^0 + f^1) = 0. \qquad (17.1.20)$$

Identifying ξ_0 and ξ_1 with the worldsheet coordinates τ and σ, respectively, we find

$$\partial_+ f^- = \partial_- f^+ = 0, \qquad (17.1.21)$$

or $f^- \equiv f^-(\sigma - \tau)$ and $f^+ = f^+(\sigma + \tau)$.

This is the reparametrization symmetry of the string theory. Since $f^\pm$ are arbitrary functions, there will be an infinite number of generators of this symmetry given by the infinite power series expansion of $f^\pm$. These generators will be identified with the elements L_n of the Virasoro algebra, to be defined shortly.

Let us now return to the discussion of string theory. For the open string, there are two more conditions that accompany eq. (17.1.11) and they are

$$x^{\mu'}(\sigma = 0, \tau) = x^{\mu'}(\sigma = \pi, \tau) = 0. \qquad (17.1.22)$$

These are Neumann boundary conditions. It is also possible that the boundary conditions are Dirichlet type, i.e., $x^\mu(0, \tau) = x^\mu(\pi, \tau) = 0$. These conditions give rise to the so called D-branes (see the book by Polchinski [1]), which have occupied a central role in the current discussion of string theories as well as their applications. For closed strings, the condition is

$$x^\mu(\sigma = 0, \tau) = x^\mu(\sigma = \pi, \tau). \qquad (17.1.23)$$

The solutions to eqs. (17.1.11) and (17.1.22) for open strings are given by

$$x^\mu(\sigma, \tau) = q^m + p^\mu \tau + \sum_{\substack{n=-\infty \\ n \neq 0}}^{+\infty} \frac{i}{n} a_{\mu,n} e^{in\tau} \cos n\sigma, \qquad (17.1.24)$$

whereas for closed strings, we have

$$x^\mu(\sigma, \tau) = q_m + P_\mu \tau + \sum_{n \neq 0} \frac{i}{2n} a_{\mu,n} e^{-2in(\tau - \sigma)} + \sum_{n \neq 0} \frac{i}{2n} \tilde{a}_{\mu,n} e^{-2in(\sigma + \tau)}. \qquad (17.1.25)$$

Reality for the space–time coordinates requires that $a_n^{\mu'} = a_{-n}^\mu$; $\tilde{a}_n^{\mu^*} = \tilde{a}_{-n}^\mu$; $\tilde{a}_n^{\mu^*} = \tilde{a}_{-n}^\mu$. Note that closed strings have twice the number of oscillators compared to open ones.

At this time, an interesting point about eqs. (17.1.8) and (17.1.10) is worth noting. Adding and subtracting the two equations in (17.1.8) we get

$$(\dot{x} \pm x')^2 = 0. \tag{17.1.26}$$

If we define the "light cone" variables in the two-dimensional world surface as

$$\xi_\pm = \sigma \pm \tau, \tag{17.1.27}$$

then we can write eq. (17.1.26) as

$$(\partial_\pm x)^2 = 0, \tag{17.1.28}$$

and we can rewrite the action in eq. (17.1.10) as

$$S = T \int d\xi_+ d\xi_- \, \partial_+ x^\mu \cdot \partial_- x_\mu. \tag{17.1.29}$$

Note that the action in eq. (17.1.29) (even though already gauge fixed) has an additional reparametrization symmetry (or conformal symmetry)

$$\xi_\pm \to \xi'_\pm(\xi_\pm). \tag{17.1.30}$$

This is a remnant of the conformal symmetry discussed earlier. Expanding $\xi'_+(\xi_+)$ in powers of ξ_+, we realize that this is an infinite parameter group, with the generators denoted by L_n, whose action on ξ_+ is given by

$$\delta_n \xi_+ \equiv \xi_+ + \varepsilon_n L_n \xi_+ = \xi_+ + \varepsilon_n \xi_+^{n+1}. \tag{17.1.31}$$

(Similar considerations hold for ξ_-, which we do not display for simplicity.) To find commutation relations between L_m and L_n, we calculate

$$\begin{aligned}
(\delta_m \delta_n - \delta_n \delta_m)\xi_+ &= \delta_m(\xi_+ + \varepsilon_n \xi_+^{n+1}) - \delta_n(\xi_+ + \varepsilon_m \xi_+^{m+1}) \\
&= (\xi_+ + \varepsilon_m \xi_+^{m+1}) + \varepsilon_n(\xi_+ + \varepsilon_m \xi_+^{m+1})^{n+1} \\
&\quad - (\xi_+ + \varepsilon_n \xi_+^{n+1}) + \varepsilon_m(\xi_+ + \varepsilon_n \xi_+^{n+1})^{m+1} \\
&= \varepsilon_n \varepsilon_m (n - m)\xi_+^{n+m+1} + O(\varepsilon^3).
\end{aligned} \tag{17.1.32}$$

This implies that, at the classical level,

$$[L_n, L_m] = (n - m)L_{n+m}. \tag{17.1.33}$$

This is known as the Virasoro algebra. It becomes anomalous once quantum corrections are included, leading to the so-called Virasoro algebra with central charges.

To understand the meaning of the generators L_m of coordinate transformations, we note that they are related to moments of the two-dimensional energy–momentum tensor, $T_{00} - T_{11}$, and T_{01}, which vanish due to the gauge fixing equation (17.1.19). In fact, we will see below that the Fourier coefficients of the left-hand side of eq. (17.1.28) have the same classical Poisson bracket as eq. (17.1.33).

Define

$$L_N = \int d\xi_+ + e^{iN\xi_+}(\partial_+ x)^2, \qquad (17.1.34)$$

and a similar $\tilde{L}_N$ for ξ_-. Let us work with the closed strings for simplicity and definiteness:

$$\partial_+ x^\mu = \tfrac{1}{2}P_\mu + \sum \tilde{a}_{\mu,n} e^{-2in\xi_+}. \qquad (17.1.35)$$

It is then easy to find that

$$L_N = -\tfrac{1}{2}\sum_{n=-\infty}^{+\infty} \tilde{a}_{\mu,n}\tilde{a}_{\mu,N-n}. \qquad (17.1.36)$$

To calculate the Poisson bracket $[L_N, L_M]_{\text{P.B.}}$, we need the Poisson bracket of a's. This follows from the basic Poisson bracket

$$[x_\mu(\sigma',\tau'), P_\nu(\sigma,\tau)]_{\text{P.B.},\tau=\tau'} = \eta_{\mu,\nu}\delta(\sigma'-\sigma). \qquad (17.1.37)$$

This is equivalent to the following Poisson brackets between a's:

$$[a_{\mu,n}, a_{\nu,m}]_{\text{P.B.}} = [\tilde{a}_{\mu,n}, \tilde{a}_{\nu,m}]_{\text{P.B.}} = -i\eta_{\mu\nu}n\delta_{m+n}, \qquad (17.1.38)$$

and

$$[P_\mu, q_\nu]_{\text{P.B.}} = -\eta_{\mu\nu}. \qquad (17.1.39)$$

Defining the Poisson bracket for any two dynamical variables A and B as follows:

$$[B, A]_{\text{P.B.}} = \frac{\partial A}{\partial q^\mu}\frac{\partial B}{\partial P_\mu} - \frac{\partial A}{\partial P^\mu}\frac{\partial B}{\partial q_\nu} + i\sum_{n\neq 0} n\frac{\partial A}{\partial a_n^\mu}\frac{\partial B}{\partial a_{-n,\mu}} + i\sum n\frac{\partial A}{\partial \tilde{a}_n^\mu}\frac{\partial B}{\partial \tilde{a}_{-n,\mu}}, \qquad (17.1.40)$$

one can easily prove that

$$[L_N, L_M]_{\text{P.B.}} = i(N - M)L_{N+M}. \qquad (17.1.41)$$

This establishes the connection between conformal transformations and the constraint equations.

Let us discuss the significance of the operators L_N, which generate the conformal transformations. Constraint equation (17.1.28) instructs us that the physical system must respect the constraints

$$L_N = 0. \qquad (17.1.42)$$

Note, in particular, that $L_0 = 0$ leads to (omitting the Lorentz indices)

$$-\tfrac{1}{8}P^2 = \sum_{n>0} a_n \cdot a_{-n} = \sum_{n>0} \tilde{a}_n \cdot \tilde{a}_n. \qquad (17.1.43)$$

This is an important formula that will play a crucial role in the applications of the string models. This will give the various energy levels of a particular string. In the quantization of the string, since the wave functions when operated on by L_N must vanish, eq. (17.1.43) will give the masses of the various quantum levels of the string.

§17.2 Light Cone Quantization and Vacuum Energy of the String

To proceed further, we use the residual conformal invariance (17.1.21) to fix the gauge such that

$$x_+ = P_+ \tau, \tag{17.2.1}$$

where we have defined $x_\pm = (x_0 \pm x_1)$, $P_\pm = P_0 \pm P_1$.

That a gauge such as this can be chosen (where the space–time variable x_+ can be identified with the time evolution parameter in the worldsheet, τ), can be seen by noting the x_+ and a transformed variable $\tilde{\tau}$ satisfying the same equation

$$\left(\frac{\partial^2}{\partial \sigma^2} - \frac{\partial^2}{\partial \tau^2} \right) \psi = 0, \tag{17.2.2}$$

where $\psi = x_+$ or $\tilde{\tau}$.

In this gauge, $a_{n,+} = 0 = \tilde{a}_{n,+} = q_+$, eqs. (17.1.32) and (17.1.27) then yield

$$a_N^- = \frac{1}{p^+} \sum_{n=-\infty}^{\infty} \mathbf{a}_{N-n} \cdot \mathbf{a}_n \tag{17.2.3}$$

where the **boldface** denotes only the transverse coordinates, which are $(D-2)$ in number, if D is the total number of space–time dimensions. (We will determine D subsequently.) The condition $L_0 = 0$ yields (since $a_n^+ = 0$)

$$-\tfrac{1}{8}P^2 = \sum_{n>0} \mathbf{a}_n \cdot \mathbf{a}_{-n} \tag{17.2.4}$$

(again here, the **boldface** denotes that only transverse coordinates enter the sum).

At this point, we note that the independent dynamical variables are q_-, $P_+ a_n^i$, and $\tilde{a}_n^i$ (i =transverse coordinates, $i = 2, \ldots, D-1$). In order to quantize the string, we therefore demand that it obey the following commutation relations:

$$[P^+, q^-] = i,$$
$$[a_n^i, a_m^j] = n\delta_{m+n,0} \cdot \delta^{ij},$$
$$[\tilde{a}_n^i, \tilde{a}_m^j] = n\delta_{m+n,0}\delta^{ij}. \tag{17.2.5}$$

We note that $a^i_{-m}(m > 0)$ are like creation operators. When we consider operators such as L_N, we must write them in the normal ordered form where creation operators appear to the left and the annihilation operators appear to the right. It is easy to see that except for L_0, all L_N can be written in normal ordered form by using the commutation relations equation (17.2.5). But as far as L_0 is concerned, we note that

$$L_0 = -\frac{1}{2} \sum_{\substack{n=-\infty \\ n \neq 0}}^{+\infty} \mathbf{a} \cdot \mathbf{a}_{-n} - \frac{1}{8} P^2 = 0, \qquad (17.2.6)$$

or

$$-\frac{1}{8} P^2 = \sum_{n>0} \mathbf{a}_{-n} \cdot \mathbf{a}_n + \frac{(D-2)}{2} \sum_{n=1}^{\infty} n, \qquad (17.2.7)$$

or

$$\frac{M^2}{8} = \Sigma : \mathbf{a}_{-n} \cdot \mathbf{a}_n : + \alpha_0.$$

Note the appearance of the constant α_0, This had the implication that the vacuum state $|0\rangle$, for which $\mathbf{a}_n|0\rangle = 0$ for $n \geq 0$, does not have zero mass; we will call this the vacuum energy. Let us try to evaluate ao using the following regularization:

$$\alpha_0 = \lim_{\beta \to 0} \frac{D-2}{2} \left(-\frac{\partial}{\partial \beta} \sum_{n=0}^{\infty} e^{-n\beta} \right) \qquad (17.2.8a)$$

$$= \lim_{\beta \to 0} \frac{D-2}{2} \left\{ -\frac{\partial}{\partial \beta} \left(\frac{1}{1-e^{-\beta}} \right) \right\}. \qquad (17.2.8b)$$

Evaluating the above expression, we find

$$\alpha_0 = \frac{D-2}{2} \lim_{\beta \to 0} \left\{ \frac{1}{\beta^2} - \frac{1}{12} + O(\beta)^2 \right\}. \qquad (17.2.9)$$

Discarding the singular piece, we obtain

$$(\alpha_0)_{\text{Reg}} = -\frac{D-2}{24}. \qquad (17.2.10)$$

This is the value of the vacuum energy. The first excited state denoted by $a^i_{-n}|0\rangle$ has mass

$$\frac{M^2}{8} = 1 - \frac{D-2}{24}. \qquad (17.2.11)$$

We note that this state has $D-2$ components and, therefore, does not transform as an irreducible representation of the $SO(D-1)$ "rotation" group, as would be required by Lorentz invariance for all massive states. Therefore, the only way that the existence of this mass level will be compatible with Lorentz invariance is if this level is massless. Equation (17.2.11) then implies that

$$D = 26. \tag{17.2.12}$$

Thus the joint requirement of conformal invariance and Lorentz invariance can only be satisfied in 26 space–time dimensions. This result holds regardless of whether the string is open or closed as long as it is bosonic. This is called the critical dimension.

§17.3   Neveu–Schwarz and Ramond Strings

The bosonic string described in the previous sections leads to excitations which are bosonic in nature. If we want strings to be relevant to the description of the real world, one must find a way to generate fermionic states out of strings. With this aim in mind, Neveu–Schwarz [3] and Ramond [3] extended the bosonic string to include fermionic space–time coordinates denoted by ψ^μ, $\mu = 0, 1, \ldots, D-1$. The action for this string in the conformal gauge can be written as follows:

$$S = T \int d\xi^0 \, d\xi^1 (\partial^\alpha x^\mu + \partial_\alpha x_\mu + \psi^{\mu^T} \gamma^0 \gamma^\alpha \partial_\alpha \psi_n), \tag{17.3.1}$$

where $\alpha = 0, 1$ are worldsheet coordinates as before and ψ^μ are two-dimensional Hermitian fermions

$$\gamma^0 = \begin{pmatrix} 0 & 1 \\ 1 & 0 \end{pmatrix}; \quad \gamma^1 = \begin{pmatrix} 0 & -1 \\ 1 & 0 \end{pmatrix} \quad \text{and} \quad \gamma_{\text{five}} = \gamma^0 \gamma^1 = \begin{pmatrix} 1 & 0 \\ 0 & -1 \end{pmatrix}. \tag{17.3.2}$$

Denoting $\psi^\mu_\pm = \frac{1}{2}(1 \pm \gamma_{\text{five}})\psi^\mu$, we see that the fermionic part of the Lagrangian can be written as

$$\mathcal{L}_F = \psi^\mu_+ \partial_- \psi_{\mu+} + \psi^\mu_- \partial_+ \psi_{\mu-}. \tag{17.3.3}$$

The field equations that follow from $\mathcal{L}_F$ are

$$\partial_\mp \psi^\mu_\pm = 0, \tag{17.3.4}$$

or ψ_+ is only a function of ξ_+ and ψ_- a function of ξ_-.

The Fourier expansion for the $\psi^\mu_\pm$ can then be given as

$$\psi^\mu_+(\xi_+) = \frac{1}{2} \sum \tilde{\psi}^\mu_n e^{-2in\xi_+}$$

and

$$\psi^\mu_-(\xi_-) = \frac{1}{2} \sum \psi^\mu_n e^{-2in\xi_-}. \tag{17.3.5}$$

We quantize the fermionic string as before by imposing the anticommutation relations

$$\{\psi^\mu_n, \psi^\nu_m\} = \eta^{\mu\nu} \delta_{m+n,0} \tag{17.3.6}$$

and a similar relation for $\tilde{\psi}$. For the closed fermionic string, there are two possible boundary conditions

(i)   Periodic: $\psi(0) = \psi(\pi)$,   Ramond string;

(ii)   Antiperiodic: $\psi(0) = -\psi(\pi)$,   Neveu–Schwarz string.

For the case of the Ramond (R) string the n in eq. (17.3.5) are integral, whereas for the Neveu–Schwarz (N–S) string the n are half-integral, so we can rewrite (17.3.5) for the N–S case

$$\psi_-^\mu(\xi_-) = \frac{1}{2} \sum_{r=-\infty} b_{r+1/2}^\mu e^{-2i(r+1/2)\xi_-}, \qquad (17.3.7)$$

and for the R case

$$\psi_-^\mu(\xi_-) = \frac{1}{2} \sum_{n=-\infty}^{+\infty} d_n^\mu e^{-2in\xi_-}. \qquad (17.3.8)$$

Let us now turn our attention to the Hilbert space of the fermionic string. As in the bosonic case, we consider the vacuum state $|0\rangle$, one-particle state $b_{-1/2}^i |0\rangle$, etc., where i goes only over the $D-2$ transverse coordinates, $i = 2, \ldots, D-1$. However, if we turn to the Ramond string, we see that

$$\{d_0^\mu, d_0^\nu\} = \eta^{\mu\nu}. \qquad (17.3.9)$$

But this is precisely the Clifford algebra and must, therefore, have $2^{D/2}$-dimensional representation in the ground state. Therefore, the ground state must transform as a spinor representation of the Lorentz group. We know that only the space–time fermions transform according to the spinor representation of the Lorentz group. Thus, unlike the bosonic or Neveu–Schwarz string that lead to bosons, the Ramond string leads to fermions. A complete description of nature must therefore involve all three types of strings.

Let us now turn to the mass spectrum. As we saw in Section 17.1, for this purpose we need the Virasoro operators L_N for the fermionic string. In analogy with the bosonic case, we evaluate the worldsheet energy momentum tensor $T_{\alpha\beta}$ and obtain its Fourier coefficients, which are nothing but L_N's:

$$T_{\alpha\beta} = \psi^{\mu T} \gamma_0 \gamma_\alpha \partial_\beta \psi_\mu. \qquad (17.3.10)$$

From this we get $T_{++} = \psi_+ \partial_+ \psi_+$ and $T_{--} = \psi_- \partial_- \psi_-$.

Our interest is in the fermionic contribution to L_0, which is given by

$$L_0^F = \frac{1}{2} \sum_{r=-\infty}^{+\infty} b_{-(r+1/2)}^\mu b_{r+1/2} \left(r + \tfrac{1}{2}\right) \qquad \text{(N–S)}, \qquad (17.3.11)$$

$$= \frac{1}{2} \sum_{n=-\infty}^{+\infty} n\, d_{-n}^\mu\, d_n \qquad \text{(R)}. \qquad (17.3.12)$$

In order to normal order this, we write (say for the Ramond case)

$$
\begin{aligned}
L_0^F &= \frac{1}{2}\sum_{n=0}^{\infty} n\, d_{-n}^i\, d_{n,i} - \frac{1}{2}\sum_{n>0} n\, d_n^i\, d_{-n,i} \\
&= \frac{1}{2}\sum : d_{-n}^i\, d_{n,i} : n - \frac{(D-2)}{2}\sum_{n=1}^{\infty} n \\
&= \frac{1}{2}\sum : d_{-n}^i\, d_{n,i} : + \frac{D-2}{24}.
\end{aligned}
\tag{17.3.13}
$$

Similar steps lead us to the vacuum energy for the Neveu–Schwarz case to be

$$
\alpha_0^{\text{N-S}} = -\frac{D-2}{2}\sum_{r=0}^{\infty}\left(r+\tfrac{1}{2}\right).
\tag{17.3.14}
$$

Using the same type of regularization as in the case of bosons, we find

$$
\alpha_0^{\text{N-S}} = -\frac{D-2}{48}.
\tag{17.3.15}
$$

Let us now suppose that we have a system of D bosonic coordinates and D fermions, then the vacuum energy is easily computed to be $-(D-2)/16$. The next excited state is $b_{-1/2}^i|0\rangle$ with mass

$$
\frac{M^2}{8} = \frac{1}{2} - \frac{(D-2)}{16}.
\tag{17.3.16}
$$

Again, since there are not enough components of the state to form an irreducible representation of the $\text{SO}(D-2)$ group, this state must have zero mass, leading to $D = 10$. Thus, the inclusion of Neveu–Schwarz fermions reduces the critical dimension from 26 to 10.

§17.4   GSO Projection and Supersymmetric Spectrum

In the previous three sections, we have seen three kinds of strings, one bosonic and two fermionic types. The bosonic string leads to a bosonic spectrum and the periodic fermionic string leads to a fermionic spectrum; but the fermionic antiperiodic or Neveu–Schwarz string leads to a spectrum which is bosonic. Furthermore, even though the string coordinates and string Lagrangian obey two-dimensional supersymmetry, the resulting spectrum is not supersymmetric. All this confusion becomes clear if we adopt the ansatz first advocated by Gliozzi, Scherk, and Olive [4]. According to this ansatz, of all the plethora of states, only those states are physical that obey the following conditions:

(i) $\qquad -(-1)^F|\text{Phys}\rangle = |\text{Phys}\rangle \quad (N-S);$

$$\sum_{n=1}^{\infty} d_{-n}d_n$$

(17.4.1)

(ii) $\Gamma_{11}(-1)^{\sum_{n=1}^{\infty} d_{-n}d_n}|\text{Phys}\rangle = |\text{Phys}\rangle.$

To see its implications, note that for a Neveu–Schwarz string, the states are

$$|0\rangle, b^i_{1/2}|0\rangle, a^i_{-1}|0\rangle, b^i_{-1/2}b^j_{-1/2}|0\rangle, a^i_{-1}b^j_{-1/2}|0\rangle, \ldots.$$

Subsequent to applying GSO projection, the allowed states, $b^i_{-1/2}b^i_{-3/2}|0\rangle$, $a^i_{-1}b^j_{-1/2}|0\rangle$ are the only ones that are allowed.

Similarly for the Ramond sector, the allowed states are $|La\rangle$, $a^i_{-1}|La\rangle$, $b^i_{-1/2}|Ra\rangle$. Here $|La\rangle$ and $|Ra\rangle$ stand for the Ramond vacuum of left or right helicity. It is important to emphasize that, even though we have adopted the GSO projection as an ansatz, it has been shown that they follow from the requirement of one- and two-loop modular invariance.

We will now show for the low-lying levels that we are considering the states allowed by GSO projection obey supersymmetry. To demonstrate this we take the example of a closed superstring. For a closed string, since all points are equivalent, the theory must be invariant under the shift $\sigma \to \sigma + \delta\sigma$. The unitary operator that implements the shift is given by $U = e^{2i\delta\sigma(L_0 - \tilde{L}_0)}$. This shift invariance implies that

$$L_0 = \tilde{L}_0. \tag{17.4.2}$$

Thus, allowed energy levels are those for which both the left- and right-movers have the same mass.

Closed Superstring

In this case, the spectrum for both the left- and right-movers are identical. We remind the reader that vacuum energies for both the left and right sectors are:

(i)   the Neveu–Schwarz case: $-\frac{1}{2}$; and

(ii)   the Ramond case: 0.

Using this information, the GSO projection [eq. (17.4.1)], and the condition (17.4.2), we find that massless states are

(a) $b^i_{-1/2}|0\rangle \otimes \tilde{b}^j_{-1/2}|0\rangle$;

(b) $|La\rangle \otimes \tilde{b}^i_{-1/2}|0\rangle, \quad b^i_{-1/2}|0\rangle \otimes |La\rangle$;

(c) $|La\rangle \otimes |La\rangle.$

From (a) we get the graviton ($h^{ij} = h^{ji}$), the antisymmetric field ($b^{ij} = -b^{ji}$), and the dilaton ϕ; (b) gives the gravitino, ψ^i_α, and the dilatino (note that the Ramond vacuum provides the fermionic character); and (c) gives rise to a set of 64 scalar bosons. This spectrum is supersymmetric since there are 128 fermionic as well as bosonic degrees of freedom. Furthermore, this theory is defined in ten space–time dimensions.

§17.5   Heterotic String

Before introducing the heterotic string, we discuss a few facts about the strings. The first point is that open strings lead only to spin-1 massless bosons that, with the addition of appropriate internal quantum numbers at the end points of the strings, can be identified with gauge fields. This string has no room for massless spin-2 particles, which would be necessary if we want the strings to describe gravitational interactions. On the other hand, massless states of closed Neveu–Schwarz strings correspond to spin-2 fields, which can be identified with the graviton. Thus, closed strings can describe gravity but apparently have no room for Yang–Mills fields. A clever combination is clearly needed to get both Yang–Mills fields as well as the graviton. This idea was proposed by Goss, Harvey, Martinec, and Rohm [5], who introduced the so-called heterotic string.

One of the essential ingredients in their work is the well-known Kaluza–Klein observation [6], that if we consider higher-dimensional gravity and compactify additional space–time dimensions, depending on the nature and isometry of the compactification manifold, new local symmetries can emerge. For instance, when the extra dimension in a five-dimensional theory of gravity is compactified on a circle, a new local U(1) symmetry emerges, which can be identified with electromagnetism.

In order to exploit this idea to get Yang–Mills theories, Gross et al. proposed a closed superstring, but one which has the left-moving string being purely bosonic so that it lives in a 26-dimensional space–time, whereas the right moving string is considered to be a superstring so that it lives in ten-dimensional space–time. The extra 16 dimensions of the left-mover can then be compactified on a torus leading to a local internal symmetry of rank 16. We will see below that the group actually is $E_8 \times E_8$ or SO(32). This being a closed string also simultaneously leads to a massless graviton in its spectrum. Thus, the heterotic string has all the basic forces; we will see in this chapter that it also has all the constituents of matter; thus, this has provided the hope that this may be the right framework for the ultimate unified theory of nature.

Let us now proceed to calculate the spectrum and the gauge group of the heterotic string. For this purpose, let us first write down the vacuum energies of the right-moving sector. Since there are fermionic coordinates in this sector, we must consider both Neveu–Schwarz and Ramond boundary conditions:

$$\text{Vacuum energy:} \quad \begin{aligned} \text{N–S} &: \quad -\tfrac{8}{24} - \tfrac{8}{48} = -\tfrac{1}{2}, \\ \text{R} &: \quad -\tfrac{8}{24} + \tfrac{8}{24} = 0. \end{aligned} \quad (17.5.1)$$

Therefore, the massless levels allowed by the GSO projection rules for this sector are

$$b^i_{-1/2}|0\rangle \quad \text{8 bosons,}$$
$$|L\rangle \quad \text{8-component left-handed spinor.} \tag{17.5.2}$$

Turning to the left-moving sector, we have got 26 bosonic coordinates, which in the light-cone gauge reduce to 24 degrees of freedom. We wish to treat 16 of these are compactified coordinates. It is convenient for later purposes to fermionize these 16 coordinates. What we mean is the following. It is well known from the work of Coleman [7] and Mandelstam [8] that in two dimensions (in our case, the worldsheet), there is an equivalence between the bosonic and fermionic systems, and one bosonic degree of freedom is equivalent to two Majorana fermions. Using this result, we can write the left-moving sector in light-cone gauge as consisting of eight bosonic coordinates denoted by $X^i_L (i = 1, \ldots, 8)$ and 32 fermionic coordinates denoted by $\lambda^A (A = 1, \ldots, 32)$. In order to study this sector, note that the fermions λ^A could have both periodic as well as antiperiodic boundary conditions. We will consider two cases:

(a) all 32 fermions obey the same boundary conditions, in this case we will show that the gauge group will be SO(32);

(b) 16 of the fermions have the same boundary conditions. This case leads to the gauge group $E_8 \times E_8$.

Case (a). Vacuum Energies.

$$\begin{aligned}(N-S): \quad & -\tfrac{8}{24} - \tfrac{32}{48} = -1, \\ (R): \quad & -\tfrac{8}{24} + \tfrac{32}{48} = +1.\end{aligned} \tag{17.5.3}$$

We see that the Ramond sector has no massless states. The states of zero mass are (tilde denotes left-movers)

$$\begin{aligned}&\tilde{a}^i_{-1}|0\rangle, \\ &\tilde{\lambda}^A_{-1/2}\tilde{\lambda}^B_{-1/2}|0\rangle.\end{aligned} \tag{17.5.4}$$

Combining eqs. (17.5.2) and (17.5.4) we get the physical particle spectrum to be

$$\begin{aligned}\tilde{a}^i_{-1}|0\rangle \otimes b^j_{-1/2}|0\rangle: \quad & \text{graviton } h^{ij}, \\ & \text{antisymmetric field } b^{ij}, \\ & \text{dilaton } \phi, \\ \tilde{a}^i_{-1}|0\rangle \otimes |L\rangle \quad & \text{gravitino, dilatino,} \\ \tilde{\lambda}^A_{-1/2}\tilde{\lambda}^B_{-1/2}|0\rangle \otimes b^i_{-1/2}|0\rangle \quad & \text{gauge bosons,} \\ \tilde{\lambda}^A_{-1/2}\tilde{\lambda}^B_{-1/2}|0\rangle \otimes |L\rangle \quad & \text{gaugino.}\end{aligned} \tag{17.5.5}$$

Looking at the gauge boson and gaugino states, we see that the gauge group is clearly SO(32). Thus, the resulting low-energy theory is a ten-dimensional $N = 1$ supergravity theory with an SO(32) Yang–Mills group.

Case (b). Vacuum Energies (v.e.) for the Left-Moving Sector.

(i) 8 bosonic, 16 (N-S) ($\tilde{\lambda}^A$), 16 (R) ($\tilde{\lambda}^P$)

$$\text{v.e.} = -\tfrac{8}{24} - \tfrac{16}{48} + \tfrac{16}{24} = 0; \tag{17.5.6}$$

(ii) 8 bosonic, 32 (N-S):

$$\text{v.e.} = -1. \tag{17.5.7}$$

The spectrum for the left-moving side is as follows.

From case (ii), we get $\tilde{\lambda}^A_{-1/2}\tilde{\lambda}^B_{-1/2}|0\rangle$ under $SO(16) \times SO(16)$, it transforms as $(120, 1)$; $\tilde{\lambda}^P_{-1/2}\tilde{\lambda}^Q_{-1/2}|0\rangle$ transforms under $SO(16) \times SO(16)$ as $(1, 120)$. Coming to case (i), the massless states are the Ramond vacuum $|\Lambda\rangle$ which is a spinor under $SO(16)$, which is 128-dimensional: Thus, this has two massless states $(128, 1)$ (denoted by $|\Lambda_A\rangle$) and $(1, 128)$ (denoted by $|\Lambda_p\rangle$). Adding these, we see that we get in the A-sector $120 \oplus 128 = 248$-dimensional representation which is the adjoint representation of E_8, and similarly in the P-sector a 248-dimensional representation of another E_8, leading to $E_8 \times E_8$, as the gauge group. As before, coming with the massless right-moving states, we get a ten-dimensional $N = 1$ supergravity theory coupled with the $E_8 \times E_8$ Yang–Mills group.

§17.6 $N = 1$ Super-Yang–Mills Theory in Ten Dimensions

The field content of this model has been described in the previous section to be

$$
\begin{aligned}
e_M^A: & \quad \text{zeinbein}\begin{cases} M = 1, \ldots, 10, \\ A = 1, \ldots, 10, \end{cases} \\
\psi_M: & \quad \text{gravitino,} \\
\phi: & \quad \text{dilaton,} \\
\lambda: & \quad \text{dilatino,} \\
& \quad B_{MN} = -B_{nm}, \\
A_M^\alpha: & \quad \text{gauge boson}, \alpha = E_8 \times E_8 \text{ index,} \\
\chi^\alpha: & \quad \text{gaugino.}
\end{aligned}
\tag{17.6.1}
$$

We have already written down the Lagrangian for this theory in Chapter 16 and we repeat it here for completeness [9, 10]:

$$
\frac{\mathcal{L}}{\tilde{e}} = -\tfrac{1}{2}R - \frac{i}{2}\bar{\psi}_M \Gamma^{MNP} D_N \psi_P + \frac{i}{2}\bar{\lambda}\Gamma^M D_M \lambda
$$

$$
+ \frac{9}{16}\left(\frac{\partial_M \phi}{\phi}\right)^2 + \frac{3}{4}\phi^{-3/2} H_{MNP} H^{MNP} + \frac{3\sqrt{2}}{8}\bar{\psi}_M \frac{\partial\!\!\!/\phi}{\phi}\Gamma^M \lambda
$$

$$+ \frac{\sqrt{2}}{16} \phi^{-3/4} H_{MNP} (i\bar{\psi}_s \Gamma^{SMNPR} \psi_R$$

$$+ 6i\bar{\psi}^M \Gamma^N \psi^P + \sqrt{2} \psi_s \Gamma^{MNP} \Gamma^s \lambda) + \cdots$$

$$- \tfrac{1}{4} \phi^{-3/4} F^{\alpha}_{MN} F^{\alpha,MN} + \frac{i}{2} \bar{\chi}^\alpha \Gamma^M (D_{M_\chi})^\alpha + \cdots$$

$$+ \frac{\sqrt{2}i}{16} \phi^{-3/4} H_{MNP} \bar{\chi}^\alpha \Gamma^{MNP} \chi^\alpha. \tag{17.6.2}$$

Before proceeding further to extract four-dimensional physics from this Lagrangian, we observe that the field ϕ appears in the Lagrangian in the denominator; therefore, for the theory to make sense, the ground state must correspond to $\langle \phi \rangle \neq 0$ and this will determine the gauge coupling constraint of the theory. Obviously, the Lagrangian written above does not lead to $\langle \phi \rangle \neq 0$, and one needs dynamical effects to determine the true vacuum state.

Another point to emphasize is that if we define a 3-form[1]

$$H = H_{MNP} \, dx^M \wedge dx^N \wedge dx^P, \tag{17.6.3}$$

$$H = dB - \omega^0_{3,y} - \omega^0_{3,L}, \tag{17.6.4}$$

where dB is the exterior derivative of the 2-form $B_{MN} \, dx^M \wedge dx^N$, and $\omega^0_{3,y}$ and $\omega^0_{3,L}$ are the Yang–Mills and Lorentz, Chern–Simons 3-forms. $\omega^0_{3,y}$ was introduced by Chapline and Manton [10] to maintain $N = 1$ supersymmetry, and $\omega^0_{3,L}$ was shown by Green and Schwarz [11] to be necessary in order to make the theory free of gravitational anomalies.

From eq. (17.6.4) we note that the exterior derivative of H is given by (using the property $d^2 = 0$)

$$dH = \text{Tr}\, R \wedge R - \tfrac{1}{30} \text{Tr}\, F \wedge F. \tag{17.6.5}$$

We will see below that eq. (17.6.5) plays a rather crucial role in the compactification.

§17.7 Compactification and the Calabi–Yau Manifold

In order to make connection with the real world, we will have to compactify six of the ten dimensions. A simple way to reduce to four dimensions is to assume the existence of a ground state for which

$$e^A_M = \begin{pmatrix} e^\alpha_\mu & e^i_\mu \\ 0 & e^j_m \end{pmatrix}, \tag{17.7.1}$$

---

[1]The notation of forms is a very powerful one to study the properties of manifolds and we will discuss some properties and uses of forms in a subsequent section of this chapter.

where

$$\mu, \alpha = 1, 2, 3, 4,$$
$$m, i = 5, \ldots, 10.$$

Here e_μ^α denotes the vierbein, e_μ^i corresponds to six vector fields, and e_i^j are 21 scalar fields. Let us look at the gravitino ϕ_μ, which is a Majorana spinor in ten dimensions and is 16-dimensional. But we see from Chapter 7 that an SO(10) spinor reduces to four SO(4) spinors; we therefore have four gravitini. This, therefore, leads to $N = 4$ supergravity. We also know from Chapter 16 that a theory with $N \geq 2$ supersymmetry has the problem of being vectorlike, and therefore cannot explain the observed chiral nature of weak interactions. We must therefore find a compactification space that automatically eliminates three of the four gravitini, leaving only one, which would lead to an $N = 1$ supergravity that is free of chirality problems. The Calabi–Yau manifold is precisely one such space, as was demonstrated by the classic paper of Candelas, Horowitz, Strominger, and Witten [12]. Furthermore, we will find that there exist a large number of spaces that are Calabi–Yau spaces. The topological characteristics of the different Calabi–Yau spaces are different, leading to different numbers of massless states (the index of the Dirac operator on the manifold).

To see how these kinds of manifolds arise from the requirement of maintaining $N = 1$ supersymmetry, we begin with the realization that the ground state must preserve supersymmetry. This means that the variation of the various fields under supersymmetry transformations must vanish on the ground state. One way all variations will vanish is if the fields vanish but, in this case, we remain in ten-dimensional space–time. The compactification requires that the curvature for six of the ten dimensions must be nonzero. Also, we want the extra six dimensions to be part of a compact manifold so that the mass spectrum is discrete. To see how this reconciliation between $N = 1$ supersymmetry and a nonvanishing curvature of the compact manifold takes place, we will write down the variation of the different fields under supersymmetry transformation in the ground state. Clearly, the variation of the Bose fields involve Fermi fields, which have vanishing vacuum expectation value due to Lorentz invariance. However, the situation is different for the fermionic fields, whose variations involve various bosonic fields as follows:

$$
\begin{aligned}
\delta\psi_\mu &= \nabla_\mu \varepsilon + \tfrac{\sqrt{2}}{32} e^{2\phi} (\gamma_\mu \gamma_5 \otimes \tilde{H}) \varepsilon, \\
\delta\psi_\mu &= \nabla_m \varepsilon + \tfrac{\sqrt{2}}{32} e^{2\phi} (\Gamma_m \tilde{H} - 12\tilde{H}_m) \varepsilon, \\
\delta\lambda &= \sqrt{2} (\Gamma_m \nabla^m \phi) \varepsilon + \tfrac{1}{8} e^{2\phi} \tilde{H} \varepsilon, \\
\delta\chi^\alpha &= -\tfrac{1}{4} e^\phi F^{\alpha,mn} \Lambda_{mn} \varepsilon,
\end{aligned}
\tag{17.7.2}
$$

where $\tilde{H} = H_{pqr} \Gamma^{pqr}$; $\tilde{H}_m = H_{mqr} T^{qr}$. The condition $\delta\psi_\mu = 0$ leads to the integrability condition

$$[\nabla_\mu, \nabla_\nu]\varepsilon = \tfrac{1}{256} e^{4\phi} (\gamma_\mu \gamma_\nu \otimes \tilde{H}^2)\varepsilon. \tag{17.7.3}$$

The left-hand side is given by

$$[\nabla_\mu, \nabla_\nu] = \tfrac{1}{2}[\Gamma^\sigma, \Gamma^\lambda]R_{\mu\nu\lambda\sigma}. \tag{17.7.4}$$

If one assumes maximally symmetric spaces, i.e.,

$$R_{\mu\nu\lambda\sigma} = \kappa(g_{\mu\lambda}g_{\nu\sigma} - g_{\mu\sigma}g_{\nu\lambda}), \tag{17.7.5}$$

eq. (17.7.3) then implies that

$$\tilde{H}^2\varepsilon \sim e^{-4\phi}\frac{\kappa}{2}\varepsilon. \tag{17.7.6}$$

The simplest solution to eq. (17.7.6) is to assume $H = 0$. This also satisfies the condition $\delta\lambda = 0$ if we further assume $\nabla^m\phi = 0$. This also implies that the four space–time dimensions are Minkowski type.

Let us now turn our attention to the compactified dimensions: $\delta\psi_m = 0$ implies

$$(\nabla^m - \beta\Gamma_{ij}H^{mij})\varepsilon = 0. \tag{17.7.7}$$

For $H = 0$, we get $\nabla^m\varepsilon = 0$, which means that ε must be a covariantly constant spinor. In order to analyze its implications, let us note that the sixteen-component ten-dimensional spinor under SO(4) ⊗ SO(6) decomposes as $\varepsilon \sim 2 \otimes 4 \oplus 2 \otimes 4^*$, and can be written as

$$\varepsilon = \begin{pmatrix} \eta \\ \eta^* \end{pmatrix}, \tag{17.7.8}$$

where $\eta = \begin{pmatrix} \eta_1 \\ \eta_2 \\ \eta_3 \\ \eta_4 \end{pmatrix}$, η_i are 2-component SO(4) spinors.

The condition of η being independent of the compactified coordinates leads to the condition

$$\Gamma_{ij}\omega_m^{ij}\begin{pmatrix} \eta \\ \eta^* \end{pmatrix} = 0, \tag{17.7.9}$$

where ω_m^{ij} are the gauge fields corresponding to the local SO(6) rotation. Since η is the four-dimensional SO(6) spinor, if ω_m^{ij} is nonvanishing for the SU(3) subgroup of SO(6), only one out of the four η_i's can be nonzero. (Of course, if we set all ω_m^{ij} as zero, the internal space would have zero curvature and is not compact, and the resulting theory would have $N + 4$ supersymmetry.) Equation (17.7.9) can be rewritten as follows:

$$\left(\begin{array}{c|c} \lambda \cdot \omega & \\ \hline & 0 \end{array} \right) \begin{pmatrix} 0 \\ 0 \\ 0 \\ \eta \end{pmatrix} = 0, \tag{17.7.10}$$

where λ are the SU(3) matrices. This space is said to have SU(3)-holonomy. Thus, $N = 1$ supersymmetry is preserved, if the compactified manifold has SU(3)-holonomy with a single covariantly constant spinor η. Equation (17.6.5) then implies that since $H = 0$,

$$\operatorname{Tr} R \wedge R = \tfrac{1}{30} \operatorname{Tr} F \wedge F. \tag{17.7.11}$$

Thus, the SU(3) subgroup of the gauge group can have nonvanishing fields F such that

$$\omega_m = A_m. \tag{17.7.12}$$

This breaks the gauge group down from $E_8 \times E_8$ to $E_6 \times E_8$ and SO(32)$\rightarrow$ SO(26) $\times$ U(1).

To study this space further, we note that, if we define a tensor

$$J_k^i = g^{ij}\bar{\eta}\Gamma_{jk}\eta, \tag{17.7.13}$$

this tensor is covariantly constant and satisfies

$$J_k^i J_l^k = -\delta_l^i. \tag{17.7.14}$$

It is this property that enables us to use J_k^i to define a complex structure in the compact manifold. A complex structure simply means that at any point in the space we can write $J_{a/b} = i\delta_{a/b}$, where $i^2 = -1$, and using this we can define complex coordinates $z_i = y_i + iy_{i+1}$, $i = 5, 7, 9$. An important question is: Can J_k^i be cast into the canonical form at every point in the manifold? This is analogous to the following situation: as is well known, the metric can be made locally flat for any space, but for it to be globally flat the curvature tensor must vanish at all points. The analogous condition to the vanishing of the curvature tensor for the complex structures is the vanishing of the Nijenhuis tensor defined as follows:

$$N_{ij}^k = J_i^l(\partial_l J_j^k - \partial_j J_l^k) - J_j^l(\partial_l J_i^k - \partial_i J_l^k). \tag{17.7.15}$$

In our case, it was shown in [12] that $N_{ij}^k = 0$. It is further possible to show that if one defines a Kähler form

$$k = g_{a\bar{b}}\, dz^a\, dz^{\bar{b}}, \tag{17.7.16}$$

then $dk = 0$ on the Kähler form is closed. In this case, the manifold is called a Kähler manifold and the metric can be written in the form

$$g_{a\bar{b}} = i\partial_a \partial_{\bar{b}}\phi. \tag{17.7.17}$$

We note that k is not an exact form [13] (i.e., k cannot be written as the exterior derivative of another lower form).

To summarize, the compact manifold of interest is a complex Kähler manifold with a metric of SU(3)-holonomy. Such manifolds are called Calabi–Yau manifolds. It was conjectured by Calabi and established by Yau that a Kähler manifold with vanishing first Chern class always admits a metric of SU(N)-holonomy. This observation is extremely important because, while it is extremely hard to write a metric of SU(3)-holonomy, it is easy to construct complex Kähler manifolds with a vanishing first Chern class. We describe this in the next section.

§17.8 Brief Introduction to Complex Manifolds

In extracting the physics content of superstring models, compactified on a Calabi–Yau manifold, several concepts from the theory of complex manifolds are essential. There exist comprehensive discussions of their properties in many places in the literature [13]. In this section we briefly touch on only some points, mainly to familiarize the reader with the forthcoming terminology.

Many topological aspects of real as well as complex manifolds can be studied by the use of homology and cohomology groups. Cohomology groups of a manifold can be studied using the properties of forms. For simplicity, let us first discuss real manifolds of dimension n. In this manifold, one can define p-forms, $p = 1, \ldots, n$, as follows:

$$p\text{-form}: \quad A^p = A_{\mu_1 \ldots \mu_p} \, dx^{\mu_1} \wedge dx^{\mu_2} \wedge \cdots dx^{\mu_p}, \qquad (17.8.1)$$

where A is totally antisymmetric in its indices and $\wedge$ denotes the wedge product, which is also totally antisymmetric. An exterior derivative of a p-form maps the p-form into a $(p + 1)$-form as follows:

$$dA^p = (\partial_{\mu_{p+1}} A_{\mu_1 \ldots \mu_p} - \partial_{\mu_1} A_{\mu_{p+1} \ldots \mu_p^+ \ldots}) \, dx^{\mu_1} \wedge dx^{\mu_p} \wedge dx^{\mu_{p+1}}. \quad (17.8.2)$$

A p-form is closed if $dA^p = 0$, and is exact if $A^p = dA^{p-1}$. Since $d^2 = 0$, an exact form is always closed. Two closed forms are said to be in the same equivalence class if they differ from each other by an exact form. The number of different, independent equivalence classes of p-forms is called the Betti number b_p, which is the dimension of the cohomology group H^p (the collection of closed p-forms). An interesting property obeyed by the Betti numbers is the so-called Poincaré duality, which says that

$$b_p = b_{n-p}, \qquad (17.8.3)$$

where n is the dimension of the manifold. Furthermore, $b_0 = 1$ for any manifold, which from eq. (17.8.3) leads to $b_n = 1$. The Euler characteristic of a manifold is defined as

$$\chi = \sum_{p=0}^{n} (-1)^p b_p. \qquad (17.8.4)$$

It often proves helpful to deal with the homology rather than the cohomology of a manifold in order to calculate the Betti numbers. To study the homology group for a compact manifold of dimension n, we consider all its closed submanifolds. Let us denote the boundary of a compact manifold B by ∂B. It is then obvious that $\partial^2 B = 0$ (since a boundary has no

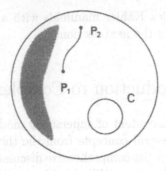

Figure 17.3.

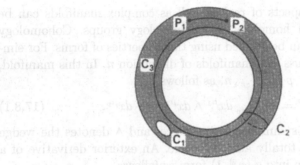

Figure 17.4.

boundary). Thus, the operation ∂ is analogous to the exterior derivative. Now, a p-dimensional closed submanifold may or may not be a boundary of a $(p + 1)$-dimensional manifold. If it is a boundary, ignore it. When two p-dimensional submanifolds $M_p^{(1)}$ and $M_p^{(2)}$ form the boundaries of a $(p + 1)$-dimensional manifold, they belong to the same equivalence class. (For example, two points on a surface constitute the boundary of a line; therefore, they belong to the same equivalence class.) The number of different equivalence classes gives the dimension of the homology group H_p and gives the Betti number. We illustrate the calculation of the Betti numbers of a 2-sphere (Fig. 17.3) and torus (Fig. 17.4) below.

For the case of the 2-sphere (Fig. 17.3) we see that all points $(p_1, p_2, \ldots)$ belong to the same equivalence class, and if a point is a zero-dimensional compact submanifold, we get $b_0 = \dim H_0 = 1$. As far as b_1 is concerned, note that all closed 1-manifolds (a typical one being c) can be expressed as boundaries, therefore $b_1 = 0$. Using Poincaré duality, we find $b_2 = b_0 = 1$. The Euler characteristic χ of the 2-sphere is 2.

To study the Betti numbers of a torus we look at Fig. 17.4 and observe that, as in the case of a 2-sphere, $b_0 = b_2 = 1$. As far as b_1 is concerned, we can identify three different equivalence classes with characteristic elements given by c_1, c_2, and c_3. Of these three, only $c_1 = \partial B$, where c_2 and c_3

are not expressible as boundaries (c_1 can be contracted to a point whereas c_2 and c_3 cannot). Therefore, we find $b_1 = 2$, and the Euler characteristic $\chi = 0$.

We can continue this procedure to obtain the Betti numbers of a 2-surface with n-handles (a torus is a 2-surface with one handle). Each time we add a new handle we increase b_1 by 2; b_0 and b_2 of course remain unchanged. Thus,

$$\chi(\text{2-surface of } n\text{-handles}) = (2 - 2n). \tag{17.8.5}$$

Next we state a very important theorem, known as the Hodge decomposition theorem, which states that for a compact manifold there is a one-to-one correspondence between a closed form and an harmonic form, which indicates the existence of a zero mode. This correspondence then enables us to obtain the number of massless modes from knowledge of the Betti numbers.

The above discussions can be extended to the case of complex manifolds with the following difference: There are now two indices to a form, typically denoted as a (p, q)-form (p denoting the number of holomorphic and q denoting the number of antiholomorphic)

$$A_{p,q} = A_{\mu_1 \ldots \mu_p \bar{\nu}_1 \ldots \bar{\nu}_q} \, dz^{\mu_1} \wedge \cdots dz^{\mu}_r \wedge d\bar{z}^{\bar{\nu}_1} \cdots \wedge d\bar{z}^{\bar{\nu}_q}. \tag{17.8.6}$$

There are, therefore, forms closed with respect to either the holomorphic or the antiholomorphic indices or both. The Betti numbers are similarly denoted by $b_{p,q}$ and the analog of Poincaré duality is

$$b_{p,q} = b_{n-p,n-q}. \tag{17.8.7}$$

Finally, as in the case of real manifolds, $b_{p,q}$ denotes the number of massless modes (or harmonic forms). For further detailed information regarding the above topological properties of the complex and real manifolds, we refer the reader to [1] and [13].

§17.9 Calabi–Yau Manifolds and Polynomial Representations for (2, 1) Forms

As discussed in the introduction, one way to obtain a four-dimensional theory with $N = 1$ supersymmetry from the heterotic string is to let the extra six dimensions compactify into a Kähler manifold with SU(3)-holonomy, or into a Kähler manifold with vanishing first Chern class known as the Calabi–Yau manifolds. In this section, we start by giving some examples of Calabi–Yau manifolds, and ways to obtain their Betti–Hodge numbers and Euler characteristics.

The simplest example of a Kähler manifold is the space $\mathbf{C}^n$ (n-tuples of complex coordinates $z_1, z_2, \ldots$). The Kähler potential ϕ is given in this case by

$$\phi = \sum_k |z^k|^2 . \tag{17.9.1}$$

This is, however, not a compact space and is therefore not useful for our purposes. We can obtain a compact space from this by excluding the origin, and identifying all points along lines passing through the origin in this space. In other words, we assume $z^k \simeq \lambda z^k$ for $k = 1, \ldots, n$ and we exclude the origin. Such a manifold is called a CP^{n-1}-manifold. A $\phi(z^k, \bar{z}^k)$ in this manifold can be written as

$$\phi(z, \bar{z}) = \ln\left(1 + \sum_{a=1}^{n-1} z^a \bar{z}^a\right). \tag{17.9.2}$$

This is still not a Calabi–Yau manifold, since it is known that the first Chern class of CP^N is nonvanishing. So we will have to choose submanifolds of CP^N. We consider submanifolds of CP^N that correspond to zero loci of homogeneous polynomials. To obtain Calabi–Yau manifolds this way, let us write the formula for Chern classes: let us suppose we have a CP^n-manifold with homogeneous polynomials $p_i = 0$ of degree q_i. The above-mentioned formula is then given by

$$\sum_{k=0}^{N+1} C_k X^k = \frac{(1+x)^{N+1}}{\prod_i (1+q_i x)}, \tag{17.9.3}$$

where C_k are the Chern classes. The Calabi–Yau condition is $C_1 = 0$. This implies

$$N + 1 = \sum_i q_i. \tag{17.9.4}$$

This formula can be generalized to include products of several CP^{N_i}-manifolds. As an example of a Calabi–Yau manifold, we consider a CP^4-space and consider a surface that satisfies the constraint

$$\sum_{i=1}^5 z_i^5 = 0. \tag{17.9.5}$$

The next question we wish to address is the Betti numbers $b_{p,q}$ for Calabi–Yau manifolds. It is given by the following array of numbers for complex dimension n of the manifold [let us choose $n = 3$ for simplicity corresponding to SU(3)-holomony]:

$$b_{0,0}$$
$$b_{0,1} \qquad b_{1,0}$$
$$b_{0,2} \qquad b_{1,1} \qquad b_{2,0}$$
$$b_{0,3} \qquad b_{1,2} \qquad b_{2,1} \qquad b_{3,0}$$
$$b_{1,3} \qquad b_{2,2} \qquad b_{3,1}$$
$$b_{2,3} \qquad b_{3,2}$$
$$b_{3,3}$$

This is called the Hodge diamond. Use of the Poincaré duality relates the various Betti numbers, e.g., $b_{0,0} = b_{3,3}$, $b_{0,2} = b_{3,1}$, $b_{0,1} = b_{3,2}$, etc. It turns out that for Calabi–Yau manifolds, $b_{0,2} = b_{0,1}$. Using these relations, we obtain only $b_{1,1} \neq 0$ and $b_{2,1} \neq 0$, and $b_{0,0} = b_{3,3} = 1$. The Euler characteristic of this manifold is then given by

$$\chi = 2(b_{1,1} - b_{2,1}). \qquad (17.9.6)$$

The Betti numbers $b_{1,1}$ and $b_{2,1}$ for a Calabi–Yau manifold can be computed as follows. For manifolds that are single CP^N, $b_{1,1} = 1$. For products of CP^N, in general, the manifold can be written as follows [14]:

$$\begin{pmatrix} n_1 & q_{11} & q_{12} & q_{1c} \\ n_2 & q_{21} & q_{22} & q_{2c} \\ \vdots & \vdots & \vdots & \vdots \\ n_k & q_{k1} & q_{k2} & q_{kc} \end{pmatrix}, \qquad (17.9.7)$$

where q_{ij} represents the power of the variable z_i occurring in the jth polynomial. The total Chern class for such a manifold is given by

$$C(M_{cr}) = \frac{\prod_{\pi=1}^{m}(1 + \chi_\pi)^{n_r+1}}{\prod_{a=1}^{n_1+n_2+n_2-3}(1 + \sum_{r=1}^{m} q_{ar}\chi_r)}. \qquad (17.9.8)$$

In order to compute the $b_{1,1}$, first ignore all polynomials that mix the different CP^N's; then

$$b_{1,1} = \sum_i (\chi_i - 2), \qquad (17.9.9)$$

where χ_i is the Euler characteristic of the different submanifolds. For a single CP^N-manifold with several polynomial constraints,

$$\chi = c_3 \cdot \prod_i q_i, \qquad (17.9.10)$$

where c_3 is the coefficient of x^3 in the expansion in (17.9.3).

Let us illustrate this with the help of the following example (which will appear later on in this chapter in our discussion of three-generation models). The manifold under consideration is [15]

$$\begin{pmatrix} 3 & \big\| & 0 & 3 & 1 \\ 3 & \big\| & 3 & 0 & 1 \end{pmatrix}.$$

Using the above formula

$$b_{1,1} = 2(\chi - 2), \tag{17.9.11}$$

where χ is the Euler characteristic of the manifold $(3||3)$. Using eq. (17.9.3), we get

$$1 + c_1 x + c_2 x^2 = \frac{(1+x)^4}{(1+3x)}. \tag{17.9.12}$$

We get $c_2 = 3$ and $\chi = 9$ leading to [16] $b_{1,1} = 14$ [17]. For a more general manifold of the type given in eq. (17.9.7), the formula for the Euler characteristic is given by

$$\chi = \left. \frac{\sum_{i=1}^{k}(1 + x_i)^{n_i + 1} \prod_{j=1}^{c} \sum_{i=1}^{k} q_{ij} x_i}{\prod_{j=1}^{c}(1 + \sum_{i=1}^{k} q_{ij} x_i)} \right|_{\pi_{i=1}^{k} x_i^{n_i}}, \tag{17.9.13}$$

where the subscript on the right-hand side means that we pick the coefficient of the term $\pi_{i=1}^{k} x_i^{n_i}$.

Let us now turn to the calculation of $b_{2,1}$. The calculation $b_{2,1}$ is done by using deformation theory [16]. It turns out that there is a one-to-one correspondence between the (2,1)-forms and the independent polynomial deformations of the manifold, which are not related to the defining polynomials by holomorphic transformations. These independent polynomials are independent complex structures of the manifold. Let us illustrate the calculation of $b_{2,1}$ for the following two manifolds of SU(3) holonomy.

Example 17.1 (CP^4 with a Quintic Polynomial, $P_5 = 0$) To count the complex structure, first note that there are

$$\frac{5 \cdot 6 \cdot 7 \cdot 8 \cdot 9}{1 \cdot 2 \cdot 3 \cdot 4 \cdot 5} = 126$$

possible polynomials of degree five. But out of them, 24 are related by SL(5,c) transformations to the polynomial P_5, and are therefore not independent. One polynomial P_5 (being the defining polynomial) vanishes, thus reducing the number of independent complex structures to 101. Thus for this manifold $b_{2,1} = 101$. Since $b_{1,1} = 1$, the Euler characteristic for this manifold is -200.

Example 17.2 Consider the manifold proposed by Tian and Yau [15]:

$$\begin{pmatrix} 3 & \| & 0 & 3 & 1 \\ 3 & \| & 3 & 0 & 1 \end{pmatrix}.$$

The total number of possible deformations in this case is

$$\frac{4 \cdot 5 \cdot 6}{1 \cdot 2 \cdot 3} + \frac{4 \cdot 5 \cdot 6}{1 \cdot 2 \cdot 3} + 4 \cdot 4 = 56.$$

Out of these, 30 are related to the original polynomials by linear (and hence holomorphic) transformations and three of the polynomials vanish.

This leaves 23 independent complex structures. Thus $b_{2,1} = 23$. We have already calculated $b_{1,1}$ for the manifold to be 14, leading to $\chi = -18$.

We now discuss some formulas and theorems from the theory of complex manifolds that are useful in obtaining the Bette–Hodge numbers for produce manifolds.

Künneth Formula

For a product of two manifolds c and c', the Betti–Hodge numbers $h^{p,q}(c \times c')$ for the produce manifold are given by

$$h^{p,q}(c \times c') = \sum_{\substack{p=r+m \\ q=s+n}} h^{r,s}(c) \cdot h^{m,n}(c'). \tag{17.9.14}$$

Lefschetz Fixed Point Theorem

Let g be a symmetry operating on the harmonic (p,q)-forms of a Kähler manifold M of complex dimension n. The Lefschetz number $L(g)$ can be defined as

$$L(g) = \sum_{n}^{n}(-1)^{p+q} \operatorname{Tr}(p,q)_M(g), \tag{17.9.15}$$

where $\operatorname{Tr}(p,q)_M(g)$ denotes the trace in the vector space $H_{\bar{\partial}}^{(p,q)}$ of the action of g on harmonic (p,q)-forms. Then

$$L(g) = \sum_{\mu_j} \chi(\mu_j), \tag{17.9.16}$$

where μ_j are the submanifolds left fixed by g, and $\chi(\mu_j)$ are their Euler characteristics.

It follows from this theorem that

$$L(\text{Identity}) = \chi(M). \tag{17.9.17}$$

This theorem is crucial to obtaining the symmetry properties of the $(1,1)$-forms or the $\overline{27}$ fields of the theory. We will consider applications of this theorem in a subsequent section, when we discuss in detail the Tian–Yau manifold.

Before concluding this section, we wish to make some general comments on the possible number of models that may emerge from these kinds of considerations. The requirement of $N = 1$ supersymmetry is satisfied by any Calabi–Yau manifold with SU(3) holonomy, i.e., a three-dimensional complex Kähler manifold with vanishing first Chern class. If we choose these manifolds to be of the algebraic variety type, given by the locus of the zeros of homogeneous polynomials in CP^N-manifolds, then they are given by the two conditions listed in eq. (17.9.7) with

$$\sum_{j=1}^{c}\sum_{i=1}^{k} q_{ij} = \sum_{l=1}^{k} n_l + k. \tag{17.9.18}$$

If we consider a single CP^N, eqs. (17.9.7) and (17.9.18) imply

$$N+1 = \sum_{i=1}^{m} q_i$$

and

$$N = m+3, \tag{17.9.19}$$

and $q_i > 1$, since $q_i = 1$ merely reduces CP^N to CP^{N-1}. It is easy to see that for $N \geq 8$, there is no solution to eq. (17.9.19) if none of the q_i is allowed to be one. Thus, the number of single CP^N-manifolds with vanishing first Chern class are indeed very few. Of course, one can consider products of CP^N; the number of manifolds that arise in this case are many more. But if we impose the phenomenological requirement of three or at most four generations, then the Euler characteristic χ must be $|\chi| = 8$ or 6 and, further the manifold not be simply connected as is required to break E_6 symmetry at the Planck scale, then the number of possibilities indeed get very limited.

§17.10   Assignment of Particles, the E_6-GUT Model [18], and Symmetry Breaking

So far we have only discussed the nature of the compactification space that leads to a low-energy theory with $N = 1$ supersymmetry. In this section, we will study the particle content of this theory. Since the theory has only $N = 1$ supergravity, it will have only one massless graviton and one massless gravitino (they will arise from e_M^A and ψ_M of Section 17.1). There will be a massless dilaton and a dilatino arising from ϕ and λ. Out of the B_{MN}, we will have an antisymmetric $B_{\mu\nu}$. But the action involving this field has a gauge invariance, and after the gauge degrees of freedom are removed, only one pseudo-scalar field remains, which can be identified as the axion.

Let us now turn to the gauge fields A_M^α and their superpartners, the gaugino χ^α in ten dimensions. Under $E_8 \times E_8$, they transform as a $(248,1) \oplus (1,248)$-dimensional representation. But as discussed earlier, compactification that respects $N = 1$ supersymmetry requires that the gauge connection corresponding to the SU(3) subgroup be identified with the spin connection for the SU(3) part of SO(6) group corresponding to the six compactified directions. Since the gauge field corresponding to an SU(3) subgroup has nonzero vacuum expectation value, it breaks the gauge group from $E_8 \times E_8$ down to $E_8 \times E_6$. [Note that E_8 has maximal sub-groups, $E_6 \times SU(3)$ or $SO(10) \times SU(4)$ or $SU(5) \times SU(5)$.] The gauginos and gauge fields corresponding to the broken E_8-group have the following transformation properties under $E_6 \times SU(3)$:

$$\{248\} \supset (27, 3) \oplus (\overline{27}, \overline{3}) \oplus (78, 1) \oplus (1, 8). \tag{17.10.1}$$

Because of the identification of the $SU(3)$ gauge connection with the spin connection, the $SU(3)$ gauge indices become indices of the compact-ified manifold. For instance, let us split the ten-dimensional space–time index M into a space–time coordinate μ and internal coordinate m. The six m-indices transform under $SU(3)$ as $3 + \overline{3}$ (denoted by indices a and $\overline{a}$). Then A_M^α has a piece in it which is $A_{\overline{a}}^a$. It is a property of Calabi–Yau manifolds with $SU(3)$-holonomy that there exists a covariantly constant, totally antisymmetric 3-tensor ε_{abc}. Multiplying by this, we can convert $A_{\overline{a}}^a$ into a $(2,1)$-form as a follows: $\varepsilon_{abc} A_{\overline{a}}^a \, dz^b \wedge dz^c \wedge d\overline{z}^{\overline{a}}$. Thus, the $(27,3)$ part of the gauge field–gaugino system corresponds to closed $(2,1)$-forms. Therefore, by our discussions of the previous section, $b_{2,1}$ counts the num-ber of $\{27\}$-dimensional superfields. Similarly, turning to the $(\overline{27}, \overline{3})$ piece of A_M^α , we can isolate a piece that transforms like $A_{\overline{a}}^{\overline{b}}$, which can be con-verted into a $(1,1)$ form as follows: $g_{a\overline{b}} A_{\overline{a}}^{\overline{b}} \, dz^a \wedge d\overline{z}^{\overline{a}}$. Thus, the number of massless $\{\overline{27}\}$-superfields are given by the Betti–Hodge number $b_{1,1}$. The net number of generations is $(b_{2,1} - b_{1,1}) = -\frac{1}{2}\chi$. Again from Section 17.3, we know that there is a one-to-one correspondence between $(2,1)$-forms and the polynomial deformations, thus we have an explicit representation for the quark and lepton fields.

Symmetry Breaking and Low-Energy Physics

We have seen in the previous sections that the starting gauge group for grand unification with Calabi–Yau type compactification is $E_6 \times E_8'$, with low-energy matter fields transforming as $\{27\}$ (the total number of which are given by $b_{2,1}$ of the corresponding manifold) and $\{\overline{27}\}$ (given by $b_{1,1}$ of the manifold). The whole system is coupled to $N = 1$ supergravity. To study symmetry breaking, we give in Table 17.1 the particle content and their transformations under the following subgroups of E_6: $SO(10) \times U(1)_\chi$ $SU(2) \times U(1)_\chi \times U(1)_\chi$, $SU(3)_c \times SU(2)_L \times U(1)_{I_{3R}} \times U(1)_{B-L} \times U(1)_\chi$. (In Table 17.1, we give the particle content of the 27-dimensional representation of E_6 and in Table 17.2, we give the E_6-weights of different particle.) From Table 17.1 we see that there are only two particles ν^c and n_0 in it, which are singlets under the standard model group. If we give vacuum expectation values to both of them, it will break E_6 down to $SU(5)$, which will remain an exact symmetry down to the electroweak scale. This is clearly phenomenologically disastrous because it implies a proton lifetime of 10^{-15} s. Therefore, to make these models phenomenologically viable, we have to

Table 17.1.

Particles	$SU(3)_C \times SU(2)_L \times U(1)_{I_{3R}}$ $\times U(1)_{B-L} \times U(1)_x$ repesentation	$SU(5) \times U(1)_{x'} \times U(1)_x$ representation	$SO(10) \times U(1)_x$ representation
$\begin{pmatrix} u \\ d \end{pmatrix}$	$(3,2,0,\frac{1}{3},1)$		
u^c	$(3^*,1,-\frac{1}{2},-\frac{1}{3},1)$	$(10,1,1)$	
e^+	$(1,1,+\frac{1}{2},+1,1)$		$(16,1)$
d^c	$(3^*,1,+\frac{1}{2},-\frac{1}{3},1)$		
$\begin{pmatrix} \nu \\ e^- \end{pmatrix}$	$(1,2,0,-1,1)$	$(\bar{5},-3,1)$	
ν^c	$(1,1,-\frac{1}{2},+1,1)$	$(1,5,1)$	
g	$(3,1,1,-\frac{2}{3},-2)$		
$\begin{pmatrix} H_u^+ \\ H_u^0 \end{pmatrix}$	$(1,2,\frac{1}{2},0,-2)$	$(5, ,-2)$	$(10,-2)$
g^c	$(3^*,1,1,\frac{2}{3},-2)$		
$\begin{pmatrix} H_d^0 \\ H_d^- \end{pmatrix}$	$(1,2,-\frac{1}{2},0,-2)$	$(\bar{5}, ,-2)$	
n_0	$(1,1,0,0,4)$	$(1,0,4)$	$(1,4)$

look for additional symmetry-breaking mechanisms. It turns out that just such a mechanism is present in the proper Calabi–Yau manifold.

To see the source of this mechanism, we recall our discussion of Calabi–Yau manifolds from Section 17.3. We saw that a useful class of such manifolds is constructed by taking CP^N-spaces or their products, and considering the zero locus of homogeneous polynomials in them. Let us call this manifold R_0. The number of fermion generations is given by $\frac{1}{2}|\chi_0| \equiv (b_{2,1} - b_{1,1})$, and is usually much larger than three (the observed number of generations). One way to reduce this number is to reduce the Euler characteristic of the manifold. This is done by choosing an appropriate discrete symmetry group H, which acts freely on the manifold R_0 (i.e., it does not leave any point in the manifold fixed); the resulting manifold R has an Euler characteristic χ given by

$$\chi = \frac{\chi_0}{\dim H}. \tag{17.10.2}$$

Table 17.2. E_6-Weights of Different Components of $\{27\}$.

u_i	(1 0 0 0 0 0)	g	(−1 1 0 0 0 0)
	(1 −1 0 0 1 0)		(−1 0 0 0 1 0)
	(1 0 0 0 0 −1)		(−1 1 0 0 0 −1)
d_i	(0 0 0 0 −1 1)	H_u^+	(0 0 1 −1 1 −1)
	(0 −1 0 0 0 1)	H_u^0	(−1 0 1 −1 0 0)
	(0 0 0 0 −1 0)	g^c	(0 0 0 −1 1 1)
u_i^e	(0 0 −1 1 0 1)		(0 1 0 −1 0 0)
	(0 1 −1 1 −1 0)		(0 0 0 −1 1 0)
	(0 0 −1 1 0 0)	H_d^0	(0 1 −1 0 1 0)
e^+	(1 −1 1 −1 0 0)	H_d^-	(−1 1 −1 0 0 1)
d^c	(0 −1 1 0 0 0)	n_0	(1 −1 0 1 −1 0)
	(0 0 1 0 −1 −1)		
	(0 −1 1 0 0 −1)		
ν	(0 0 0 1 0 −1)		
e^-	(−1 0 0 1 −1 0)		
ν^c	(1 0 −1 0 0 1)		

Let us give an example for the $CP^3 \times CP^3$ bases manifolds discussed in Section 17.3 and specify the three polynomials as follows:

$$P_1 = \tfrac{1}{3}(x_0^3 + x_1^3 + x_2^3 + x_3^3) = 0,$$
$$P_2 = \tfrac{1}{3}(y_0^3 + y_1^3 + y_2^3 + y_3^3) = 0, \qquad (17.10.3)$$
$$P_3 = (x_0 y_0 + x_1 y_1 + x_2 y_2 + x_3 y_3) = 0.$$

This manifold R_0 has $\chi_0 = -18$, which predicts nine generations. To reduce the number of generations, we consider the Z_3-group that acts freely on the manifold (i.e., no point of the manifold is invariant under it), and let it operate on the coordinates x_i and y_i $(i = 0, 1, 2, 3)$ as follows:

$$(x_0, x_1, x_2, x_3) \to (x_0, \alpha^2 x_1, \alpha x_2, \alpha x_3),$$
$$(y_0, y_1, y_2, y_3) \to (y_0, \alpha y_1, \alpha^2 y_2, \alpha^2 y_3), \qquad (17.10.4)$$

where $\alpha^3 = 1$.

This reduces χ to -6 which gives only three generations of fermions. In fact, it has been argued [19] that this is the only manifold that can lead to three generations of quarks and leptons.

We will now argue that the discrete symmetry also plays another role in reducing the size of the gauge group via a dynamical symmetry breaking, first advocated by Hosotani [20] and later on in the superstring context by Witten [21]. To see this, we observe that a function ψ acting on R is equivalent to a function on R_0 provided

$$\psi(h(x)) = \psi(x), \qquad h \in H. \qquad (17.10.5)$$

However, since the Lagrangian is E_6-invariant, a weaker condition than the above is sufficient:

$$\psi(h(x)) = U_h\psi(x), \qquad (17.10.6)$$

where U_h is an element of the E_6-group that provides a homomorphism of H into the group E_6. Since under a gauge rotation, $\psi \to V\psi$, eq. (17.10.6) requires that

$$[V, U_h] = 0. \qquad (17.10.7)$$

The surviving group below the compactification scale is the one that commutes with the embedding of H in E_6 and is, therefore, a smaller group as would be necessary for good phenomenology.

A concrete way to realize this kind of symmetry breaking is as follows: we note that in the process of compactification, all E_6-gauge field strengths in ten dimensions must vanish (i.e., $F_{MN}^\alpha = 0$ for α being an E_6-index). In particular, for the indices in the compact manifold, they must vanish too. But does the vanishing of field strength imply that A^α (the gauge potential $m = 5, \ldots, 10$) vanishes? In a simply connected manifold, $F_{MN} = 0$ will imply $A_m = 0$ (up to gauge transformations), but on a multiply connected manifold this is not true, since

$$\oint A_m\, dx^m \neq \int_A F_{mn}\, dA^{mn}. \qquad (17.10.8)$$

A well-known example is that of a 2-sphere that encloses a monopole in standard electrodynamics. Since A_m transforms as the adjoint representation of E_6, $\langle A_m \rangle \neq 0$ reduces the gauge group without reducing the rank. To embed this in E_6, we must consider functions of A_m (which are unitary operators), and the obvious choice is the Wilson loop integral

$$W = P\exp\left(-i\int_\gamma A_m\, dx^m\right), \qquad (17.10.9)$$

where

$$A_m = \theta^a A_m^a, \qquad a = E_6\text{ index,}$$

and γ is a closed loop in the multiply connected compactified space. When $\langle A_m^a \rangle \neq 0$, $\langle W \rangle \neq 0$, and it leads to breaking of the E_6-symmetry. Below we discuss how this symmetry-breaking pattern can be determined for different manifolds. We will concentrate here only on Z_n-groups. For a Z_n-group, it follows that $W^n = 1$, since W is the image of an element of this group. Since W is also an element of an E_6-group and Z_n is abelian, we can parametrize W as

$$W = e^{2\pi i \sum_{a=1}^{b} \lambda_a \theta_a}, \qquad (17.10.10)$$

where $\theta_a (a = 1, \ldots, 6)$ are elements of the Cartan subalgebra of the E_6-group and λ_a are real parameters. $W^n = 1$ implies that $n\lambda_a = 0$. The next question is how do we know the λ_a?

First we note that the mass of a gauge boson, whose weight vector is given by α, is proportional to $\alpha \cdot \lambda$ (α is a six-component vector for the E_6-group). When $\alpha \cdot \lambda = 0$, we obtain an unbroken symmetry. Using this symmetry and using the root vectors of E_6, we find that if we want to keep $SU(3)_c \times SU(2)_L$ unbroken, we must have λ in the following form [1]:

$$\lambda = \{-c, c, a, b, c, 0\}. \tag{17.10.11}$$

For arbitrary a, b, and c, in general, we would obtain the unbroken gauge group to be $SU(3)_c \times SU(2)_L \times U(1)$. If, however, they are related to each other by some additional constraint, the group becomes larger.

Let us again illustrate these ideas for the specific case of the three-generation model. In this case, since the discrete symmetry is Z_3 we have $3a = 3b = 3c = 0$. In Table 17.3, we list $\alpha \cdot \lambda$ for all the positive roots of E_6. The following conclusions can be drawn for the symmetry-breaking patterns in the following cases:

	Choice of parameters	Unbroken subgroup
(i)	$a = 0, b = -c$	E_6
(ii)	$a \neq 0, b = -c$	$SU(6)_I \times U(1)$
(iii)	$a = b + c$	$SU(6)_{II} \times U(1)$
(iv)	$a = 0, b, c$ arbitrary	$SU(6)_{III} \times U(1)$
(v)	$a + b = 2c$	$[SU(3)]^3$

We thus see that the local symmetry group is now considerably reduced below the compactification scale. The rest of the symmetry can be broken for the $[SU(3)]^3$ case to the standard model. We will argue in the next section that $[SU(3)]^3$ breaking to a standard model must be achieved in general at an intermediate scale of the order 10^{14} GeV.

It is also worth pointing out here that, while we have considered only abelian groups to break E_6-symmetry at the Planck scale, a quite interesting class of models emerges, if the discrete symmetry group is chosen to be non-abelian. The freely acting discrete symmetry group in the parent manifold must be non-abelian. An advantage of the non-abelian case is that flux breaking at the Planck scale also reduces the rank of the group. Therefore, in this class of models, one could perhaps avoid problems with proton decay, although the neutrino mass problem still has to be dealt with.

Necessity of Intermediate Scale

In this subsection, we will argue that for E_6-type superstring models to be phenomenologically viable, one needs at least one intermediate scale [22] but possibly two ($\sim 10^{12} - 10^{14}$ GeV). We will then show how one can generate such scales in realistic Calabi–Yau type models. There are three

Table 17.3.

Root α	$\lambda \cdot \alpha$	α	$(\lambda \cdot \alpha)$
(0 0 0 0 0 1)	0	(0 1 0 0 −1 0)	0
(0 −1 0 0 1 1)	0	(1 0 0 0 1 −1)	0
(−1 1 0 0 1 − 1)	$3c$	(−2 1 0 0 0 0)	$3c$
(0 −1 1 1 −1 1)	$a+b-2c$	(0 0 −1 2 −1 0)	$2b-a-c$
(0 −1 2 −1 0 −1)	$2a-b-c$	(1 −1 1 −1 1 0)	$a-b-c$
(1 0 1 −1 0 −1)	$a-b-c$	(1 −1 1 −1 1 − 1)	$a-b-c$
(0 −1 1 −1 0 1)	$a-b-c$	(0 0 1 −1 −1 0)	$a-b-c$
(0 −1 1 −1 0 0)	$a-b-c$	(0 0 1 0 0 −1)	a
(0 −1 1 0 1 −1)	a	(0 0 1 0 0 −2)	a
(−1 0 1 0 −1 0)	a	(−1 −1 1 0 0 0)	a
(−1 0 1 0 −1 − 1)	a	(−1 0 0 1 0 0)	$b+c$
(−1 1 0 1 1 −1)	$b+c$	(−1 0 0 1 0 −1)	$b+c$
(0 1 0 −1 1 0)	$2c-b$	(0 0 0 −1 2 0)	$2c-b$
(0 1 0 −1 1 −1)	$2c-b$	(−1 1 0 −1 0 1)	$2c-b$
(−1 0 0 −1 1 1)	$2c-b$	(−1 1 0 −1 0 0)	$2c-b$
(−1 0 1 −1 1 0)	$a-b+2c$	(−1 1 1 −1 0 −1)	$a-b+2c$
(−1 0 1 −1 1 − 1)	$a-b+2c$	(−1 1 −1 0 1 1)	$3c-a$
(−1 2 −1 0 0 0)	$3c-a$	(−1 1 −1 0 1 0)	$3c-a$

phenomenological requirements that a grand unified theory of E_6-type must satisfy in order to be realistic:

(i) Since grand unified theories contain baryon-violating gauge and Higgs interactions, there must be a mechanism in the theory that will suppress the observable baryon-violating processes to the appropriate level.

(ii) E_6-type theories being left–right symmetric contain the right-handed neutrino, which pairs up with the left-handed neutrino to give a mass of the order of the typical charged lepton mass of the corresponding generation. Since there are many orders of magnitude of difference between m_{v_l} and m_l, the theory must have a built-in mechanism to explain this mass hierarchy. In superstring models, we are not allowed to have representations other than those transforming as 27 or $\overline{27}$; as a result, the simple "seesaw" mechanism (Chapter 6) does not work here. However, as we will see, an extended mechanism [23, 24] can do the job in these models, if there exist appropriate E_6-singlet field couplings to the 27 and $\overline{27}$'s.

(iii) The third requirement has to do with the unification of coupling constants. As is well known, the evolution of weak and electromagnetic coupling constants depends on the β function for the gauge theory. Also, we know that the existence of extra fermions always slows down the evolution of the coupling constants. As we saw from the previous section, E_6-supergravity models not only have gluinos, winos, etc., but also an additional $SU(2)_L \times U(1)_Y$ singlet quark, the g quark. In the presence of all these additional fermions (for masses $\lesssim m_W$), the QCD coupling remains fixed beyond the electroweak scale if the number of generations is three, whereas the $SU(2)_L$ coupling starts to grow, thereby "sabotaging" the whole unification program.

We will see now that, to meet the first and the third requirements, we must have at least one intermediate scale. The neutrino mass problem does not strictly require a new intermediate scale, but if one exists, a solution that does not rely on the E_6-singlet fields can be given [23].

To illustrate these points, let us assume a compactification manifold, where $E_6 \rightarrow SU(3)_c \times SU(2)_L \times SU(2)_R \times U(1)_{B-L} \times U(1)_X$. The most general superpotential invariant under this group is

$$W = \lambda_1 QQ^c\phi + \lambda_2 LL^c\phi + \lambda_3(QQg + Q^cQ^cg^c)$$
$$+ \lambda_4(QLg^c + Q^cL^cg) + \lambda_5 \operatorname{Tr}\phi^2 n_0 + \lambda_6 gg^c n_0, \quad (17.10.12)$$

where

$$Q = \begin{pmatrix} u \\ d \end{pmatrix}; \qquad L = \begin{pmatrix} v \\ e \end{pmatrix},$$

$$\Phi = \begin{pmatrix} H_u^0 & H_u^+ \\ H_d^- & H_d^0 \end{pmatrix}.$$

Let us now turn to the proton decay problem. There are the usual gauge interaction mediated graphs, which are highly suppressed if E_6 breaks down at the compactification scale $\sim M_{pl}$. We will therefore discuss only the baryon-violating interactions present in eq. (17.10.12), i.e., the couplings λ_3 and λ_4. We see from eq. (17.10.12) that the g quark acquires a mass, if $\langle n_0 \rangle \neq 0$; this will give rise to $\Delta B = 1$ decay of the type, $p \rightarrow \pi^0 e^+, \ldots$.

The strength of this effective $QQQL$-type Hamiltonian is given by

$$G_{\Delta B=1} \simeq \frac{\lambda_3\lambda_4 M_{\tilde{G}}\alpha_3}{4\pi M_a M_{\tilde{Q}}^2}. \quad (17.10.13)$$

If we assume that

$$\lambda_3 \simeq g\left(\frac{m_W}{m_W}\right) \simeq 10^{-5} \quad \text{and} \quad \lambda_4 \simeq \frac{gm_s}{m_W} \simeq 10^{-3}$$

(as would be the case in $p \rightarrow K^+\nu$ decay),

$$G_{\Delta B=1} \simeq \frac{10^{-16}}{(M_g \text{ in GeV})}.$$

This implies that if $M_g \geq 10^{14}$ GeV, then the $p \to K^+\nu$ type decay will be in agreement with the observed upper limits on such processes. Thus, one must have $\langle n_0 \rangle \geq 10^{14}$ GeV, i.e., at least one intermediate scale [22].

This also meets the third requirement, since the g quark starts contributing to the evolution of coupling constants only above 10^{14} GeV. There is another class of baryon-violating interactions that can arise when g^c and d^c mix after symmetry breaking. These couplings have dimension 4 and will lead to catastrophic rates for proton decay unless naturally prohibited. It has been suggested by Bento, Hall, and Ross [25] that, if the product of a discrete symmetry of the Calabi–Yau manifold and a discrete subgroup of E_6 remains unbroken at the intermediate scale, the spectrum of particles will split in such a way as to forbid all dimension-4 terms (involving light g quarks) that violate the baryon number. The resulting symmetry has been called matter parity in [25].

Let us now turn to neutrino masses. The problem here is that of the $(27)^3$ couplings present in the fundamental Lagrangian, after $SU(2)_L$ breaking leads to a Dirac mass $m_{\nu D}$ coupling ν^c and ν, $m_{\nu D}$ is expected to be of order of the changed lepton masses of the theory due to weak isospin symmetry. The experimental limits on neutrino masses are smaller than the m_l by orders of magnitude. The presently popular way to understand the smallness of neutrino masses is via the so-called seesaw mechanism where ν^c acquires a large mass M_N associated with the breaking of $B - L$ symmetry. This makes both ν and ν^c into Majorana particles, with $m_\nu \simeq m_{\nu D}^2/M_N$ and $m_{\nu^c} \simeq M_N$. The generation of a large Majorana mass, however, requires the existence of a Higgs boson with $B - L$ quantum number 2; no such Higgs bosons are present in the theory. Therefore, new ways must be devised to cure the neutrino mass problem, and several such mechanisms are known [23, 24].

(i) Generalized "Seesaw" Mechanism

According to this proposal ([23]), by adjoining a two-component chiral neutral "lepton" S to each pair of ν, ν^c, one can write a 3×3 mass matrix of type

$$
\begin{array}{c}
\begin{array}{ccc} \nu & \nu^c & \quad S \end{array} \\
\begin{array}{c} \nu \\ \nu^c \\ S \end{array}
\begin{pmatrix}
O & m_\nu D & O \\
m_\nu D & O & V_{BL} \\
O & V_{BL} & \mu
\end{pmatrix}.
\end{array}
\qquad (17.10.14)
$$

Note that in order to generate the $\nu^c S$ entry in eq. (17.10.14), we need a Higgs boson with the $SU(2)_R \times U(1)_{B-L}$ quantum number $(\frac{1}{2}, 1)$, which is part of the {27}-dimensional representation of E_6. The seesaw–Majorana mass could arise from a supersymmetry-breaking effect. One can then calculate the eigenvalues and eigenvectors of this matrix. The lightest eigenvectors and eigenvalues are

$$\left| v \right\rangle + \frac{m_{\nu_D}}{V_{BL}} \left| \nu^c \right\rangle, \qquad m_\nu \simeq \frac{m_{\nu_D}^2 \mu}{V_{BL}^2}, \qquad (17.10.15)$$

and $\nu^c S$ forms a pseudo-Dirac particle with Dirac mass V_{BL}. Thus, we see that for moderate values of V_{BL} (such as 1 TeV for $\mu \simeq 10 - 100$ GeV), m_ν becomes acceptably small. As far as the particle S is concerned, it could either be the $B - L$ gaugino λ_{B-L} or any of the E_6-singlet fields present in the theory. This mechanism does not need the existence of an intermediate scale, although its existence would not pose any problem.

(ii) Higher-Dimensional Terms

A second mechanism to solve the neutrino mass problem relies on the existence of an intermediate mass scale around 10^{12} GeV or more [24]. The basic assumption here is that the string excitations may lead to effective dimension 4 terms in the low-energy superpotential of the form

$$W^{(4)} = \frac{h}{M_{P1}} (27)^2 (\overline{27})^2. \qquad (17.10.16)$$

When broken down into its components, one obtains terms in the effective Lagrangian of the type

$$\mathcal{L} = \frac{h}{M_{P1}} \nu^{cT} C^{-1} \nu^c \tilde{\bar{\nu}}^c \tilde{\bar{\nu}}^c. \qquad (17.10.17)$$

Therefore, when the $\tilde{\bar{\nu}}^c$ acquires a v.e.v. at the intermediate scale, i.e., $\langle \tilde{\bar{\nu}}^c \rangle = M_I$, it leads to a Majorana mass for ν^c of the order $h(M_I^2/M_{P1})$. The ν, ν^c mass matrix then effectively becomes of the seesaw type, leading to a small m_ν,

$$m_\nu \simeq \frac{m_D^2}{M_I^2} M_{P1}. \qquad (17.10.18)$$

Having discussed the need for an intermediate mass scale, let us now discuss the mechanisms that might generate such an intermediate scale. This mechanism was first proposed by Dine et al. [22]. This also requires the existence of dimension-4 terms in the superpotential of the type shown in (17.10.16). This can lead to dimension-6 terms in the superpotential of the form

$$V^{(6)} = \frac{h^2}{M_{P1}^2} \cdot \tilde{\nu}^{c4} \tilde{\bar{\nu}}^c + \ldots. \qquad (17.10.19)$$

Since we want supersymmetry unbroken at the intermediate scale, the D terms must vanish, requiring that $\langle \tilde{\nu}^c \rangle = \langle \tilde{\bar{\nu}}^c \rangle$. To analyze the minimum of this potential, we include the supersymmetry-breaking mass term, radiatively extrapolated down to the TeV scale. In that case, we obtain (setting

$\tilde{\nu}^c = \tilde{\bar{\nu}}^c$ as required by the vanishing of the D-term)

$$V = -\mu^2 \tilde{\nu}^{c^2} + \frac{h^2}{M_{\text{Pl}}^2} \tilde{\nu}^{c^6}. \qquad (17.10.20)$$

Minimizing this, we find

$$\langle \tilde{\nu}^c \rangle \simeq \sqrt{\frac{M_{\text{Pl}} \mu}{h}}. \qquad (17.10.21)$$

Since we expect it, $\mu \approx \mu_{3/2} \simeq$ TeV, $\langle \tilde{\nu}^c \rangle \simeq 10^{11}$ GeV for $h = 1$. It has been argued that [26] h may vanish perturbatively and arise solely from the nonperturbative effects of worldsheet instantons. In such a case, we expect $h \sim e^{-\alpha'/R^2}$, where R is the radius of the compactification manifold. It is therefore reasonable to assume that $h \ll 1$, in which case $\tilde{\nu}^c$ can be much larger than 10^{11} GeV. Similar results also seem to follow from the study of blown-up orbifolds [27].

Another lesson that we learn from the above analysis is that, if in a particular field direction, $V^{(6)}$ vanishes at the minimum, then dimensions 5 [terms of the form $(27)^3 27\,\overline{27}$] or dimension $6[(27)^3(\overline{27})^3]$ terms may determine the intermediate scale; a simple calculation convinces one that M_I becomes higher if higher dimension terms dominate.

It is also important to point out that, while an intermediate scale solves many problems of superstring models, it also has the potential to create a new problem, i.e., all $SU(2)$, doublets H_u and H_d may pair up to acquire mass M_I, leaving no massless Higgs doublets to break $SU(2)_L$ symmetry. This situation can be avoided if there are discrete symmetries which prevent such terms. For discussion of these and other phenomenological issues of the Calabi–Yau type models, see [22–28].

§17.11   Supersymmetry Breaking

In Section 17.6 we showed that compactification on the Calabi–Yau manifold preserves $N = 1$ supersymmetry. There is, of course, no trace of supersymmetry in the observed low-energy spectrum (e.g., there is no evidence for a selectron $\tilde{e}$ with a mass of half on Mev, etc.). Therefore, physics below the compactification scale must break supersymmetry and this must occur in such a way that the cosmological constant vanishes. Furthermore, the supersymmetry breaking must be communicated to the low-energy sector. In this section, we address this question.

It has been suggested [29] that supersymmetry breaking could arise from a nonvanishing v.e.v. of H_{mnp} since, under supersymmetry transformation [eq. (17.7.2)],

$$\delta\lambda = \tfrac{1}{8} e^{2\phi} \tilde{H}\varepsilon \qquad (\nabla_m \phi = 0). \qquad (17.11.1)$$

Dine et al. [29] then observed that if the hidden sector E_8' gauginos also form a condensate at the same scale where $\langle H_{mnp} \rangle \neq 0$, then since H_{mnp} always appears in linear combination with H [see eq. (17.6.2)] the nonsupersymmetric ground state will correspond to a vanishing cosmological constant. This is a very desirable feature. We, therefore, explore the implications of this situation for low-energy physics [30].

For this purpose, we need to know the Kähler potential and superpotential in the four dimensions. We start from the ten-dimensional Lagrangian in eq. (17.6.2) and compactify to four-dimensional by the following choice of the metric:

$$g_{MN} = \begin{pmatrix} e^{-3\sigma} g_{\mu\nu} & 0 \\ 0 & e^{\sigma} \delta_{mn} \end{pmatrix}. \tag{17.11.2}$$

Here e^{σ} parametrizes the size of the compact manifold. From this, one finds $\det e_A^M = e^{-3\sigma} \cdot \det e_\nu^\mu$. We, therefore, need a rescaling of the four-dimensional gravitino field $\psi_\mu \to e^{3\sigma/4} \psi_\mu$. Using this in eq. (17.6.2), a mass term results for the gravitino field of the form

$$-e e^{G/2} \bar{\psi}_{\mu L}^c \gamma^{\mu\nu} \gamma^5 \psi_{\nu L}. \tag{17.11.3}$$

This G is the Kähler potential that appears in the kinetic energy terms of the four-dimensional supergravity Lagrangian. From eq. (17.6.2), we can identify G as

$$G = \ln(\tfrac{1}{16} e^{-6\sigma} \varphi^{-3/2}) + \ln\left|\sqrt{\tfrac{1}{2}} \langle \Gamma^{mnp} \rangle H_{mnp}\right|^2. \tag{17.11.4}$$

Here σ and φ are real scalar fields; but they combine with the pseudo scalar fields θ and η that emerge from the fields $B_{\mu\nu}$ and B_{mn} upon compactification, i.e.,

$$\varepsilon^{\alpha\mu\nu\lambda} H_{\mu\nu\lambda} = \frac{1}{M} \partial_\alpha \theta$$

and

$$B_{mn} = \eta \varepsilon_{mn}, \tag{17.11.5}$$

to lead to

$$\begin{aligned} S &= \varphi^{-3/4} e^{3\sigma} + i\theta, \\ T &= e^\sigma \varphi^{-3/4} + |c|^2 + i\eta. \end{aligned} \tag{17.11.6}$$

In terms of S, T, and C, which represent the scalar partners of $\{27\}$-fields, they arise from the ten-dimensional Yang–Mills potentials A_M^α, where M takes values on the compact manifold.

To see the contribution of the second term in eq. (17.11.4), we see that only the Yang–Mills, Chern–Simons term contributes and has the form $d_{xyz} c^x c^y c^z$, where x, y, and z run over the $\{27\}$-dimensional representation of E_6. Combining all these, we get the Kähler potential to be

$$G = -\ln(s + s^*) - 3\ln(T + T^* - 2|c|^2). \tag{17.11.7}$$

This is the Kähler potential used in the so-called no-scale models [31]. The Higgs potential can be inferred from eq. (17.11.7) by the following formula:

$$V_B = e^G \left\{ \sum_i \left| \frac{\partial W}{\partial z_i} + \frac{\partial G}{\partial z_i} \cdot W \right|^2 - 3|W|^2 \right\} + \frac{g^2}{2} D^\alpha D^\alpha, \qquad (17.11.8)$$

where

$$W = d_{xyz} C^x C^y C^z \equiv (27)^3. \qquad (17.11.9)$$

Using eq. (17.11.7), we get

$$V_B = \frac{1}{(S+S^*)(T+T^* - 2|C|^2)} \left(|W|^2 + \frac{T+T^* - 2|C|^2}{6} \left| \frac{\partial W}{\partial C} \right|^2 \right). \qquad (17.11.10)$$

The minimum of this potential corresponds to $\langle C \rangle = 0$, with $\langle S \rangle$ and $\langle T \rangle$ undetermined. Let us now include the effects of supersymmetry breaking. As mentioned earlier, we assume that this is caused by condensation of the shadow E_8 gauginos in vacuum. The scale of gaugino condensation is then related to the Planck scale as follows:

$$\mu \sim M_{\text{P}1} e^{-2\sigma} \exp(-1/2b_0 g_A^2), \qquad (17.11.11)$$

where the gauge coupling g_q^2 is given by $g_4^2 = \varphi^{3/4} e^{-3\sigma}$. In order to find the net contribution of this supersymmetry breaking to the superpotential, we note that $2/g_4^2 = S + S^*$ and remember that superpotential is an analytic function. Furthermore, we also give nonzero v.e.v. to H_{mnp}, i.e.,

$$\langle H_{mnp} \rangle = c M_{\text{P}1}^3 \varepsilon_{mnp}. \qquad (17.11.12)$$

Since $\bar{\lambda}\lambda$ and H_{mnp} occur as a complete square in the ten-dimensional Lagrangian, we conclude that W receives a new contribution of the form

$$W(S) = M_{\text{P}1}^3 (c + h e^{-3S/2b_0}). \qquad (17.11.13)$$

Adding this to the superpotential changes the potential V_B to $V_B + \delta V_B$ where

$$\delta V_B = \frac{1}{(S+S^*)(T+T^* - 2|C|^2)} \left| (S+S^*) \frac{\partial W}{\partial S} - W(S) - W(C) \right|^2. \qquad (17.11.14)$$

Minimization of the potential now says that $\langle c \rangle = 0$, and $\langle s \rangle$ is determined by a W

$$\left\langle (S+S^*) \frac{\partial W}{\partial S} - W(s) \right\rangle = 0, \qquad (17.11.15)$$

and $\langle T \rangle$ remains undetermined. Since supersymmetry has now been broken, we get a nonzero gravitino mass, but the scalar matter fields still have no mass. This presents a rather awkward situation for superstring theories.

In order to see if radiative corrections improve the situation at all, a one-loop correction to the tree level potential has been calculated in [31] and the situation does not improve, unless the value of the cutoff is interpreted in a particle manner. It is perhaps fair to summarize the situation as unsatisfactory and in need of clarification or new ideas.

§17.12   Cosmological Implications of the Intermediate Scale

The scenario for intermediate scales outlined in the preceding sections has very interesting cosmological implications [32–35]. To see this, let us write down the zero-temperature potential in symbolic notation

$$V = -\mu^2 \phi^2 + \frac{\lambda}{M_{P1}^2} \phi^6, \qquad (17.12.1)$$

where $\mu \simeq m_W$. At high temperatures, the form of the above potential changes to

$$V = (-\mu^2 + \lambda' T^2)\phi^2 + \frac{\lambda \phi^6}{M_{P1}^2}. \qquad (17.12.2)$$

It is now clear from eq. (17.4.2) that if $\lambda' \approx 1$, the sign of the (mass)^2 term for ϕ remains positive until we reach temperatures down to $T \approx m_W$, and only below the scale of electroweak transition does the (mass)^2 term become negative. This means that as the universe evolves, the phase transitions corresponding to the intermediate scale ($\sim 10^{14}$ GeV or so) are delayed down to the electroweak scale; this is a first-order phase transition. Once the phase transition takes place, there is enormous entropy release, which would dilute any previously generated baryon asymmetry. Therefore, if one assumes the conventional picture where the baryon to photon ratio is generated at the GUT scale, then the low-temperature entropy release would spoil one of the major successes of the grand unified theories. New mechanisms for baryogenesis at low temperature must be considered seriously [32–35]. Several of these mechanisms imply severe constraints on the particle spectrum of the model. For instance, if baryogenesis is supposed to occur via the decay mode $\nu^c \to 3q$, as in [35], the constraint that baryogenesis must occur before the era of nucleosynthesis implies that the mass of the signlet g quarks must be less than 10^8 GeV. The proton decay constraints then imply that several $\Delta B \neq 0$ couplings in acceptable superstring models must either be highly suppressed or vanish.

§17.13   A Real Superstring Model with Four Generations

So far we have been discussing the general mathematical framework, as well as the physics expectations, of superstring models compactified on a suitable Calabi–Yau manifold. In the remainder of this chapter, we wish to discuss some examples of Calabi–Yau manifolds that lead to four and three generations, and then study their viability for the description of known low-energy quark–lepton physics. In this section, we would like to discuss a four-generation model first proposed by Candelas, Horowitz, Strominger, and Witten [12]. We will calculate the Yukawa couplings [36] and discuss the phenomenology [37, 38].

The defining manifold for this model is CP^4 with a vanishing homogeneous quintic polynomial. Using the formulas in eq. (17.9.3), it is easy to see that its first Chern class vanishes. Since it is a single CP^N-manifold, $b_{1,1} = 1$. To calculate $b_{2,1}$, we calculate the total number of holomorphically nonequivalent deformations of a quintic polynomial in CP^4. This number is given by

$$\frac{1 \cdot 2 \cdot 3 \cdot 4 \cdot 5}{5 \cdot 6 \cdot 7 \cdot 8 \cdot 9} - 24 - 1 = 101.$$

In the above calculation, 24 denotes the number of complex linear transformations in five dimensions, and 1 arises from stands for the quintic polynomial that vanishes to define the hypersurface. The Euler characteristic of this manifold in -200. Using our previous discussion, we conclude that this model, without further ado, predicts 100 generations. Furthermore, the manifold is simply connected so that it will not lead to any symmetry breaking at the Planck symmetry, thus making it unsuitable for model building.

We, therefore, now use a discrete symmetry G that acts freely (i.e., leaves no fixed points) to contract the manifold from K_0 to $K \equiv K_0/G$. To show this, let us denote the intrinsic coordinates of this manifold by z_i $(i = 1, \ldots, 5)$, and choose the defining polynomial to be P_5:

$$P_5 = \frac{1}{5} \sum_{i=1}^{5} z_i^5 + c z_1 z_2 z_3 z_4 z_5. \tag{17.13.1}$$

Owing to the nature of P_5, we can define the following freely acting symmetries on this manifold:

$$\begin{aligned} Z_5 &: z_i \to \alpha^i z_i, \qquad \alpha = e^{2\pi/5}, \\ Z_5' &: z_i \to z_{i+1}. \end{aligned} \tag{17.13.2}$$

One can check that these symmetries do not leave any point of the manifold fixed. It is easy to see that the only solution to $\lambda z_i = \alpha^i z_i (i = 1, \ldots, 5)$ is $\lambda = 1$, $z_k = 0 \, (k = 1, \ldots, 4)$, but eq. (17.13.1) implies $z_5 = 0$; but the point $(0, 0, 0, 0, 0)$ is outside the manifold. Thus, no point in the manifold is left fixed by Z_5. If we choose $G = Z_5 \times Z'_5$, the Euler characteristic of K reduces to -8 leading to a model with four generations.

To discuss further details of the models, we write down the various deformations of the model, which are Z'_5 invariant and transform differently under Z_5. For this purpose, we use the following shorthand notation employed in [43].

$$(abcde) = (z_1^a z_2^b z_3^c z_4^d z_5^e + z_z^a z_3^b z_4^c z_5^d z_1^e + z_3^a z_4^b z_5^c z_1^d z_z^e$$
$$+ z_4^a z_5^b z_1^c z_2^d z_3^e + z_5^a z_1^b z_2^c z_3^d z_4^e). \tag{17.13.3}$$

At this point we are ready to identify the various polynomial deformations with the various particles of the E_6-models. To do this, let us embed the Z_5-subgroup into E_6. The E_6-group can break down via the flux-breaking mechanism to many subgroups, depending on the nature of background gauge field configuration denoted by the parameters a, b, and c. To see these subgroups, we note the phenomenological constraint that, in order to have an intermediate scale, we must have either the pair $(\nu^c, \bar{\nu}^c)$ or $(n_0, \bar{n}_0)$ light at the intermediate scale. We then demand that the light Higgs $\mathbf{H}$ must satisfy the condition $\boldsymbol{\lambda} \cdot \mathbf{H} = 0$, where $\boldsymbol{\lambda}$ is given in eq. (17.10.11). In each case, the unbroken subgroup below the Planck scale will be different.

$(\nu^c, \bar{\nu}^c)$ *as the Light Pair:* $(b = -c)$

The unbroken subgroups in this case are:

(a) $SU(3)_c \times SU(2)_L \times U(1)^3$;

(b) $SU(4) \times SU(2)_L \times U(1)^2$;

(c) $SU(5) \times SU(2)_R \times U(1)$.

$(n_0, \bar{n}_0)$ *as the Light Pair:* $(b = 3c)$

In this case, the unbroken subgroups are:

(a) $SU(3)_c \times SU(2)_L \times SU(2)_R \times U(1)^3$;

(b) $SU(5) \times SU(2)_L \times U(1)$;

(c) $SU(4) \times SU(2) \times U(1)^2$.

For the sake of illustration, we consider here only case (ii)(a), which leads to the left–right symmetric group at low energies. In this case, we have the fields transforming as shown in Table 17.5.

Now, matching the Z_5-transformation properties between Table 17.4 and 17.5, we can identify the different fields with the different polynomials; this is a major step toward evaluating Yukawa couplings. For instance, if we want to calculate the Yukawa coupling $Q\phi Q^c$ for any particular generation, we see that we have to evaluate the following integral:

$$h_{Q\phi Q^c} \equiv \int d^3 z \, d^3 \bar{z} \, (21110)(31001)(20111) \tag{17.13.4}$$

Table 17.4.

Polynomials	Transformation under Z_5
(50000) λ_0	
(31001) λ_{+1}	
(30110) λ_{-1}	α^0
(12002) λ_{+2}	
(10220) λ_{-2}	
(21110)(30020)(30101)	
(12200)	α'
(21101)(32000)(30011)	
(20210)	α^2
(21011)(23000)(03110)	
(02010)	α^3
(20111)(30200)(03101)	
(00221)	α^4

Table 17.5.

Fields	Transformation under Z_5
$\phi \equiv (H_u, H_d)n_0$	α^0
Q	α
$\begin{pmatrix} \nu \\ e \end{pmatrix}, g^c$	α^2
$\begin{pmatrix} \nu^c \\ e^c \end{pmatrix}, g$	α^3
Q^c	α^4

(using the notation in Table 17.4). At this point, the generation label is arbitrary since we are just below the Planck scale. The identification of different generations will follow only after the $SU(2)_L \times U(1)$ breaking and the diagonalization of the mass matrix.

Let us now proceed to the actual evaluation of the Yukawa couplings. It is important for this purpose to note that the "parent manifold" K_0 defined by eq. (17.13.1) has the symmetry

$$Z_i \rightarrow \alpha^{n_i} Z_i. \qquad (17.13.5)$$

Using the complex projective character, we can choose $\sum n_i = 0$ ($\alpha = e^{2\pi/5}$). Since $d^3z \, d_{\bar{z}}^3$ is invariant under this symmetry, for a Yukawa cou-

Table 17.6.

	Invariant polynomials	Value of the corresponding integral
I_1	$(z_1 z_2 z_3 z_4 z_5)$	μ
I_2	$(z_1 z_2 z_3 z_4 z_5)^2 z_i^5$	$c\mu$
I_3	$(z_1 z_2 z_3 z_4 z_5) z_i^5 z_j^5$	$c^2 \mu$
I_4	z_i^{15}	$c^3 \mu$
I_5	$z_i^{10} z_j^5$	$c^3 \mu$
I_6	$z_i^5 z_j^5 z_k^5$	$c^3 \mu$

pling to be nonzero the integrand in eq. (17.13.5), which is a 15th-order polynomial, must be invariant under the symmetry in eq. (17.13.5). Thus, the integrand must have one of the polynomials [36] (Table 17.6) to yield a nonzero Yukawa coupling.

We see from Table 17.6 that all the integrals are given in terms of only two parameters: μ and c. To see how they all get related, note that a particular polynomial P_1 is equivalent to another polynomial P_2 if they satisfy the following relation [39]:

$$P_1 \simeq P_2 + z_i \frac{\partial P}{\partial z_i}. \tag{17.13.6}$$

Using eq. (17.13.6), we see that

$$z_i^5 \simeq c z_1 z_2 z_3 z_4 z_5. \tag{17.13.7}$$

Use of eq. (17.13.7) leads the relations between various Yukawa coupling integrals displayed in Table 17.6. Using Table 17.6, we can compute all the Yukawa couplings. We give a few examples:

$$(21110)(31001)(20111) = (72222) \equiv I_2. \tag{17.13.8}$$

Therefore the "raw" Yukawa coupling in question is $c\mu$. As another example, consider

$$(21110)(12002)(20111) \equiv \int (10211)(12002)(11120)$$

$$= \int (3333) = \mu. \tag{17.13.9}$$

On the other hand,

$$\int (30101)(31001)(30200) = 0, \tag{17.13.10}$$

since the integrand cannot be written as any of the invariants listed in Table 17.6.

Table 17.7.

Polynomial type	Normalization
$z_1 z_2 z_3 z_4 z_5$	N_0
$z_i^3 z_j^2 \ (i \neq j)$	N_1
$z_i^3 z_j z_k \ (i \neq j \neq k)$	N_2
$z_i^4 z_j \ (i \neq j)$	N_3
$z_i^2 z_j^2 z_k \ (i \neq j \neq k)$	N_4
$z_i^2 z_j z_k z_l$	$N_3/|c|^2$

These integrands give only the raw Yukawa couplings. To get the actual Yukawa couplings, one must take into account the normalizations defined as follows:

$$\int d^3z \, d^3\bar{z} \varphi \varphi^* = N_\varphi. \tag{17.13.11}$$

Evaluation of these requires the explicit form of the metric and cannot therefore be done. In this example, there are only four normalizations as given in Table 17.7.

One can extend these considerations to the study of higher-dimensional terms in the superpotential ($d > 3$), which could arise as a result of compactification. These terms play an essential role in determining the nature of the intermediate scale needed for phenomenological consistency. To be specific, let us consider the polynomials in Table 17.4 invariant under z_5, and find the dimension-4 term in the superpotential involving them [38].

$$P_4 = \frac{c_4}{M_{\text{P1}}}[N_0^2 \bar{\lambda}_0^2 \lambda_0^2 + N_2^2 \bar{\lambda}_0^2 \lambda_2 \lambda_{-2} + N_1^2 c \bar{\lambda}_0^2 \lambda_1 \lambda_{-1}]. \tag{17.13.12}$$

Note that for $c = 0$, P_4 does not have any term proportional to $\lambda_{\pm 1}$; therefore, one has an F- and D-flat direction along $\langle \bar{\lambda}_0 \rangle \neq 0$ and $\langle \lambda_\pm \rangle \neq 0$, leading to a phenomenologically viable intermediate scale. However, this property is destroyed for $c \neq 0$.

Finally, we illustrate briefly how global symmetries of the four-dimensional quark–lepton models emerge in superstring models with Calabi–Yau compactification using this CP^4 example. First, we note that the original manifold K_0 has the permutation symmetry S_5, since eq. (17.13.7) is invariant under it. The manifold is also invariant under $z_i \rightarrow \alpha^{n_i} z_i$ with $\sum n_i = 0$. The total number of elements of the combined group is denoted by G_0 and is $120 \times 5^4 = 75{,}000$. A subgroup of this, $G = z_5 \times z_5'$, defined in (17.13.2) is used to reduce the manifold K_0 to $K \equiv K_0/G$. To find symmetries of K, we denote elements of z_5 and z_5' by A and B, respectively. Obviously, those elements of G_0 that commute with

A and B are symmetries of K; however, since $A^k B^l z = a$ for $Z \subset K$, all elements of U of G_0 satisfying

$$UAU^{-1} = A^k B^l$$

and

$$UBU^{-1} = A^m B^n \tag{17.13.13}$$

are also symmetries of K from this; it can then be concluded that the symmetries of K are given by

$$B : z_i \to \alpha^{2i^2} z_i,$$
$$Y : z_i \to z_{2i}. \tag{17.13.14}$$

The presence of these symmetries imposes further restrictions on the Yukawa couplings of the theory.

§17.14 String Theories, Extra Dimensions, and Gauge Coupling Unification

In this section, we would like to explore the question of mass scales in string theories and their implications for gauge coupling unification. Two main and related areas we will explore are the weakly and strongly coupled string theories. In the latter case, the very interesting possibility emerges that there may be extra large hidden dimensions in nature.

17.14.1 Weakly Coupled Heterotic String, Mass Scales, and Gauge Coupling Unification

As a typical theory let us consider the Calabi–Yau compactification which begins with the 10-dimensional super-Yang–Mills theory coupled to supergravity based on the gauge group $E_8 \times E'_8$. The Lagrangian for the massless states of the theory is fixed by the above symmetry requirement and writing only the first two bosonic terms, we have

$$S = \frac{4}{(\alpha')^3} \int d^{10}x e^{-2\phi} \left[\frac{R}{\alpha'} + \frac{1}{4} F^{\mu\nu} F_{\mu\nu} + \ldots \right]. \tag{17.14.1}$$

Compactifying to 4-dimensions, it is easy to derive the relations:

$$G_N = \frac{\alpha'^4 e^{2\phi}}{64\pi V^{(6)}}; \alpha_U = \frac{\alpha'^3 e^{2\phi}}{16\pi V^{(6)}} \tag{17.14.2}$$

leading to the relation $G_N = \frac{1}{4}\alpha' \alpha_U$. Thus for typical values of the unified gauge coupling (say $\simeq 1/24$), $M_{str} \simeq 0.1 M_{P\ell}$ and the string scale is larger than the GUT scale by a factor roughly of 20.

How does one view this? One possible attitude is to say that at the GUT scale a grand unified group emerges so that between M_U and M_{str}, the coupling of the grand unifying group evolves and presumably remains perturbative. This makes a heavy demand on the string theory that one must search for a vacuum which has a GUT group [say SO(10)] with three generations and the appropriate Higgs fields so that the theory remains perturbative.

Another way to look at this is that we have a puzzle and new ideas must be looked for to understand this scale discrepancy. We will see that there exist some very interesting possibilities in strongly coupled string theories.

Regardless of whether there is a grand unifying gauge group present at the the scale M_U or not, string models do have gauge coupling unification as is apparent from the above equation. In cases where there are different gauge groups below the string scale, one has the more general relation

$$G_N = \frac{1}{4}k_i\alpha_i\alpha' \qquad (17.14.3)$$

where k_i are the Kac–Moody level of the theory. For the simple heterotic construction with Calabi–Yau compactification, all k_i's are unity (in the proper normalization for the hypercharge). In more general theories however, k_i's could be different from one and a more general unification occurs.

17.14.2 Strongly Coupled Strings, Large Extra Dimensions, and Low String Scales

It has been realized that once one goes to the strong-coupling limit of string theories, many new possibilities emerge. To have an overall picture of how this happens, let us recall the basic relation between string coupling and the observable couplings such as Newton's constant and the gauge couplings:

$$G_N = \frac{\alpha'^4}{V^{(6)}64\pi\alpha_{10}}; \qquad k_i\alpha_i = \frac{\alpha'^3}{V^{(6)}16\pi\alpha_{10}} \qquad (17.14.4)$$

$$k_i\alpha_i = \alpha_{str}$$

where α_{10} and α_{str} are the 10- and 4-dimensional string couplings, respectively. If we identify $V^{(6)}|sim M_{str}^{-6}$, we can derive the following relations among the couplings:

$$G_N \sim \frac{\alpha_{str}^{4/3}}{M_{str}^2\alpha_{10}^{1/3}}. \qquad (17.14.5)$$

Clearly, if we now want to identify the α_{str} and M_{str} with α_U and M_U, the smallness of G_N can be understood only if α_{10} is very large, i.e., we are in the strong coupling limit of strings. This indicates that the profile of mass scales can be considerably different in the strongly couped string theories [40]. The first realization of these ideas in a concrete string model was presented by Horava and Witten [40] in the context of the M-theory whose low-energy limit is given by an 11-dimensional supergravity compactified on an $S_1 \times Z_2$ orbifold. In this case the picture is that of a 11-dimensional bulk bounded by 10-dimensional walls where in resides the gauge groups with one E_8 on each wall. The separation between the two walls (or the compactification radius in the 11th dimension) R_{11} is related in these modsels to the string coupling as $R_{11} \sim (\alpha')^{1/2} \alpha_{10}^{1/3}$. This means that in the weak-coupling limit the two walls sit on top of each other and we have the usual weakly coupled picture described in the previous subsection. On the other hand, as the string becomes larger, the two walls separate. The effective Lagrangian in this case is given by

$$S = \frac{1}{(\kappa)^2} \int d^{11}x \sqrt{g} \left[-\frac{R}{2} - \frac{1}{4\pi(4\pi\kappa^2)^{2/3}} \int d^{10}x \sqrt{g} \frac{1}{4} F^{\mu\nu} F_{\mu\nu} + \ldots \right].$$

$$(17.14.6)$$

Upon compactification, we get the following relations between the four-dimensional couplings:

$$G_N \sim \frac{\kappa^2}{8\pi R_{11} V^{(6)}}; \qquad (17.14.7)$$

$$\alpha_U \sim \frac{(4\pi\kappa^2)^{2/3}}{V^{(6)}}.$$

It is now clear that adjusting R_{11}, we can get the correct string scale while keeping both the string scale and the compactification scales at the M_U. One gets $R_{11} \sim (10^{12}\ GeV)^{-1}$. Thus in strongly coupled string theories, the string scale can be lower. This idea was carried another step further in Refs. 41 and 42 in the proposal that perhaps the string scale can be as low as a TeV. In such a scenario, one has the following general relation between the various mass scales:

$$M_{P\ell}^2 = M_{str}^{n+2} R_1 R_2 \cdots R_n. \qquad (17.14.8)$$

It is then clear that one of the compactification radii could be quite large [43] and indeed it was suggested in Ref. 43 that this would change Newton's law at submillimeter distances. As it turns out, validity of Newton's inverse square law has been checked only above millimeter distances. This is exciting for experimentalists. Sinilarly, the fact that the string scale can be as low as a TeV implies that string states could be excited in colliders and many detailed studies of this has been carried out.

17.14.3 Effect of Extra Dimensions on Gauge Coupling Unification

Once it is accepted that the compactification scales as well as the string scales could be arbitrary, it is clear that the presence of Kaluza–Klein excitations will affect the evolution of couplings in gauge theories and may alter the whole picture of unification of couplings. This question was first studied by Dienes, Dudas and Gherghetta [44]. The formula for this evolution above the compactification scale μ_0 was derived in Ref. 44 on the basis of an effective (4-dimensional) theory approach and the general result at one-loop level is given by

$$\alpha_i^{-1}(\mu_0) = \alpha_i^{-1}(\Lambda) + \frac{b_i - \tilde{b}_i}{2\pi} \ln\left(\frac{\Lambda}{\mu_0}\right) + \frac{\tilde{b}_i}{4\pi} \int_{r\Lambda^{-2}}^{r\mu_0^{-2}} \frac{dt}{dt} \quad (17.14.9)$$

with Λ as the ultraviolet cut-off, δ the number of extra dimensions, and R the compactification radius identified as $1/\mu_0$. The Jacobi theta function

$$\vartheta(\tau) = \sum_{n=-\infty}^{\infty} e^{i\pi\tau n^2} \quad (17.14.10)$$

reflects the sum over the complete (infinite) Kaluza–Klein (KK) tower. In eq. (17.14.9) b_i are the beta functions of the theory below the μ_0 scale, and $\tilde{b}_i$ are the contribution to the beta functions of the KK states at each excitation level. Besides, the numerical factor r in the former integral could not be deduced purely from this approach. Indeed, it is obtained assuming that $\Lambda \gg \mu_0$ and comparing the limit with the usual renormalization group and also assuming that the number of KK states below certain energy μ between μ_0 and Λ is well approximated by the volume of a δ-dimensional sphere of radius μ/μ_0:

$$N(\mu, \mu_0) = X_\delta \left(\frac{\mu}{\mu_0}\right)^\delta, \quad (17.14.11)$$

with $X_\delta = \pi^{\delta/2}/\Gamma(1+\delta/2)$. The result is a power-law behavior of the gauge coupling constants given by

$$\alpha_i^{-1}(\mu) = \alpha_i^{-1}(\mu_0) - \frac{b_i - \tilde{b}_i}{2\pi} \ln\left(\frac{\mu}{\mu_0}\right) - \frac{\tilde{b}_i}{2\pi} \cdot \frac{X_\delta}{\delta} \left[\left(\frac{\mu}{\mu_0}\right)^\delta - 1\right].$$
$$(17.14.12)$$

It however turns out that for MSSM the energy range between μ_0 and Λ—identified as the unification scale—is relatively small due to the steep behavior in the evolution of the couplings. For instance, for a single extra dimension the ratio Λ/μ_0 has an upper limit of the order of 30, which substantially decreases for higher δ to be less than 6. It is therefore unclear whether the power-law approximation is a good description of the coupling evolution. This question has been recently examined in Ref. 45.

In general, the mass of each KK mode is well approximated by

$$\mu_n^2 = \mu_0^2 \sum_{i=1}^{\delta} n_i^2. \quad (17.14.13)$$

Therefore, at each mass level μ_n there are as many modes as solutions to Eq. (17.14.13). KK level will have two KK states that match each other, with the exception of the zero modes which are not degenerate and correspond to (some of) the particles in the original (4-dimensional) theory manifest below the μ_0 scale. In this particular case, the mass levels are separated by units of μ_0. In higher extra dimensions the KK levels are no longer regularly spaced. Indeed, as it follows from Eq. (17.14.13).

Combining all these equations together is straightforward to get

$$\alpha_i^{-1}(\mu) = \alpha_i^{-1}(\mu_0) - \frac{b_i}{2\pi} \ln\left(\frac{\mu}{\mu_0}\right) - \frac{\tilde{b}_i}{2\pi} \cdot 2\left[n \ln\left(\frac{\mu}{\mu_0}\right) - \ln n!\right].$$

(17.14.14)

which explicitly shows a logarithmic behavior just corrected by the appearance of the n thresholds below μ.

Using Stirling's formula $n! \approx n^n e^{-n} \sqrt{2\pi n}$ valid for large n, the last expression takes the form of the power law running

$$\alpha_i^{-1}(\mu) = \alpha_i^{-1}(\mu_0) - \frac{b_i - \tilde{b}_i}{2\pi} \ln\left(\frac{\mu}{\mu_0}\right) - \frac{\tilde{b}_i}{2\pi} \cdot 2\left[\left(\frac{\mu}{\mu_0}\right) - \ln\sqrt{2\pi}\right].$$

(17.14.15)

In the DDG paper [44], it was concluded that for MSSM, unification can essentially occur for arbitrary values of the M_U starting all the way from a TeV to 10^{16} GeV if one puts the gauge bosons in the bulk but leaves the chiral fermions in the brane; however, the value of α_{strong} increases as M_U is lowered. This can be corrected in many ways [45,46].

There are however certain immediate issues that come up in models with GUT scale as low as a TeV. The two main issues are that of proton decay and neutrino masses. It has been conjectured that in strongly coupled theories there are U(1) symmetries that will help to stabilize the proton. One must therefore show that string vacua exist with such properties. As far as neutrino mass goes, there is no way to implement the seesaw mechanism now unless one generates the neutrino Dirac masses radiatively. So completely new approaches have been tried [47] where the postulated bulk neutrinos form Dirac masses with the known neutrinos. These ideas will work only if the string scale is low. There is a new way to circumvent the constraint of low string scale if one considers the left–right symmetric models in the bulk [48].

These problems become moot if one considers high string scale models but with large extra dimensions so that the interesting gravity remains. One particular result of interest in this connection is the way that high-scale seesaw mechanism emerges from higher-dimensional unification. Recall that the minimal SUSY left–right model with the seesaw mechanism resisted grand unification with the minimal particle content. It was noted in Ref. 44 that in the presence of higher dimensions, if all the gauge bosons are in the bulk and matter in the brane, then the left–right model unifies with a left–right seesaw scale around 10^{13} GeV and the KK scale for one dimension slightly above it.

§17.15   Conclusion

String theories clearly provide one of the most elegant and the only consistent framework for unifying gravity with the other forces of nature. For this alone it deserves most serious consideration. It offers more—it limits the choice for the direction of unification and provides a new view of the very, very early universe.

On the down side, the plethora of vacua in the theory plague its uniqueness as does its apparent lack of solid testable predictions. The recent upsurge of activities in the low-scale string theories may have brought some relief on the latter score, whereas the problems such as proton decay and neutrino masses that accompany them keep one from being completely "wild" with enthusiasm in favor in them. Most important, however, are the new ideas they bring to the fore, out of which, some day the final theory unification that Einstein dreamed about, may become a reality.

Exercises

17.1. A class of discrete symmetries of the manifold known as the R-symmetries do not impose any restrictions on the Yukawa couplings. Under the R-symmetry, the holomorphic (3, 0)-form is not invariant and the superpotential W, which transforms the same way as a (3, 0)-form, is also not invariant. Consider the following T–Y manifold:

$$\sum_i x_i^3 = 0,$$

$$x_0 y_0 + x_1 y_1 + c x_3 y_3 = 0,$$

$$\sum_i y_i^3 = 0.$$

Show that the symmetry $B_x = \mathrm{diag}(1, 1, \alpha, 1)x \otimes \mathrm{diag}(1, 1, 1, 1)y$ is an R symmetry using the Lefschetz fixed point theorem.

17.2. Write the torus as a $CP^1 \times CP^1$ manifold with polynomial constraints.

17.3. Consider the $Y_{7;2,2,2,2}$ space that is defined as a subspace of CP^7 with four transversely intersecting quadratics set to zero. What is its Euler characteristic? What are the $b_{2,1}$ and $b_{1,1}$ for this manifold?

17.4. Write down the supersymmetry transformations that leave the La-grangian for a string with one bosonic and fermionic variable invariant.

17.5. Write down a free string Lagrangian with $N = 2$ worldsheet supersymmetry. Calculate the critical dimension for this case.

17.6. Write down the algebra for $(1, 1)$ and $(2, 2)$ superconformal symmetry in two dimensions.

References

[1] For a complete survey of the developments up to 1986, see
M. Green, J. H. Schwarz, and E. Witten, *Superstring Theories*, vols. I and II, Cambridge University Press, Cambridge, 1986;
M. Kaku, *Introduction to Superstrings*, Springer-Verlag, New York, 1988;
J. Polchinski, TASI Lectures, 1995 and *String Theories, I, II*, Cambridge University Press (1998).

[2] Y. Nambu, in *Symmetries and Quark Model* (edited by R. Chand), Gordon and Breach, New York, 1970, p. 269;
T. Goto, *Prog. Theor. Phys.* **46**, 1560 (1971).

[3] A. Neveu and J. Schwarz, *Nucl. Phys.* **B31**, 86 (1971);
P. Ramond, *Phys Rev.* **D3**, 2415 (1971)

[4] F. Gliozzi, J. Scherk, and D. Olive, *Phys. Lett.* **65B**, 282 (1976); *Nucl. Phys.* **B22**, 253 (1977).

[5] D. Gross, J. Harvey, E. Martinec, and R. Robin, *Nucl. Phys.* **B256**, 253 (1985).

[6] For a thorough discussion of higher-dimensional theories, see
T. Appelquist, A. Chodos, and P. Freund, *Modern Kaluza–Klein Theories*, Benjamin-Cumings, New York, 1988.

[7] S. Coleman, *Phys. Rev.* **D11**, 2088 (1975).

[8] S. Mandelstam, *Phys. Rev.* **D11**, 3026 (1975).

[9] E. Bergshoeff, M. de Roo, B. de Wit, and P. van Nieuenhuizen, *Nucl. Phys.* **B195**, 97 (1982).

[10] G. Chapline and N. S. Manton, *Phys. Lett.* **120B**, 105 (1983).

[11] M. B. Green and J. H. Schwarz, *Phys. Lett.* **149B**, 117 (1982).

[12] P. Candelas, G. Horowitz, A. Strominger, and E. Witten, *Nucl. Phys.* **B258**, 46 (1985).

[13] An excellent review of differential geometry relevant to our discussion can be found in [1] as well as
T. Eguchi, R. Gilkey, and A. Hanson, *Phys. Rep.* **66**, 213 (1980).

[14] T. Hübsch, *Comm. Math. Phys.* **108**, 291 (1987);
see also T. Hübsch, University of Maryland Ph.D. Thesis (1987).

[15] G. Tian and S. T. Yau, *Proceedings of the Argonne Symposium on "Anomalies, Geometry and Topology"* (edited by W. Bardeen et al.), World Scientific, Singapore, 1985.

[16] K. Kodaira, *Complex Manifolds and Deformations of Complex Structures*, Springer-Verlag, New York, 1985.

[17] B. Greene, K. Kirklin, P. Miron, and G. G. Ross, *Nucl. Phys.* **B278**, 667 (1986).

[18] For consideration of E_6-GUT, see
F. Gursey, P. Sikivie, and P. Ramond, *Phys. Lett.* **60B**, 177 (1976);
F. Gursey and M. Serdaroglue, *Nuovo Cimento* **65A**, 337 (1981);
Y. Achiman and B. Stech, *Phys. Lett.* **77B**, 389 (1978).

[19] R. Aspinwall, B. Greene, K. Kirklin, and P. Miron, *Nucl. Phys.* **B294**, 1983 (1987);
P. Candelas, A. Dale, C. Lutken, and R. Schimmrigk, *Nucl. Phys.* **B298**, 493 (1988).

[20] Y. Hosotani, *Phys. Lett.* **126B**, 303 (1983).

[21] E. Witten, *Nucl. Phys.* **B258**, 75 (1985);
see G. Segre, "Schladming Lectures" (1986) for a review;
A. Sen, *Phys. Rev. Lett.* **55**, 33 (1985).

[22] M. Dine, V. Kaplunovsky, M. Mangano, C. Nappi, and N. Seiberg, *Nucl. Phys.* **B259**, 519 (1985).

[23] R. N. Mohapatra, *Phys. Rev. Lett.* **56**, 561 (1986);
R. N. Mohapatra and J. W. F. Valle, *Phys. Rev.* **D34**, 1642 (1986).

[24] J. P. Deredinger, L. Ibanez, and H. P. Nilles, *Nucl. Phys.* **B267**, 365 (1986);
S. Nandi and U. Sarkar, *Phys. Rev. Lett.* **56**, 564 (1986).
For other solutions to the neutrino mass problem, see
A. Masiero, D. Nanopoulos, and A. Sanda, *Phys. Rev. Lett.* **57**, 663 (1986);
E. Ma, *Phys. Rev. Lett.* **58**, 969 (1987).

[25] M. Bento, L. Hall, and G. G. Ross, *Nucl. Phys.* **B292**, 400 (1987).

[26] M. Dine, N. Seiberg, X. Wen, and E. Witten, *Nucl. Phys.* **B278**, 769 (1986).

[27] M. Cvetič, *Phys. Rev. Lett.* **59**, 1795 (1987).

[28] J. Ellis, K. Enquist, D. Nanopoulos, and K. Olive, *Phys. Lett.* **188B**, 415 (1987);
G. Costa, F. Feruglio, F. Gabbiani, and F. Zwirner, *Nucl. Phys.* **B286**, 325 (1986).

[29] J. P. Deredinger et al., Ref. [24];
M. Dine, N. Seiberg, R. Rohm, and E. Witten, *Phys. Lett.* **156B**, 55 (1985).

[30] J. Breit, B. Ovrut, and G. Segre, *Phys. Lett.* **162B**, 303 (1985);
P. Binetruy and M. K. Gaillard, *Phys. Lett.* **168B**, 347 (1986);
J. Ellis. C. Gomez, and D. V. Nanopoulos, *Phys. Lett.* **171B**, 302 (1986);
M. Quiros, *Phys. Lett.* **173B**, 265 (1986);

Y. J. Ahn and J. Breit, *Nucl. Phys.* **B273**, 253 (1986);
P. Binetruy, S. Dawson, and I. Hinchliffe, *Phys. Lett.* **179B**, 262 (1986).

[31] J. Ellis, C. Kounras, and D. Nanopoulos, *Nucl. Phys.* **B241**, 406 (1984);
Nucl. Phys. **B247**, 373 (1984);
N. Chang, S. Ouvry, and X. Wu, *Phys. Rev. Lett.* **5**, 327 (1983).

[32] K. Yamamoto, *Phys. Lett.* **B168**, 341 (1986).

[33] G. Lazaridis, C. Panagiotakopoulos, and Q. Shafi, *Phys. Rev. Lett.* **56**, 557 (1986).

[34] R. N. Mohapatra and J. W. F. Valle, *Phys. Lett.* **B186**, 303 (1987).

[35] K. Yamamoto, *Phys. Lett.* **B194**, 390 (1987).

[36] S. Kalara and R. N. Mohapatra, *Phys. Rev.* **D35**, 3143 (1987).

[37] S. Kalara and R. N. Mohapatra, *Z. Phys.* **C37**, 395 (1988);
see also F. del Aguila, M. Daniel, M. Blair, and G. G. Ross, *Nucl. Phys.* **B272**, 413 (1986).

[38] M. Matsuda, T. Matsuoka, H. Mino, D, Suematsu, and Y. Yamada, *Prog. Theor. Phys.* **79**, 174 (1988).

[39] P. Candelas, *Nucl. Phys.* **B298**, 458 (1988).

[40] P. Horava and E. Witten, *Nucl. Phys.* **B 460**, 506 (1996).

[41] J. Lykken, *Phys. Rev.* **D 54**, 3693 (1996).

[42] I. Antoniadis, *Phys. Lett.* **B 246**, 377 (1990);
I. Antoniadis, K. Benakli and M. Quirós, *Phys. Lett.* **B331**, 313 (1994).

[43] N. Arkani-Hamed, S. Dimopoulos and G. Dvali, *Phys. Lett.* **B429** (1998) 263; Phys. Rev. **D59**, 086004 (1999);
I. Antoniadis, S. Dimopoulos and G. Dvali, *Nucl. Phys.* **B 516**, 70 (1998);
N. Arkani-hamed, S. Dimopoulos and J. March-Russell, hep-th/9809124.

[44] K.R. Dienes, E. Dudas and T. Gherghetta, *Phys. Lett.* **B436**, 55 (1998) ; *Nucl. Phys.* **B537**, 47 (1999);
for an earlier discussion see T. Taylor and G. Veneziano, *Phys. Lett.* **B212**, 147 (1988).

[45] A. Perez-Lorenzana and R. N. Mohapatra, hep-ph/9904504.

[46] D. Ghilencea and G.G. Ross, *Phys. Lett.* **B442**, 165 (1998);
Z. Kakushadze, hep-ph/9811193; C.D. Carone, hep-ph/9902407;
A. Delgado and M. Quirós, hep-ph/9903400;
P. H. Frampton and A. Räsin, hep-ph/9903479;
D. Dumitru and S. Nandi, hep-ph/9906514.

[47] K. Dienes, E. Dudas and T. Gherghetta, *Nucl. Phys.* **B557**, 25 (1999);
N. Arkani-Hamed, S. Dimopoulos, G. Dvali and J. March-Russell, hep-ph/9811448.

[48] R. N. Mohapatra, S. nandi and A. Perez-Lorenzana, *Phys. Lett.* **B 466**, 115 (1999);
R. N. Mohapatra and A. Perez-Lorenzana, *Nucl. Phys.* **B 576**, 466 (2000).

"I thought that my voyage had come to its end at the last limit of my power—that, the path before me was closed, that provisions were exhausted and the time come to take shelter in a silent obscurity.

But I find that thy will knows no end in me. And when old words die out on the tongue, new melodies break forth from the heart; and where the old tracks are lost, new country is revealed with its wonders."

GITANJALI, RABINDRA NATH TAGORE,
NOBEL LAUREATE IN LITERATURE

Index